OXFORD REVIEWS OF REPRODUCTIVE BIOLOGY

Volume 10
1988

OXFORD REVIEWS OF REPRODUCTIVE BIOLOGY

EDITED BY
J. R. CLARKE

Volume 10
1988

OXFORD UNIVERSITY PRESS
1988

Oxford University Press, Walton Street, Oxford OX2 6DP
Oxford New York Toronto
Delhi Bombay Calcutta Madras Karachi
Petaling Jaya Singapore Hong Kong Tokyo
Nairobi Dar es Salaam Cape Town
Melbourne Auckland

and associated companies in
Berlin Ibadan

OXFORD is a trade mark of Oxford University Press

Published in the United States
by Oxford University Press, New York

© Oxford University Press, 1988

ISBN 0 19 857648-X

British Library Cataloguing in Publication Data
Oxford reviews of reproductive biology.—
Vol. 10 (1988)—
1. Animals. Reproductive system. Serials
591.1'6'05
ISBN 0 19 857648-X

Set by Eta Services (Typesetters) Ltd, Beccles, Suffolk

Printed in Great Britain
at the University Printing House, Oxford
by David Stanford
Printer to the University

Contents

vi   Contents

Contributors

Lizbeth A. Adams: Department of Physiology and Biophysics, University of Washington, Seattle, Washington 98195, USA.

Robert A. Steiner: Department of Obstetrics & Gynaecology, Physiology and Biophysics, and Zoology, University of Washington, Seattle, Washington 98195, USA.

S. Maddocks: Department of Animal Sciences, Waite Agricultural Research Institute, University of Adelaide, Glen Osmond, South Australia 5064.

B. P. Setchell: Department of Animal Sciences, Waite Agricultural Research Institute, University of Adelaide, Glen Osmond, South Australia 5064.

Michael D. Griswold: Program in Biochemistry, Washington State University, Pullman, WA 99164, USA.

Carlos Morales: Department of Anatomy, McGill University, Montreal, Quebec, Canada H3A 2B2.

Steven R. Sylvester: Program in Biochemistry, Washington State University, Pullman, WA 99164, USA.

Anne McLaren: MRC Mammalian Development Unit, Wolfson House (University College London), 4 Stephenson Way, London NW1 2HE, UK.

John W. Moore: Department of Clinical Endocrinology, Imperial Cancer Research Fund Laboratories, Lincoln's Inn Fields, London WC2A 3PX, UK.

Richard D. Bulbrook: Imperial Cancer Research Fund Laboratory, Department of Surgery, Western Infirmary, Glasgow G11 6NT, UK.

William M. Pardridge: Department of Medicine, Division of Endocrinology, UCLA School of Medicine, Los Angeles, California 90024, USA.

J. F. Savouret: Unité de Recherches Hormones et Reproduction (INSERM

U 135), Faculté de Medecine de Bicêtre, Université Paris-Sud, 78 rue du Général Leclerc 94275 Le Kremlin-Bicêtre Cedex, France.

M. Misrahi: Unité de Recherches Hormones et Reproduction (INSERM U 135), Faculté de Medecine de Bicêtre, Université Paris-Sud, 78 rue du Général Leclerc 94275 Le Kremlin-Bicêtre Cedex, France.

E. Milgrom: Unité de Recherches Hormones et Reproduction (INSERM U 135), Faculté de Medecine de Bicêtre, Université Paris-Sud, 78 rue du Général Leclerc 94275 Le Kremlin-Bicêtre Cedex, France.

N. Einer-Jensen: Department of Physiology, Odense University, DK-5230 Odense M., Denmark.

C. P. Sibley: Action Research Placental and Perinatal Unit, Department of Child Health and Physiological Sciences, University of Manchester, St. Mary's Hospital, Hathersage Road, Manchester M13 0JH, UK.

R. D. H. Boyd: Action Research Placental and Perinatal Unit, Department of Child Health, University of Manchester, St. Mary's Hospital, Hathersage Road, Manchester M13 0JH, UK.

R. E. Garfield: Department of Neurosciences, and Obstetrics and Gynaecology, McMaster University Health Sciences Centre, 1200 Main Street West, Hamilton, Ontario, Canada L8N 3Z5.

M. G. Blennerhassett: Department of Neurosciences, McMaster University Health Sciences Centre, 1200 Main Street West, Hamilton, Ontario, Canada L8N 3Z5.

S. M. Miller: Department of Neurosciences, McMaster University Health Sciences Centre, 1200 Main Street West, Hamilton, Ontario, Canada L8N 3Z5. (Present address: Department of Physiology and Biophysics, Mayo Medical School, Rochester, MN 55904, U.S.A.)

Hilary Dobson: Department of Veterinary Clinical Science, University of Liverpool, Leahurst, Neston, South Wirral, L64 7TE, UK.

1   Puberty

LIZBETH A. ADAMS AND ROBERT A. STEINER

I   Introduction

The ability of an animal to reproduce depends upon the coordination and integration of the endocrine components of the reproductive system. These components are the hypothalamus, producing the decapeptide gonadotrophin-releasing hormone (GnRH); the anterior pituitary, producing luteinizing hormone (LH) and follicle-stimulating hormone (FSH); and the gonads (testis or ovary), producing mature gametes (sperm or oocytes) capable of fertilization, as well as the sex hormones (androgens, progestins, oestrogens and inhibins). These organ systems must develop to a functional state and interact in the correct manner to produce mature gametes; while they mature to some degree during fetal life, they undergo final maturation and synchronization during postnatal development. The time course of this maturation varies profoundly among species, depending not only on the length of the life cycle, but also on the particular reproductive strategies used by the species in question.

The period of time referred to as *puberty* falls somewhere along this continuum of development. It is regarded as the transitional stage from childhood to adulthood, and is somewhat difficult to define explicitly or to delimit temporally. Aspects of the reproductive axis on which authors have focussed in their attempts to define puberty include: (1) phenotypic changes: "[Puberty] . . . is marked by the occurrence of those constitutional changes whereby the two sexes become fully differentiated. It is at this period that the secondary sexual characters first become conspicuous, and the essential organs of reproduction undergo a great increase in size." (Marshall 1922; Donovan and Van der Werff ten Bosch 1965); (2) neuroendocrine events: ". . . The time when the hypothalamo-hypophysial unit is competent to respond to an effective increment in circulating oestrogens by a preovulatory surge of LH and FSH." (Dierschke, Karsch, Weick, Weiss, Hotchkiss, and Knobil 1974); and (3) a combination of physiological and behavioural occurrences: "Puberty is the stage of individual development characterized mainly by the ability to elaborate functional gametes and by the ability and the desire to play the appropriate role in sexual congress" (Crew 1931).

This chapter is not meant to be an exhaustive survey of the process of puberty as it occurs throughout the animal kingdom. Rather, it emphasizes the principles and strategies of pubertal maturation employed by a few groups of animals: rodents, domestic ruminants, and primates. It is divided into two main parts. The first part is devoted to a description of puberty in rats, sheep, non-human primates, and human beings. The second part describes the maturation of components of the hypothalamo-pituitary axis during the pubertal process in these same species, and addresses the roles of gonadal steroids and central neurotransmitters such as catecholamines, melatonin, serotonin and endogenous opiates in the maturation process. Included in this section is a discussion of the role of nutritional cues in timing the onset of puberty.

II   Descriptive aspects of puberty

1   RATS

The process of sexual maturation has been studied extensively in the laboratory rat, in which weaning occurs at the end of the third week of life, or postnatal day (D) 20–21, and full reproductive capabilities are established within the following two or three weeks. The initiation of a mature rate of gameto-genesis occurs near the time of weaning, and during the weeks between weaning and the end of puberty, animals undergo the greatest increase in body weight and sexual organ weight (Watanabe and McCann 1969; Swerdloff, Walsh, Jacobs, and Odell 1971). Although the Sertoli cell blood-testis barrier in males is formed and primary spermatocytes appear around D 21, mature sperm are not released from the germinal epithelium until the end of puberty (D 49), a delay representing the length of the spermatogenic cycle in this species. Similarly, follicular maturation in the female is sufficiently advanced by the time of weaning to support ovulation under conditions of extraordinary exogenous stimulation, although the animal does not spontaneously ovulate until D 40. The time between birth and the onset of puberty may be divided into the neonatal period (D 0–7), the early phase (D 7–21), the late juvenile phase (D 21–40) and puberty (D 40–50) (Ramaley 1979). The endocrine events that characterize these stages of development are described below.

i Males

During the neonatal period the reproductive axis in male rats exhibits a transient phase of increased activity before entering the relatively quiescent phase characteristic of the early juvenile stage. Circulating plasma LH levels are approximately 50 per cent of adult values on D 5, decline to low levels by D 15 (Miyachi, Nieschlag, and Lipsett 1973), and increase from weaning onward to reach adult values by around D 70 (Miyachi *et al.* 1973; Negro-Vilar, Krulich, and McCann 1973*a*; Gupta, Rager, Zarzycki, and Eichner 1975). The pituitary LH content during these developmental stages reflects a different pattern to plasma levels, increasing gradually from very low levels during the neonatal period to peak levels which are 400 per cent of adult values on D 60, and declining thereafter to reach adult levels by D 75 (Dupon and Schwartz 1971).

Plasma FSH levels are similar in the neonatal and the adult male rat, but exhibit a pronounced increase just prior to the onset of puberty (D 35) to levels which are much higher than at any other time (Miyachi *et al.* 1973; Negro-Vilar *et al.* 1973*a*; Gupta *et al.* 1975; Chappel and Ramaley 1985). Following this pronounced prepubertal increase, FSH levels rapidly decline to adult values by D 40. The pituitary FSH content increases steadily from D 10 onwards, reaching adult levels by D 30 and maintaining those levels thereafter (Kragt and Ganong 1968, Chappel and Ramaley 1985). From D 21 onward, the pituitary begins to synthesize a greater proportion of FSH molecules of high biological activity, so that both the amount and the biological impact of the FSH molecule increases prior to puberty (Chappel and Ramaley 1985). The neonatal elevation of plasma FSH levels may play a role in triggering the onset of spermatogenesis (Ramaley 1979; Ojeda, Andrews, Advis, and White 1980) and the prepubertal (D 35) FSH elevation is most likely important in inducing the formation of LH receptors on Leydig cells (Ketelslegers, Hetzel, Sherins, and Catt 1978), which in turn increases testosterone production in response to LH.

Prolactin levels, which are low during the early juvenile phase, rise dramatically around D 25 to levels approximating 50 per cent of adult levels. These levels are sustained until D 50 when another rise in prolactin occurs, followed by a third rise around D 70. The initial elevation in prolactin is correlated temporally with the beginning of a rapid growth phase of the testes, seminal vesicles, and prostate (Negro-Vilar *et al.* 1973*a*), and is also coincident with the beginning of the major prepubertal FSH peak. Secondary sex organs exhibit rapid growth while prolactin levels remain high (even in the face of stable or diminishing gonadotrophin levels), with the highest rate of weight increase between D 50–D 60, concurrent with the secondary elevation of prolactin levels. Theories have been advanced that prolactin stimulates growth of the accessory sex organs by a direct effect on the tissues themselves (Golder, Boyns, Harper, and Griffiths 1972) or by synergism with LH and testosterone (Bartke 1980).

The relative proportions of the various androgens produced by the testis changes during development. In fetal, neonatal, and adult animals, testosterone is the main conversion product of progesterone; this conversion takes place in the testis. However, during the late juvenile phase (D 20–D 35), androstanediol is the predominant circulating androgen, achieving levels 6–8 times those of testosterone (Moger 1977). At D 30 the shift back to testosterone production is mirrored by an increase in plasma testosterone and dihydrotestosterone levels (Gupta *et al.* 1975), while androstanediol levels decline. Testosterone and dihydrotestosterone levels achieve adult values by D 40 (Gupta *et al.* 1975). Although the biological consequences of this pattern of androgen production are not fully understood, androstanediol may be important in initiating spermatogenesis, and particularly in supporting maturation of sperm in the epididymis (Lubicz-Nawrocki 1976). Also, androstanediol decreases the pituitary responsiveness to GnRH stimulation (Epstein, Lunenfeld, and Kraiem 1977), and thus may play a role in maintaining low circulating levels of gonadotrophins in the prepubertal animal. Leydig cell maturation and spermatogenesis precede the late juvenile increase in testosterone levels (Knorr, Vanha-Perttula, and Lipsett 1970), suggesting that this steroid is not directly responsible for these early developmental processes. However, testosterone does stimulate the growth of the accessory organs during the late juvenile period.

Oestrogen levels in the male rat are elevated during the first two days of post-natal life, then decline only to increase again between D 9 and D 19. The peak levels at day 19 are fivefold higher than normal adult levels of this hormone (Döhler and Wuttke 1975). During the neonatal and juvenile periods, Sertoli cells produce oestrogen in response to stimulation by FSH. Oestrogen may suppress testosterone production by Leydig cells until, around D 20,

Sertoli cell sensitivity to FSH declines (Tsai-Morris, Aquilano, and Dufau 1985) and the testicular concentration of oestradiol falls. This decrease in oestrogen titre may in turn release Leydig cells from suppression, allowing them to respond to LH stimulation with the production of testosterone (Keel and Abney 1985).

The GnRH content of the hypothalamus increases from birth onward (Chappel and Ramaley 1985), and the responsiveness of the pituitary to GnRH stimulation increases with age, particularly just prior to the onset of puberty (Debeljuk, Arimura, and Schally 1972*a*).

ii Females

The stages of prepubertal development in the female rat are delineated similarly to those in the male. During the neonatal phase (D 0–D 7) the ovaries are insensitive to the effects of gonadotrophin stimulation and circulating steroid hormone levels are low (Ben-Or 1963). During the early juvenile phase (around D 9) gonadotrophin receptors are first detected in the ovaries (Peluso, Stegert, and Hafez 1976; Smith-White and Ojeda 1981), the ovaries begin to respond to FSH, and steroid titres rise (Andrews and Ojeda 1981). The late juvenile stage begins around the time of weaning and it is at this point that ovulation may be induced by extraordinary stimuli (Andrews and Ojeda 1981). Ovulation may occur spontaneously around D 26–D 28 as the animal enters the peripubertal state, but vaginal opening, true ovulation, and the onset of 4–5 day cycles begins between D 35–D 40. During the late juvenile phase the body weight of female rats increases almost twofold, and there is a concurrent increase in the weights of the uterus and ovaries (Watanabe and McCann 1969).

LH levels during the neonatal period are approximately twice those in adult females in the dioestrus phase of the oestrous cycle, and twice those in the neonatal male (Goldman, Grazia, Kamberi, and Porter 1971). During the next two weeks, LH levels are elevated markedly and appear to undergo pulsatile and circadian fluctuations (Döhler and Wuttke 1975; Döhler and Wuttke 1976). After D 23, LH levels decrease dramatically until D 40, when they rise again to adult levels and begin to show characteristic 4- or 5-day cyclic variations (Döhler and Wuttke 1975). There is a peripubertal shift in the nature of the LH molecule synthesized such that a more biologically active form is produced after puberty (Buckingham and Wilson 1985). Pituitary content of LH increases until D 26 and remains at these levels until the onset of puberty around D 35, then falls coincidently with vaginal opening and first ovulation and remains low thereafter (Dupon and Schwartz 1971).

Circulating FSH is also elevated in the neonatal female to values above those in the neonatal male or the adult female (Goldman *et al.* 1971). FSH continues to increase until around D 17, when it declines rapidly to values characteristic of the adult animal (Döhler and Wuttke 1975). Pituitary FSH content declines markedly between D 25 and D 35 (Watanabe and McCann 1969), coincidently with an increased growth rate of the ovaries and uterus, and just preceding the time of vaginal opening. The pattern of prolactin secretion in females is similar to that in male rats. Prolactin levels are low during the neonatal and early juvenile periods, then increase markedly at the time of weaning to reach adult levels around D 40 (Döhler and Wuttke 1975). Prolactin probably increases the LH receptor content of the granulosa cells in the ovary, thereby enhancing the production of progesterone and oestrogen in response to gonadotrophin stimulation (Ojeda, Urbanski, and Ahmed 1986).

As is the case in the male, the pattern of gonadal steroid production in the female demonstrates a biphasic pattern. The neonatal and adult ovary predominately converts progesterone to androstenedione and testosterone, but during the late juvenile period mainly 5-alpha-reduced products are formed (for example, androsterone). Near the time of puberty, the ovary produces oestradiol in increased amounts, because of an enhancement of aromatase activity and a diminution of 5-alpha-reductase activity. Gonadotrophins are likely responsible for the peripubertal changes in enzyme activity, since these events occur both during normal puberty and during precocious pseudopuberty induced with exogenous gonadotrophin. Progesterone levels are low until D 14, increase slightly between D 15–D 30, and increase dramatically at the time of vaginal opening around D 40 (Parker and Mahesh 1976).

As in the male, the GnRH content of the hypothalamus increases from birth onwards in female rats (Hompes, Vermes, Tilders, and Schoemaker 1982). In contrast to the male, the responsiveness of the pituitary to GnRH stimulation decreases just prior to the onset of puberty (Debeljuk et al. 1972b; Nash, Brawer, Robaire, and Ruf 1986).

2 SHEEP

Mammals living in climates where food supplies wax and wane over the course of the year generally give birth to their young during a season when food is plentiful. To time births correctly, animals must restrict their breeding activities to some period of time before the season of births; the timing of the conception season is, of course, dictated by the length of gestation in that particular species. Animals whose reproductive activity is strongly influenced by the season of the year are referred to as seasonal breeders. Conversely, mammals living near the equator, where the food supply remains fairly constant, and mammals whose food supply is otherwise stabilized, such as human beings and species which are protected by man (for example, the laboratory rat) do not exhibit pronounced seasonal variations in reproductive activity. Of the many environmental cues used by seasonal breeders, it is clear that day length or photoperiod is one of the most powerful, and constitutes a signal which profoundly influences the activity of the reproductive axis in sheep. The mechanism whereby photoperiod acts on the reproductive axis will be discussed in a later section.

i Males

A working definition for the time of puberty in male sheep is the time at which reproduction first becomes possible. Many authors view puberty as a continuum between the activation of the endocrine function of the testes (when they become responsive to pituitary gonadotrophins and begin to secrete androgens) and the release of mature spermatozoa, which constitutes the final stages of puberty. As mentioned above, the fact that one particular event is not designated as a marker of puberty renders precise timing of the attainment of puberty difficult; this problem is not confined to the study of male domestic ruminants, but complicates the study of male mammals in general. Using the presence of mature spermatozoa in the reproductive tract as a criterion for timing pubertal maturation in sheep does not obviate this problem, since there is considerable variability among breeds in just this single parameter. For example, Suffolk, Dorper, and Merino rams all first show mature spermatozoa in the reproductive tract around D 110 and a body weight of 27 kg (Watson, Sapsford, and McCance 1956), whereas rams of the Namaqua breed of South Africa do not show mature sperm until a later age (D 196) and a heavier weight (32 kg; Skinner 1970; Dyrmundsson 1973). In many breeds, the attainment of a critical testis weight of 10 g (occurring around D 42) is a prerequisite for the onset of spermatogenesis, although the latter event does not occur until around D 60–D 70; completion of the spermatogenic cycle requires an additional 49 days (Skinner, Booth, Rowson, and Karg 1968; Skinner 1970; Skinner 1971). The following discussion of the ontogeny of endocrine function focuses on data taken from breeds which develop relatively early.

Mean circulating levels of LH are low in the neonatal ram lamb, and increase between the second and fourth months of life, before returning to low levels by the fifth postnatal month (Crim and Geschwind 1972a; Foster, Mickelson, Ryan, Coon, Drongowski, and Holt 1978). The onset of pulsatile LH discharge occurs as early as the first week of postnatal life; the exact timing is related to the rate of growth of the young ram, with faster growing animals exhibiting LH pulses at an earlier age. Furthermore, the ontogeny of pulsatile LH release is reflected in the time at which elevations in mean LH levels are first observed. LH pulse frequency increases dramatically until the eighth week of life, and decreases slightly over the ensuing eight weeks (Foster *et al.* 1978). At all ages studied, a pulse of LH is followed by a rise in the circulating levels of testosterone. Pituitary LH content increases from birth to peak levels around D 84 and remains high through the end of puberty (Skinner *et al.* 1968). Circulating FSH levels in male lambs do not appear to vary over prepubertal development, when assessed on a monthly basis (Crim and Geschwind 1972*a*), although pituitary FSH content shows much the same pattern over development as does LH.

Prolactin levels are measurable by one week after birth, and remain fairly constant until the completion of puberty, with the exception of a pronounced peak between the tenth and twelfth weeks of life (Ravault and Courot 1975). This peak in prolactin concentration is coincident with the onset of a rapid phase of testicular growth, as is true in the male rat, and with the onset of spermatogenesis (Skinner *et al.* 1968).

Testosterone is the predominant androgen present in the testes (Skinner *et al.* 1968) and in the circulation (Crim and Geschwind 1972*b*) of the ram throughout prepubertal development, although androstenedione is also present in the testes in very small amounts (Skinner *et al.* 1968). Androgen content of the testes increases sharply after D 42, preceding the onset of spermatogenesis by a few weeks (Skinner *et al.* 1968); the most pronounced increase in circulating testosterone levels occurs slightly later, near the time of the onset of spermatogenesis (Crim and Geschwind 1972*b*; Schanbacher, Gomes, and VanDemark 1974). The predominance of testosterone production throughout prepubertal development distinguishes male sheep from other male mammals, in which androstenedione and other 5-alpha-reduced products are preferentially secreted during juvenile life. As mentioned, high circulating levels of 5-alpha-reduced androgens may suppress reproductive function prepubertally in rats; the low levels of these compounds in sheep suggest that some other factor, for example photoperiod or nutritional status, may play a more important role in timing the onset of puberty in this species.

ii   Females

The female sheep is reproductively active during the six month breeding season extending from late summer to early winter, and exhibits regular 16-day oestrous cycles during this time. Beginning with the increasing day-lengths of late winter, females enter a six month period (anoestrus) during which oestrous cycles cease and the animals are reproductively quiescent. Puberty in the female lamb is defined as the time of first ovulation and first oestrus, and occurs only during the breeding season. Female lambs born in the late winter and spring grow quite rapidly and enter puberty during the breeding season of the following autumn. By this time they are 25–35 weeks old and have attained a body weight of approximately 40 kg. If a female lamb were born too late in the year and were unable to attain this critical weight, or if the animal were deprived of adequate nutrition, she would not come into puberty until the breeding season of the following year (see Foster, Karsch, Olster, Ryan, and Yellon 1986 for review).

Mean plasma LH levels in female lambs are uniformly low during the first

3 or 4 weeks of life. In some animals, pulsatile release of LH begins to occur at the age of 4 weeks, and is observed in virtually all animals by 9 to 11 weeks of age (Foster, Jaffe, and Niswender 1975a; Foster, Lemons, Jaffe, and Niswender 1975b). The pulsatile mode of release results in mean LH levels that exceed the baseline values found in the adult female. The high levels of LH extant during the maturation process persist through the oestrous cycles of the first breeding season, then LH secretion declines to levels characteristic of the cycling adult. In contrast, mean plasma FSH levels are relatively constant over development, do not show evidence of pulsatile release, and remain within the range of values found in the adult female (Foster *et al.* 1975a; Foster *et al.* 1975b).

Oestradiol levels are low throughout the juvenile period in female lambs (Foster *et al.* 1986) and increase at the time of the first preovulatory surge and ovulation. Progesterone levels rise from low prepubertal levels to high levels during the luteal phase of the first oestrous cycle and show cyclic changes in the adult female thereafter (Foster *et al.* 1986). The weights of the ovaries and uterus of the developing lamb peak between 4–8 weeks of life, then decline until the onset of puberty (Kennedy, Worthington, and Cole 1974).

A comparison of rats and sheep shows similarities and disparities in the details of their postnatal development. The males of both species, and the female rat, exhibit a transient increase in plasma LH levels during the early postnatal period. The female sheep, but not the female rat, sustains a relatively high level of plasma LH from the initiation of LH secretion until the end of the first breeding season. Sheep neither demonstrate a prepubertal peak in plasma FSH levels, nor go through a period during juvenile life when 5-alpha-reduced steroid products predominate in the plasma, as do rats. In the males of both species, increases in plasma prolactin levels are correlated temporally with increased growth of sex organs. In females of both species, marked elevations of oestradiol are delayed until near the time of first ovulation.

3 NON-HUMAN PRIMATES

Non-human primates such as macaques, baboons, and apes have a much longer life cycle than sheep or rats. Old World monkeys (macaques and baboons) live to an age of 30–40 years in captivity, whereas apes (chimpanzees, gorillas, and orangutans) can live to be 50 years old (Bowden and Jones 1979), and the length of time spent in a state of reproductive quiescence during juvenile life is commensurately longer than that in the rat or sheep. Macaques and baboons are weaned from their mothers at 4 to 6 months of age, and do not enter puberty until 3 years of age. The apes have an even more protracted "childhood", entering puberty at 7 to 10 years of age (Butler

1974). The following discussion focuses on studies performed in macaques, from which the largest body of literature pertaining to reproductive endocrinology has been derived.

i   Males

Pubertal maturation has been less extensively studied in the macaque than in the rat, human, or sheep. In the male macaque, plasma LH levels are slightly elevated during the first 6 months of life (Frawley and Neill 1979), then decrease to very low levels which persist over the next 2 to 3 years (Steiner and Bremner 1981; Plant 1985). The peripubertal period extends from about 2 to 4 years of age and is heralded by nocturnal elevations in plasma LH levels; daytime elevations of LH are observed shortly thereafter (Plant 1985). LH is probably released in a pulsatile fashion throughout the life cycle, since analyses of serial nighttime blood samples reveal that infantile, peripubertal, and adult macaques demonstrate episodic LH release, and approximately half of the juvenile animals release LH in pulses (Steiner and Bremner 1981; Plant 1983).

FSH levels follow a similar pattern to those of LH: a small postnatal elevation, sustained over the first few months of life (Frawley and Neill 1979; Fuller, Faiman, Winter, Reyes, and Hobson 1982), is followed by low to undetectable levels for the next 2 or 3 years. The peripubertal nocturnal reinitiation of FSH secretion appears to lag behind that of LH (Plant 1985); however, monkey FSH radioimmunoassays are notoriously poor and may fail to detect small amounts of FSH secreted during the earliest phases of reinitiation. After the first year, prolactin begins to exhibit long term fluctuations in plasma concentrations which occur with a periodicity of approximately one year (Plant 1985). The appearance of elevated plasma androgen levels at night, signalling reactivation of gonadotrophin secretion, is correlated temporally with low or declining prolactin levels.

Testosterone levels mirror those of gonadotrophins over development, being high in neonates, declining through late infancy, and reaching nadir values during juvenile life (Meusy-Dessolle and Dang 1985). Nocturnal elevations in androgen levels mark the peripubertal reactivation of the reproductive axis, preceding slightly the nocturnal increments in LH secretion (Plant 1985). Testosterone secretion during development appears to be episodic in nature, with testosterone pulses being coincident with or preceded by an LH pulse (Steiner and Bremner 1981).

Somatic changes occurring in conjunction with these endocrine events are a low and stable rate of increase of body weight to 36 months, then accelerating rates of weight gain as the animal enters puberty. The rate of increase of testis size is low and stable until 36 months, then undergoes a rapid increase (Steiner and Bremner 1981).

ii Females

The female macaque undergoes menarche, or the first shedding of menstrual blood, around 30 months of age. Animals raised in a light- and temperature-controlled environment do not ovulate until nearly 18 months later, and do not achieve full reproductive competence until 60 months of age. The period between first ovulation and reproductive competence is characterized by irregular ovulation and, quite often, short luteal phases (Terasawa, Bridson, Nass, Noonan, and Dierschke 1984). First ovulation in spring-born animals raised outdoors is restricted to the late fall and early winter months, occurring either shortly after menarche (at around 31 months of age), or 11 months later at 42 months of age (Wilson, Gordon, and Collins 1986).

Basal LH levels are low during the prepubertal period (10–20 months of age) and increase during the early pubertal phase (20–30 months). Circadian fluctuations are evident during this time, with nocturnal secretion exceeding daytime release. LH pulses, occurring at a frequency of about 1 to 3 per hour and superimposed on diurnal fluctuations in basal levels, are first noted during the early pubertal phase. The period between menarche and first ovulation (30–48 months) is designated the midpubertal stage and is characterized by further increases in basal LH levels and maximal amplitude in circadian fluctuations. After first ovulation the late pubertal state ensues, when the animal is incapable of reproducing as an adult because she is not experiencing regular ovulatory cycles. During the late pubertal state, basal LH declines somewhat and the amplitude of circadian fluctuations diminishes (Terasawa *et al.* 1984). Basal FSH levels increase around 22 months of age and are maximal at 40 months, declining to adult levels after the first ovulation. In the individual animal, the rise in FSH precedes that of LH and oestradiol (by about 5 months), and menarche (by 6–8 months).

Plasma oestradiol levels increase around 27 months and rise steadily to a maximum just before the first ovulation. In individual animals, oestradiol increases 4–5 months before menarche, coincidently with an increase in nipple size. Progesterone concentrations remain low until after the first ovulation. Somatic changes occurring during pubertal maturation include an increase in nipple size and an increase in body weight. As mentioned above, nipple size increases coincidently with oestradiol levels (at around 27 months) and continues to increase to maximum size around 45 months, just before the first ovulation. The most rapid rate of increase of body weight occurs during the first 10 months of life, and rapid weight gain is also observed between 18 and 31 months and between 38 and 47 months (the periods preceding menarche and first ovulation, respectively). This is true in individual animals as well as group means.

4 HUMAN BEINGS

Human beings, with an expected lifespan of more than 50 years, have a protracted childhood of at least 10 years, similar in duration to that experienced by apes. During the prepubertal years social and intellectual skills are developed which prepare the child for an autonomous existence, and body weight and physical skills increase to the extent that a pregnancy can be sustained by a female, and protection of offspring can be offered by both males and females. The enormous amount of information which needs to be assimilated by young human beings living in technologically advanced societies creates a situation in which the length of time spent living with adults is prolonged and the age at which offspring can be provided for is delayed. The benefits of delayed reproduction have not, it appears, been communicated to the reproductive axis, which awakens some 10 years "too early" by these societal standards, perhaps contributing to the emotional conflicts experienced by many human beings during the second decade of life. This situation contrasts that of non-human primate troops, in which young females produce offspring within a few years of attaining reproductive maturity, and males often leave their troop of origin when they reach sexual maturity and seek breeding opportunities within other troops.

Detailed descriptions of somatic changes occurring during puberty have been collected for hundreds of years, although characterization of the endocrine correlates of these changes have been a much more recent addition to the scientific literature. The earliest recorded longitudinal study of growth rates in the human was made in the 18th century, by Count Philibert Gueneau de Montbeillard upon his son (Tanner 1962). In these early data from an individual boy, and from other individual and group data collected since then, it is evident that birth occurs during the time of peak growth velocity of the fetus/neonate. Subsequently, the most rapid growth occurs during the adolescent growth spurt between 12 and 16 years of age in boys, and between 10 and 14 years of age in girls. During this time, most elements of the muscular, skeletal, and reproductive systems grow at an accelerated rate and marked changes in motor skills and intellectual function occur (Donovan and Van der Werff ten Bosch 1965). Puberty is divided into Stages 1 to 5, distinguished by the development of secondary sexual characters (Tanner 1962).

During normal development the adrenal glands also mature and secrete increasing amounts of androgens. This process, referred to as adrenarche, occurs several years before the onset of gonadal maturation (gonadarche) and appears to be unique to the anthropoid apes, exhibited only by human beings and chimpanzees. Androgens secreted by the adrenal glands stimulate development of some secondary sex characteristics (for example, pubic hair) and, when converted to oestrogen, stimulate breast budding, Still, the precise relationship between adrenarche and the timing of puberty is unclear. Gonadarche and adrenarche can be activated independently of each other, and adrenarche is probably not a necessary precursor to the onset of puberty in primates (Cutler and Loriaux 1980; Wierman, Beardsworth, Crawford, Crigler, Mansfield, Bode, Boepple, Kushner, and Crowley 1986).

i   Males

The growth of the testes and thinning of the scrotum, occurring at an average age of 12 years and preceding the growth of pubic hair and the penis, signals that puberty is imminent (Tanner 1962). Testicular growth occurs under the influence of FSH and is attributable to an increased seminiferous tubule volume resulting from a thickening of the tubular lining, the formation of a lumen, and the differentiation and growth of Sertoli and Leydig cells. Mature spermatozoa first appear in the urine at a median age of 13 years (Nielsen, Skakkebaek, Richardson, Darling, Hunter, Jorgensen, Nielsen, Ingerslev, Keiding, and Muller 1986).

Studies performed in the early 1970s with (then) recently developed gonadotrophin radioimmunoassays demonstrated that mean LH levels rise steadily from low prepubertal values to adult values by the end stages of puberty (Burr, Sizonenko, Kaplan, and Grumbach 1970; Faiman and Winter 1974), although subsequent studies indicate that the most rapid increase in LH levels occurs during stages P2–P4 (Lee and Migeon 1975; Nottlemann, Susman, Dorn, Inoff-Germain, Loriaux, Cutler, and Chrousos 1987). The pubertal augmentation of LH secretion, driven by GnRH release from the hypothalamus, was shown by Boyar, Kulin, and others (Boyar, Finkelstein, Roffwang, Kapen, Weitzman, and Hellman 1972; Boyar, Rosenfeld, Kapen, Finkelstein, Roffwarg, Weitzman, and Hellman 1974; Kulin, Moore, and Santner 1976) to occur first during sleep, and in the later stages of puberty to occur during the waking hours as well. In early puberty, the sleep-related augmentation of LH is accompanied by an increase in testosterone levels, but in later pubertal boys and adult men the temporal coincidence of LH and testosterone pulses is not as clearcut. When serum LH levels are measured by bioassay, both basal and GnRH-stimulated peak LH levels increase over the course of puberty (Burstein, Schaff-Blass, Blass, and Rosenfield 1985); as measured by some assay systems, the increase in bioactive LH is much more dramatic than that of immunoreactive LH (Montanini, Celani, Baraghini, Carani, Marrama 1984*a*; Montanini, Celani, Carani, Cioni, Negri, Morabito, Marrama 1984*b*).

Mean plasma FSH levels increase early in puberty (P1–P2) then increase again between P4 and P5 (Nottelmann *et al.* 1987). The early increment in FSH levels is mirrored by a parallel elevation in plasma testosterone values and coincides with a pronounced increase in testicular volume, presumably resulting from a stimulatory effect of FSH on seminiferous tubule development (Burr *et al.* 1970; Lee and Migeon 1975). The predominant circulating gonadal steroids in the pubertal boy are testosterone and androstenedione. The most pronounced increment in testosterone levels occurs between stages P2 and P3 (Grumbach, Roth, Kaplan, and Kelch 1974, Nottlemann *et al.* 1987), while that of androstenedione occurs between stages P3 and P4 (Lee and Migeon 1975).

Prolactin levels, measured over a 24 hour period, do not differ between prepubertal and adolescent boys, although at all ages mean plasma prolactin concentrations are higher during sleep than during wakefulness (Finkelstein, Kapen, Weitzman, Hellman, and Boyar 1978). Progesterone levels are low during the early stages of puberty, and show a transient peak during stage P4. Oestradiol is similarly low during the early stages of puberty and rises gradually through the end of puberty (P5; Nottelmann et al. 1987).

ii   Females

Females experience the adolescent growth spurt earlier than males: the earliest sign of sexual development occurs at around age 7, at the time of adrenarche. The adrenal gland produces increasing amounts of androgens, which are converted to oestrogens and initiate breast development. Breast growth begins in P2, when the breast and papilla are elevated as a small mound above the chest, but is not complete until 18 to 20 years of age. Approximately 6 months after the start of breast growth, axillary and pubic hair appear. Menarche occurs at an average age of 13 years, approximately one year after the apex of the height spurt (Tanner 1962). Females are not fully reproductively competent at the time of menarche, however. As in macaques, human females have irregular and often anovulatory cycles for about a year after menarche. Although Graafian follicles are present at all ages before puberty, ovulation of mature ova does not occur until late in puberty. The ovary undergoes progressive enlargement from 3 years of age onward, whereas the uterus only demonstrates a rapid growth phase beginning around 10 years of age (Tanner 1962).

Early radioimmunoassays revealed that LH levels change very little over P1–P4, but are higher in menstruating (P5) compared to nonmenstruating girls (Sizonenko, Burr, Kaplan, and Grumbach 1970). LH levels measured by bioassay increase dramatically during puberty (P1–P5), with a peak of LH bioactivity occurring about one year after menarche (Lucky, Rich, Rosenfield, Fang, and Roche-Bender 1980; Reiter, Beitins, Ostrea, and Gutai 1982). That LH bioactivity declines over the next few years is consistent with the observation that LH bioactivity is similar in fertile adult women and prepubertal girls (Marrama, Zaidi, Montanini, Celani, Cioni, Carani, Morabito, Resentini, Bonati, and Baraghini 1983). As is true in boys, early prepubertal girls have augmented LH secretion during sleep, presumably as a result of pulsatile GnRH release (Boyar et al. 1972).

FSH levels are high in the human female between 0 and 2 years of age, then decline only to increase two-fold in early puberty (P1–P3). Subsequently, FSH levels are elevated during the follicular phase of the menstrual cycle in menstruating girls compared to nonmenstruating girls (Sizonenko et al. 1970; Nottelmann et al. 1987). The early increase in FSH levels in girls is in contrast to the relatively late rise in FSH levels in pubertal boys. The increment in FSH during P2 and P3 is accompanied by an increase in circulating levels of oestradiol, testosterone, dihydrotestosterone, and androstenedione (Nottelmann *et al.* 1987). Circulating progesterone levels increase dramatic-ally between P3 and P4 (Winter, Faiman, Reyes, and Hobson 1978), and pro-lactin levels increase slightly during the later stages of puberty. Normal girls experiencing menarche relatively early have a larger, more sustained increase in oestradiol, and a larger increase in FSH at the time of puberty than girls who experience menarche at a late age (Apter and Vihko 1985).

5   COMPARATIVE ASPECTS OF PUBERTAL MATURATION

All of the species discussed above demonstrate a transient neonatal activa-tion of the hypothalamo-pituitary axis, followed by a period of juvenile de-velopment when the reproductive axis is relatively quiescent. The length of time spent in a state of reproductive quiescence is correlated with the length of the life cycle (longer-living species having a more protracted prepubertal development), and in seasonal breeders such as sheep, is further influenced by environmental factors such as daylength. The onset of puberty occurs coincident with or just following a rapid increase in body weight and (in rodents and primates) sexual organ weight. The pubertal increase in the amount of gonadotrophins secreted is accompanied by a shift in the quality of gonadotrophins such that the LH and FSH molecules have greater bio-logical activity. The females of each of these species go through a period between the onset of puberty and the attainment of true reproductive fertility and regular ovulatory cycles which may last for days (rats), weeks (sheep), or years (primates). A feature of prepubertal patterns of gonadotrophin secre-tion common to rats, female macaques and human beings is an early eleva-tion of FSH levels; this has not been demonstrated in sheep or male macaques.

III   Mechanisms of puberty onset

The hiatus of reproductive axis activity during juvenile life in a variety of spe-cies has been described above. The question of what holds the reproductive system in check during this time and what causes its reactivation has not been answered, despite receiving considerable attention. In the 1930s, it was thought that the quiescent state of the system was imposed by the "im-mature" state of one or more of the components of the reproductive axis, that is hypothalamus, pituitary, and gonads. Subsequently, various com-ponents were assessed for functional competence and the pituitary and the gonads were shown to be capable of responding to appropriate stimulation with responses characteristic of the adult state. These observations suggested that neither the pituitary nor the gonads were limiting to the onset of puberty, but that some central process imposes a restraint, either actively by direct inhibition or passively by virtue of being in an immature state: evidence supporting this hypothesis is presented below.

Immunoreactive GnRH is present in the hypothalamus of male and female rats by day 12 of gestation (Aubert, Begeot, Winiger, Morel, Sizonenko, and Dubois 1985), at which time it appears to be required for differentiation of the pituitary anlage into gonadotrophs capable of secreting LH and FSH (Begeot, Morel, Rivest, Aubert, Dubois, and Dubois 1984). GnRH in the human hypothalamus is detected with immunocytochemistry (King, Anthony, Fitzgerald, and Stopa 1985) and radioimmunoassay (Aubert, Grumbach, and Kaplan 1977; Paulin, Dubois, Barry, and Dubois 1977; Aksel and Tyrey 1977; Siler-Khodr and Khodr 1978), and is present by 4.5 weeks of gestation (Winters, Eskay, and Porter 1974). By 9 weeks of human gestation, GnRH perikarya with axonal connections terminating on the median eminence can be distinguished; by 12 weeks the pituitary can produce gonadotrophins measurable in the fetal circulation (Winter 1982). The appearance of high affinity binding sites on fetal rat pituitary and the first appearance of radioimmunoassayable LH in the pituitary occur concurrently around D 17 (Aubert et al. 1985). Injection of synthetic GnRH into 17 or 18 day fetal rats results in an elevation of circulating levels of LH (Salisbury, Dudley, and Weisz 1982; Daikoku, Adachi, Kawano, and Wakabayashi 1981). These data demonstrate that GnRH is synthesized and that the pituitary is capable of responding to GnRH stimulation during fetal life.

Not only is GnRH present in the hypothalamus from fetal life onward, but it also appears to be released, and released in a pulsatile fashion, throughout postnatal development (Foster et al. 1975a; Foster et al. 1978; Steiner and Bremner 1981; Bourguignon and Franchimont 1984; Ojeda, Urbanski, and Ahmed 1986). It may be that the amount of GnRH released is just too small to sustain an adult level of reproductive activity. When GnRH neurons are activated by administration of a neuroexcitatory amino acid analog, N-methyl-DL-aspartic acid (NMA), gonadotrophin release ensues in monkeys (Gay and Plant 1987) and the onset of puberty is advanced in female rats (Urbanski and Ojeda 1987). These data suggest that in the prepubertal animal, GnRH neurons are relatively quiescent only because they lack sufficient excitatory stimulation. Morphological changes in GnRH neurons occurring during postnatal development may enable the neurons at puberty to integrate diverse incoming signals into a net excitatory signal and to respond with pulsatile release of GnRH (Rodriquez-Sierra 1986; Wray and Hoffman 1986; Gerocs, Rethelyi, and Halasz 1986); these changes may be steroid-mediated.

The pituitary is capable of releasing gonadotrophins in response to GnRH administration during postnatal development (Brook, Stanhope, Hindmarsh, and Adams 1986), but it exhibits striking variations in its responsiveness to GnRH treatment over the course of sexual maturation. In male rats, plasma LH levels increase rapidly following GnRH administration to immature ($<$ 31 days) and pubertal ($>$ 40 days) animals; however, pubertal animals release more LH in response to repeated doses of GnRH than do immature animals (Nazian 1983). In macaques, the striking increases in plasma gonadotrophin levels observed following GnRH administration to neonates are lost or severely attenuated after 3 months of age, and become evident again sometime after 1.5 years of age (Steiner and Bremner 1981; Monroe, Yamamoto, and Jaffe 1983). However, in animals younger than 1.5 years of age, hourly administration of GnRH over a period of days can stimulate pituitary responsiveness and result in gonadotrophin release not unlike that of adult animals (Wildt, Marshall, and Knobil 1980); indeed, normal 28 day menstrual cycles are initiated in prepubertal macaques by continued exposure to hourly pulses of GnRH. In human beings, an increased pituitary sensitivity to GnRH is evident in pubertal compared to prepubertal children (Reiter, Root, and Duckett 1976): a standardized dose of GnRH elicits an increase in LH levels which is almost 10 fold higher in pubertal than in prepubertal subjects. The LH response in pubertal but not prepubertal children is characterized by a biphasic release pattern suggestive of a maturity related increase in the pituitary LH reserve. One explanation for the pubertal increase in gonadotrophin release in response to exogenous GnRH stimulation is that endogenous GnRH has begun to be released in sufficient quantities to prime the pituitary, thereby effectively increasing pituitary sensitivity and therefore responsiveness. The trophic effect of GnRH on its own pituitary receptors and on pituitary stores of gonadotropins has been well documented and is referred to as a "priming effect" (Nazian 1985).

The capacity of the gonads of immature animals to respond to the appropriate stimulus is evident from studies performed in human beings (Brook *et al.* 1986), macaques (Sopelak, Collins, and Hodgen 1986; Wildt *et al.* 1980), sheep (Foster and Ryan 1981), and rats (Schwartz 1974). These studies demonstrate that the ovaries of immature animals will undergo cyclic changes and that ovulation can be effected by the appropriate regimen of GnRH or gonadotrophin treatment.

1 STEROID DEPENDENT AND STEROID INDEPENDENT MECHANISMS

It is clear from the above discussion that the pituitary is capable of responding to stimulation by GnRH and that the brain produces GnRH during prepubertal life. What is it that holds the system in abeyance? Is GnRH released in insufficient quantities to initiate gonadotrophin secretion? Is the pituitary relatively unresponsive to the GnRH which is released? One hypothesis is that the prepubertal hypothalamic-pituitary axis is exquisitely sensitive to suppression by gonadal steroids. The basis of this so-called gonadostat theory is that during the prepubertal years, the set point of the negative feedback system is much lower than during the adult years; that is, much less steroid is required to suppress the axis to a low level of activity. The earliest suggestion of this was made by Hohlweg and Dohrn in 1932, who stated that a central nervous system (CNS) "Sexualzentrum" exists which regulates gonadotrophin secretion and which undergoes a change in sensitivity to circulating sex steroids around the time of puberty.

Direct tests of this hypothesis came later when the suppressive effects of a given amount of steroid were tested in children and animals at various stages of pubertal development. In one type of experiment, animals are castrated at various postnatal ages and different doses of gonadal steroids (either testosterone or oestradiol) are administered beginning at the time of gonadectomy. The dose of steroid required to prevent the postcastration rise in gonadotrophins is taken as an index of the "sensitivity" of the hypothalamic-pituitary axis to the suppressive effects of steroids. When this is performed in male rats, it is shown that testosterone is effective in lowering gonadotrophin levels following castration, but that much higher doses are required to achieve this in the adult animal than in the immature animal (< 28 days); this is true both when testosterone is administered as daily injections and in Silastic capsules which maintain constant plasma levels of the hormone (Ramirez and McCann 1965; Negro-Vilar, Ojeda, and McCann 1973b; Smith, Damassa, and Davidson 1977). Similar results have been obtained in female rats and mice given oestradiol replacement following ovariectomy: more steroid is necessary to suppress gonadotrophin levels in the adult than in the juvenile animals, whether administered peripherally (Ramirez and McCann 1963; Eldridge, McPherson, and Mahesh 1974; Steele and Weisz 1974; Steele 1977; Bronson 1981) or centrally as implants into the hypothalamic ventromedial arcuate region (Docke, Rohde, Tonjes, Lange, and Dorner 1978).

Another way to examine developmental changes in steroid sensitivity is to assess the magnitude of the postcastration rise in gonadotrophins and GnRH in animals of different ages. When male rats are castrated at various times from birth to maturity, it is clear that the postcastration rise in LH levels increases during development, reaching a maximum around the time of puberty (Negro-Vilar et al. 1973b; Ojeda and Ramirez 1973). Females demonstrate a smaller increase in postcastration LH levels during development (Ojeda and Ramirez 1973), but the maximum postcastration increase in FSH levels occurs well before puberty (around day 20) in both male and female castrates (Ojeda and Ramirez 1973).

Unlike rats, juvenile primates do not have an immediate postcastration gonadotrophin rise, as do adult monkeys, but that rise is delayed until near the age at which the animal would normally go through puberty. However, once the postcastration rise has occurred, less oestradiol is required to suppress gonadotrophins in the immature compared to the adult female monkey (Karsch, Dierschke, Weick, Yamaji, Hotchkiss, and Knobil 1973). Although castration and steroid replacement experiments clearly are not performed in humans, there is nevertheless evidence for a change in the "set point" of the negative feedback mechanism over the course of pubertal development in humans. When oestradiol is administered to children at different stages of maturation, a dose which suppresses urinary FSH to undetectable levels in prepubertal children, only slightly suppresses FSH in older pubertal subjects, and does not suppress FSH at all in adult subjects (Kulin and Reiter 1972; Kelch, Kaplan, and Grumbach 1973).

Thus it seems clear that as puberty proceeds, there is a change in the sensitivity of the hypothalamo-pituitary axis to the suppressive effects of gonadal steroids. The question then becomes whether this is an obligatory event which is principally responsible for the final stages of pubertal maturation, or whether it is merely a consequence of the maturation of CNS components which finally override the suppressive effects which gonadal steroids had exerted. To illustrate: in prepubertal primates GnRH and gonadotrophin secretion is miniscule in amount. The difficulty of stopping the release of gonadotrophins in such a low volume system by the negative feedback action of gonadal steroids might be likened to the problem of stopping a small leak in a garden hose—a little bit of tape might suffice. However, in the adult in which GnRH and LH secretion is much amplified, stopping the flow would be more like stemming the flow of a river—either a great deal of tape or a dam of considerable size would be required!

Arguing against the gonadostat theory are data from animals in which the effects of gonadal steroids are removed at birth or shortly thereafter. These animals demonstrate a gonad-independent maturation of the hypothalamic-pituitary axis and increased gonadotrophin secretion at around the time when puberty normally occurs in intact animals. Female lambs ovariectomized prepubertally show an immediate postcastration rise in LH levels, and a secondary increase at the time of first ovulation in intact lambs (Foster and Ryan 1979). Both male and female rhesus macaques gonadectomized neonatally exhibit a postnatal rise in gonadotrophin levels similar in duration but larger in amplitude than that of age matched, gonad intact animals. There follows the juvenile hiatus in gonadotrophin secretion characteristic of the intact animal, and near the time at which puberty occurs in intact animals, gonadotrophin levels rise in the gonadectomized animals to castrate adult levels (Terasawa *et al.* 1984; Plant 1985; Winter, Ellsworth, Fuller, Hobson, Reyes, and Faiman 1987). The increase in LH levels occurs slightly earlier in ovariectomized (OVX) compared to intact or OVX, oestradiol-treated female macaques, suggesting that oestradiol can delay the maturation of the gonad-independent mechanism (Wilson *et al.* 1986). Patients with gonadal dysgenesis have elevated LH and FSH levels postnatally, experience a juvenile hiatus in gonadotrophin secretion, and demonstrate a marked increase in gonadotrophin production at an age when puberty normally occurs. As in macaques, the time course of developmental changes in gonadotrophin secretion is similar in normal children and children with gonadal dysgenesis, but the levels of gonadotrophins reached during the neonatal and pubertal phases are much higher in the latter group (Conte, Grumbach, and Kaplan 1975; Conte, Grumbach, Kaplan, and Reiter 1980; Ross, Loriaux, and Cutler 1983). These data suggest that gonadal factors do not drive the pubertal reinitiation of gonadotrophin secretion, but are important in modulating the level of activity of the axis once it is activated.

Gonad-independent maturation of the hypothalamic pituitary axis can be demonstrated in castrated animals given exogenous steroids. For example, rats and lambs gonadectomized prepubertally respond to castration with an increase in LH levels, and this increase can be reversed by treatment with exogenous steroids. When steroid titres are maintained at a constant level in castrates by infusions or by Silastic capsules, an increase in LH levels occurs at the time of puberty as in intact controls (Steele and Weisz 1974; Foster and Ryan 1979; Matsumoto, Karpas, Southworth, Dorsa, and Bremner 1986). Similarly, in female rhesus macaques ovariectomized shortly after menarche and given replacement oestradiol capsules, oestradiol reverses the postcastration rise in LH levels until the age at which ovulation first begins, at which point tonic LH secretion escapes from inhibition and LH levels rise (Rapisarda, Bergman, Steiner, and Foster 1983). Thus, it appears that in rats, sheep, and monkeys the final stages of pubertal maturation are characterized by a decrease in the sensitivity of the hypothalamo-pituitary axis to the suppressive effects of gonadal steroids.

The maturation of the hypothalamo-pituitary axis and the concomitant decrease in effective suppression of that axis by gonadal steroids may be the result of a number of factors. Hypothalamic production of GnRH may increase. That this may be true is suggested by studies in male rats that are castrated at various ages and examined for changes in hypothalamic GnRH content. Adult animals show a pronounced decrease in hypothalamic GnRH content following castration, but prepubertal animals accumulate increasing stores of GnRH for up to three weeks after castration, even in the face of markedly elevated gonadotrophin levels (Bourguignon, Gerard, and Franchimont 1984). This suggests that a gonad-independent increase in GnRH synthesis and storage occurs prior to puberty. Hypothalamic GnRH content declines only at the time of puberty in animals castrated prepubertally; again, while gonadotrophin levels remain elevated. These data suggest that a gonad-independent increase in GnRH release occurs around the time of puberty. Further support for this suggestion may be found in studies in sheep, ovariectomized prepubertally and treated chronically with oestradiol. In the prepubertal animals there is an absence of LH pulses, but concomitantly with the escape from oestradiol inhibition and the elevation of mean LH levels, LH pulses (presumably reflecting GnRH pulses) occur on an hourly basis (Foster and Ryan 1981). Finally, the pituitary may become more responsive to endogenous GnRH. In rats, more LH is released in response to a GnRH challenge if the animals have been pretreated with GnRH; this GnRH self-priming effect can only be demonstrated in pubertal rats and castrated prepubertal rats, but is absent in intact prepubertal rats. Thus, it may be that gonadal steroids act prepubertally to suppress the responsiveness of the pituitary to GnRH, and that this in concert with low levels of endogenous GnRH release, acts to keep gonadotrophin secretion low.

In addition to the negative feedback effects of gonadal steroids described above, there exists a positive feedback mechanism whereby oestradiol can elicit the release of a large quantity (a "surge") of gonadotrophins from the pituitary resulting in ovulation. This important aspect of pubertal development first is manifest at various times during postnatal life in the species under consideration here: relatively early (after 3 weeks of age) in sheep (Foster *et al.* 1986), not before D 15 in rats (Andrews, Mizejewski, and Ojeda 1981), and relatively late (after 34 months) in macaques (Terasawa *et al.* 1983). The maturation of the positive feedback system in rats has been described in a recent review by Ojeda (Ojeda, Urbanski, and Ahmed 1986). In truncated form, the course of events will be outlined here.

Circulating levels of oestradiol are low throughout the infantile and early juvenile stages of life. This is because early on the ovary is unresponsive to stimulation by gonadotrophins, and later because high circulating levels of alpha fetoprotein (AFP), to which oestradiol binds avidly, keep free oestradiol levels low. After D 16, decreasing AFP levels allow serum oestradiol titres to rise. (In the macaque, free oestradiol is maintained at low levels by a combination of a high sex hormone binding globulin [SHBG] capacity, a high metabolic clearance rate of oestradiol [MCR-E2], and a low production rate of oestradiol [PR-E2]. SHBG levels decrease, PR-E2 increases, and MCR-E2 decreases as the female matures, resulting in more E2 being available to act on target tissues [Hotchkiss 1983, Hotchkiss 1985]). The rise in serum oestradiol amplifies the signal from a central gonad-independent mechanism regulating prolactin release (Ojeda and McCann 1974; Andrews and Ojeda 1977) such that afternoon surges of prolactin become prominent by the end of the juvenile period. Increasing prolactin levels facilitate the stimulatory actions of gonadotrophins on the ovaries, and levels of ovarian steroids rise. Oestradiol then potentiates two important processes in GnRH-releasing neurons in the median eminence: the stimulatory effect of noradrenaline on prostaglandin E_2 (PGE_2) synthesis, and the stimulatory effect of PGE_2 on GnRH release. These processes appear to be oestradiol-dependent (Ojeda, Urbanski, Katz, and Costa 1986). The activation of a protein kinase C-dependent pathway is also probably necessary for the positive feedback mechanism to function. Finally, oestradiol may increase pituitary responsiveness to GnRH on the day before the surge occurs.

In summary, there appear to be both steroid- dependent and steroid-independent mechanisms acting at the level of the brain to influence the onset of puberty. The steroid dependent mechanisms encompass those processes which lead to a change in sensitivity of the hypothalamo-pituitary axis to the suppressive effects of gonadal steroids, and allow the system to produce

GnRH and gonadotrophins in patterns and amounts commensurate with adult reproductive function. That the GnRH "pulse generator" in the brain can be activated independently of gonadal steroids is evident from studies performed in gonadectomized animals and patients with gonadal dysgenesis. The fact that the augmentation of pulse generator activity occurs at about the same time in gonadectomized and gonad intact animals suggests that the time course of pubertal maturation is a gonad-independent event. The steroid-independent initiation of pulsatile GnRH secretion probably consti- tutes the fundamental activation of the system, but the output of the system is adjusted or fine-tuned by the magnitude of feedback effects exerted by gon- adal steroids.

2   NEUROTRANSMITTERS

i   Catecholamines

The catecholamines (CA) noradrenaline (NA) and dopamine (DA) play an important role in regulating the reproductive system (see Barraclough and Wise 1982; Ramirez, Feder, and Sawyer 1984; Ramirez, Kim, and Dluzen 1985 for reviews) and perhaps in timing the onset of puberty as well. The neuroanatomic basis for the interaction between GnRH neurons and CA neurons has been shown in rat, mouse, and macaque with immunostaining for GnRH and the CA synthesizing enzyme tyrosine hydroxylase (TH), and histofluorescence for localizing catecholamines. Elements (perikarya, fibre varicosities, or terminals) of catecholamine neurons and GnRH neurons coexist in the same regions of the rat and mouse hypothalamus and they occur together in especially high concentrations in the median eminence (McNeill and Sladek 1977), the arcuate nucleus, and the periventricular region (Hoffman, Wray, and Goldstein 1982). In juvenile and adult rhesus macaque brains, CA terminal innervation is the most dense at the caudal as- pect of the median eminence, and extensive innervation is also seen in the periventricular, paraventricular, and supraoptic regions (Ishikawa and Tanaka 1977; Schofield and Everitt 1981). Large numbers of varicosities are seen in the preoptic, supraoptic, arcuate, and periventricular areas of the rhe- sus brain (Hoffman, Felten, and Sladek 1977). These observations provide the anatomic correlates for possible physiological interactions of catechol- amines and GnRH, and the functional relationship of CA and GnRH neuro- nal systems has been described in a number of reviews (cited above). Briefly, the mechanism by which NA is thought to stimulate GnRH release is the fol- lowing: NA activates alpha adrenergic receptors (Ojeda, Negro-Vilar, and McCann 1982) which in turn stimulates PGE_2 synthesis from arachidonic acid. PGE_2 stimulates GnRH release (Ojeda, Negro-Vilar, and McCann 1979) probably via a cAMP dependent pathway (Ramirez et al. 1985).

One approach to the question of CA involvement in the timing of puberty is to study developmental changes in CA biosynthesis during puberty, Such analyses generally involve the measurement of CA content or concentration and/or CA biosynthetic enzyme activity in tissue homogenates. For example, it was established that NA content and TH activity in the whole diencephalon of the rat increase markedly during postnatal development (Porcher and Heller 1972; Coyle and Axelrod 1972), but this information did not elucidate changes in the specific regions involved in the regulation of gonadotrophin secretion. Examining the hypothalamus *per se* of the female rat, Weiner and Ganong demonstrated that adult concentrations of NA are achieved by postnatal day 30, but decline between then and the time of vaginal opening at day 37 (Weiner and Ganong 1972). More recently it has been demonstrated that hypothalamic CA levels increase during puberty in male rats (Matsumoto *et al.* 1986), and that TH content in the rat median eminence increases between prepubertal and adult life (Porter 1986).

Because measurements of content or concentration alone do not indicate whether neural activity is increased or decreased, they are most useful when coupled with measurements of turnover rates, in that high turnover rates are associated with a high degree of neural activity. An examination of CA turnover rates or turnover indices (TI) in the hypothalami of female rats during the first oestrous cycle at the onset of puberty reveals that an increase in NA-TI and a decrease in DA-TI precede the first LH surge at the onset of puberty (Advis, Simpkins, Chen, and Meites 1978). Individual hypothalamic nuclei, microdissected from female rat brains and assayed for both CA-TI and content, exhibit the highest NA-TI and concentration in the medial preoptic nucleus (MPOA) and the mediobasal hypothalmus (MBH) of 35–40 day old animals. Peak DA content and TI occurs in these same areas in 25–35 day old animals. The MPOA and MBH are known to contain high levels of GnRH (Hohn and Wuttke 1979). An interesting observation is that NA-TI in the anterior hypothalamus (containing the MPOA) and in the residual hypothalamus (containing the MBH) increases following castration in prepubertal female rats, that the magnitude of this rise diminishes as the animal matures (Raum, Glass, and Swerdloff 1980), and that the postcastration rise in LH in prepubertal animals can be blocked by a specific inhibitor of NA synthesis (Cocchi, Fraschini, Jalanbo, and Muller 1974). These data, in conjunction with evidence that NA is thought, in general, to be stimulatory and DA is thought to be inhibitory to gonadotrophin release (see reviews cited above), suggest that NA is involved in the stimulation of gonadotrophin release at the time of puberty, and perhaps that DA is involved in the restraint of gonadotrophin secretion during the late juvenile stage. Furthermore, gonadal steroids might exert a suppressive effect of NA activity in prepubertal animals, but this effect may be attenuated as the animal approaches puberty, thereby allowing an increased production of NA and gonadotrophins.

Another approach to this issue is to determine whether blockade of CA biosynthesis influences the timing of puberty. Injection of alpha-methyl-para-tyrosine (MPT), an inhibitor of CA synthesis, to prepubertal female rats does not delay vaginal opening, although it does appear to induce periods of prolonged dioestrus in the animals postpubertally (Weiner and Ganong 1971). The interpretation of studies in which CA biosynthesis is blocked is difficult for a number of reasons: (1) CA affect a number of neuronal systems in addition to GnRH neurons, and may cause generalized perturbations of CNS function; (2) if the CA system were impaired prepubertally, there could be compensatory mechanisms which become functional and initiate puberty at a normal time. Also, in reference to the study cited above, the use of vaginal opening as an index of the timing of puberty may mask CNS events which more accurately reflect puberty onset.

Alternatively, one can examine the effects of agents which alter the timing of puberty on various indices of CA function. Precocious pseudopuberty or premature ovulation can be induced in immature female rats by the injection of pregnant mare serum gonadotrophins (PMSG); an LH surge usually occurs on the afternoon of the second day, and ovulation on the morning of the third day following PMSG injection. When changes in CA biosynthesis are examined in PMSG-treated animals, the following observations are made: NA-TI increases dramatically in the median eminence (ME) and MPOA just prior to the LH surge (Lofstrom, Agnati, and Fuxe 1977a); hypothalamic DA concentrations increase within 18 hours after the injection (Parker and Soliman 1983); but DA-TI decreases in the ME just prior to the LH surge (Lofstrom et al. 1977a). DA agonists, agents which deplete brain NA, and NA receptor blockers administered just before the LH surge all block ovulation in PMSG treated animals (Coppola, Leonardi, and Lippmann 1966; Lofstrom, Agnati, Fuxe, and Hokfelt 1977b). These results suggest that NA plays a stimulatory role and DA plays an inhibitory role in the LH surge and subsequent ovulation induced by PMSG in the immature rat. It is intriguing to speculate that the function of DA-mediated suppression of GnRH release during the intervals between GnRH pulses is to allow the accumulation of relatively large intraneuronal stores of the decapeptide, ensuring that an adequate amount of gonadotropins are available for release during a "pulse". Finally, alpha adrenergic receptors in the hypothalamus decrease over two days following PMSG injections, then increase again just prior to first ovulation (Wilkinson, Herdon, Pearce, and Wilson 1979): the authors suggest that PMSG causes an increase in NA turnover, which in turn causes a desensitization of the receptors and a subsequent resensitization by the time of first ovulation. Evidence that desensitization-resensitization may be important in the dopaminergic regulation of puberty onset will be discussed below. Although the PMSG treated immature animal provides a useful experimental model, a caveat to interpreting the above studies is that the events resulting from PMSG treatment may not represent the events which actually occur during the onset of "natural" puberty. PMSG acts downstream from the brain, at the gonads, to stimulate production of gonadal steroids. These steroids may then act at the brain and the pituitary to influence CA activity and to effect or affect other maturational events which simulate those occurring during natural puberty.

An interesting model for the role of CA in the timing of puberty in the female rat has been proposed by Wuttke (Wuttke, Honma, Lamberts, and Hohn 1980). The salient points of this model are: (1) An increase in the NA-TI in the hypothalamus after day 20 stimulates prolactin release. Prolactin in turn stimulates hypothalamic DA turnover and may also inhibit LH release at the pituitary. The increased DA levels inhibit GnRH release, an effect to which the prepubertal animal may be particularly sensitive (Beck, Hancke, and Wuttke 1978). (2) Continued exposure of hypothalamic DA receptors to DA ultimately desensitizes them, and GnRH escapes DA-mediated inhibition. Receptor desensitization can be demonstrated by the fact that a DA receptor-stimulating drug suppresses LH in chronically ovariectomized animals for a period of a few days, but LH levels rise despite continued treatment with the drug (Beck *et al*. 1978). Furthermore, highly sensitive DA receptors produce more cAMP in response to DA treatment than do desensitized receptors. With cAMP production as an index of receptor sensitization, it can be shown that DA receptors do become desensitized in rats between D 20 and puberty (Wuttke *et al*. 1980). Between D 20 and D 30, elevated levels of prolactin may enhance gonadal maturation and prepare the system for activation once GnRH release is initiated. Thus, manipulations resulting in increased DA receptor stimulation lead to precocious puberty, perhaps by causing an earlier desensitization of the DA receptors, whereas those which delay prolactin increases, delay puberty (Wuttke *et al*. 1980).

Catecholaminergic neuronal systems are clearly implicated in the control of the onset of puberty in a number of species. NA, acting through PGE_2 and cAMP dependent pathways, most likely stimulates GnRH releasing mechanisms in the hypothalamus; an increase in noradrenergic neurotransmission constitutes one means of increasing GnRH output during pubertal development. The increased drive to the GnRH neurons may override inhibitory inputs which had held GnRH secretion in abeyance. Conversely, DA most likely inhibits GnRH release. A decrease in dopaminergic input would also increase the effective drive to the GnRH releasing system, and contribute to the pubertal increase in GnRH output. In the above discussion, the issue of catecholaminergic receptor subtypes has been ignored. Noradrenergic receptor subtypes include alpha 1 and 2, and beta 1 and 2, and dopaminergic subtypes are classified as D1 and D2 (Cooper, Bloom, and Roth 1986). Interpretations of CA studies are confounded by the fact that CA agonists or antagonists are rarely "pure", and exert effects at more than one receptor subtype. Furthermore, different receptor subtypes are regulated differently, and become desensitized and resensitized at different rates. Finally, one receptor subtype may be stimulatory to GnRH release while another may be inhibitory or be without effect. Alpha adrenergic receptors are most likely stimulatory to GnRH release, while beta receptors are probably inhibitory or ineffective (Kalra and Kalra 1983; Ramirez *et al.* 1985; Kaufman, Kesner, Wilson, and Knobil 1985).

ii   Pineal indoleamines

The pineal gland has been linked to the reproductive system for nearly three centuries. As early as the 17th century, an acceleration of pubertal growth was linked to a pineal tumor, and similar reports appeared in the medical literature over the next 200 years. The pineal gland produces the indole-amines serotonin and melatonin. These compounds have each been impli-cated in the control of gonadotrophin secretion and their respective functions will be outlined below.

One approach towards understanding the role of the pineal in the control of gonadotrophin secretion and puberty is to remove the pineal gland and observe the consequences. When this operation is performed in neonatal female laboratory rats (which are not considered to be seasonal breeders), the onset of puberty (as indicated by vaginal opening, the time when the positive feedback effects of oestradiol are manifest, and the time when diminished sensitivity to the negative feedback effects of oestradiol become apparent) is advanced compared to sham-lesioned controls (Relkin 1971; Faigon, Cardi-nali, and Moguilevsky 1982). Conversely, there is little evidence that the pineal gland plays a role in restraining the reproductive system during juven-ile life in the seasonally breeding rhesus macaque: when neonatally orchidec-tomized animals are pinealectomized or left intact, the juvenile hiatus in gonadotrophin secretion occurs at the same time in the two groups (Plant and Zorub 1986). Removal of the pineal gland interferes with the production of both melatonin and serotonin, and studies of this sort do not allow a dis-tinction to be made between the effects of the two indoleamines. The effects of each of these compounds will be considered separately in the sections below.

● *Serotonin.* The indoleamine serotonin (5-HT) is believed to play a role in the control of reproduction, although its exact functions have not been fully characterized. It is present in high concentrations in the suprachiasmatic nucleus (SCN), arcuate nucleus, and median eminence of the hypothalamus of the adult male rat (Saavedra, Palkovits, Brownstein, and Axelrod 1974), and the 5-HT content of the hypothalamus and pituitary of the female rat increases markedly between days 20 and 35 of age (Moguilevsky, Faigon, Rubio, Scacchi, and Szwarcfarb 1985a). Binding sites for 5-HT are present in the female rat hypothalamus and MPOA by D 5, and the number of sites and affinity for the ligand show a large increase between D 16 and D 37 (Becu-Villalobos and Libertun 1986).

As with many neurotransmitters, serotonin has both inhibitory (see Gallo 1980 for review) and facilitatory (Wilson and Endersby 1979; Moguilevsky, Faigon, Scacchi, and Szwarcfarb 1985*b*; Vitale, Parisi, Chiocchio, and Tramezzani 1986) effects on gonadotrophin release, depending on the age of the animal and the prevailing steroid milieu. In ovariectomized adult rats, electrical stimulation of the arcuate nucleus or the midbrain dorsal raphe nucleus (DRN, known to contain cell bodies synthesizing 5-HT whose fibre tracts ascend to the hypothalamus) results in an inhibition of pulsatile LH release; however, in serotonin-depleted animals this inhibition does not occur, and in serotonin-replaced animals, some of the inhibitory effect is restored (Gallo and Moberg 1977; Arendash and Gallo 1978, Gallo 1980). In animals with high levels of steroids, serotonin or its agonists might enhance the positive feedback effects of steroids on LH release, since: (1) pretreatment of ovariectomized rats with gonadal steroids reverses the suppressive effect of arcuate stimulation (Gallo and Osland 1976); (2) in ovariectomized steroid-primed rats the preovulatory LH surge is completely abolished in serotonin-depleted animals but is restored when 5-HT levels are restored (Chen, Sylvester, Ieiri, and Meites 1981); and (3) central administration of 5-HT potentiates the ability of PMSG to induce ovulation in immature rats (Wilson and Endersby 1979).

5-HT may be involved in timing the development of the positive feedback effects of ovarian steroids in the prepubertal rat. The positive feedback effects of oestrogen-progesterone treatment are manifest after D 20 in normal female rats, but destruction of serotonin neurons with p-chloroamphetamine (PCA) advances the onset of the positive feedback mechanism such that it is apparent in 20 day old animals (Moguilevsky *et al.* 1985*b*). Administration of a precursor of serotonin (5-hydroxytryptophan or 5-HTP) to female rats 20 days of age or younger induces LH release, but this stimulatory effect disappears around the time when oestrogen positive feedback normally matures (Moguilevsky 1985*b*), and in adult rats 5-HTP exerts an inhibitory effect on LH release (Moguilevsky 1985*a*). These latter data suggest an inhibitory role for serotonin in the control of gonadotrophin secretion during the late juvenile phase of development in the rat, when it may delay the maturation of steroid positive feedback and contribute to the prolongation of the prepubertal hiatus in reproductive activity.

5-HT may also influence the onset of puberty by stimulating prolactin release. 5-HT administration results in prolactin release in both adult (Clemens, Sawyer, and Cerimele 1977) and prepubertal female rats, and in the latter group its stimulatory effects increase gradually between D 12 and the onset of puberty. Hypothalamic 5-HT-TI increases concomitantly with the prolactin peak observed during the first oestrous cycle at the onset of puberty (Advis *et al.* 1978). The same model for the involvement of prolactin in puberty onset that was described above may be invoked to describe the role of 5-HT in the onset of puberty.

• *Melatonin.* One way to approach the question of the role of melatonin in reproductive development is to administer melatonin to immature animals and assess the effects of this treatment on the onset of puberty. Melatonin injected daily either peripherally or centrally to immature male or female rats delays sexual maturation as judged by vaginal opening, reproductive organ weights, and plasma levels of gonadotrophins and gonadal steroids (Wurtman, Axelrod, and Chu 1963; Collu, Fraschini, and Martini 1971; Lang, Aubert, Rivest, Vinas-Bradtke, and Sizonenko 1985). However sexual maturation resumes after cessation of melatonin treatments, and if treatments are prolonged there is an escape phenomenon such that sexual maturation proceeds despite daily melatonin injections (Lang *et al.* 1985). Thus, although melatonin may be able to delay sexual maturation in the rat, it cannot do so indefinitely.

To examine the role of this compound in human puberty, changes in melatonin levels over the course of pubertal development have been examined. Although some authors can detect no change in daytime plasma melatonin levels over pubertal development (Lenko, Lang, Aubert, Paunier, and Sizonenko 1982; Waldhauser, Frisch, Waldhauser, Weiszenbacher, Zeitlhuber, and Wurtman 1984), others report a decline in boys during the earliest stages of puberty (Tanner stages 1–2), coincident with an increase in plasma gonadotrophins and testosterone (Silman, Leone, and Hooper 1979). A decline in nocturnal plasma levels and a concomitant increase in urinary excretion of melatonin occurs in children in the early stages of puberty (Tetsuo, Poth, and Markey 1982; Penny 1982; Waldhauser *et al.* 1984). Although in general it seems that mean plasma melatonin levels decrease over puberty, there have been reports that they increase (Penny 1985) and that the nocturnal elevation in LH is accompanied by an elevation of melatonin levels (Fevre, Segel, Marks, and Boyer 1978). Boys with precocious puberty have daily plasma melatonin patterns indistinguishable from those of normal boys in Tanner stages P1–P3 (Ehrenkranz, Tamarkin, Comite, Johnsonbaugh, Bybee, Loriaux, and Cutler 1982; Gupta 1986). However, boys with delayed puberty have significantly higher nocturnal levels of melatonin than do normal, age-matched boys (Gupta 1986), and show the greatest decrease in melatonin levels between mid and late puberty rather than in early puberty as do normal subjects (Cohen, Hay, Annesley, Beastall, Wallace, Spooner, Thomson, Eastwold, and Klee 1982).

The true role of melatonin in the onset of puberty in nonseasonal breeders is still obscure, but evidence from *in vitro* studies suggests that melatonin could exert effects at a number of points in the hypothalamo-pituitary-gonadal axis. It has been argued that melatonin exerts a central effect on the frequency of GnRH release (Rivest, Aubert, Lang, and Sizonenko 1986), and in fact melatonin does stimulate the release of GnRH from perifused fragments of rat medial basal hypothalamus (Kao and Weisz 1977). Melatonin inhibits the release of gonadotrophins in response to GnRH by cultured pituitary cells from very young ($<$ D 15) but not older ($\geqslant$ D 20) animals (Martin, Engel, and Klein 1977; Martin and Sattler 1979). Also, melatonin stimulates progesterone and androstenedione synthesis by the human ovary (MacPhee, Cole, and Rice 1975). These gonadal steroids may in turn feed back on the hypothalamus and/or pituitary to influence gonadotrophin production. Because clear correlations between serum melatonin and serum gonadotrophin levels have not been established *in vivo* (Lisoni, Resentini, Mauri, De Medici, Morabito, Esposti, Di Bella, Esposti, Rossi, Parravicini, Legname, and Fraschini 1986) it has been suggested that changes in melatonin levels during puberty are chronologically, but not causally, related to changes in gonadotrophin levels (Waldhauser and Dietzel 1985; Sizonenko, Lang, Rivest, and Aubert 1985; Gupta 1986).

It is easier to defend the importance of melatonin in the control of reproduction in seasonal breeders, such as sheep, in which photoperiod is an environmental clue used to time reproductive activity. Synthesis and release of melatonin are tightly coupled to the light–dark cycle, such that plasma and pineal levels of melatonin are high at night and low during the day (Rollag and Niswender 1976).

In adult sheep, evidence for a direct photoperiodic drive on GnRH secretion comes from the ovariectomized ewe in which LH pulse patterns follow those of intact animals. In the long day nonbreeding season (anoestrus), large amplitude, slow frequency pulses prevail whereas during the short day breeding season (oestrus) faster frequency, lower amplitude pulses are dominant. Oestradiol exerts a profound suppressive effect on GnRH pulse frequency during anoestrus but not during oestrus. Pinealectomized ewes do not respond to either the inductive effects of short daylength or the inhibitory effects of long daylength, but these responses can be restored with nightly infusions of melatonin (Bittman, Dempsey, and Karsch 1983*a*; Bittman, Karsch, and Hopkins 1983*b*). The data indicate that melatonin can influence reproduction in the adult sheep, but how does the sensitivity of the young ewe to photoperiod time the onset of puberty? In general it appears that photoperiod exerts its effect by influencing tonic LH secretion. As mentioned earlier, lambs born in the spring normally come into puberty 30 weeks later, as the long days of summer change into the short days of fall; it appears that this sequence of daylengths is critical in the timing of puberty. Lambs maintained from birth on either long or short day photoperiods demonstrate a delay in the onset of puberty, whereas those which are exposed to long days then short days go through puberty at the normal time (Yellon and Foster 1985). Denervation of the pineal gland of young ewes delays the onset of puberty (Foster, Yellon, and Olster 1985*b*), as does continuous administration of melatonin to very young ($<$ 8 weeks) ewes raised in natural photoperiod (Kennaway and Gilmore 1984). Continuous melatonin treatment initiated at 19 weeks advances the onset of puberty, again suggesting that once exposure to long daylength has occurred, a shift to short daylength will trigger puberty

(Nowak and Rodway 1985). The lamb becomes sensitive to photoperiod around 10–20 weeks of age; that is, the animal can establish melatonin rhythms which correspond to the prevailing daylength. It may be that under chronic short day treatment the (as yet unidentified) target tissue(s) of melatonin become refractory to the short day stimulus, whereas long day melatonin patterns establish responsiveness to subsequent short day stimuli. Furthermore, the lamb may be born refractory to short days, and either long days or the attainment of a certain physiological age may be necessary to override this refractoriness. The ultimate effect of inhibitory photoperiods appears to be to inhibit tonic LH secretion by maintaining a high degree of responsiveness to oestradiol negative feedback (Foster *et al.* 1985*b*).

iii   Endogenous opioids

Endogenous opioid peptides (EOP) were identified in 1975 (Hughes, Smith, Kosterlitz, Fothergill, Morgan, and Morris 1975), and subsequently characterized and grouped into three families: endorphins, enkephalins, and dynorphins. Each family comprises the cleavage products of a unique precursor molecule; for example, the endorphins derive from the precursor proopiomelanocortin (POMC) (Roberts and Herbert 1977). Over the last decade, a considerable effort has been expended to elucidate the physiological actions of EOPs, including their role in regulating the onset of puberty.

It has been established that EOPs are involved in the control of GnRH secretion. The anatomical basis for this interaction has been demonstrated in the human (Bloch, Bugnon, Fellman, and Lenys 1978; Leonardelli and Tramu 1979) and the rat (Bloom, Battenberg, Rossier, Ling, and Guillemin 1978). In the fetal human hypothalamus, β-endorphin and GnRH neurons are co-localized in the arcuate nucleus, the lamina terminalis, and the pre-optic area, and immunoreactivity for both β-endorphin and GnRH can be detected in some neurons in the arcuate nucleus. Cell bodies containing immunoreactive β-endorphin in the rat are located predominantly in the arcuate nucleus.

Physiological studies demonstrating the suppressive effects of opiates on reproductive function date back three decades to the studies of Barraclough and Sawyer, who showed that morphine inhibits ovulation in the rat (Barraclough and Sawyer 1955). Subsequently, morphine was shown to suppress plasma LH levels in male rats (Bruni, Van Vugt, Marshall, and Meites 1977; Cicero, Badger, Wilcox, Bell, and Meyer 1977), whereas the opiate receptor antagonist naloxone produces an elevation in plasma LH and FSH levels in humans (Morley, Baranetsky, Wingert, Carlson, Hershman, Melmed, Levin, Jamison, Weitzman, Chang, and Varner 1980; Delitala, Devilla, and Arata 1981) and rats (Bruni *et al.* 1977). Furthermore, microinjections of anti-endorphin antibodies into the arcuate nucleus of female rats elicit an increase in plasma LH levels (Schulz, Wilhelm, Pirke, Gramsch, and Herz 1981).

In the rat, EOPs do not appear to act directly on the anterior pituitary to suppress gonadotrophin release. The anterior pituitary has very few opiate binding sites (Simantov and Snyder 1977); neither morphine (Cicero *et al.* 1977) nor naloxone (Cicero, Schainker, and Meyer 1979; Detlitala *et al.* 1981) alter the pituitary response to GnRH stimulation; and opiate receptor antagonists increase the frequency of LH pulses (Veldhuis, Rogol, Johnson, and Dufau 1983; Delitala *et al.* 1981). These data suggest that EOPs exert their effect at the level of the hypothalamus to reduce GnRH secretion. However, in prepubertal boys chronic administration of naltrexone, an opiate antagonist with weak agonist properties, reduces the LH response to GnRH administration (Mauras, Veldhuis, and Rogol 1986), which may reflect a direct effect of naltrexone on the pituitary. Further support for the contention that EOPs act primarily on GnRH neurons in the hypothalamus comes from *in vitro* studies in which perifused hypothalamic tissue from humans (Rasmussen, Liu, Wolf, and Yen 1983) and rats (Leadem, Crowley, Simpkins, and Kalra 1985; Wilkes and Yen 1981) are used. In these studies, infusion of naloxone elicits an increase in GnRH release, suggesting that there is a tonic inhibition of GnRH release effected by EOPs.

The inhibition of GnRH and gonadotrophin release by EOPs appears to require the presence of gonadal steroids. Following gonadectomy of male or female rats, the ability of naloxone to stimulate and of an exogenous opiate peptide to inhibit LH secretion is lost; these effects are manifest by 7 days following gonadectomy (Bhanot and Wilkinson 1983*a*; Bhanot and Wilkinson 1984; Almeida, Nikolarakis, Schulz, and Herz 1987). When male rats are castrated prepubertally (on D 26), opiate inhibition of LH declines only transiently and is reinstated around the time of normal puberty in males (Almeida *et al.* 1987), but if prepubertal castrates are given testosterone at castration they remain unresponsive as adults. The authors suggest that the presence of testosterone during a 'critical window' of time prepubertally is an important determinant of the opiate sensitivity of the adult (Almeida *et al.* 1987). When long term gonadectomized rats are given steroid replacement therapy, gonadotrophin responsiveness to naloxone and opiates is reinstated (Bhanot and Wilkinson 1984).

Changes in plasma opiate levels and changes in the sensitivity of the reproductive axis to the effects of opiates over the course of sexual maturation have been examined. Plasma levels (which may not accurately reflect brain levels) of two opioid peptides, β-lipotropin and β-endorphin, increase from low levels during infancy in humans to reach adult levels at the earliest stages of puberty (P1–P2), and do not change significantly as pubertal maturation proceeds (Genazzani, Facchinetti, Petraglia, Pintor, Bagnoli, Puggioni, and Corda 1983). However, the sensitivity of the hypothalamic-pituitary axis to the effects of opiate infusion or opiate receptor blockade changes over pubertal development. Peripheral injection of a synthetic opiate to 48-hour gonadectomized rats at various ages reveals that more opiate is required to suppress LH levels in pubertal and adult animals than in prepubertal animals (Wilkinson and Bhanot 1982; Bhanot and Wilkinson 1983*b*). There is a sexual difference in the ontogeny of this EOP-mediated suppression of gonadotrophin secretion: whereas morphine suppresses LH levels at an earlier age in males than in females, naloxone infusions result in increased LH secretion at a younger age in female than in male rats (Blank, Panerai, and Friesen 1979; Cicero, Schmoeker, Meyer, Miller, Bell, Cytron, and Brown 1986), and the delayed effect seen in males can be shown to be dependent on neonatal exposure to androgens (Sylvester, Sarkar, Briski, and Meites 1985). A decreased sensitivity to the inhibitory effects of opiates during pubertal development also may occur in human beings. Chronic administration of naltrexone to prepubertal and pubertal boys results in a reduction in LH secretion in the former group and an augmentation of LH secretion in the latter group. The authors suggest that the hypothalamic-pituitary axis is highly sensitive to opiate substances in the prepubertal child, and that in this setting, the weak agonist properties of naltrexone are manifest. In pubertal boys, in whom the sensitivity to opiates is diminished, the weak agonist effects of naltrexone are overridden by its antagonist properties. In other studies in which naloxone is administered to children of different ages, naloxone is shown to stimulate LH secretion only in children in mid-puberty or late puberty (Sauder, Case, Hopwood, Kelch, and Marshall 1984; Petraglia, Bernasconi, Iughetti, Loche, Romanini, Facchinetti, Marcellini, and Genazzani 1986). These observations are consistent with the hypothesis that gonadal steroids are necessary for the suppressive effects of opiates on LH secretion to be manifest. Another explanation for the attenuation of opiate inhibition of LH secretion during puberty may be in differential processing of the β-endorphin molecule: the adult rat hypothalamus contains more of an inactive proteolytic breakdown product than that of the juvenile (Martensz 1985).

In the female rat, chronic naloxone treatment during the neonatal period (P1–10), but not during later stages of prepubertal development, advances the onset of puberty (Sirinathsinghji, Motta, and Martini 1985). The authors propose that chronic opiate receptor blockade during the neonatal period alters synaptic connections between opioid and GnRH neurons, diminishing the interaction between the two systems and allowing an earlier maturation of GnRH releasing mechanisms. Other authors have shown that naloxone administration to prepubertal animals advances the onset of the positive feedback action of oestradiol (Faigon, Szwarcfarb, Scacchi, and Moguilevsky 1987).

In summary, EOPs are implicated as one of the neurotransmitter systems involved in the onset of puberty in mammals. The decreased inhibitory effect of EOPs on gonadotrophin secretion during pubertal maturation may reflect the diminution of a tonic inhibition of GnRH secretion. The interaction between gonadal steroids and EOPs suggests that the developmental change in sensitivity of the hypothalamic-pituitary axis to suppression by steroids may be mediated by the EOP system.

3 NUTRITIONAL AND METABOLIC FACTORS

The influence of nutritional status on reproductive function has been noted for some time but has proved to be a difficult relationship to assess. Malnutrition seems to impair reproductive function, since children who are malnourished during childhood experience a delay in pubertal maturation as judged by increases in gonadotrophin secretion, the development of secondary sex characteristics, and the onset (in girls) of menarche (Kulin, Bwibo, Mutie, and Santner 1984; Galler, Ramsey, and Solimano 1985). Other factors, such as intensive athletic training, which alter the rate of growth or alter body composition in prepubertal children, can also influence the onset of puberty. Young athletes who begin intensive training prior to menarche often have a delayed menarche and/or primary amenorrhea, whereas those who begin training after menarche may experience secondary amenorrhea (Feicht, Johnson, Martin, Sparkes, and Wagner 1978; Frisch, Wyshak, and Vincent 1980; Frisch, Gotz-Welbergen, McArthur, Albright, Witschi, Bullen, Birnholz, Reed, and Hermann 1981; Warren 1980). Anorectic and nonanorectic women with pronounced weight loss often become amenorrheic and have a blunted pituitary response to GnRH administration (Warren, Jewelewicz, Dyrenfurth, Ans, Khalaf, and Vande Wiele 1975; Vigersky, Andersen, Thompson, and Loriaux 1977; Miles and Wright 1984). Frisch and Revelle proposed that a critical body weight had to be reached before menarche could occur (Frisch and Revelle 1970) and subsequently amended this to suggest that the critical factor is attainment of a minimum percentage of body fat (Frisch and McArthur 1974; see also Frisch 1984 for review). Other authors have suggested that the high calorific demands of the brain relative to the body influence the timing of puberty (Winterer, Cutler, and Loriaux 1984). Thus it seems evident that metabolic factors strongly influence the development and maintenance of the reproductive system, but the nature of the metabolic signal(s) and the exact site at which it acts have not been identified.

Studies which attempt to ascertain the effects of limited nutrition on various reproductive parameters and to identify the metabolic factor(s) have been carried out in rats, mice, sheep and macaques. Female rats maintained on a limited amount of food or on a diet lacking a single essential amino acid, either before (Kennedy and Mitra 1963), or after weaning, have delayed vaginal opening (Widdowson and Cowen 1972; Ronnekliev, Ojeda, and McCann 1978; Pau and Milner 1984, Holehan and Merry 1985). In adult females fasting appears to interfere with the positive feedback actions of oestradiol (McClure and Saunders 1985). Male rats deprived of protein postweaning have decreased pituitary and plasma levels of LH and (especially) FSH, and lower accessory sex organ weights than normally fed controls

(Srebnik and Nelson 1962; Stewart, Kopia, and Gawlak 1975; Salem, Coward, Lunn, and Hudson 1984; Glass, Anderson, Herbert, and Vigersky 1984; Glass, Herbert, and Anderson 1986). Glass refutes the critical body weight or critical body fat hypothesis, especially as applied to males, and argues that the timing of underfeeding is an important determinant of the reproductive outcome of nutritional stress (Glass and Swerdloff 1980, Glass *et al.* 1986). He suggests that undernutrition begun before weaning has a more dramatic impact on reproductive function in the male rat than does undernutrition begun after weaning, and that the male rat may respond to a decrease in feeding postweaning with an accelerated onset of sexual function. In addition, there might well be a difference in the response of laboratory-selected species and wild species; relatively wild deer mice respond to a mild reduction of food intake with significant reductions in plasma LH and testosterone, and deranged spermatogenesis, whereas house mice under the same conditions fail to show any alterations in the function of the pituitary-testicular axis (Blank and Desjardins 1984). As Blank and Desjardins point out in their discussion of this issue, "domestication involves countless generations of artificial selection for maximal reproductive performance under optimum energy availability" (Blank and Desjardins 1985). They suggest that each species has a particular response to food restriction which is shaped by the reproductive strategies that have evolved by natural selection. Gender-related differences in the response to nutritional stress, specifically that females tend to respond to food restriction with a more significant impairment of reproductive function than males, are easily interpreted on the basis of the greater amount of energy stores required by females to sustain pregnancy and lactation.

Studies performed in the lamb reinforce the concept that undernutrition retards the maturation of the reproductive system and provide evidence that the effect is exerted primarily at the hypothalamic level to prolong the period of hypersensitivity to steroid negative feedback and prevent the increase in GnRH pulse frequency necessary for the onset of cyclicity (Foster and Olster 1985). Lambs raised on low nutrition during early life have retarded growth and remain anovulatory during the first breeding season, but refeeding leads to rapid catchup growth and the initiation of cycles (Foster and Olster 1985). That underfed lambs are hypersensitive to the negative feedback effects of gonadal steroids is suggested by the ability of refed lambs to escape the inhibitory effects of oestradiol and to produce GnRH pulses at a high enough frequency to support follicular development and ovulation. Starvation of rats interferes with stimulation of LH release by both PGE_2 and naloxone, suggesting a direct effect on GnRH neurons (Kuderling, Dorsch, Warnhoff, and Pirke 1984).

Metabolic factors also modulate the level of activity of the reproductive axis in non-human primates. Castrated adult male rhesus macaques respond to a restricted calorie diet with depressed levels of LH, FSH, insulin, triiodothyronine (T3), and thyroxine (T4), reduced body weight, and elevated cortisol levels; GnRH administration restores plasma gonadotrophins to normal levels, suggesting that underfeeding initially had interrupted the intermittent discharge of GnRH (Dubey, Cameron, Steiner, and Plant 1986). Compared to adults, juvenile macaques undergo a more rapid transition from a fed to a fasted state, as indicated by lower levels of insulin, glucose, and large neutral amino acids, and more rapid increases in plasma ketone body concentrations in juveniles following a meal (Cameron, Koerker, and Steiner 1985). Juvenile animals may exist in a more chronically fasted state, perhaps because of a greater sensitivity to the effects of insulin (Bloch, Clemons, and Sperling 1987), and for that reason may be limited in some metabolic substrate. When levels of that as yet unidentified substrate become high enough, the hypothalamic pulse generator may produce GnRH in a pulsatile mode characteristic of the adult state. Indeed, when prepubertal macaques are given a constant infusion of dextrose and amino acids over a 4–6 week period, some of the animals have elevated LH levels at the end of that time and in at least two of six animals there is evidence of pulsatile LH release (Steiner, Cameron, McNeill, Clifton, and Bremner 1983; Cameron *et al.* 1985).

Although the relative importance of the various metabolic substrates in signalling to the hypothalamus that the organism is ready to reproduce is as yet unclear, it is likely that insulin plays an important role. During the transition into puberty, prolonged postprandial elevations in plasma insulin levels (Cameron *et al.* 1985) may be integrated over time into an elevated cerebrospinal fluid-insulin signal. Insulin binds to the arcuate nucleus and median eminence of the hypothalamus (Van Houten, Posner, Kopriwa, and Brawer 1980; Baskin, Porte, Guest, and Dorsa 1983), where it may provide an important signal to the brain relating body size or metabolic demand to reproductive function. This signal may take the form of increased activity of hypothalamic neurons, or alterations in steroid binding to brain regions (see Steiner *et al.* 1983 for discussion), or may be manifest at the pituitary as an increase in the sensitivity of the gonadotrophs to GnRH stimulation (Adashi, Hseuh, and Yen 1981). Insulin may also affect the availability of precursors of neurotransmitters to the brain; if the synthesis of neurotransmitters regulating GnRH neurons is affected, then GnRH secretion might be altered (see Steiner *et al.* 1983, Steiner 1987 for discussions).

The interaction among opiates, feeding behaviour, and reproduction may also be important. In human beings and other mammals, opiate agonists enhance feeding behaviour, particularly of foods that are high in fat content or are especially palatable, and may also coordinate food-seeking behaviour (see Morley 1987 for review). It is not surprising, then, that fasting activates an opioid pathway in rats: and it appears that this pathway is inhibitory to LH release (Dyer, Mansfield, Corbet, and Dean 1985). Might the chronically semi-fasted state of the juvenile activate this opioidergic pathway, and thus restrain pubertal maturation?

IV   Future work

Although a great deal has been elucidated about the neuroendocrine events which underlie the pubertal process, much remains to be clarified. A potentially fruitful area of inquiry is that of the control of GnRH and gonadotrophin synthesis at the molecular level. Using the currently available techniques of molecular biology, investigators may now address many questions about the transcriptional and translational regulation of the GnRH, LH, and FSH genes and their protein products. Is the juvenile hiatus in GnRH secretion the result of a decreased synthetic capacity of hypothalamic GnRH neurons? This can be assessed using the technique of *in situ* hybridization to measure levels of GnRH mRNA in individual neurons in the basal forebrain and hypothalamus of prepubertal compared to adult animals. A decreased amount of GnRH mRNA in neurons of prepubertal animals would suggest that less peptide can be produced by these animals. One significant advantage of using the *in situ* hybridization technique is that the level of transcription of the GnRH gene in different subpopulations of neurons can be assessed as a function of development. Used in conjunction with retrograde transport assays, it could be ascertained, for example, whether gene expression changes only in those cells which project to the median eminence and which therefore probably control anterior pituitary function. If the amount of GnRH mRNA in these neurons is the same in the prepubertal and adult states, then it could be argued that the apparent decrease in GnRH release during the prepubertal years is caused by a posttranscriptional control mechanism: perhaps the mRNA present is not translated into protein, or the protein product is biologically inactive, or the product is not released from nerve terminals.

Given the limitations of studies which examine changes in tissue levels of neurotransmitters, an elucidation of the ways in which neurotransmitter receptors are regulated would be helpful. Recently, information has been made available pertaining to second messenger systems and the interactions of steroid hormones with neurotransmitters at the cellular level, providing a foundation for future work. More extensive knowledge about the synthesis of receptors (catecholaminergic, indoleaminergic, and opioidergic), the regulation of receptor number, the (intracellular) processes of sensitization and desensitization of these receptors, and the interaction of the various receptor subtypes will facilitate the interpretation of the roles of neurotransmitters in controlling the onset of puberty.

References

Adashi, E. Y., Hsueh, A. J. W., and Yen, S. S. C. (1981). Insulin enhancement of luteinizing hormone and follicle-stimulating hormone release by cultured pituitary cells. *Endocrinology* **108**, 1441–1449.

Advis, J. P., Simpkins, J. W., Chen, H. T., and Meites, J. (1978) Relation of biogenic amines to onset of puberty in the female rat. *Endocrinology* **103**, 11–16.

Almeida, O. F. X., Nikolarakis, K. E., Schulz, R., and Herz, A. (1987). A 'window of time' during which testosterone determines the opiatergic control of LH release in the adult male rat. *J. Reprod. Fert.* **79**, 299–305.

Andrews, W. W., Mizejewski, G. J., and Ojeda, S. R. (1981). Development of oestradiol-positive feedback on luteinizing hormone release in the female rat: a quantitative study. *Endocrinology* **109**, 1404–1413.

—— and Ojeda, S. R. (1977). On the feedback actions of oestrogen on gonadotrophin and prolactin release in infantile female rats. *Endocrinology* **101**, 1517–1523.

—— —— (1981). A quantitative analysis of the maturation of steroid negative feedbacks controlling gonadotropin release in the female rat: the infantile-juvenile periods, transition from an androgenic to a predominantly estrogenic control. *Endocrinology* **108**, 1313–1320.

Apter, D. and Vihko, R. (1985). Premenarcheal endocrine changes in relation to age at menarche. *Clin. Endocr.* **22**, 753–760.

Arendash, G. W. and Gallo, R. V. (1978). Serotonin involvement in the inhibition of episodic luteinizing hormone release during electrical stimulation of the midbrain dorsal raphe nucleus in ovariectomized rats. *Endocrinology* **102**, 1199–1206.

Askel, S. and Tyrey, L. (1977). Luteinizing hormone-releasing hormone in the human foetal brain. *Fert. Steril.* **28**, 1067–1071.

Aubert, M. L., Grumbach, M. M., and Kaplan, S. L. (1977). The ontogenesis of human foetal hormones. IV. Somatostatin, luteinizing hormone releasing factor, and thyrotropin releasing factor in hypothalamus and cerebral cortex of human foetuses 10–22 weeks of age. *J. clin. Endocr. Metab.* **44**, 1130–1141.

——, Begeot, M., Winiger, B. P., Morel, G., Sizonenko, P. C., and Dubois, P. M. (1985). Ontogeny of hypothalamic luteinizing hormone-releasing hormone (GnRH) and pituitary GnRH receptors in foetal and neonatal rats. *Endocrinology* **116**, 1565–1576.

Barraclough, C. A. and Sawyer, C. H. (1955). Inhibition of the release of pituitary ovulatory hormone in the rat by morphine. *Endocrinology* **57**, 329–337.

—— and Wise, P. M. (1982). The role of catecholamines in the regulation of pituitary luteinizing hormone and follicle-stimulating hormone secretion. *Endocr. Rev.* **3**, 91–119.

Bartke, A. (1980). Role of prolactin in reproduction in male mammals. *Fedn. Proc.* **39**, 2577–2581.

Baskin, D. G., Porte, Jr., D., Guest, K., and Dorsa, D. M. (1983). Regional concentrations of insulin in the rat brain. *Endocrinology* **112**, 898–903.

Beck, W., Hancke, J. L., and Wuttke, W. (1978). Increased sensitivity of dopaminergic inhibition of luteinizing hormone release in immature and castrated female rats. *Endocrinology* **102**, 837–843.

Becu-Villalobos, D. and Libertun, C. (1986). Ontogenesis of [^{3}H] serotonin binding sites in the hypothalamus of the female rat: Relation to serotonin-induced LH release in moxestrol-pretreated rats. *Dev. Brain Res.* **25**, 111–116.

Begeot, M., Morel, G., Rivest, R. W., Aubert, M. L., Dubois, M. P., and Dubois, P. M. (1984). Influence of gonadoliberin on the differentiation of rat gonadotrophs: An *in vivo* and *in vitro* study. *Neuroendocrinology* **38**, 217–225.

Ben-Or, Sarah (1963). Morphological and functional development of the ovary of the mouse: I. Morphology and histochemistry of the developing ovary in normal conditions and after FSH treatment. *J. Embryol. exp. Morph.* **2**, 1–11.

Bhanot, R. and Wilkinson, M. (1983a). Opiatergic control of LH secretion is eliminated by gonadectomy. *Endocrinology* **112**, 399–401.

———— (1983b). Opiatergic control of gonadotropin secretion during puberty in the rat: A neurochemical basis for the hypothalamic "gonadostat?" *Endocrinology* **113**, 596–603.

———— (1984). The inhibitory effect of opiates on gonadotrophin secretion is dependent upon gonadal steroids. *J. Endocr.* **102**, 133–141.

Bittman, E. L., Dempsey, R. J., and Karsch, F. J. (1983a). Pineal melatonin secretion drives the reproductive response to daylength in the ewe. *Endocrinology* **113**, 2276–2283.

Bittman, E. D., Karsch, F. J., and Hopkins, J. W. (1983b). Role of the pineal gland in ovine photoperiodism: Regulation of seasonal breeding and negative feedback effects of oestradiol upon luteinizing hormone secretion. *Endocrinology* **113**, 329–336.

Blank, J. L. and Desjardins, C. (1984). Spermatogenesis is modified by food intake in mice. *Biol. Reprod.* **30**, 410–415.

———— (1985). Differential effects of food restriction on pituitary-testicular function in mice. *Am. J. Physiol.* **248**, R181–R189.

Blank, M. S., Panerai, A. E., and Friesen, H. G. (1979). Opioid peptides modulate luteinizing hormone secretion during sexual maturation. *Science, N.Y.* **203**, 1129–1131.

Bloch, B., Bugnon, C., Fellman, D., and Lenys, D. (1978). Immunocytochemical evidence that the same neurons in the human infundibular nucleus are stained with anti-endorphins and antisera of other related peptides. *Neurosci. Lett.* **10**, 147–152.

Bloch, C. A., Clemons, P., and Sperling, M. A. (1987). Puberty decreases insulin sensitivity. *J. Pediatr.* **110**, 481–487.

Bloom, F., Battenberg, E., Rossier, J., Ling, N., and Guillemin, R. (1978). Neurons containing β-endorphin in rat brain exist separately from those containing enkephalin: Immunocytochemical studies. *Proc. natn. Acad. Sci. U.S.A.* **75**, 1591–1595.

Bourguignon, J. P. and Franchimont, P. (1984). Puberty-related increase in episodic LHRH release from rat hypothalamus *in vitro*. *Endocrinology* **114**, 1941–1943.

——, Gerard, A., and Franchimont, P. (1984). Age-related differences in the effect of castration upon hypothalamic LHRH content in male rats. *Neuroendocrinology* **38**, 376–381.

Bowden, D. M. and Jones, M. L. (1979). Aging research in nonhuman primates. In *Aging in Nonhuman Primates*. (ed. D. M. Bowden). Van Nostrand Reinhold Co., New York.

Boyar, R., Finkelstein, J., Roffwang, H., Kapen, S., Weitzman, E., and Hellman, L. (1972). Synchronization of augmented luteinizing hormone secretion with sleep during puberty. *New Engl. J. Med.* **287**, 582–586.

——, Rosenfeld, R. S., Kapen, S., Finkelstein, J. W., Roffwang, H. P., Weitzman, E. D., and Hellman, L. (1974). Human puberty: simultaneous augmented secretion of luteinizing hormone and testosterone during sleep. *J. clin. Invest.* **54**, 609–618.

Bronson, F. H. (1981). The regulation of luteinizing hormone secretion by oestrogen: relationships among negative feedback, surge potential, and male stimulation in juvenile, peripubertal, and adult female mice. *Endocrinology* **108**, 506–516.

Brook, C. G. D., Stanhope, R., Hindmarsh, P., and Adams, J. (1986). The control of the onset of puberty. *Acta endocr.* Suppl **279**, 202–6.

Bruni, J. F., Van Vugt, D., Marshall, S., and Meites, J. (1977). Effects of naloxone, morphine and methionine enkephalin on serum prolactin, luteinizing hormone, follicle stimulating hormone, thyroid stimulating hormone and growth hormone. *Life Sci.* **21**, 461–466.

Buckingham, J. C. and Wilson, C. A. (1985). Peripubertal changes in the nature of LH. *J. Endocr.* **104**, 173–177.

Burr, I. M., Sizonenko, P. C., Kaplan, S. I., and Grumbach, M. M. (1970). Hormonal changes in puberty. I. Correlation of serum luteinizing hormone and follicle stimulating hormone with stages of puberty, testicular size, and bone age in normal boys. *Pediat Res.* **4**, 25–35.

Burstein, S., Schaff-Blass, E., Blass, J., and Rosenfield, R. L. (1985). The changing ratio of bioactive to immunoreactive luteinizing hormone (LH) through puberty principally reflects changing LH radioimmunoassay dose-response characteristics. *J. clin. Endocr. Metab.* **61**, 508–513.

Butler, H. (1974). Evolutionary trends in primate sex cycles. *Contributions to Primatology* **3**, (ed. W. P. Luckett) pp. 2–35, S. Karger, Basel.

Cameron, J. L., Koerker, D. J., and Steiner, R. A. (1985). Metabolic changes during maturation of male monkeys: Possible signals for onset of puberty. *Am. J. Physiol.* **249**, E385–E391.

Chappel, S. C. and Ramaley, J. A. (1985). Changes in the isoelectric focusing profile of pituitary follicle-stimulating hormone in the developing male rat. *Biol. Reprod.* **32**, 567–573.

Chen, H. T., Sylvester, P. W., Ieiri, T., and Meites, J. (1981). Potentiation of luteinizing hormone release by serotonin agonists in ovariectomized steroid-primed rats. *Endocrinology* **108**, 948–952.

Cicero, T. J., Badger, T. M., Wilcox, C. E., Bell, R. D., and Meyer, E. R. (1977). Morphine decreases luteinizing hormone by an action on the hypothalamic pituitary axis. *J. Pharmac. exp. Ther.* **203**, 548–555.

——, Schmoeker, P. F., Meyer, E. R., Miller, B. T., Bell, R. D., Cytron, S. M., and Brown, C. C. (1986). Ontogeny of the opioid-mediated control of reproductive endocrinology in the male and female rat. *J. Pharmac. exp. Ther.* **236**, 627–633.

——, Schainker, B. A., and Meyer, E. R. (1979). Endogenous opioids participate in the regulation of the hypothalamic-pituitary-luteinizing hormone axis and testosterone's negative feedback control of luteinizing hormone. *Endocrinology* **104**, 1286–1291.

Clemens, J. A., Sawyer, B. D., and Cerimele, B. (1977). Further evidence that serotonin is a neurotransmitter involved in the control of prolactin secretion. *Endocrinology* **100**, 692–698.

Cocchi, D., Fraschini, F., Jalanbo, H., and Muller, E. E. (1974). Role of brain catecholamines in the postcastration rise in plasma LH of prepuberal rats. *Endocrinology* **95**, 1649–1657.

Cohen, H. N., Hay, I. D., Annesley, T. M., Beastall, G. H., Wallace, A. M., Spooner, R., Thomson, J. A., Eastwold, P., and Klee, G. G. (1982). Serum immunoreactive melatonin in boys with delayed puberty. *Clin. Endocr.* **17**, 517–521.

Collu, R., Fraschini, F., and Martini, L. (1971). The effect of pineal methoxyindoles on rat vaginal opening time. *J. Endocr.* **50**, 679–683.

Conte, F. A., Grumbach, M. M., and Kaplan, S. L. (1975). A diphasic pattern of gonadotropin secretion in patients with the syndrome of gonadal dysgenesis. *J. clin. Endocr. Metab.* **40**, 670–674.

—— —— —— and Reiter, E. O. (1980). Correlation of luteinizing hormone-releasing factor-induced luteinizing hormone and follicle-stimulating hormone release from infancy to 19 years with the changing pattern of gonadotropin secretion in agonadal patients: Relation to the restraint of puberty. *J. clin. Endocr. Metab.* **50**, 163–168.

Cooper, J. R., Bloom, F. E., and Roth, R. H. (1986). *The biochemical basis of neuropharmacology*, fifth edition. Oxford University Press, New York.

Coppola, J. A., Leonardi, R. G., and Lippman, W. (1966). Ovulatory failure in rats after treatment with brain norepinephrine depletors. *Endocrinology* **78**, 225–228.

Coyle, J. T. and Axelrod, J. (1972). Tyrosine hydroxylase in rat brain: Developmental characteristics. *J. Neurochem.* **19**, 1117–1123.

Crew, F. A. E. (1931). Puberty and maturity. *Proc. 2nd. Intern. Congr. Sex Res.* Oliver and Boyd, London.

Crim, L. W. and Geschwind, I. I. (1972*a*). Patterns of FSH and LH secretion in the developing ram: The influence of castration and replacement therapy with testosterone propionate. *Biol. Reprod.* **7**, 47–54.

—————(1972*b*). Testosterone concentration in spermatic vein plasma of the developing ram. *Biol. Reprod.* **7**, 42–46.

Cutler, G. B., Jr., and Loriaux, D. L. (1980). Adrenarche and its relationship to the onset of puberty. *Fedn. Proc.* **39**, 2384–2390.

Daikoku, S., Adachi, T., Kawano, H., and Wakabayashi, K. (1981). Development of the hypothalamic-hypophysial-gonadotrophic activities in foetal rats. *Experientia* **37**, 1346–1347.

Debeljuk, L., Arimura, A., and Schally, A. V. (1972*a*). Studies on the pituitary responsiveness to luteinizing hormone-releasing hormone (LH-RH) in intact male rats of different ages. *Endocrinology* **90**, 585–588.

————— (1972*b*). Pituitary responsiveness to LH-releasing hormone in intact female rats of different ages. *Endocrinology* **90**, 1499–1502.

Delitala, G., Devilla, L., and Arata, L. (1981). Opiate receptors and anterior pituitary hormone secretion in man. Effect of naloxone infusion. *Acta endocr. Copenh.* **97**, 150–156.

Dierschke, D. J., Karsch, F. J., Weick, R. F., Weiss, G., Hotchkiss, J., and Knobil, E. (1974). In *Control of the Onset of Puberty* (eds M. M. Grumbach, G. D. Grave, and F. E. Mayer) pp. 104–114, John Wiley and Sons, New York.

Docke, F., Rohde, W., Tonjes, R., Lange, T., and Dorner, G. (1978). Experimental studies on the puberal desensitization to oestrogen. In *Hormones and Brain Development*, (eds G. Dorner and M. Kawakami) pp. 361–368, Elsevier/North-Holland Biomedical Press.

Döhler, K. D. and Wuttke, W. (1975). Changes with age in levels of serum gonadotropins, prolactin, and gonadal steroids in prepubertal male and female rats. *Endocrinology* **97**, 898–907.

————— (1976). Circadian fluctuations of serum hormone levels in prepubertal male and female rats. *Acta endocr. Copenh.* **83**, 269–279.

Donovan, B. T. and Van der Werff ten Bosch, J. J. (1965). Physiology of puberty. *Monographs of the Physiological Society*. No. 15. (eds H. Barcroft, H. Davson, and W. D. M. Paton) Arnold, London.

Dubey, A. K., Cameron, J. L., Steiner, R. A., and Plant, T. M. (1986). Inhibition of gonadotropin secretion in castrated male rhesus monkeys (*Macaca mulatta*) induced by dietary restriction: Analogy with the prepubertal hiatus of gonadotropin release. *Endocrinology* **118**, 518–525.

Dupon, C. and Schwartz, N. B. (1971). Pituitary LH patterns in prepuberal normal and testosterone-sterilized rats. *Neuroendocrinology* **7**, 236–248.

Dyer, R. G., Mansfield, S., Corbet, H., and Dean, A. D. P. (1985). Fasting impairs LH secretion in female rats by activating an inhibitory opioid pathway. *J. Endocr.* **105**, 91–97.

Dyrmundsson, O. R. (1973). Puberty and early reproductive performance in sheep. II. Ram lambs. *Animal Breeding Abstracts* **41**, 419–430.

Ehrenkranz, J. R. L., Tamarkin, L., Comite, F., Johnsonbaugh, R. E., Bybee, D. E.,

Loriaux, D. L., and Cutler, G. B. Jr. (1982). Daily rhythm of plasma melatonin in normal and precocious puberty. *J. clin. Endocr. Metab.* **55**, 307–310.

Eldridge, J. C., McPherson, III, J. C. and Mahesh, V. B. (1974). Maturation of the negative feedback control of gonadotropin secretion in the female rat. *Endocrinology* **94**, 1536–1540.

Epstein, Y., Lunenfeld, B., and Kraiem, Z. (1977). The effects of testosterone and its 5-α-reduced metabolites on pituitary responsiveness to gonadotrophin-releasing hormone (GnRH). *Acta endocr. Copenh.* **86**, 728–732.

Faigon, M. R., Cardinali, D. P., and Moguilevsky, J. A. (1982). Pinealectomy advances the time of development of steroid feedback on luteinizing hormone release in immature female rats. *Brain Res.* **241**, 366–369.

——, Szwarcfarb, B., Scacchi, P., and Moguilevsky, J. A. (1987). Effect of naloxone on the development of the positive feed-back action of oestrogen-progesterone on LH secretion in rats. *Acta endocr. Copenh.* **115**, 16–20.

Faiman, C. and Winter, J. S. D. (1974). In *Control of the Onset of Puberty.* (eds M. M. Grumbach, G. D. Grave, and F. E. Mayer.) pp. 32–55. John Wiley and Sons, New York.

Fevre, M., Segel, T., Marks, J. F., and Boyar, R. M. (1978). LH and melatonin secretion patterns in pubertal boys. *J. clin. Endocr. Metab.* **47**, 1383–1386.

Feicht, C. B., Johnson, T. S., Martin, B. J., Sparkes, K. E., and Wagner, Jr., W. W. (1978). Secondary amenorrhoea in athletes. *Lancet 1*, 1145–1146.

Finkelstein, J. W., Kapen, S., Weitzmann, E. D., Hellman, L., and Boyar, R. M. (1978). Twenty-four-hour plasma prolactin patterns in prepubertal and adolescent boys. *J. clin. Endocr. Metab.* **47**, 1123–1128.

Foster, D. L., Jaffe, R. B., and Niswender, G. D. (1975a). Sequential patterns of circulating LH and FSH in female sheep during the early postnatal period: Effect of gonadectomy. *Endocrinology* **96**, 15–22.

—— Lemons, J. A., Jaffe, R. B., and Niswender, G. D. (1975b). Sequential patterns of circulating luteinizing hormone and follicle-stimulating hormone in female sheep from early postnatal life through the first estrous cycles. *Endocrinology* **97**, 985–994.

—— Mickelson, I. H., Ryan, K. D., Coon, G. A., Drongowski, R. A., and Holt, J. A. (1978). Ontogeny of pulsatile luteinizing hormone and testosterone secretion in male lambs. *Endocrinology* **102**, 1137–1146.

—— and Ryan, K. D. (1979). Endocrine mechanisms governing transition into adulthood: A marked decrease in inhibitory feedback action of oestradiol on tonic secretion of luteinizing hormone in the lamb during puberty. *Endocrinology* **105**, 896–903.

—— —— (1981). Endocrine mechanisms governing transition into adulthood in female sheep. *J. Reprod. Fert.* Suppl. **30**, 75–90.

—— and Olster, D. H. (1985). Effect of restricted nutrition on puberty in the lamb: Patterns of tonic luteinizing hormone (LH) secretion and competency of the LH surge system. *Endocrinology* **116**, 375–381.

——, Yellon, S. M., and Olster, D. H. (1985). Internal and external determinants of the timing of puberty in the female. *J. Reprod. Fert.* **75**, 327–344.

——, Karsch, F. J., Olster, D. H., Ryan, K. D., and Yellon, S. M. (1986). Determinants of puberty in a seasonal breeder. *Recent Prog. Horm. Res.* **42**, (ed. R. O. Greep) pp. 331–384. Academic Press, Orlando.

Frawley, L. S., and Neill, J. D. (1979). Age related changes in serum levels of gonadotropins and testosterone in infantile male rhesus monkeys. *Biol. Reprod.* **20**, 1147–1151.

Frisch, R. E. and Revelle, R. (1970). Height and weight at menarche and a hypothesis of critical body weights and adolescent events. *Science, N.Y.* **169**, 397–399.

—— and McArthur, J. W. (1974). Menstrual cycles: Fatness as a determinant of minimum weight for height necessary for their maintenance or onset. *Science, N.Y.* **185**, 949–951.

——, Wyshak, G., and Vincent, L. (1980). Delayed menarche and amenorrhea in ballet dancers. *New Engl. J. Med.* **303**, 17–19.

——, Gotz-Welbergen, A. V., McArthur, J. W., Albright, T., Witschi, J., Bullen, B., Birnholz, J., Reed, R. B., and Hermann, H. (1981). Delayed menarche and amenorrhea of college athletes in relation to age of onset of training. *J. Am. med. Ass.* **246**, 1559–1563.

—— (1984). Body fat, puberty and fertility. *Biol. Rev.* **59**, 161–188.

Fuller, G. B., Faiman, C., Winter, J. S. D., Reyes, F. I., and Hobson, W. C. (1982). Sex-dependent gonadotropin concentrations in infant chimpanzees and rhesus monkeys. *Proc. Soc. exp. Biol. Med.* **169**, 494–500.

Galler, J. R., Ramsey, F., and Solimano, G. (1985). A follow-up study of the effects of early malnutrition on subsequent development. I. Physical growth and sexual maturation during adolescence. *Pediat. Res.* **19**, 518–527.

Gallo, R. V. (1980). Neuroendocrine regulation of pulsatile luteinizing hormone release in the rat. *Neuroendocrinology* **30**, 122–131.

—— and Moberg, G. P. (1977). Serotonin mediated inhibition of episodic luteinizing hormone release during electrical stimulation of the arcuate nucleus in ovariectomized rats. *Endocrinology* **100**, 945–954.

—— and Osland, R. B. (1976). Electrical stimulation of the arcuate nucleus in ovariectomized rats inhibits episodic luteinizing hormone (LH) release but excites LH release after oestrogen priming. *Endocrinology* **99**, 659–668.

Gay, V. L. and Plant, T. M. (1987). N-methyl-D, L-aspartate elicits hypothalamic gonadotrophin-releasing hormone release in prepubertal male rhesus monkeys (*Macaca mulatta*). *Endocrinology* **120**, 2289–2296.

Genazzani, A. R., Facchinetti, F., Petraglia, F., Pintor, C., Bagnoli, F., Puggioni, R., and Corda, R. (1983). Correlations between plasma levels of opioid peptides and adrenal androgens in prepuberty and puberty. *J. steroid Biochem.* **19**, 891–895.

Gerocs, K., Rethelyi, M. and Halasz, B. (1986). Quantitative analysis of dendritic protrusions in the medial preoptic area during postnatal development. *Dev. Brain Res.* **26**, 49–57.

Glass, A. R., Herbert, D. C., and Anderson, J. (1986). Fertility onset, spermatogenesis, and pubertal development in male rats: effect of graded underfeeding. *Pediat. Res.* **20**, 1161–1167.

—— and Swerdloff, R. S. (1980). Nutritional influences on sexual maturation in the rat. *Fedn. Proc.* **39**, 2360–2364.

——, Anderson, J., Herbert, D., and Vigersky, R. A. (1984). Relationship between pubertal timing and body size in underfed male rats. *Endocrinology* **115**, 19–24.

Golder, M. P., Boyns, A. R., Harper, M. E., and Griffiths, K. (1972). An effect of prolactin on prostatic adenylate cyclase activity. *Biochem. J.* **128**, 725–727.

Goldman, B. D., Grazia, Y. R., Kamberi, I. A., and Porter, J. C. (1971). Serum gonadotropin concentrations in intact and castrated neonatal rats. *Endocrinology* **88**, 771–776.

Grumbach, M. M., Roth, J. C., Kaplan, S. L., and Kelch, R. P. (1974). In *Control of the Onset of Puberty* (eds M. M. Grumbach, G. D. Grave, and F. E. Mayer) pp. 115–181. John Wiley and Sons, New York.

Gupta, D. (1986). The pineal gland in relation to growth and development in children. *J. Neural. Transm.* Suppl **21**, 217–232.

——, Rager, K., Zarzycki, J., and Eichner, M. (1975). Levels of luteinizing hormone, follicle-stimulating hormone, testosterone and dihydrotestosterone in the circulation of sexually maturing intact male rats and after orchidectomy and experimental bilateral cryptorchidism. *J. Endocr.* **66**, 183–193.

Hoffman, G. E., Felten, D. L., and Sladek, Jr., J. R. (1977). Monoamine distribution in primate brain. *Am. J. Anat.* **147**, 501–514.

——, Wray, S., and Goldstein, M. (1982). Relationship of catecholamines and LHRH: Light microscopic study. *Brain Res. Bull.* **9**, 417–430.

Hohn, K. G. and Wuttke, W. (1979). Ontogeny of catecholamine turnover rates in limbic and hypothalamic structures in relation to serum prolactin and gonadotropin levels. *Brain Res.* **179**, 281–293.

Holehan, A. M. and Merry, B. J. (1985). The control of puberty in the dietary restricted female rat. *Mechanisms of Ageing and Development* **32**, 179–191.

Hompes, P. G. A., Vermes, I., Tilders, F. J. H., and Schoemaker, J. (1982). *In vitro* release of LHRH from the hypothalamus of female rats during prepubertal development. *Neuroendocrinology* **35**, 8–12.

——(1983). The metabolic clearance rate and the production rate of estradiol in sexually immature and adult female rhesus monkeys. *J. clin. Endocr. Metab.* **56**, 979–984.

Hotchkiss, J. (1985). Changes in sex hormone-binding globulin binding capacity and percent free oestradiol during development in the female rhesus monkey (*Macaca mulatta*): relation to the metabolic clearance rate of oestradiol. *J. clin. Endocr. Metab.* **60**, 786–792.

Hughes, J., Smith, T. W., Kosterlitz, H. W., Fothergill, L. A., Morgan, B. A., and Morris, H. R. (1975). Identification of two related pentapeptides from the brain with potent opiate agonist activity. *Nature, Lond.* **258**, 577–579.

Ishikawa, M. and Tanaka, C. (1977). Morphological organization of catecholamine terminals in the diencephalon of the rhesus monkey. *Brain Res.* **119**, 43–55.

Kalra, S. P. and Kalra, P. S. (1983). Neural regulation of luteinizing hormone secretion in the rat. *Endocr. Rev.* **4**, 311–351.

Kao, L. W. L. and Weisz, J. (1977). Release of gonadotrophin-releasing hormone (Gn-RH) from isolated, perifused medial-basal hypothalamus by melatonin. *Endocrinology* **100**, 1723–1726.

Karsch, F. J., Dierschke, D. J., Weick, R. F., Yamaji, T., Hotchkiss, J., and Knobil, E. (1973). Positive and negative feedback control, by oestrogen, of luteinizing hormone secretion in the rhesus monkey. *Endocrinology* **92**, 799–804.

Kaufman, J-M., Kesner, J. S., Wilson, R. C., and Knobil, E. (1985). Electrophysiological manifestation of luteinizing hormone-releasing hormone pulse generator activity in the rhesus monkey: influence of α-adrenergic and dopaminergic blocking agents. *Endocrinology* **116**, 1327–1333.

Keel, B. A. and Abney, T. O. (1985). Oestrogenic regulation of testicular androgen production during development in the rat. *J. Endocr.* **105**, 211–218.

Kelch, R. P., Kaplan, S. L., and Grumbach, M. M. (1973). Suppression of urinary and plasma follicle stimulating hormone by exogenous oestrogens in prepubertal and pubertal children. *J. clin. Invest.* **52**, 1122–1128.

Kennaway, D. J. and Gilmore, T. A. (1984). Effects of melatonin implants in ewe lambs. *J. Reprod. Fert.* **70**, 39–45.

Kennedy, G. C. and Mitra, J. (1963). Body weight and food intake as initiating factors for puberty in the rat. *J. Physiol., Lond.* **166**, 408–418.

Kennedy, J. P., Worthington, C. A., and Cole, E. R. (1974). The post-natal development of the ovary and uterus of the Merino lamb. *J. Reprod. Fert.* **36**, 275–282.

Ketelslegers, J. M., Hetzel, W. D., Sherins, R. J., and Catt. K. J. (1978). Developmental changes in testicular gonadotropin receptors: plasma gonadotropins and plasma testosterone in the rat. *Endocrinology* **103**, 212–222.

King, J. C., Anthony, E. L. P., Fitzgerald, D. M., and Stopa, E. G. (1985). Luteinizing hormone-releasing hormone neurons in human preoptic/hypothalamus: Differential intraneuronal localization of immunoreactive forms. *J. clin. Endocr. Metab.* **60**, 88–97.

Knorr, D. W., Vanha-Perttula, T., and Lipsett, M. B. (1970). Structure and function of rat testis through pubescence. *Endocrinology* **86**, 1298–1304.

Kragt, C. L. and Ganong, W. F. (1968). Pituitary FSH content in male rats at various ages. *Proc. Soc. exp. Biol. Med.* **128**, 965–967.

Kuderling, I., Dorsch, G., Warnhoff, M., and Pirke, K-M. (1984). The actions of prostaglandin E$_2$, naloxone and testosterone on starvation-induced suppression of luteinizing hormone-releasing hormone and luteinizing-hormone secretion. *Neuroendocrinology* **39**, 530–537.

Kulin, H. E., Bwibo, N., Mutie, D., and Santner, S. J., (1984). Gonadotropin excretion during puberty in malnourished children. *J. Pediat.* **105**, 325–328.

——, Moore, R. G., and Santner, S. J. (1976). Circadian rhythms in gonadotropin excretion in prepubertal and pubertal children. *J. clin. Endocr. Metab.* **42**, 770–773.

——and Reiter, E. O. (1972). Gonadotropin suppression by low dose estrogen in men: evidence for differential effects upon FSH and LH. *J. clin. Endocr. Metab.* **35**, 836–839.

Lang, U., Aubert, M. L., Rivest, R. W., Vinas-Bradtke, J. C., and Sizonenko, P. C. (1985). Inhibitory action of exogenous melatonin, 5-methoxytryptamine, and 6-hydroxymelatonin on sexual maturation of male rats: Activity of 5-methoxytryptamine might be due to its conversion to melatonin. *Biol. Reprod.* **33**, 618–628.

Leadem, C. A., Crowley, W. R., Simpkins, J. W., and Kalra, S. P. (1985). Effects of naloxone on catecholamine and LHRH release from the perifused hypothalamus of the steroid-primed rat. *Neuroendocrinology* **40**, 497–500.

Lee, P. A. and Migeon, C. J. (1975). Puberty in boys: Correlation of plasma levels of gonadotropins (LH, FSH), androgens (testosterone, androstenedione, dehydro-epiandrosterone and its sulfate), oestrogens (estrone and estradiol) and progestins (progesterone and 17-hydroxyprogesterone). *J. clin. Endocr. Metab.* **41**, 556-562.

Lenko, H., Lang, U., Aubert, M. L., Paunier, L., and Sizonenko, P. C. (1982). Hormonal changes in puberty. VII. Lack of variation of daytime plasma melatonin. *J. clin. Endocr. Metab.* **54**, 1056–1058.

Leonardelli, J. and Tramu, G. (1979). Immunoreactivity for β-endorphin in LH-RH neurons of the foetal human hypothalamus. *Cell Tissue Res.* **203**, 201–207.

Lisoni, P., Resentini, M., Mauri, R., De Medici, C., Morabito, F., Esposti, D., Di Bella, L., Esposti, G., Rossi, D., Parravicini, L., Legname, G., and Fraschini, F. (1986). Effect of an acute injection of melatonin on the basal secretion of hypophyseal hormones in prepubertal and pubertal healthy subjects. *Acta endocr. Copenh.* **111**, 305–311.

Lofstrom, A., Agnati, L. F., and Fuxe, K. (1977*a*). Evidence for an inhibitory dopaminergic and stimulatory noradrenergic hypothalamic influence of PMS-induced ovulation in the immature rat. I. Catecholamine turnover changes in relation to the critical period. *Neuroendocrinology* **24**, 270–288.

—— —— ——and Hokfelt, T. (1977*b*). Evidence for an inhibitory dopaminergic and stimulatory noradrenergic hypothalamic influence on PMS-induced ovulation in the immature rat. II. A pharmacological analysis. *Neuroendocrinology* **24**, 289–316.

Lubicz-Nawrocki, C. M. (1976). The effects of metabolites of testosterone on the development of fertilizing ability by spermatozoa in the epididymis of castrated hamsters. *J. exp. Zool.* **197**, 89–96.

Lucky, A. W., Rich, B. H., Rosenfield, R. L., Fang, V. S., and Roche-Bender, N. (1980). LH bioactivity increases more than immunoreactivity during puberty. *J. Pediat.* **97**, 205–213.

MacPhee, A. A., Cole, F. E., and Rice, B. F. (1975). The effect of melatonin on steroidogenesis by the human ovary *in vitro. J. clin. Endocr. Metab.* **40**, 688–696.

Marrama, P., Zaidi, A. A., Montanini, V., Celani, M. F., Cioni, K., Carani, C., Morabito, F., Resentini, M., Bonati, B., and Baraghini, G. F. (1983). Age and sex related variations in biologically active and immunoreactive serum luteinizing hormone. *J. endocr. Invest* **6**, 427–433.

Marshall, F. H. A. (1922). *The Physiology of Reproduction*, 2nd ed. Longmans, Green, and Co., London.

Martensz, N. D. (1985). Changes in the processing of β-endorphin in the hypothalamus and pituitary gland of female rats during sexual maturation. *Neuroscience* **16**: 625–640.

Martin, J. E., Engel, J. N., and Klein, D. C. (1977). Inhibition of the *in vitro* pituitary response to luteinizing hormone-releasing hormone by melatonin, serotonin, and 5-methoxytryptamine. *Endocrinology* **100**, 675–680.

—— and Sattler, C. (1979). Developmental loss of the acute inhibitory effect of melatonin on the *in vitro* pituitary luteinizing hormone and follicle-stimulating hormone responses to luteinizing hormone-releasing hormone. *Endocrinology* **105**, 1007–1012.

Matsumoto, A. M., Karpas, A. E., Southworth, M. B., Dorsa, D. M., and Bremner, W. J. (1986). Evidence for activation of the central nervous system-pituitary mechanism for gonadotrophin secretion at the time of puberty in the male rat. *Endocrinology* **119**, 362–369.

Mauras, N., Veldhuis, J. D., and Rogol, A. D. (1986). Role of endogenous opiates in pubertal maturation: Opposing actions of naltrexone in prepubertal and late pubertal boys. *J. clin. Endocr. Metab.* **62**, 1256–1263.

McClure, T. J. and Saunders, J. (1985). Effects of withholding food for 0–72 h on mating, pregnancy rate and pituitary function in female rats. *J. Reprod. Fert.* **74**, 57–64.

McNeill, T. H. and Sladek, Jr., J. R. (1977). Fluorescence-immunocytochemistry: Simultaneous localization of catecholamines and gonadotropin-releasing hormone. *Science, N.Y.* **200**, 72–74.

Meusy-Dessolle, N. and Dang, D. C. (1985). Plasma concentrations of testosterone, dihydrotestosterone, Δ_4-androstenedione, dehydroepiandrosterone and oestradiol-17β in the crab-eating monkey (*Macaca fascicularis*) from birth to adulthood. *J. Reprod. Fert.* **74**, 347–359.

Miles, S. W. and Wright, J. J. (1984). Psychoendocrine interaction in anorexia nervosa, and the retreat from puberty: A study of attitudes to adolescent conflict, and luteinizing hormone response to luteinizing hormone releasing factor, in refed anorexia nervosa subjects. *Br. J. med. Psych.* **57**, 49–56.

Miyachi, Y., Nieschlag, E., and Lipsett, M. B. (1973). The secretion of gonadotropins and testosterone by the neonatal male rat. *Endocrinology* **92**, 1–5.

Moger, W. H. (1977). Serum 5α-androstane-3α, 17 β-diol, androsterone, and testosterone concentrations in the male rat. Influence of age and gonadotropin stimulation. *Endocrinology* **100**, 1027–1032.

Moguilevsky, J. A., Faigon, M. R., Rubio, M. C., Scacchi, P., and Szwarcfarb, B. (1985*a*). Sexual differences in the effect of serotonin on LH secretion in rats. *Acta endocr. Copenh.* **109**, 320–325.

—— —— —— —— —— (1985*b*). Effect of the serotoninergic system on luteinizing hormone secretion in prepubertal female rats. *Neuroendocrinology* **40**, 135–138.

Monroe, S. E., Yamamoto, M., and Jaffe, R. B. (1983). Changes in gonadotrope responsivity to gonadotropin releasing hormone during development of the rhesus monkey. *Biol. Reprod.* **29**, 422–431.

Montanini, V., Celani, M. F., Baraghini, G. F., Carani, C., and Marrama, P. (1984*a*). Effects of acute stimulation with luteinizing hormone-releasing hormone (LRH) on biologically active and immunoreactive serum luteinizing hormone (LH) in pubertal boys. *Acta endocr. Copenh.* **107**, 289–294.

————Carani, C., Cioni, K., Negri, R., Morabito, F., and Marrama, P. (1984*b*). LH bioactivity in male puberty and senescence. *J. endocr. Invest.* **7**, 43–52.

Morley, J. E., Baranetsky, N. G., Wingert, T. D., Carlson, H. E., Hershman, J. M., Melmed, S., Levin, S. R., Jamison, K. R., Weitzman, R., Chang, R. J., and Varner, A. A. (1980). Endocrine effects of naloxone-induced opiate receptor blockade. *J. clin. Endocr. Metab.* **50**, 251–257.

———— (1987). Neuropeptide regulation of appetite and weight. *Endocr. Rev.* **8**, 256–287.

Nash, S. J., Brawer, J., Robaire, B., and Ruf, K. B. (1986). Changes in the dynamics of luteinizing hormone-releasing hormone-stimulated secretion of luteinizing hormone during sexual maturation of female rats. *Biol. Reprod.* **34**, 549–557.

Nazian, S. J. (1983). Serum and pituitary luteinizing hormone and serum androgens during luteinizing hormone releasing hormone self-priming in immature and pubertal male rats. *Biol. Reprod.* **29**, 912–918.

———— (1985). Induction of the self-priming effect of luteinizing hormone releasing hormone during the sexual maturation of the male rat. *Proc. Soc. exp. Biol. Med.* **179**, 348–351.

Negro-Vilar, A., Krulich, L., and McCann, S. M. (1973*a*). Changes in serum prolactin and gonadotropins during sexual development of the male rat. *Endocrinology* **93**, 660–664.

————, Ojeda, S. R., and McCann, S. M. (1973*b*). Evidence for changes in sensitivity to testosterone negative feedback on gonadotropin release during sexual development in the male rat. *Endocrinology* **93**, 729–735.

Nielsen, C. T., Skakkebaek, N. E., Richardson, D. W., Darling, J. A. B., Hunter, W. M., Jorgensen, M., Nielsen, A., Ingerslev, O., Keiding, N., and Muller, J. (1986). Onset of the release of spermatozoa (spermarche) in boys in relation to age, testicular growth, pubic hair, and height. *J. clin. Endocr. Metab.* **62**, 532–535.

Nottelmann, E. D., Susman, E. J., Dorn, L. D., Inoff-Germain, G., Loriaux, D. L., Cutler, G. B., and Chrousos, G. P. (1987). Developmental processes in early adolescence. *J. adolesc. Health Care* **8**, 246–260.

Nowak, R. and Rodway, R. G. (1985). Effect of intravaginal implants of melatonin on the onset of ovarian activity in adult and prepubertal ewes. *J. Reprod. Fert.* **74**, 287–293.

Ojeda, S. R., Andrews, W. W., Advis, J. P., and White, S. S. (1980). Recent advances in the endocrinology of puberty. *Endocr. Rev.* **1**, 228–257.

———— and McCann, S. M. (1974). Development of dopaminergic and oestrogenic control of prolactin release in the female rat. *Endocrinology* **95**, 1499–1505.

————, Negro-Vilar, A., and McCann S. M. (1982). Evidence for involvement of α-adrenergic receptors in norepinephrine-induced prostaglandin E_2 and luteinizing hormone-releasing hormone release from the median eminence. *Endocrinology* **110**, 409–412.

———— ———— ———— (1979). Release of prostaglandin Es by hypothalamic tissue: evidence for their involvement in catecholamine-induced luteinizing hormone-releasing hormone release. *Endocrinology* **104**, 617–624.

—— and Ramirez, V. D. (1973/74). Short-term steroid treatment on plasma LH and FSH in castrated rats from birth to puberty. *Neuroendocrinology* **13**, 100–114.

——, Urbanski, H. F., and Ahmed, C. E. (1986). The onset of female puberty: studies in the rat. *Recent Prog. Horm. Res.* **42**, 385–442.

—— —— Katz, K. H., and Costa, M. E. (1986). Activation of oestradiol-positive feedback at puberty: oestradiol sensitizes the LHRH-releasing system at two different biochemical steps. *Neuroendocrinology* **43**, 259–265.

Parker, Jr., C. R. and Mahesh, V. B. (1976) Hormonal events surrounding the natural onset of puberty in female rats. *Biol. Reprod.* **14**, 347–353.

Parker, V. D. and Soliman, K. F. A. (1983). Brain dopamine variations in gonado-atropin-treated immature rat. *Experientia* **39**, 215–217.

Pau, M. and Milner, J. A. (1984). Dietary arginine deprivation and delayed puberty in the female rat. *J. Nutr.* **114**, 112–118.

Paulin, C., Dubois, M. P., Barry, J., and Dubois, P. M. (1977). Immunofluorescence study of LH-RH producing cells in the human fetal hypothalamus. *Cell Tissue Res.* **182**, 341–345.

Peluso, J. J., Steger, R. W., and Hafez, E. S. E. (1976). Development of gonadotropin-binding sites in the immature rat ovary. *J. Reprod. Fert.* **47**, 55–58.

Penny, R. (1985). Episodic secretion of melatonin in pre- and post-pubertal girls and boys. *J. clin. Endocr. Metab.* **60**, 751–756.

—— (1982). Melatonin excretion in normal males and females: Increase during puberty. *Metabolism* **31**, 816–823.

Petraglia, F., Bernasconi, S., Iughetti, L., Loche, S., Romanini, F., Facchinetti, F., Marcellini, C., and Genazzani, A. R. (1986). Naloxone-induced luteinizing hormone secretion in normal, precocious, and delayed puberty. *J. clin. Endocr. Metab.* **63**, 1112–1115.

Plant, T. M. (1983). Ontogeny of gonadotropin secretion in the rhesus macaque (*Macaca mulatta*). In *Neuroendocrine Aspects of Reproduction*. (ed. R. L. Norman) Academic Press, New York.

—— (1985). A study of the role of the postnatal testes in determining the ontogeny of gonadotropin secretion in the male rhesus monkey (*Macaca mulatta*). *Endocrinology* **116**, 1341–1350.

—— and Zorub, D. S. (1986). Pinealectomy in agonadal infantile male rhesus monkeys (*Macaca mulatta*) does not interrupt initiation of the prepubertal hiatus in gonadotropin secretion. *Endocrinology* **118**, 227–232.

Porcher, W. and Heller, A. (1972). Regional development of catecholamine biosynthesis in rat brain. *J. Neurochem.* **19**, 1917–1930.

Porter, J. C. (1986). Relationship of age, sex, and reproductive status to the quantity of tyrosine hydroxylase in the median eminence and superior cervical ganglion of the rat. *Endocrinology* **118**, 1426–1432.

Ramaley, J. A. (1979). Development of gonadotropin regulation in the prepubertal mammal. *Biol. Reprod.* **20**, 1–31.

Ramirez, V. D., Feder, H. H., and Sawyer, C. H. (1984). The role of brain catecholamines in the regulation of LH secretion: A critical inquiry. *Frontiers in Neuroendocrinology* **8**, 27–84.

—— Kim, K., and Dluzen, D. (1985). Progesterone action on the LHRH and the nigrostriatal dopamine neuronal systems: *in vitro* and *in vivo* studies. *Recent Prog. Horm. Res.* **41**, 421–472.

—— and McCann, S. M. (1963). Comparison of the regulation of luteinizing hormone (LH) secretion in immature and adult rats. *Endocrinology* **72**, 452–464.

—— —— (1965). Inhibitory effect of testosterone on luteinizing hormone secretion in immature and adult rats. *Endocrinology* **76**, 412–417.

Rapisarda, J. J., Bergman, K. S., Steiner, R. A., and Foster, D. L. (1983). Response to oestradiol inhibition of tonic luteinizing hormone secretion decreases during the final stage of puberty in the rhesus monkey. *Endocrinology* **112**, 1172–1179.

Rasmussen, D. D., Liu, J. H., Wolf, P. L., and Yen, S. S. C. (1983). Endogenous opioid regulation of gonadotropin-releasing hormone release from the human foetal hypothalamus *in vitro*. *J. clin, Endocr. Metab.* **57**, 881–884.

Raum, W. J., Glass, A. R., and Swerdloff, R. S. (1980). Changes in hypothalamic catecholamine neurotransmitters and pituitary gonadotropins in the immature female rat: Relationships to the gonadostat theory of puberty onset. *Endocrinology* **106**, 1253–1258.

Ravault, J. P. and Courot, M. (1975). Blood prolactin in the male lamb from birth to puberty. *J. Reprod. Fert.* **42**, 563–566.

Reiter, E. O., Beitins, I. Z., Ostrea, T., and Gutai, J. P. (1982). Bioassayable luteinizing hormone during childhood and adolescence and in patients with delayed pubertal development. *J. clin. Endocr. Metab.* **54**, 155–161.

—— and Grumbach, M. M. (1982). Neuroendocrine control mechanisms and the onset of puberty. *A. Rev. Physiol.* **44**, 595–613.

—— Root, A. W., and Duckett, G. E. (1976). The response of pituitary gonadotropes to a constant infusion of luteinizing hormone-releasing hormone (LHRH) in normal prepubertal and pubertal children and in children with abnormalities of sexual development. *J. clin. Endocr. Metab.* **43**, 400–411.

Relkin, R. (1971). Relative efficacy of pinealectomy, hypothalamic and amygdaloid lesions in advancing puberty. *Endocrinology* **88**, 415–418.

Rivest, R. W., Aubert, M. L., Lang, U., and Sizonenko, P. C. (1986). Puberty in the rat: modulation by melatonin and light. *J. neural Transm.* **21**, 81–108.

Roberts, J. L. and Herbert, E. (1977). Characterization of a common precursor to corticotropin and β-lipotropin: Identification of β-lipotropin peptides and their arrangement relative to corticotropin in the precursor synthesized in a cell-free system. *Proc. natn. Acad. Sci. U.S.A.* **74**, 5300–5304.

Rodriguez-Sierra, J. F. (1986). Extended organizational effects of oestrogen at puberty. *Ann. N. Y. Acad. Sci.* **474**, 293–307.

Rollag, M. D. and Niswender, G. D. (1976). Radioimmunoassay of serum concentrations of melatonin in sheep exposed to different lighting regimens. *Endocrinology* **98**, 482–489.

Ronnekleiv, O. K., Ojeda, S. R., and McCann, S. M. (1978). Undernutrition, puberty and the development of oestrogen positive feedback in the female rat. *Biol. Reprod.* **19**, 414–424.

Ross, J. L., Loriaux, D. L., and Cutler, Jr. G. B. (1983). Developmental changes in neuroendocrine regulation of gonadotropin secretion in gonadal dysgenesis. *J. clin. Endocr. Metab.* **57**, 288–293.

Saavedra, J. M., Palkovits, M., Brownstein, M. J., and Axelrod, J. (1974). Serotonin distribution in the nuclei of the rat hypothalamus and preoptic region. *Brain Res.* **77**, 157–165.

Salem, S. I., Coward, W. A., Lunn, P. G., and Hudson, G. J. (1984). Response of the reproductive system of male rats to protein and zinc deficiency during puberty. *Ann. Nutr. Metab.* **28**, 44–51.

Salisbury, R. L., Dudley, S. D., and Weisz, J. (1982). Effect of gonadotrophin-releasing hormone on circulating levels of immunoreactive luteinizing hormone in foetal rats. *Neuroendocrinology* **35**, 265–269.

Sauder, S. E., Case, G. D., Hopwood, N. J., Kelch, R. P., and Marshall, J. C. (1984). The effects of opiate antagonism on gonadotropin secretion in children and in women with hypothalamic amenorrhea. *Pediat. Res.* **18**, 322–328.

Schanbacher, B. D., Gomes, W. R., and VanDemark, N. L. (1974). Developmental changes in spermatogenesis, testicular carnitine acetyltransferase activity and serum testosterone in the ram. *J. Anim. Sci.* **39**, 889–892.

Schofield, S. P. M. and Everitt, B. J. (1981). The organization of catecholamine-containing neurons in the brain of the rhesus monkey (*Macaca mulatta*). *J. Anat.* **132**, 391–418.

Schulz, R., Wilhelm, A., Pirke, K. M., Gramsch, C., and Herz, A. (1981). *β*-endorphin and dynorphin control serum luteinizing hormone level in immature female rats. *Nature, Lond.* **294**, 757–759.

Schwartz, N. B. (1974). The role of FSH and LH and of their antibodies on follicle growth and on ovulation. *Biol. Reprod.* **10**, 236.

Siler-Khodr, T. M. and Khodr, G. S. (1978). Studies in human fetal endocrinology. I. Luteinizing hormone-releasing factor content of the hypothalamus. *Am. J. Obstet. Gynecol.* **130**, 795–800.

Silman, R. E., Leone, R. M., and Hooper, R. J. L. (1979). Melatonin, the pineal gland and human puberty. *Nature, Lond.* **282**, 301–303.

Simantov, R. and Snyder, S. H. (1977). Opiate receptor binding in the pituitary gland. *Brain Res.* **124**, 178–184.

Sirinathsinghji, D. J. S., Motta, M., and L. Martini (1985). Induction of precocious puberty in the female rat after chronic naloxone administration during the neonatal period: the opiate "brake" on prepubertal gonadotrophin secretion. *J. Endocr.* **104**, 299–307.

Sizonenko, P. C., Burr, I. M., Kaplan, S. L., and Grumbach, M. M. (1970). Hormonal changes in puberty. II. Correlation of serum luteinizing hormone and follicle stimulating hormone with stages of puberty and bone age in normal girls. *Pediat. Res.* **4**, 36–45.

——, Lang, U., Rivest, R. W., and Aubert, M. (1985). The pineal and pubertal development. In *Photoperiodism, Melatonin, and the Pineal* (eds D. Evered and S. Clark) *Ciba. Found. Symp.* **117**, 208–230. Pitman, London.

Skinner, J. D., Booth, W. D., Rowson, L. E. A., and Karg, H. (1968). The post-natal development of the reproductive tract of the Suffolk ram, and changes in the gonadotrophin content of the pituitary. *J. Reprod. Fert.* **16**, 463–477.

——, (1970). The post-natal development of the male reproductive tract in two indigenous fat-tailed South African sheep, the Pedi and the Namaqua. *Proc. S. Afr. Soc. anim. Prod.* **9**, 195–196.

——, (1971). Post-natal development of the reproductive tract in the Dorper ram. *Agroanimalia* **3**, 7–12.

Smith, E. R., Damassa, D. A., and Davidson, J. M. (1977). Feedback regulation and male puberty: Testosterone-luteinizing hormone relationships in the developing rat. *Endocrinology* **101**, 173–180.

Sopelak, V. M., Collins, R. L., and Hodgen, G. D. (1986). Follicular stimulation *versus* ovulation induction in juvenile primates: Importance of gonadotropin-releasing hormone dose. *J. clin. Endocr. Metab.* **62**, 557–562.

Srebnik, H. H. and Nelson, M. M. (1962). Anterior pituitary function in male rats deprived of dietary protein. *Endocrinology* **70**, 723–730.

Steele, R. E. and Weisz, J. (1974). Changes in sensitivity of the oestradiol-LH feedback system with puberty in the female rat. *Endocrinology* **95**, 513–520.

—— (1977). Role of the ovaries in maturation of the oestradiol-luteinizing hormone negative feedback system of the pubertal rat. *Endocrinology* **101**, 587–597.

Steiner, R. A. (1987). Nutritional and metabolic factors in the regulation of reproductive hormone secretion in the primate. *Proc. Nutr. Soc.* **46**, 159–175.

—— and Bremner, W. J. (1981). Endocrine correlates of sexual development in the male monkey, *Macaca fascicularis*. *Endocrinology* **109**, 914–919.

——, Cameron, J. L., McNeill, T. H., Clifton, D. K., and Bremner, W. J. (1983). Metabolic signals for the onset of puberty. In *Neuroendocrine Aspects of Reproduction*, pp. 183–227 (ed. R. L. Norman). Academic Press, New York.

Stewart, S. F., Kopia, S., and Gawlak, D. L. (1975). Effect of underfeeding, hemigonadectomy, sex and cyproterone acetate on serum FSH levels in immature rats. *J. Reprod. Fert.* **45**, 173–176.

Swerdloff, R. S., Walsh, P. C., Jacobs, H. S., and Odell, W. D. (1971). Serum LH and FSH during sexual maturation in the male rat: Effect of castration and cryptorchidism. *Endocrinology* **88**, 120–128.

Sylvester, P. W., Sarkar, D. K., Briski, K. P., and Meites, J. (1985). Relation of gonadal hormones to differential LH response to naloxone in prepubertal male and female rats. *Neuroendocrinology* **40**, 165–170.

Tanner, J. M. (1962). *Growth at Adolescence*, 2nd Ed. Blackwell Scientific Publishers, Oxford.

Terasawa, E., Bridson, W. E., Nass, T. E., Noonan, J. J., and Dierschke, D. J. (1984). Developmental changes in the luteinizing hormone secretory pattern in peripubertal female rhesus monkeys: Comparisons between gonadally intact and ovariectomized animals. *Endocrinology* **115**, 2233–2240.

—— Nass, T. E., Yeoman, R. R. Loose, M. D., and Schultz, N. J. (1983). Hypothalamic control of puberty in the female rhesus macaque. In *Neuroendocrine Aspects of Reproduction* (ed. Reid L. Norman) pp. 149–182. Academic Press, Inc. New York.

Tetsuo, M., Poth, M., and Markey, S. P. (1982). Melatonin metabolite excretion during childhood and puberty. *J. clin. Endocr. Metab.* **55**, 311–313.

Tsai-Morris, C-H., Aquilano, D. R., and Dufau, M. L. (1985). Cellular localization of rat testicular aromatase activity during development. *Endocrinology* **116**, 38–46.

Urbanski, H. F. and Ojeda, S. R. (1987). Activation of luteinizing hormone-releasing hormone release advances the onset of female puberty. *Neuroendocrinology* **46**, 273–276.

Van Houten, M., Posner, B. I., Kopriwa, B. M., and Brawer, J. R. (1980). Insulin binding sites localized to nerve terminals in rat median eminence and arcuate nucleus. *Science, N.Y.* **207**, 1081–1083.

Veldhuis, J. D., Rogol, A. D., Johnson, M. L., and Dufau, M. L. (1983). Endogenous opiates modulate the pulsatile secretion of biologically active luteinizing hormone in man. *Am. Soc. clin. Invest.* **72**, 2031–2040.

Vigersky, R. A., Andersen, A. E., Thompson, R. H., and Loriaux, D. L. (1977). Hypothalamic dysfunction in secondary amenorrhea associated with simple weight loss. *New Engl. J. Med.* **297**, 1141–1145.

Vitale, M. L., Parisi, M. N., Chiocchio, S. R., and Tramezzani, J. H. (1986). Serotonin induces gonadotrophin release through stimulation of LH-releasing hormone release from the median eminence. *J. Endocr.* **111**, 309–315.

Waldhauser, F. and Dietzel, M. (1985). Daily and annual rhythms in human melatonin secretion: role in puberty control. *Ann. N.Y. Acad. Sci.* **453**, 205–214.

——, Frisch, H., Waldhauser, M., Weiszenbacher, G., Zeitlhuber, U., and Wurtman, R. J. (1984). Fall in nocturnal serum melatonin during prepuberty and pubescence. *Lancet*, Feb. 18, 362–365.

Warren, M. P. (1980). The effects of exercise on pubertal progression and reproductive function in girls. *J. clin. Endocr. Metab.* **51**, 1150–1157.

——, Jewelewicz, R., Dyrenfurth, I., Ans, R., Khalaf, S., and Vande Wiele, R. L. (1975). The significance of weight loss in the evaluation of pituitary response to

LH-RH in women with secondary amenorrhea. *J. clin. Endocr. Metab.* **40**, 601–611.

Watanabe, S. and McCann, S. M. (1969). Alterations in pituitary follicle-stimulating hormone (FSH) and hypothalamic FSH-releasing factor (FSH-RF) during puberty. *Proc. Soc. exp. Biol. Med.* **132**, 195–201.

Watson, R. H., Sapsford, C. S., and McCance, I. (1956). The development of the testis, epididymis, and penis in the young Merino ram. *Aust. J. agric. Res.* **7**, 574–590.

Weiner, R. I. and Ganong, W. F. (1971). Effect of the depletion of brain catecholamines on puberty and the estrous cycle in the rat. *Neuroendocrinology* **8**, 125–135.

—— and Ganong, W. F. (1972). Norepinephrine concentration in the hypothalamus, amygdala, hippocampus, and cerebral cortex during postnatal development and vaginal opening. *Neuroendocrinology* **9**, 65–71.

White, S. S. and Ojeda, S. R. (1981). Changes in ovarian luteinizing hormone and follicle-stimulating hormone receptor content and in gonadotropin-induced ornithine decarboxylase activity during prepubertal and pubertal development of the female rat. *Endocrinology* **109**, 152–161.

Widdowson, E. M. and Cowen, J. (1972). The effect of protein deficiency and calorie deficiency on the reproduction of rats. *Br. J. Nutr.* **27**, 85–95.

Wierman, M. E., Beardsworth, D. E., Crawford, J. D., Crigler, J. F. Jr., Mansfield, M. J., Bode, H. H., Boepple, P. A., Kushner, D. C., and Crowley, W. F. Jr. (1986). Adrenarche and skeletal maturation during luteinizing hormone-releasing hormone analogue suppression of gonadarche. *J clin. Invest.* **77**, 121.

Wildt, L., Marshall, G., and Knobil. E. (1980). Experimental induction of puberty in the infantile female rhesus monkey. *Science, N.Y.* **207**, 1373–1375.

Wilkes, M. M. and Yen, S. S. C. (1981). Augmentation by naloxone of efflux of LRF from superfused medial basal hypothalamus. *Life Sci.* **28**, 2355–2359.

Wilkinson, M. and Bhanot, R. (1982). A puberty-related attenuation of opiate peptide-induced inhibition of LH secretion. *Endocrinology* **110**, 1046–1048.

——, Herdon, H., Pearce, M., and Wilson, C. (1979). Precocious puberty and changes in α- and β-adrenergic receptors in the hypothalamus and cerebral cortex of immature female rats. *Brain Res.* **167**, 195–199.

Wilson, C. A. and Endersby, C. A. (1979). The stimulatory effect of 5IIT and the role of the paraventricular nucleus on PMS induced ovulation in the immature rat. *Neuroendocrinology* **28**, 415–424.

Wilson, M. E., Gordon, T. P., and Collins, D. C. (1986). Ontogeny of luteinizing hormone secretion and first ovulation in seasonal breeding rhesus monkeys. *Endocrinology* **118**, 293–301.

Winter, J. S. D. (1982). Hypothalamic-pituitary function in the foetus and infant. *Clin. Endocr. Metab.* **11**, 41–55.

—— Ellsworth, L., Fuller, G., Hobson, W. C., Reyes, F. I., and Faiman, C. (1987). The role of gonadal steroids in feedback regulation of gonadotrophin secretion at different stages of primate development. *Acta endocr. Copenh.* **114**, 257–268.

——, Faiman, C., Reyes, F. I., and Hobson, W. C. (1978). Gonadotrophins and steroid hormones in the blood and urine of prepubertal girls and other primates. *Clin. Endocr. Metab.* **7**, 513–559.

Winters, A. J., Eskay, R. L., and Porter, J. C. (1974). Concentration and distribution of TRH and LRH in the human fetal brain. *J. clin. Endocr. Metab.* **39**, 960–963.

Winterer, J., Cutler, Jr., G. B., and Loriaux, D. L. (1984). Caloric balance, brain to body ratio, and the timing of menarche. *Med. Hypoth.* **15**, 87–91.

Wray, S. and Hoffman, G. (1986). Postnatal morphological changes in rat LHRH neurons correlated with sexual maturation. *Neuroendocrinology* **43**, 93–97.

Wurtman, R. J., Axelrod, J., and Chu, E. W. (1963). Melatonin, a pineal substance: Effect on the rat ovary. *Science, N.Y.* **141**, 277–278.

Wuttke, W., Honma, K., Lamberts, R., and Hohn, K. G. (1980). The role of monamines in female puberty. *Fedn. Proc.* **39**, 2378–2383.

Yellon, S. M. and Foster, D. L. (1985). Alternate photoperiods time puberty in the female lamb. *Endocrinology* **116**, 2090–2097.

2   The physiology of the endocrine testis

S. MADDOCKS AND B. P. SETCHELL

I   Introduction

The testes of mammals have two interrelated functions: to produce the male
gametes, the spermatozoa, and to produce hormones. The hormones are primarily required for the production of the spermatozoa, but they have other
important secondary roles in regulating behaviour and development, structure and function of the accessory sex organs, and in controlling, by negative
feedback, the secretion of gonadotrophic hormones by the pituitary. The
spermatozoa are produced in the seminiferous tubules, which are long con-
voluted cylindrical two-ended structures, opening at each end into the rete

testis, from whence the spermatozoa and the fluid in which they are suspended are carried into the epididymis. Inside these tubules are the various germ cells from which the spermatozoa develop, the spermatogonia, spermatocytes and spermatids, together with the somatic Sertoli cells, but there are no blood or lymph vessels or nerves there. The walls of the tubules are formed by peritubular myoid cells, lying between two or more layers of noncellular material. Between the tubules is the interstitial tissue, comprising all the blood vessels, lymphatic vessels and nerves, together with the steroid-producing Leydig cells, macrophages, and possibly other cell types.

LH and FSH are undoubtedly the two hormones which have the greatest influence on testicular function. However, a number of hormones from other tissues have been implicated in modulating the endocrine and paracrine activity in the testis. Prolactin, arginine vasotocin-like and LHRH-like factors, insulin-like growth factor, epidermal growth factor, catecholamines and benzodiazapine are some of the compounds for which specific receptors have been found on the Leydig cell membrane, and their possible roles have been addressed in other reviews (see Sharpe 1982; Tähkä 1986).

In this review, we will concentrate our attention on the hormone production by the testis and only consider sperm production insofar as it affects hormone secretion. In the past, a great deal of attention has been directed towards the two cell types in the testis most directly involved in hormone production, Leydig cells and Sertoli cells, and in most of these studies, the two cell types have been examined in isolation. There have also been many investigations involving analyses of whole testes or peripheral blood for specific hormones. While the findings from these projects have provided us with a great deal of important information, we believe that the time is now appropriate for a consideration of the physiological factors and cell interactions involved in hormone transport and secretion in the testis, and this is the intention of this review.

II   The Leydig cells

Franz Leydig described in 1850 some characteristic cells in the interstitial tissue between the seminiferous tubules of several species of mammals. An endocrine function was suggested for these cells only much later, and was not at first generally accepted (see Setchell 1984). It is now agreed by most authorities that the Leydig cells are the major site of *de novo* steroid synthesis in the testis, although the tubules may transform steroids synthesized elsewhere.

1   ANATOMY

The mammalian Leydig cell is a relatively large, polyhedral, epithelioid cell. It is surrounded by a typical plasma membrane which is frequently folded with microvilli. It usually has a single, eccentrically located nucleus, although binucleated cells are not uncommon. The nucleus is spherical or ovoid and distinctly vesicular, and contains one to three large nucleoli. The remaining chromatin is present as granules distributed predominantly around the nuclear membrane giving this membrane an appearance of exaggerated thickness when seen with the light microscope (Hooker 1970). No other cell of the interstitial tissue has a nucleus with these characteristics, making identification relatively easy. The cytoplasm is usually abundant, with a prominent amount of smooth endoplasmic reticulum. This is particularly so in guinea pigs, opossums, mice and boars. There are also scattered patches of rough endoplasmic reticulum which interconnect with the smooth ER. The mitochondria are of moderate size and number and contain cristae that commonly appear in the characteristic lamellar form, although many are tubular. However tubular cristae are not particularly distinct in the human, guinea pig or rat (Christensen 1975). The Golgi complex is well developed and often found at one pole of the nucleus. The cytoplasm of the mature cell contains lipid globules of different size and number depending on the species. They are particularly abundant in the cat, and common in humans, guinea pigs and mice but are rare in rats and opossums. These globules were mentioned in Leydigs' original description and Loisel (1903) guessed that the secretion of these cells may be lipid in nature. Reinke crystals, microtubules, and microfilaments are also found in the cytoplasm along with primary lysosomes, digestive vacuoles (secondary lysosomes) and residual bodies (late secondary lyosomes). These residual bodies often take the form of lipofuscin granules when lipid droplets are present, but are usually rare in species in which the cells lack lipid droplets such as rat or opossum. Boar Leydig cells do not show lipid droplets, however they do possess residual bodies containing an unusual reddish pigment. Leydig cells range in size from about 10 μm diameter in the human to 30 μm in the boar, and constitute anything from 2 per cent (rat) to 37 per cent (boar) of the testicular volume. The number of Leydig cells in the adult boar is extreme and they virtually fill the interstitial tissue (Christensen 1975).

Most seasonal breeders also show Leydig cells of differing appearance depending on the season. In such animals, Leydig cells appear to transform from an "undifferentiated" interstitial cell just before the breeding season, into an active Leydig cell with abundant smooth ER, possessing mitochondria with tubular cristae, and possibly a number of medium sized lipid droplets. Following the period of activity, they regress, and lose the specific Leydig cell characteristics (Wing and Lin 1977; Hochereau-de Reviers and Lincoln 1978).

Two generations of Leydig cells have been described for most mammals studied, although three populations have been found in the human (Mancini, Vilar, Alvarez, and Seiguer 1965; Pelliniemi and Niemi 1969) and pig (van Straaten and Wensing 1978; van Vorstenbosch, Colenbrander, and Wensing

1984). The first is termed the "fetal" generation because this is when they first appear. Their function is to secrete the androgens responsible for differentiation of the male reproductive tract but not the regression of the paramesonephric Müllerian ducts (see Josso and Picard 1986). They function independently of gonadotrophins and regress at birth or in the early postnatal period, the extent of regression depending on the species. In the human and the pig, a second population of Leydig cells has been described which are gonadotrophin dependent, transient, and present in the perinatal period. Within a few months after birth, the interstitial tissue of the human testis becomes essentially devoid of Leydig cells and remains in this condition throughout childhood until puberty. On the other hand, there is apparently little involution postnatally in the guinea pig or in the mouse. The rat undergoes some decrease in Leydig cell number postnatally, however there remains a population of Leydig cells which continue to secrete low, but measurable amounts of testosterone in plasma. This apparently initiates neural changes important for subsequent development of male behaviour. The final generation of all species appear to be initiated post-natally at the time of puberty. They differentiate from interstitial cells to acquire the characteristics of the "adult" Leydig cell. Exactly which cells are the precursors for this population has yet to be determined. Fibroblast-like or mesenchymal cells (Mancini *et al.* 1965; Kerr and Sharpe 1985), macrophages (Clegg and MacMillan 1965*a*), and endothelial cells (Laws 1985) have been suggested as Leydig cell precursors, as have dedifferentiated fetal Leydig cells remaining in the testis (Prince 1984). Proliferation of those organelles involved in manufacturing secretory products takes place (mitochondria, smooth ER, Golgi apparatus) with nuclear changes and an increase in the microvilli processes on the cell surface (see Burgos, Vitale-Calpe, and Aoki 1970; Christensen 1975). During maturation, the number of Leydig cells and their size increases (Knorr, Vanha-Pertula, and Lipsett 1970; Tapanainen, Kuopio, Pelliniemi, and Huhtaniemi 1984) along with LH receptor numbers (Ketelslegers, Hetzel, Sherins, and Catt 1978). In the sexually mature animal, the plasma membrane of the Leydig cell has many specializations including junctional complexes, projections and surface indentations that are of both the smooth and coated variety (Connell and Christensen 1975). Mitosis of fully differentiated Leydig cells may also occur to a lesser extent and Christensen and Peacock (1980) and Amat, Paniagua, Nistal, and Martin (1986) have shown that Leydig cell division and differentiation is still possible in mature rats.

Much of the available information on Leydig cell function and metabolism has been derived from *in vitro* studies using isolated cells in culture. Mechanical and enzymatic digestion of rat and mouse testes yields an interstitial cell preparation of which the Leydig cells are but a minor component (Mendelson, Dufau, and Catt 1975; Schumacher, Schafer, Holstein, and Hilz 1978; Payne, O'Shaughnessy, Chase, Dixon, and Christensen 1982; Shaw, Georgopoulos, and Payne 1979). Enriched populations of Leydig cells can be obtained by fractionation of the cells obtained on Metrizamide or Percoll gradients (Browning, D'Agata, and Grotjan 1981; Dehejia, Nozu, Catt, and Dufau 1982) or by centrifugal elutriation (Aquilano and Dufau 1984). Such studies have led to reports of heterogeneity of isolated Leydig cells (Payne, Downing, and Wong 1980; Chen, Lin, Murono, Osterman, Cole, and Nankin 1981; Cooke, Magee-Brown, Golding, and Dix 1981) with two populations being found in the adult rat testis and classified as LH responsive or LH unresponsive (see Sharpe 1982). However, this subject remains controversial. The use of histochemical criteria for identification of isolated cell populations may be of questionable use, since some of the reagents used may only enter disrupted or fragmented Leydig cells (see Aldred and Cooke 1982; Molenaar, Rommerts, and Van der Molen 1983). Aquilano and Dufau (1984) report the active Leydig cell population to consist functionally of only one population of Leydig cells with comparable LH receptor numbers, steroidogenic activity and susceptibility to gonadotrophic desensitization. Kerr, Robertson, and de Kretser (1985) have studied the morphology of two populations of hCG-binding cells obtained from adult mice testes; while one group contained 60–80 per cent morphologically intact Leydig cells, the second group contained large numbers of indeterminate connective cell types with less than 10 per cent morphologically intact Leydig cells. These reports suggest that great care is needed in identification of isolated cell types, prior to their use in biological investigations.

2   RELATIONSHIP TO BLOOD VESSELS, LYMPH VESSELS AND SEMINIFEROUS TUBULES

In most species Leydig cells occur as clusters in the triangular intertubular space, and as strands between closely opposed tubules (see Burgos et al. 1970; Fawcett, Leak, and Heidger 1970; Fawcett, Neaves, and Flores 1973; Christensen 1975; Neaves 1975). Surrounding clusters of Leydig cells is a basal lamina of variable width and disposition that appears to function as a supporting structure in conjunction with collagen fibres located around the Leydig cells and basal lamina. These collagen fibres are few in the rat and are usually found in bundles of several fibres, which loop out over the Leydig cells and extend into the fluid space (Clark 1975).

The relationship between the walls of the tubules, the Leydig cells, and the blood and lymph vessels varies amongst different species. In 1973, Fawcett et al. described the intertubular lymphatics in 14 species. The abundance of Leydig cells, the amount of intertubular connective tissue and the location and degree of development of the lymphatics allowed these authors to define three main categories of interstitial tissue organization. The following descriptions, and Figure 2.1, are from Fawcett (1973) who increased this classification to four groups. The first group (Fig. 2.1A) is characterized by the guinea pig, in which the Leydig cells comprise a small fraction of the testicu-

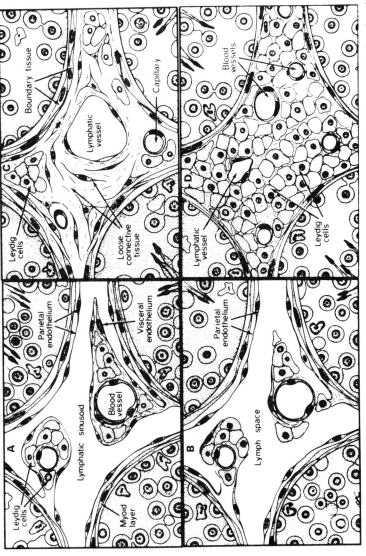

Fig. 2.1. Diagram showing the variation in the anatomy of the interstitial tissue in several species. (A) Guinea pig, showing Leydig cells clustered around blood vessels, with the whole groups of cells completely surrounded by endothelial cells and floating in a lymphatic sinusoid, the contents of which also bathe the walls of the seminiferous tubules. (B) Rat, similar to the guinea pig except that the groups of Leydig cells are surrounded by an incomplete layer of endothelial cells. (C) Ram, showing the Leydig cells either in groups near a capillary or in separate clusters embedded in loose connective tissue which also contains lymph vessels and other blood vessels. (D) Boar, showing the interstitial space crammed with closely packed Leydig cells with a few small blood and lymph vessels. (From Fawcett, 1973).

lar volume, and occur in clusters closely applied to blood vessels. The greater part of the interstitium is occupied by extensive lymphatic sinusoids of irregular outline. These are bounded by a "visceral" (or interstitial; Clark 1975) layer of attenuated endothelium covering the vessels and their associated Leydig cell clusters, and a "parietal" (or peritubular; Clark 1975) layer of endothelium closely applied to the myoid layer of the seminiferous tubules. The endothelium of the lymphatic sinusoids in this space is generally continuous. Narrow sheets of collagen bounded by two layers of endothelium attach the vascular-endocrine complexes to the tubules. In this species therefore, Leydig cells are interposed between the walls of the blood vessels and the endothelium of the lymphatic sinusoids.

In the second group (Fig. 2.1B), characterized by the rat and mouse, interstitial organization is basically the same as in the first group. However the visceral layer of endothelium is discontinuous over wide areas so that Leydig cells are directly bathed in lymph.

The majority of larger mammals, including the bull, ram, elephant, monkey and man, have a very different interstitial organization (Fig. 2.1C). In these species, Leydig cells do not have such an obvious association with blood vessels, occurring in clusters of varying size, scattered in an oedematous loose connective tissue which is drained by conspicuous lymph vessels located centrally or eccentrically in each intertubular area.

The fourth group (Fig. 2.1D) includes the domestic boar, warthog, zebra and naked mole rat. Closely packed Leydig cells occupy large intertubular spaces and comprise up to 50 per cent of the volume of the testis. There is very little interstitial connective tissue in these species, and small lymphatic vessels are infrequently encountered.

Given that the Leydig cell is the most important source of androgen in the testis, these species differences in interstitial organization have obvious implications for how androgen is partitioned between the vascular and intratesticular circulation. Fawcett *et al.* (1973) hypothesized that in the rodents (Groups 1 and 2), movement of testosterone from the Leydig cells could be envisioned as a release of androgens into the blood capillaries and into the protein-rich extracellular fluids that move from the blood vascular system into the lymphatic sinusoids surrounding the seminiferous tubules. These authors suggested that the lymphatic route of testosterone transport might be the primary source of androgens for the seminiferous tubules in these animals. In the larger mammals of Group 3, blood capillaries are in much closer contact with the tubules and therefore may directly supply testosterone to the tubules, as suggested many years ago by Halley (1960). Lymphatics in these species are believed to be more involved in return of extracellular fluid to the general circulation, than in distribution of testosterone. The large mass of Leydig cells in Group 4 is probably able to provide testosterone directly to the tubules without the need for any indirect system such as the lymphatics or blood vascular system (Connell and Connell 1977). The high density of

Leydig cells in these species may be related to the production of other hormones such as oestrogens and pheromones (for example, "boar taint", 5α-androst-16-en-3-one, see Setchell 1978).

Leydig cells have been found in the boundary tissue of the seminiferous tubules in guinea pigs (Fawcett *et al.* 1970) and man (Fawcett and Burgos 1960) in association with the myoid cells and fibroblasts of this tissue. These Leydig cells may be important in Leydig cell–Sertoli cell interactions and Bergh (1982; 1983; 1985*a*) has demonstrated that peritubular Leydig cells in the rat testis change in size according to the stage of the spermatogenic cycle in the adjacent seminiferous tubules. The largest Leydig cells are found adjacent to tubules in ante-release stages (VII–VIII) and are larger than perivascular cells. Leydig cells adjacent to tubules in post-release (IX–XII) and ante-meiosis stages (XIII–XIV) were similar in size to perivascular Leydig cells, and both peritubular and perivascular cells in abdominal testes were of comparable size (Bergh 1983). This suggests that Sertoli cells may be able to regulate Leydig cell function. It has also been suggested that newly differentiated Leydig cells may arise from peritubular cells and subsequently leave the lamina propria to gain access to the interstitial tissue (Kerr, Donachie, and Rommerts 1985). During pubertal maturation, new Leydig cells apparently differentiate from peritubular interstitial cells (Christensen 1975; de Kretser 1967; Fawcett 1973; van Straaten and Wensing 1978), and this region is also the site of development of numerous Leydig cells 4 weeks after EDS treatment (Kerr *et al.* 1985).

3   RELATIONSHIP TO OTHER CELLS IN THE INTERSTITIAL TISSUE

While the Leydig cell is widely recognized as a major cell type of the interstitial compartment of the mature testis (Niemi and Ikonen 1973), a number of cells of connective tissue origin have been reported in this region including fibroblasts, macrophages, mast cells and lymphocytes (Ross 1967; Christensen 1975). Testicular macrophages in the rat may comprise up to 25 per cent of the interstitial cell population (Ewing, Zirkin, Cochran, Kromann, Pefers, and Ruiz-Bravo 1979; Niemi, Sharpe, and Brown 1986) or one macrophage per four Leydig cells (Bergh 1985*b*). Macrophages have frequently been found in close association with Leydig cells (Connell and Christensen 1975; Wing and Lin 1977; Miller, Bowman, and Rowland 1983; Niemi *et al.* 1986) and morphological changes in Leydig cells (caused by gonadotrophin withdrawal, cryptorchidism or seasonal dysfunction) are associated with morphological changes in the macrophages (see Wing and Lin 1977; Gondos, Rao, and Ramachandran 1980; Bergh 1985*b*). The importance of this association is unclear although Bergh (1985*b*) indicated that the morphology of the two cell types is correlated so that a functional inter-relationship is likely.

Because macrophages are so closely associated with Leydig cells, this can lead to considerable contamination of Leydig cell preparations. Similarities in function between the two cell types has led Niemi *et al.* (1986) to suggest that this may in fact account for "unwarranted" suggestions of Leydig cell heterogeneity (see Cooke *et al.* 1981 and **Section II.1**).

Functionally, Milewich, Chen, Lyons, Tucker, Uhr, and McDonald (1982) found peritoneal macrophages to be steroidogenic, and Molenaar, Rommerts, and Van der Molen (1984) have shown that macrophages possess surface antigens in common with Leydig cells. Wahlstrom, Huhtaniemi, Hovatta, and Seppala (1983) and Hovatta, Huhtaniemi, and Wahlstrom (1986) have found macrophages which contain FSH demonstrable by immunohistochemistry in the human testis, and Yee and Hutson (1985) have found high affinity hFSH receptors on macrophages isolated from the rat testis, which suggests that these cells may also respond to gonadotrophins. Miller *et al.* (1983) found that portions of Leydig cells may be endocytosed by macrophages, supporting the suggestion that testicular macrophages are involved in regulating Leydig cell function. Yee and Hutson (1985) demonstrated that conditioned medium from cultures of testicular macrophages stimulated testosterone production when added to Leydig cells *in vitro*. On the other hand, Niemi *et al.* (1986) observed macrophages in prepubertal animals which may imply other functions such as an immunological role for these cells. Most of these studies have been carried out with rats and the situation in other species is yet to be fully investigated. The macrophage population in the sheep testis appears to be quite different from that in the rat (Pollanen and Maddocks 1988). Macrophages have also been reported in the peritubular region of the rat testis (Ross 1967) and Leydig cells which lie adjacent to the seminiferous tubules are often in close association with the peritubular or parietal endothelium (Clark 1975). This may be of consequence in determining distribution of hormones produced by these cells.

Mast cells are a cell-type usually found free in connective tissues, and their presence has been confirmed in interstitial testicular and epididymal tissues, as well as the testis mediastinum and tunica albuginea (Hermo and Lalli 1978; Nistal, Paniagua, Abaurra, and Pallardo 1980; Nykanem 1980; Maseki, Miyake, Mitsuya, Kitamura, and Yamada 1981). Nistal, Santamaria, and Panigua (1984) have shown that mast cells of the testis and epididymis are similar to those found in other connective tissues. Mast cells are a major source of histamine (Riley and West 1953; Schayer 1956) and in the rat and mouse also contain serotonin (5-hydroxytryptamine) and can form and store dopamine (see Goth and Johnson 1975). Histamine has been shown to affect vascular permeability in a number of tissues (Williamson, Chang, Kilzer, Marvel, and Kilo 1984) but although Assaykeen and Thomas (1965) have found histamine in the testis, it appears to have little effect on vascular permeability in this organ (Williamson, personal communication), and is not involved in mediating the testicular vascular response to hCG

(Sowerbutts, Jarvis, and Setchell 1986). However these authors have demonstrated large numbers of mast cells in the tissue surrounding the testicular artery, and found ketanserin (an antagonist to 5-HT) to block partially the hCG-mediated vascular response in the testis. Krishna and Terranova (1985) have also reported mast cells in the vicinity of blood vessels entering and leaving the hamster ovary, and have correlated the gonadotrophin surge on the day of pro-oestrus with histamine release by ovarian mast cells; and a similar relationship may exist between gonadotrophins and mast cells in the testis. The number of mast cells increases in infancy, decreases in childhood, and increases again at puberty. During adulthood, mast cell numbers progressively decline in all testicular and epididymal connective tissues, although Maseki *et al* (1981) reported an increase in numbers (mastocytosis) in patients with idiopathic male infertility.The increase in mast cell numbers in the normal individual correlates with development of testicular connective tissues, and mast cell degranulation and heparin release is concomitant with collagen synthesis (see Nistal *et al.* 1984). At the beginning of puberty, proliferation of active fibroblasts occurs, coinciding with increased gonadotrophin levels and Leydig cell differentiation. The increase of mast cell numbers during this time may be of importance for developing Leydig cells given their inevitable association in the interstitium.

4   HORMONES PRODUCED BY THE LEYDIG CELLS

The evidence that the Leydig cell is the predominant source of testosterone in the mammalian testis has already been discussed in this review. However a variety of steroids are synthesized in the testis from either cholesterol, formed elsewhere in the body and transported to the testis in blood; or acetate, derived from the blood or formed as acetyl-coenzyme A during metabolism of glucose (see Setchell 1978). There is great variation amongst mammals in the identity of the steroids secreted by the testis, and in the concentrations achieved. While testosterone is the steroid secreted in the greatest amounts by most mammalian testes, the human testis secretes a number of sulphated steroids into venous blood, including dehydroepiandrosterone sulphate (DHAS), 5-androstene-3β,17β-diol sulphate, and testosterone sulphate (Leinonen, Ruokonen, Kontturi, and Vihko 1981; Ruokonen, Lukkarinen, and Vihko 1981). The pig secretes as much 5α-androsterone (boar taint) as testosterone (Gower, Harrison, and Heap 1970; Claus and Hoffman 1980) and almost as much DHAS and oestrone sulphate into venous blood (Claus and Hoffman 1980; Setchell, Laurie, Flint, and Heap 1983), while the horse secretes appreciably greater amounts of oestrone sulphate than testosterone into venous blood (Setchell and Cox 1982). In the pig and horse, testicular lymph appears to be a more important route than venous blood for the secretion of conjugated steroids (Setchell and Cox 1982; Setchell *et al.* 1983), and it is possible that the detection of such steroids in other species will occur when the appropriate fluids can be analysed. Ewing and co-workers have also studied steroid secretion in the maximally-stimulated perfused testis of several species. While testosterone remains the major steroid secreted under these conditions, the rabbit testis also secretes significant amounts of 5α-reduced androgens and the rat testis produces almost as much pregnenolone and 17α-hydroxypregnenolone as it does testosterone (Ewing, Brown, Irby, and Jardine 1975; Chubb and Ewing 1979a,b; Ewing et al. 1979; Zirkin, Ewing, Kromann, and Cochran 1980).

Prepubertal and fetal Leydig cells appear to produce a different balance of steroids from the adult Leydig cell. Hsueh, Schreiber, and Erickson (1981) have reported that isolated cells from the prepubertal rat testis secrete androsterone and 5α-androstene-3α,17β-,diol whereas testosterone was the major product in adult cells (Hsueh 1980). Likewise, the Leydig cell in the sexually mature animal is responsible for testicular oestrogen production (Payne and Valladares 1980) whereas the Sertoli cell performs this function in the immature animal (Dorrington and Armstrong 1975). This has obvious implications for studies on isolated cells since age differences must be carefully considered before any conclusions are drawn about cell function.

Apart from the androgenic products, it is now evident that the Leydig cell synthesizes other bioactive substances that are probably involved in the paracrine and autocrine regulation of testicular function. Early studies by Niemi and Kormano (1965) demonstrated a possible role for oxytocin in mediating contractility of the seminiferous tubules, and since then, oxytocin and vasopressin have been found in small amounts in testicular material from rats, men and bulls (Wathes 1984) together with neurophysin (Nicholson, Swann, Burford, Wathes, Porter, and Pickering 1984) suggesting local biosynthesis. Recently Guldenaar and Pickering (1985) have demonstrated immunocytochemically that oxytocin is present in rat Leydig cells, although they found no evidence of vasopressin or neurophysin. Other substances which cells in the testis seem to be capable of synthesizing and which may regulate endocrine function of the testis by autocrine or paracrine actions include prostaglandins (Ellis, Sorensen, and Buhrley 1975; Greazissis and Dray 1977a,b, 1979; Haour, Mather, Saez, Kouznetzova, and Dray 1979; Carpenter, Manning, Robinson, and To 1978; Sairam 1979), renin (Parmentier, Inagami, Pochet, and Desclin 1983; Pandey, Melner, Parmentier, and Inagami 1984) and β-endorphin and other pro-opiomelanocortin (POMC)-derived peptides (Margioris, Liotta, Vaudry, Bardin, and Krieger 1983; Bardin, Shaha, Mather, Salamon, Margioris, Liotta, Gerendai, Chen, and Krieger 1984; Shaha, Liotta, Krieger, and Bardin 1984; Fabbri, Tsai-Morris, Luna, Fraioli, and Dufau 1985). Parmentier, Inagami, Pochet, and Desclin (1983) found renin-like immunoreactivity in the testis, and Pandey, Melner, Parmentier, and Inagami (1984) have found similar activity in purified Leydig cell preparations, but its function is unknown.

III   The Sertoli cells

In 1865, Enrico Sertoli described columnar cells with cytoplasmic processes extending from the basement membrane to the lumen of the seminiferous tubules, and enveloping the neighbouring germ cells to provide physical support and a "nursing" function. While germ cells continue to divide, differentiate and be released into the tubular lumen eventually to pass out of the testis, the Sertoli cells are a stable population of cells, which do not divide in the adult animal (Clegg 1963; Steinberger and Steinberger 1971; Nagy 1972; Orth 1982, 1984). The importance of Sertoli cells in regulating and maintaining the process of spermatogenesis has been emphasized by numerous investigators, although their precise functions and significance for the developing germ cells are still not fully understood and their endocrinological significance is still a matter for debate.

1   ANATOMY

The Sertoli cells rest upon the limiting basement membrane of the seminiferous tubule and can be easily distinguished by their shape and nuclear and cytoplasmic characteristics. The basal portion of Sertoli cells are recognized in tangential section as polygonal areas. The nucleus is infolded in all species studied, being elaborately lobulated in some (Burgos *et al.* 1970; Fawcett 1975). It often lacks both the fibrous lamina that reinforces the inner surface of the nuclear envelope in many cell types, and the karysome and peripheral clumps of heterochromatin found in the nuclei of most other somatic cell types. Specialized junctional complexes are formed between neighbouring Sertoli cells, with opposed membranes approaching to within 2 nm in some areas (Dym and Fawcett 1970). These junctional complexes divide the germinal epithelium into "basal" and "adluminal" (or "central" and "peripheral", Bellvé 1979) compartments and provide a structural basis for the maintenance of a different milieu in each of these compartments which may be crucial during specific stages of cell development.

The cytoplasm of the Sertoli cells contain long mitochondria. While randomly orientated in the basal cytoplasm, they tend to be orientated parallel to the cell axis in the supranuclear columnar portion of the cell (Fawcett 1975). Again unlike other cell types, the Golgi apparatus appears to consist of multiple separate Golgi elements scattered throughout the basal cytoplasm and occasionally in the supranuclear region.

Numerous membrane-limited dense bodies are found in the cytoplasm, and include primary lysosomes, autophagic and heterophagic vacuoles, and deposits of lipochrome pigment (Fawcett 1975). In addition to normal housekeeping lysosomal activities in a cell, the Sertoli cells digest those germ cells that normally degenerate during spermatogenesis along with residual spermatid cytoplasm left behind after the release of the spermatozoa (Lacy 1962; Clegg and McMillan 1965b; Black 1971; Russell 1980).

Rough and smooth endoplasmic reticulum is found in the Sertoli cell cytoplasm, with rough or granular reticulum predominantly in the basal region. This granular form is usually tubular, and while the population of attached ribosomes is large in the monkey, it is relatively sparse in other species (Dym 1973; Fawcett 1975). The smooth endoplasmic reticulum is generally more abundant than the rough, but again Fawcett (1975) describes significant species differences, with ruminants possessing more extensive quantities than the laboratory rat. The ram Sertoli cell contains aggregates of smooth reticulum in the basal cytoplasm often closely associated with lipid. Localization also occurs in cytoplasm immediately surrounding developing acrosomes of associated spermatids (Russell 1980). Crystalline inclusions have also been described in the Sertoli cell, although they are seldom found in laboratory and domestic animals (Fawcett 1975).

Lipid content is highly variable between species, in both total amount and size of droplets. They are generally localized near the base of cells and appears to vary with the stage of the spermatogenic cycle in some species (Lacy 1967). After heat, X-irradiation or oestrogen treatment, all of which cause germ cell degeneration, the amount of lipid increases and Lacy (1967) has speculated that phagocytosis by Sertoli cells of germ cell residual bodies may contribute to this lipid pool, which in turn might be utilized by Sertoli cells for steroid synthesis. The abundance of smooth endoplasmic reticulum suggests that Sertoli cells may have the capacity for steroid biosynthesis (Steinberger and Steinberger 1977).

At certain stages of the spermatogenic cycle, microtubules are abundant in the Sertoli cell cytoplasm. While considered to be mainly cytoskeletal elements providing support for the columnar portions of the cell, they may also be actively involved in movements of cytoplasm involved in displacing late spermatids (Fawcett 1975; Means, Dedman, Tindall, and Welsh 1978).

Unlike the Leydig cell population, which is primarily established in the adult animal at puberty; the mitotic activity of the Sertoli cell is most pronounced in the fetal and early postnatal period (Orth 1984). In the rat, Sertoli cell proliferation is maximal on day 20 postconception, that is, one day before birth (Orth 1982) and falls steadily after parturition, ceasing between the twelfth and fifteenth day after birth (Clermont and Perey 1957; Hilscher and Makoski 1968; Orth 1982, 1984). While it is widely accepted that the number of Sertoli cells per testis appears to remain unchanged in rats after the second postnatal week, there is contrary evidence in other species. Johnson and Thompson (1983) have reported a 36 per cent increase in the number of Sertoli cells in the testis of the adult horse during the breeding season, and Hochereau-de Reviers and Lincoln (1978) found a similar increase in the number of Sertoli cells in the testis of the adult red deer in the breeding season, although the latter change was not statistically significant. While increases in the Sertoli cell population have also been reported to occur in adult animals during recovery from X-irradiation (Nebel and Murphy 1960), artificial cryptorchidism (Clegg 1963) and during FSH replacement treatment following hypophysectomy (Murphy 1965), the lack of statistical significance (Hochereau-de Reviers and Lincoln 1978) and the possible shrinkage of seminiferous tubules after various treatments (Steinberger and Steinberger 1977) has meant that such reports have not gained wide acceptance. Some seasonal breeders can dramatically alter the number of germ cells per Sertoli cell, with only the Sertoli cells and younger germinal cells remaining in the nonbreeding season (Neaves 1973). However, in situations where seasonal breeders have a limited capacity to alter the number of germinal cells per Sertoli cell (as reported for the horse; Johnson and Thompson 1983), the ability to alter Sertoli cell numbers would possibly be of importance.

Sertoli cells are now routinely kept in culture following isolation from the testes of immature rats (Dorrington and Fritz 1975; Dorrington and Armstrong 1975; Dorrington, Roller, and Fritz 1975; Welsh and Wiebe 1975; Steinberger, Heindel, Lindsay, Elkinton, Sanborn, and Steinberger 1975; Fritz, Rommerts, Louis, and Dorrington 1976; Tung and Fritz 1977; Lacroix, Smith, and Fritz 1977; Wilson and Griswold 1979; Wright, Musto, Mather, and Bardin 1981; Kissinger, Skinner, and Griswold 1982; Le Gac and de Kretser 1982; Sharpe, Fraser, Cooper, and Rommerts 1982) and also from young pigs (Vazeille and Chevalier 1979; Perrard, Saez, and Dazord 1985), calves (Francis, Triche, Brown, Brown, and Bercu 1981; Vigieri, Picard, Campargue, Forest, Heyman, and Josso 1985; Coombs, Woods, Ellison, and Jenkins 1986; Hayes 1986) and immature monkeys (Lee, Pineda, Spiliotis, Brown, and Bercu 1983). The cells do not divide in culture, which is probably not surprising as division of these cells has already ceased *in vivo* at the age at which they are usually isolated. However, several cell lines with Sertoli cell characteristics have been established (Mather 1980; Mather, Zhuang, Perez-Infante, and Phillips 1982). Studies with cells isolated from adult animals are complicated by the presence of greater numbers of germ cells (rat: Steinberger *et al.* 1975; DeMartino, Marcante, Floridi, Citro, Bellocci, Cantaforca, and Natali 1977; Cameron and Merkwald 1981; monkey: Lee *et al.* 1983) and contamination with peritubular cells has also been a problem (see Tung, Skinner, and Fritz 1984). The possibility of contamination with endothelial cells has apparently not been considered although recent evidence showed that γ-glutamyl transpeptidase, an enzyme originally claimed to be specific to Sertoli cells (Hodgen and Sherins 1973; Lu and Steinberger 1977), could be localized histochemically in the endothelial cells of the arterioles of the testis (Niemi and Setchell 1986).

These problems of contamination with cells other than Sertoli cells are of obvious concern when it comes to interpreting data on cell products obtained from *in vitro* studies. However a further complication arises from the process of culture itself. Sertoli cells grown on culture plates in defined medium lose their columnar form and adopt a flattened morphology (Suarez-Quian, Hadley, and Dym 1984). This implies important cell-cell relationships as we shall discuss shortly. However it has been of concern that cells in such a state may not continue to perform the same functions that columnar cells perform *in vivo*. Androgen binding protein (ABP) secretion of Sertoli cells cultured in plastic dishes declines immediately after isolation and culture, and transferrin production declines after 5–6 days of culture (Skinner and Griswold 1982; Rommerts, Kruger-Sewnarain, Van Woerkom-Blik, Grootesgoed, and Van der Molen 1978; Byers, Hadley, Djakiew, and Dym 1986). Coating the plastic with a reconstituted basement membrane preparation maintains high levels of secretion of both products for at least 14 days (Hadley, Byers, Suarez-Quian, Kleinman, and Dym 1985). Dym and co-workers have recently cultured Sertoli cells on a reconstituted basement membrane extract in bicameral culture chambers, and in such cultures the columnar form of the cells persists (Suarez-Quian *et al.* 1984; Hadley *et al.* 1985; Byers *et al.* 1986; Djakiew, Hadley, Byers, and Dym 1986). Using a similar system, Janecki and Steinberger (1987) have reported secretion of ABP and transferrin at relatively high and stable rates, for much longer time periods than achieved by cells cultured on plastic.

2   RELATIONSHIP TO THE BLOOD-TESTIS BARRIER AND OTHER CELLS

A number of substances of widely varying molecular size introduced into the blood stream rapidly appear in testicular lymph, but not in the fluid collected from the cannulated rete testis (Setchell, Voglmayr, and Waites 1969; Setchell 1970*a*, 1978, 1980). Such studies demonstrated the existence of a blood-testis barrier, analogous to the blood-brain barrier discovered several decades previously. However, unlike the blood-brain barrier which has been found to reside in cell-cell junctions in the walls of cerebral capillaries (Reese and Karnovsky 1967), the blood-testis barrier is found principally in the walls of the seminiferous tubules. The specialized junctions between pairs of Sertoli cells above the spermatogonia, but below the spermatocytes, divide the seminiferous epithelium into basal and adluminal compartments, and it is these junctional complexes that are a principle component of the blood-testis barrier to larger hydrophilic substances such as proteins. These sites of restricted permeability have been defined by studies using electron opaque markers (see Fawcett 1975). They are symmetrical specializations of adjoining Sertoli cells, confined to the basal third of the epithelium (see Fig. 2.2) where overarching lateral processes of the Sertoli cells meet just above the spermatogonia. In these junctions subsurface cisternae and bundles of filaments develop, and the opposing membranes (normally some 15–20 nm apart) approach to within 2 nm to form the junctions. Multiple sites of obliteration of the intercellular cleft are apparent, and it is at these sites that

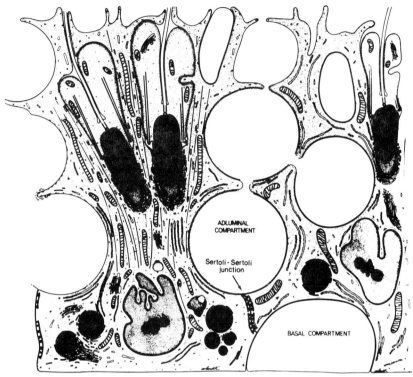

Fig. 2.2.   Drawing illustrating the occluding junctions between Sertoli cells. These divide the seminiferous epithelium into basal and adluminal compartments, and form the principle component of the blood-testis barrier. (Adapted from Fawcett 1975).

penetrating electron-opaque markers abruptly stop (Fawcett 1975). In the immature testis, occluding junctions are absent, and while typical gap junctions are common, they gradually disappear (Vitale-Calpe, Fawcett, and Dym 1973; Gilula, Fawcett, and Aoki 1976). The development of the specialized junctions between Sertoli cells would seem to be under hormonal control since the blood-testis barrier does not appear until the onset of spermatogenesis at puberty (Kormano 1967a). The unique cell-cell junctions between Sertoli cells are an effective intra-epithelial component of the blood-testis barrier. However, the Sertoli cell has important interactions with other cell types, and the epitheloid cells in the peritubular contractile layer of the seminiferous tubules are now thought to provide an adventitial, albeit incomplete, component to the blood-testis barrier. This is particularly so in rodents, but Fawcett (1975) suggests that in primates, including man, the extensive gaps in the multiple layers of adventitial cells mean that the blood-testis barrier depends exclusively on the Sertoli cell junctions. In the ram testis, the limiting membrane of the tubules probably provides a more complete layer, since it contains up to fifteen layers of condensed material lying

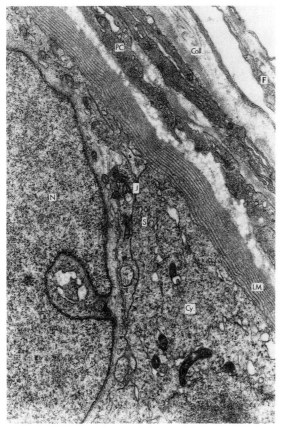

Fig. 2.3.   Electron micrograph (× 30 000) of the boundary tissue of a seminiferous tubule from the testis of a ram. LM—Inner noncellular layer; PC—peritubular contractile cells of the inner cellular layer; Coll—collagen fibres of outer noncellular layer; F—fibroblast of outer cellular layer; N—nucleus and Cy—cytoplasm of cells inside the seminiferous tubule (S); J—junction between two adjacent cells. Kindly supplied by Dr. G. H. McIntosh, CSIRO Division of Human Nutrition, Adelaide, and Dr. B. Morris, Department of Experimental Pathology, Australian National University, Canberra.

beneath a layer of modified fibroblasts, with the internal surface of this membrane lying directly against the cell membranes of the adjacent spermatogonia or Sertoli cells (Morris and McIntosh 1971; Fig. 2.3).

The division of the Sertoli cell epithelium by the junctional complexes is of major consequence for partitioning of all substances; those reaching the base of the epithelium from the blood have more or less direct access to cells in the basal compartment. However the occluding junctions mean that substances must pass through the Sertoli cell cytoplasm to reach germ cells in the adluminal compartment. Likewise the fluid of the tubular lumen must be derived from material passed selectively through, or synthesized by the Sertoli cell (Setchell 1970, 1980). A conspicuous feature of the composition of the fluid of the seminiferous tubule is a high content of potassium (40 mM in the rat; Tuck, Setchell, Waites, and Young 1970), and this is probably produced by the Sertoli cells (see Setchell and Waites 1975; Muffly, Turner, Brown, and Hall 1985). To achieve this, these cells would require a conventional coupled sodium-potassium exchange pump on the basal or peripheral side of the cell, with these membranes having a low permeability to sodium and potassium (Tuck *et al.* 1970; Waites and Gladwell 1982). The luminal surface however would have relatively high permeability to these ions, and a potassium gradient develops from the base to the apex of the Sertoli cell (Muffly *et al.* 1985).

The function of Sertoli cells can be influenced by interaction with other cell types. The potassium concentration in tubular fluid is similar to plasma in testes from which the germ cells had been eliminated by X-irradiation or busulphan treatment (see Setchell, Davies, Gladwell, Hinton, Main, Pilsworth, and Waites 1978). The secretion of ABP by cultured Sertoli cell monolayers stimulated with FSH can be further increased if pachytene spermatocytes are allowed to adhere (Galdieri, Monaco, and Stefanini 1984).

In the intact testis, the mesenchymal cells (peritubular myoid cells and fibroblasts) in the boundary tissue of the seminiferous tubules are adjacent to the epithelial-type Sertoli cells, separated by a basal lamina (Fawcett 1975) with which only 3 per cent of the Sertoli cell is in contact (Weber, Russell, Wong, and Peterson 1983). These two cell types have been observed to interact specifically with each other in a complex manner, both *in vivo* and *in vitro* (Tung and Fritz 1986). Peritubular myoid cells of rodents are thought to be derived from fibroblast-like cells in the fetal gonadal interstitium under the influence of adjacent Sertoli cells (Bressler and Ross 1973). Peritubular cells and Sertoli cells in co-culture interact in a variety of ways resulting in prolonged mutual survival in a medium containing no added protein, under conditions in which neither cell type would otherwise survive in monoculture (Tung and Fritz 1980). Peritubular cells in co-culture with Sertoli cells sustain the production of ABP, a recognized marker for Sertoli cell function (Tung and Fritz 1980; Hutson and Stocco 1981), and the addition of peritubular cell conditioned medium to Sertoli cells in culture increases ABP and transferrin formation (Skinner and Fritz 1985a). Indeed, peritubular cells are known to secrete a 70 kD non-mitogenic, paracrine factor termed P-Mod-S which maximally stimulates ABP production in cultured Sertoli cells (Skinner and Fritz 1986). This indicates that a protein secreted by peritubular cells may be involved in mesenchymal-epithelial cell interactions between peritubular and Sertoli cells.

These two cell types also co-operate to form components of the extracellular matrix (ECM) and basal lamina of the tubular boundary layer. Cultured peritubular cells secrete collagen and fibronectin, and Sertoli cells in monoculture produce type IV collagen and laminin (Tung *et al.* 1984; Tung, Skinner, and Fritz 1985; Skinner, Tung, and Fritz 1985). Pollanen, Kallajoki, Ristelli, Ristelli, and Suominen (1985) have also demonstrated laminin and type IV collagen immunohistochemically in myoid cell layers, and laminin in Sertoli cells of both normal and pathological human testes. The finding by Hermo and Lalli (1978) of mast cells and monocytes as regular components of the walls of seminiferous tubules may relate to collagen synthesis in this region. Proteoglycans constitute important components of the ECM, and are thought to play an essential role in ECM deposition. Skinner and Fritz (1985b) have demonstrated that cultured Sertoli cells synthesize and secrete proteoglycans which contain both chondroitin and heparin glycosaminoglycan chains (GAG chains), while cultured peritubular cells produce proteoglycans of higher molecular mass containing chondroitin GAG but not heparin GAG chains.

As culture and particularly co-culture systems are further developed and improved, the physiological significance of much of this work will become clearer. While the physiological role of peritubular cells, collagen fibrils and other ECM components in the boundary wall in limiting the passage of cells and macromolecules and preventing the penetration of blood vessels is recognized (Dym and Fawcett 1970), they may also play an important role in shielding the immunologically-foreign haploid germ cells from immune surveillance. Hermo and Lalli (1978) reported that in the limiting membrane of the seminiferous tubules in human testicular biopsies, leukocytes were found subjacent to Sertoli germ cells in contact with the basal lamina, but never in the seminiferous epithelium.

3   HORMONES PRODUCED BY SERTOLI CELLS

The possibility that the seminiferous tubules might possess steroidogenic functions was suggested by the occurrence of tumours believed to originate from Sertoli cells, which appeared to secrete oestrogens (Huggins and Moulder 1945). The abundance of smooth endoplasmic reticulum in the Sertoli cells, similar to that found in other steroid producing cells, made this suggestion plausible on a morphological basis (Christensen and Fawcett 1961). However, there has been some controversy over whether hormones are actually produced. While Christensen and Mason (1965) found significant androgen biosynthesis by isolated seminiferous tubules *in vitro*, it seems likely that these preparations were contaminated by interstitial cells as the rate of synthesis was much higher in the latter. Cooke, de Jong, Van der Molen, and Rommerts (1972) incubated interstitial tissue *in vitro* and found the content of testosterone increased with time, whereas the amount of testosterone in isolated tubules decreased during incubation or was unchanged. However, de Jong, Hey, and Van der Molen (1974) found the oestradiol content of separated tubules increased during incubation, while that in interstitial tissue did not change. Sertoli cells isolated from the testes of prepubertal rats formed oestrogens from testosterone and other steroid precursors. However, this activity falls off rapidly with age and no detectable oestrogen is formed by cells isolated from rats older than 40 days (Armstrong, Moon, Fritz, and Dorrington 1975; Dorrington and Armstrong 1975; Dorrington, Fritz, and Armstrong 1976), that is, before the first spermatozoa are shed. It should also be remembered that isolated tubules can form from progesterone or pregnenolone a number of other steroids which are not formed by the Leydig cells (Bell, Vinson, and Lacy 1971; Lacy 1973). Steroids from the tubules must therefore be considered as possible regulators of Leydig cell function, and Leydig cells have been shown to have oestrogen receptors (Steinberger and Steinberger 1977).

However, the hormone produced by Sertoli cells which probably has the most chequered history of all reproductive hormones is inhibin. While its existence was postulated over 60 years ago, it has only recently been isolated and characterized. The name was coined by McCullagh (1932) for a non-steroidal substance that feeds back on the anterior pituitary specifically to decrease the secretion of FSH (see reviews by Setchell and Main 1974; Baker, Bremner, Burger, de Kretser, Dulmanis, Eddie, Hudson, Keogh, Lee, and Rennie 1976; Setchell, Main, and Davies 1977; Setchell 1980; Steinberger 1983; de Jong and Robertson 1985; de Jong 1987). By its inhibiting effects on FSH secretion, it may alter spermatogenesis (see de Jong, Welschen, Hermans, Smith, and Van der Molen 1978) and while this probably remains the major role for inhibin in the male, it may exert other effects within the testis. Franchimont, Croze, Demoulin, Bologne, and Hutson (1981) found a dose-dependant inhibitory effect on the division of germ cells (see also Demoulin, Hustin, Lambotte, and Franchimont 1981). Nagendranath, Jose, Sheth, and Juneja (1982) report that dihydrotestosterone inhibits the release of inhibin by Sertoli cells, and Hurkaldi, Arbatti, Mehte, and Sheth (1984) suggest that testosterone concentrations may regulate inhibin levels. Inhibin has recently been shown to be a glycoprotein composed of an interlinked α- and β-subunit; Vale, Rivier, Vaughan, McClintock, Corrigan, Woo, Karr, and Spiess (1986) and Ling, Ying, Ueno, Shimasaki, Esch, Hotta, and Guilleman (1986) have shown that dimers of the β-subunit can form (which complicates any immunoassays of this hormone where the antisera is directed against the β-subunit), and that these dimers are potent *stimulators* of FSH synthesis and secretion, which now raises the possibility of dual control of FSH from within the gonads (Tsonis and Sharpe 1986). Just how this dimer interacts at the level of the pituitary with LHRH remains to be determined.

A number of possible paracrine regulators in the testis have been considered (see Sharpe 1984, 1986; Täkhä 1986). There is evidence that an LHRH-like peptide can be formed by the Sertoli cells and may act on the Leydig cells (Sharpe, Fraser, Cooper, and Rommerts 1981; Sharpe *et al.* 1982). Two other peptides from the Sertoli cells which influence the Leydig cells have recently been described (Benhamed, Grenot, Tabone, Sanchez, and

Morera 1985), and the Leydig cells may influence the Sertoli cells by factors other than testosterone (Perrard-Saperi, Chatelain, Vallier, and Saez 1986). A peptide growth factor stimulating mitoses in quiescent fibroblasts can be isolated from Sertoli cells (Feig, Klagsbrun, and Bellvé 1983) and a similar but not identical peptide is secreted into rete testis fluid (Brown, Blakeley, Henville, and Setchell 1982), suggesting that the Sertoli cells may be influencing cells in the epididymis directly.

IV   Mechanism of transport of hormones out of and across cells

1   SECRETION OF HORMONES BY CELLS

Cholesterol is produced by the smooth endoplasmic reticulum of the Leydig cell and passes to the mitochondria for side-chain cleavage and conversion to pregnenolone, which then returns to the smooth endoplasmic reticulum for conversion to testosterone (Ewing and Zirkin 1983, see Fig. 2.4). How the testosterone then leaves the cell is unknown (see Christensen 1975). The mode of secretion will depend on whether the newly synthesized testosterone is released into the cytoplasm outside the tubules of the smooth reticulum, into the hydrophobic portion of the reticulum membrane itself, or into the cavity of the smooth reticulum. Christensen (1975) discusses the implications of each of these possible mechanisms on the mode of testosterone secretion. It is unlikely that simple diffusion is the only process involved, and it is more than likely that a carrier protein is present there, possibly in association with the Golgi bodies.

It was proposed several years ago that the cells of the corpus luteum sequester progesterone in membrane-bound granules containing a progesterone-binding protein. Evidence was presented that these granules were exocytosed, releasing progesterone and the binding-protein into the extracellular fluid (Gemmell, Stacy, and Thorburn 1974). More recent evidence suggests that oxytocin, and not progesterone, is released from these granules (Rice and Thorburn 1985; Theodosis, Wooding, Sheldrick, and Flint 1986; Hirst, Rice, Jenkin, and Thorburn 1986). Lipid droplets are common in the cytoplasm of the Leydig cells of some species, although not in the adult rat (Christensen and Gillim 1969). It is not known if they are exocytosed, although in view of the high lipid solubility of most steroids, it could be one way of secretion. Fat droplets are pinched off at the surface of acinar cells in the mammary glands (see Linzell and Peaker 1971) and the endothelial cells of the mammary gland can engulf chylomicra on their vascular surface, and release their content into the tissue (Schofl and French 1968). The endothelial cells in the ram testis certainly contain significant numbers of lipid inclusions similar to those found in Leydig cells (McIntosh 1969), and while their roles are yet to be identified, the possibility that they are involved in steroid secretion is of great interest.

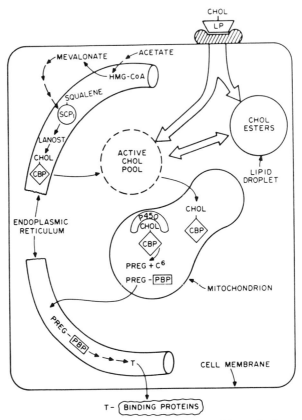

Fig. 2.4.   The spatial organization of the testosterone biosynthetic apparatus in a Leydig cell. Cholesterol (CHOL) in the metabolically active pool is derived from de novo synthesis, cholesterol esters in lipid droplets, or in blood plasma. Cholesterol probably is transported from the blood plasma into the Leydig cell by lipoprotein (LP), which may bind to Leydig cell membrane receptors. Cholesterol biosynthesis takes place in the cytoplasm and endoplasmic reticulum. Cholesterol binding proteins (CBP) may exist to transport cholesterol to the mitochondria for conversion to pregnenolone. Pregnenolone binding proteins (PBP) are then thought to return the pregnenolone to the endoplasmic reticulum for conversion to testosterone. (From Ewing, Davis, and Zirkin 1980.)

Protein hormone secretion by testicular cells is likely to be similar to the mechanism employed by other protein secreting cells. Studies of the pancreas (Jamieson and Palade 1967a,b; Orci 1986) demonstrated that after protein synthesis on the rough endoplasmic reticulum, the product passes promptly into the cavity of the endoplasmic reticulum and is then transported in solution to the periphery of the Golgi complex. There small membrane vesicles bud off the reticulum and carry the protein into the Golgi region where they are fused with condensing vacuoles. Water withdrawal from the vacuoles

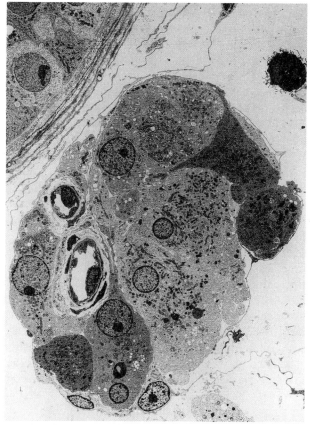

Fig. 2.5.   A cluster of Leydig cells in the interstitial tissue of a human testis, showing their close association with blood vessels. A portion of a seminiferous tubule is shown in the top left corner ($\times$ 2970). (From Schulze 1984).

concentrates the protein therein. The resulting secretory granule is transported to the surface of the cell where its content is released into extracellular space (Christensen 1975).

2   FATE OF HORMONES AFTER LEAVING THE SECRETORY CELLS

The fate of a hormone molecule after it has been released from a cell depends on the anatomical relationships between that cell and other cells, and the composition of the fluid surrounding the cell. Several authors have suggested that the Leydig cells may have a close association with endothelial cells of the testicular blood vessels (Fawcett et al. 1973; McIntosh 1969; Schulze 1984, Fig. 2.5). While this arrangement apparently has not been systematically investigated by electron microscopists, it seems possible that hormones may be secreted by the Leydig cells directly into the blood vessels, or at least into their immediate vicinity, so that a higher concentation is achieved around and in the lumen of the vessels than in the interstitial extracellular fluid. This does seem to happen under some circumstances (see **Section V**). Similarly, because some of the Leydig cells are so close to the seminiferous tubules they could produce localized high concentrations of testosterone next to the tubular wall, leading to higher concentrations inside the tubules than would result from secretion into the extracellular fluid. In fact, the concentration of testosterone in tubular fluid appears to be similar, or slightly lower than in extracellular interstitial fluid in normal and hCG-stimulated rats (Comhaire and Vermeulen 1976). If secretion is into the extracellular fluid, then the fate of the hormone will depend on the composition of this fluid, particularly with respect to large hydrophilic molecules, such as binding proteins, which could affect the physiochemical properties of the steroid molecule.

In the case of the Sertoli cells, the fate of a secreted molecule would depend initially on whether it was secreted by all parts of the Sertoli cell membrane, or only by that part above (on the luminal side) or below (on the basal side) the specialized junctions between pairs of these cells. There appears to be no evidence that the cell membrane is different between the two zones of the cell, and therefore it is probably safe to assume that secreted molecules would leave the Sertoli cell both below and above the specialized junctions. A substance being secreted above the junctions would find itself in the fluid surrounding the later germ cells or in the luminal fluid; one being secreted below the junctions would pass into the space between the Sertoli cells and the spermatogonia or the peritubular tissue, through which it must pass to reach the extracellular interstitial fluid. The proportion leaving by these two routes, assuming that secretion is evenly distributed around the cell surface, would depend on the proportion of the cell surface above and below the junctions, and this would appear to be 90.36 to 4.74 or about 20 to 1 (Weber *et al.* 1983).

3   FACTORS AFFECTING THE ROUTE OF TRANSPORT OUT OF THE TESTIS

Hormones secreted into the interstitial extracellular fluid leave the testis by one of three routes: by entering a capillary, by passing into a lymph vessel, or by traversing the peritubular tissue to enter a seminiferous tubule. As discussed earlier, the great species variation in interstitial anatomy probably affects the relative partitioning of hormone between lymph and blood. In those species possessing large lymphatic sinusoids (rodents), the lymphatic route is envisaged as the primary source of androgen supplied for the seminiferous tubules, while those animals with more clearly defined lymphatic vessels (sheep, humans) have blood capillaries which are in much closer contact with the tubules and which may directly supply testosterone to them (Fawcett *et al.* 1973). Testicular lymph in the ram, stallion and boar contains much more testosterone than testicular arterial blood (Lindner 1963, 1967, 1969; Setchell and Cox 1982; Setchell *et al.* 1983) but blood is the major route of transport for testosterone from the testis to the rest of the body in these species. This is primarily because of the much greater flow rate of blood compared with lymph. Blood in the deferential vein also contains more testosterone than arterial blood (Pierrepont, Davies, Millington, and John 1975), and blood from arteries below the spermatic cord may contain slightly more testosterone than arterial blood elsewhere in the body because of transfer from vein to artery (Free and Jaffe 1975; Free and Tillson 1975; Free 1977).

A significant proportion of testosterone in blood is bound to plasma protein, including albumin and corticosteroid-binding globulin (Eik-Nes, Schellman, Lumry, and Samuels 1954; Eik-Nes 1970). A sex hormone-binding globulin (SHBG) distinct from cortisol-binding globulin (CBG) is present in the plasma of several species including man, but is not found in the rat or boar (Corval and Bardin 1973; Bardin, Musto, Gunsalus, Kotite, Cheng, Larreau, and Becker 1981). The affinities of physiologically important steroids for CBG or SHBG are two to three orders of magnitude higher than those for albumin (Giorgi 1980). However, because it is so abundant, albumin is important in determining the magnitude of the non-protein bound or "free" fraction of steroid in plasma (Siiteri, Murai, Hammond, Nisker, Raymoure, and Kuhn 1982). It used to be thought that only the free steroid was available to the target tissues, but more recent evidence obtained using the uptake of intra-arterially injected radioactive testosterone compared with the uptake of simultaneously injected butanol shows clearly that albumin-bound hormone can be taken up by a variety of tissues (see Pardridge this volume). Another technique, the single injection multiple tracer method (Crone 1963; Yudilevich and Mann 1982), the reverse of the Oldendorf-Pardridge technique, measures radioactivity in blood or perfusate leaving the organ when a mixture of isotopes is injected into the inflow. This technique has been used with the isolated perfused rat testis to study capillary permeability (Bustamante and Setchell 1981) and amino acid transport (Bustamante and Setchell 1982), and was also applied to the study of testosterone transport. However such a high proportion of the testosterone was removed, in comparison with mannitol as a reference compound (Bustamante and Setchell, unpublished observations), that quantitative estimates of transport could not be made.

The possibility of a specific transport mechanism for testosterone was suggested by the observation that the passage of testosterone from blood to seminiferous tubular fluid (STF) or rete testis fluid (RTF) in rats is much faster than that of dihydrotestosterone (Cooper and Waites 1975; Setchell and Main 1975; Main and Setchell 1978) although the latter steroid is marginally more lipid soluble. Furthermore, the entry of radioactive testosterone into the STF or RTF can be reduced by raising the concentration of non-radioactive testosterone in the blood, either by injections of the steroid or by pretreatment with human chorionic gonadotrophin (hCG) (Setchell, Laurie, Main, and Goats 1978, Main and Setchell 1978). However, when measurements were made of the entry of testosterone and dihydrotestosterone into isolated seminiferous tubules *in vitro*, both steroids entered at the same very fast rate (Setchell, unpublished observations). If radioactive testosterone was infused into a single seminiferous tubule for 1 hour in anaesthetized rats, radioactivity increased in the blood during the infusion, and the rate of rise was decreased if non-radioactive testosterone was added to the infusate. However, a decrease was also produced by ligating the efferent ducts before starting the infusion, and if testosterone was then added to the infusate, only a small further decrease was observed, suggesting that this technique was principally measuring transport from luminal fluid to blood in the epididymis (Burrow and Setchell 1980; Burrow, Pholpramool, and Setchell 1981). In this context, it is worth noting that in general, free steroids enter cells in proportion to their lipid solubility (Giorgi 1980; Giorgi and Stein 1981), although there is some evidence for an androgen transporting system in the prostate (Giorgi 1976), and the entry of testosterone into the uterine lumen can be reduced by a high level of progesterone (Eiler 1986). It is also worth noting that the concentration of non-radioactive testosterone in rete testis fluid and seminiferous tubule fluid does not change as much as the concentration in testicular venous blood in rats given hCG or injected with testosterone (Setchell *et al.* 1978; Setchell 1980), or in rams injected with LH (Voglmayr, Roberson, and Musto 1980).

Lymph is an important route for the secretion of conjugated steroids and some proteins. In pigs and horses, the concentrations of dehydroepiandrosterone sulphate (DHAS) and oestrone sulphate in lymph are up to 40 times higher than in venous blood and as much as 80 per cent of testicular production of these hormones is secreted via the lymph (Setchell and Cox 1982; Setchell *et al.* 1983). More than 70 per cent of radioactive iodinated albumin injected directly into the testis of rams and pigs was cleared in testicular lymph (Galil, Laurie, Main, and Setchell 1981). In rats, of the ^{125}I recovered in thoracic duct lymph and blood after an intratesticular injection of iodinated human serum albumin mixed with rat serum, 56.8 ± 2.1 per cent (n = 21) was in the lymph (see Fig. 2.6); when ^{3}H-DHAS was included in the mixture, the ^{3}H/^{125}I ratio in the lymph samples with the highest counts (usually those between 15 and 60 min after injection for both isotopes) was 75.6 ± 3.6 per cent (n = 6) of the ratio in the dose, whereas at the same time the ratio in blood plasma was 39.5 ± 6.9 per cent. This indicates that DHAS behaves very similarly to albumin, despite the large difference in molecular weight. In contrast, if ^{3}H-testosterone was substituted for the DHAS (Fig. 2.6), the ^{3}H/^{125}I ratio in thoracic duct lymph was much lower than that in the dose (2.96 ± 0.05 per cent, n = 3), and this ratio was not changed if serum from women with high levels of SHBG or rat RTF with high levels of ABP

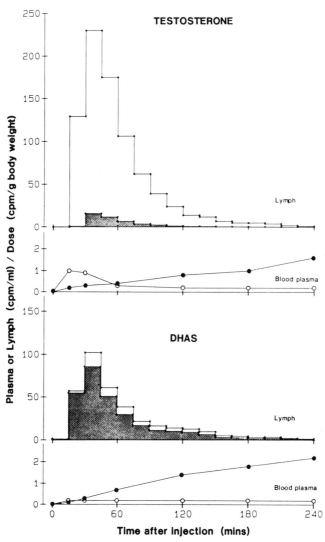

Fig. 2.6.   The appearance of radioactivity in thoracic duct lymph and right atrial blood plasma following the injection of a mixture of ^{125}I-iodinated human serum albumin and either ^{3}H-testosterone or ^{3}H-dihydroepiandrosterone-sulphate in 25 µl of serum from a hypophysectomized rat, into the left testis of anaesthetized rats. ^{125}I—closed circles, open columns; ^{3}H—open circles, stippled columns. Unpublished data of B. P. Setchell and J. L. Zupp, showing one typical experiment of four conducted with each steroid.

were substituted for the rat serum (2.38 ± 0.33, n = 6 and 1.74 ± 0.62, n = 3 respectively); at the same time the ratio in blood plasma was similar to that in the dose. This suggests that testosterone, even when combined with a binding protein, is separated from that protein and secreted into the blood stream (Setchell and Zupp 1988).

The protein hormone inhibin is also found in higher concentrations in testicular lymph than in venous blood (see Hudson, Baker, Eddie, Higginson, Burger, de Kretser, Dobos, and Lee 1979; Au, Robertson, and de Kretser 1984), and presumably, the same would apply to all moderately large and large peptides. It has not yet been determined what is the "cut-off" point for molecular weight at which a hydrophilic molecule passes preferentially into the lymph. The concentration of inhibin in seminiferous tubular fluid can be calculated from the data of Au *et al.* (1984) if one assumes that the extra inhibin in the efferent duct-ligated testis is all in the fluid trapped in the tubules and that ligation has no effect on the inhibin in the rest of the testis, using the "difference" calculation for the composition of total secreted fluid described by Setchell, Hinton, Jacks, and Davies (1976). This calculation gives values of 724 ± 36 U/ml for tubular fluid, compared with 120 U/ml for "interstitial" fluid collected post mortem and levels of less than 2 U/ml in testicular venous blood plasma. From these concentrations, values for lymph flow from Setchell and Sharpe (1981) and estimates of fluid secretion made by Au *et al.* (1984) using the technique originally described by Setchell (1970*b*), it can be calculated that 6 U inhibin leaves the testis per hour in the lymph and 18 U per hour in the tubular fluid.

There is some interesting new evidence on the fate of androgen-binding protein (ABP) which is secreted by the Sertoli cells of the rat testis. ABP is a protein with a molecular weight of about 90,000 (Hansson, Ritzen, French, and Nayfeh 1975; Bardin *et al.* 1981) and it appears to be secreted at both the basal and luminal surfaces of the Sertoli cells with between 70 and 80 per cent being released into the tubules *in vivo* (Gunsalus, Musto, and Bardin 1980; Turner, Jones, Howards, Ewing, Zegeye, and Gunsalus 1984). Similar proportions (80:20) are secreted across the luminal and basal membranes of Sertoli cells grown in a bicameral culture system (Hadley, Djakiew, Byers, and Dym 1987). There are two forms of ABP, judged by its binding to concanavalin A, and in adult rats, the ratio of Form I to Form II is about 1:1.5 in testicular and epididymal cytosols, but about 1:5 in serum. This difference is non-existent in 22 day-old rats (Cheng, Gunsalus, Musto, and Bardin 1984), before the functional blood-testis barrier is established (Setchell, Laurie, and Jarvis 1981; Setchell, Pollanen, and Zupp 1987). The ratio of Form I to Form II and the total concentration of ABP in thoracic duct lymph and peripheral blood are similar (1:4.6 and 1:4.4, and 228 and 114 µl eq/ml respectively), and while the total concentration is higher in "interstitial fluid" collected post mortem (862 µl eq/ml), the ratio of Form I to Form II (1:3.1) was not significantly different from that in thoracic duct lymph or blood plasma, although it was slightly lower (Cheng, Gunsalus, Morris, Turner, and Bardin 1986). These results presumably reflect the fact that testicular lymph contributes only a small proportion of the lymph which can be collected from the thoracic duct (0.7 µl/min from the testis (Setchell 1986) compared with 10 to 20 µl/min from the thoracic duct), and there must also be significant doubts about any results obtained with their technique for collecting "interstitial fluid" (see **Section V**). There is also some doubt about the purity of their tubular fluid, as the protein content is about 5 times higher than that reported in pure samples collected by micropuncture (Setchell *et al.* 1978; Hinton and Keefer 1983). Nevertheless, the results indicate that either a different mix of proteins is being secreted from the two faces of the Sertoli cells, or that the protein once in the interstitial fluid is being changed in its binding affinity for concanavalin A.

Transferrin is another protein, a unique form of which is secreted by Sertoli cells into seminiferous tubular fluid, but because the assay used does not distinguish between testicular and circulating transferrin, and because of the high concentrations in blood plasma and testicular interstitial fluid (Sylvester and Griswold 1984), no estimate can be made of the ratio of luminal to basal secretion *in vivo*. However using Sertoli cells in a bicameral culture system, luminal secretion is about twice basal secretion, a lower ratio than for ABP (Hadley *et al.* 1987).

Once in a capillary, a molecule presumably passes into the general circulation, unless it passes back from a vein into the testicular or epididymal arteries in the spermatic cord (Jacks and Setchell 1973; Free 1977) and returns by a short-circuit to the testis or is transported directly to the epididymis. A hormone in the lymph would also eventually reach the general circulation, but probably passes through at least one lymph node en route. Some authors (Engeset 1959) have suggested that the testicular lymphatics are unusual in that a significant proportion of the testicular lymph enters the blood without passing through a lymph node, but others do not agree with this suggestion (Tilney 1971; McCullough 1975). A hormone such as inhibin secreted into seminiferous tubule fluid would pass out through the rete testis. However it is impossible to predict its subsequent fate since the total protein concentration is reduced in the rete to 20 per cent (Setchell *et al.* 1978; Hinton and Keefer 1983) and its protein spectrum altered (Koskimies and Kormano 1973; Olson and Hinton 1985), but the concentration of ABP is unchanged or rises (Turner *et al.* 1984; Cheng *et al.* 1986). The rete testis fluid then passes into the epididymis where most of it is reabsorbed (Setchell and Hinton 1981; Setchell and Brooks 1988). What happens to the ABP, transferrin and inhibin is not clear. At least some of the ABP is taken up by the cells of the head of the epididymis (Attramadal, Bardin, Gunsalus, Musto, and Hansson 1981; Pelliniemi, Dym, Gunsalus, Musto, Bardin, and Fawcett 1981), and the concentration of ABP in fluid from the cauda is much less than that in RTF (Danzo, Cooper, and Orgebin-Crist 1977; Jegou, Dacheux, Garnier, Terqui, Colas, and Courot 1979). The concentration of transferrin rises and then falls again as the fluid passes along the epididymis, but significant resorption occurs all along the epididymis if fluid reabsorption is calculated from the spermatocrits (Djakiew, Griswold, Lewis, and Dym 1987). No estimates appear to have been made of the concentrations of inhibin along the epididymal duct, although Le Lannou, Chambon and Le Calve (1979) found inhibin-like activity in the head of the epididymis that was not present in the cauda. There remains some uncertainty as to the endocrinological significance of the inhibin entering the epididymis. Trapping the tubular fluid in the testis for 24 hours by ligation of the efferent ducts or diverting it to the exterior through a catheter in the rete testis has no effect on the concentrations of FSH in the blood of that animal (Davies, Main, and Setchell 1978; Walton, Evans, and Waites, 1978; Le Lannou et al. 1979). However plasma levels of FSH are significantly greater at 36 hours after removal of the testis (orchidectomy) if the epididymis is also removed with the testis (orchido-epididymectomy; Le Lannou et al. 1979). These authors were also able to demonstrate significant reductions in circulating levels of FSH in castrated male rats following an intraperitoneal injection of an epididymal extract obtained from normal rats, and suggested that inhibin in the tubular fluid was being reabsorbed in the caput epididymis and passed into the bloodstream to act ultimately on the pituitary or hypothalamus.

V   Composition and function of interstitial fluid

By now it should be apparent that the extracellular or "interstitial" space in the testis is critically important for continued normal endocrine function. The fluid in this region not only bathes the interstitial cells of the testis but also performs the vital role of providing an avenue for communication both between cells within the testis, and between the testis and the rest of the body. It is always assumed that interstitial fluid is formed by filtration of a fluid of low protein concentration at the arterial end of the testicular capillaries where capillary hydrostatic pressure exceeds colloid osmotic pressure. Most of this fluid but not the protein is resorbed at the venous end of the capillaries where hydrostatic pressure is less than colloid osmotic pressure. The protein and the rest of the fluid leaves the testis as lymph. While direct evidence for this process has been demonstrated for other tissues (see Yoffey and Courtice 1970) we can only assume that the system is similar in the testis, where regulation of tissue pressure by contraction of the capsule may also be of considerable significance (Setchell 1986).

For some years now, it has been recognized that all substances entering or leaving the tubules and probably the Leydig cells, must pass through some interstitial fluid. The close association of some Leydig cells with the walls of capillaries may mean that they have direct access to the blood stream. Indeed, the finding that the concentration of testosterone was higher in the spermatic vein blood than in lymph might suggest a closer association of Leydig cells with blood vessels than with lymphatics. McIntosh (1969) has shown by electron microscopy that Leydig cells can be intimately associated with either, but there are more blood capillaries than lymph capillaries and therefore more Leydig cells associated with blood capillaries (see also Yoffey and Courtice 1970). Likewise, certain Sertoli cell products may be directly secreted into the tubular lumen. However, the fluid in the interstitial region remains an important avenue for cellular communication, and thus a potentially useful window on what is happening within the testis. Not surprisingly, a number of attempts have been made to monitor the interstitial region.

Practically all *in vivo* studies have relied on collecting fluids leaving the testis (blood, lymph, rete testis fluid), although as may be appreciated by now, the composition of these fluids does not necessarily reflect the composition of fluids within the testis. Lymphatic vessels in the spermatic cord are closely associated with blood vessels, particularly veins, and there is a strong possibility that transfer of material can occur between the various fluids in the cord (Free 1977; Setchell 1986). This is particularly so for compounds such as testosterone which is produced in the testis itself and has important actions there.

A method for obtaining interstitial fluid was reported by Pande, Chowdhury, Dasgupta, Chowdhury, and Kar in 1966, using a method that might be termed the "Drip Collection". This method involves removing the testes from the animal, and making a

... "small incision at a suitable site through the tunica albuginea avoiding injury to blood vessels and seminiferous tubules. The organs were then placed in a test tube with a small perforation at the bottom and containing a few glass beads. The testes were suspended above the glass beads which prevented blocking of the perforation. The test tube was put inside a conical bottom centrifuge tube into which the fluid from the testes trickled down through the perforation. The operation was carried out at a temperature of 2°C under sterile conditions."...

The collected fluid was centrifuged and the cell free supernatant taken for estimations. No mention is made in the paper as to how long collection continued, but in presenting similar data for the human testis, Pande, Dasgupta, and Kar (1967) report collection for 15–17 hours. Since these reports, a number of researchers have used this technique or variations of it (for example, centrifugation of the whole apparatus for 15–20 min at 54 g instead of gravity collection for 15–17 hours) to obtain interstitial fluid for analysis, or for addition to *in vitro* cultures (see Hagenas, Ritzen, and Suginami 1978; Sharpe 1979; Sharpe and Cooper 1983; Sharpe, Doogan, and Cooper 1983; Turner *et al.* 1984; Widmark, Damber, and Bergh 1986). However the reported composition of fluid obtained by this method is somewhat unusual for a fluid supposedly formed by blood filtration. While Pande *et al.* (1966, 1967) discuss their results in comparison with blood serum, they make no attempt to explain any of the differences found. In terms of actual ratios the following differences can be highlighted:

	Ratio of Interstitial fluid/Serum		
Substance measured	Human[1]	Rat[2]	Immature rats[3] given hCG
Protein	1/2	2.3/1	1/1
Lactic dehydrogenase	400/1	3.5/1	
Glucose-6-phosphatase	3/1	2/1	
Glucose-6-phosphate dehydrogenase	5000/1	1000/1	
Acid phosphatase	112/1	65/1	
Ascorbic acid	30/1	15/1	
Glucose	1/3.5	1/7	
Glycogen	7/1	5/1	
Total Lipids	1/3	1/3.5	
Sodium ion	183/146	165/152	140/141
Potassium ion	7.7/5.1	7/5.6	14.7/5.7
Chloride ion	139/115	125/115	

[1] Pande *et al.* 1967   [2] Pande *et al.* 1966   [3] Sharpe 1979

A number of the enzymes listed are usually recognized as intracellular enzymes, and these values together with the high potassium levels are suggestive of cellular damage, particularly as potassium levels in lymph from the spermatic cord are similar to those in serum (Wallace and Lascelles 1964; Setchell 1982). The low glucose and the high lactate and glycogen levels might be indicative of post-mortem anaerobic metabolism. The possibility that isolating an organ from its blood supply and collecting fluid from it for many hours at near freezing temperatures might be unphysiological was never addressed.

Interstitial extracellular fluid, as well as tubular luminal fluid, can also be collected by micropuncture techniques from anaesthetized rats (Tuck, Setchell, Waites, and Young 1970; Comhaire and Vermeulen 1976). This requires removal of part of the tunical albuginea, which may affect fluid movements in the testis, and the small samples obtained limit the possible analyses. Potassium levels in this fluid were again about 50 per cent higher than blood plasma (Tuck *et al.* 1970). Comhaire and Vermeulen (1976) measured testosterone in micropuncture samples of interstitial and tubular fluid, but unfortunately did not measure venous blood concentrations.

Testosterone concentrations in fluid collected by the "drip" technique and venous blood have been reported by a number of workers (Hagenas *et al.* 1978; Sharpe *et al.* 1983; Turner *et al.* 1984; Turner, Ewing, Jones, Howard, and Zegeye 1985; Widmark *et al.* 1986). It appears from these results that interstitial fluid contains appreciably higher concentrations of testosterone than testicular venous blood plasma (Table 2.1). However, none of the

Table 2.1

Testosterone levels in testicular interstitial fluid obtained by various techniques, and testicular venous blood of male rats. Values are presented as means with the standard errors of the means, and the number of observations

Method	Testosterone levels (ng/ml) Testicular interstitial fluid	Testicular venous blood plasma	Reference
Micropuncture	150 ± 27 (17)	—	Comhaire and Vermeulen 1976*
Drip/centrifuge	137 ± 25 (10)	—	Hagenas *et al.* 1978
	73 ± 5 (26)	28 ± 5.3 (10)	Turner *et al.* 1984, 1985
Drip/gravity	315 ± 29 (4)	—	Sharpe *et al.* 1983
	590 ± 20 (8)	90 ± 20 (8)	
Push-pull cannula	11.6 ± 4.2 (9)	23.4 ± 5.2 (7)	Maddocks and Setchell 1986

* These authors unfortunately did not analyse venous blood, but report values of 91 ± 14 (15) for seminiferous tubular fluid also collected by micropuncture and a mean STF:IF ratio of 0.94 ± 0.24 (14).

authors comments on this point or discuss how such concentration differences might be achieved or maintained within the specific environs of the testicular interstitium. What is also apparent in Table 2.1, and yet never discussed, is that the faster the method of recovery of fluid (centrifugation/micropuncture vs gravity over 16–20 hours) the lower the level of testosterone. This must surely raise further questions about the validity of these methods of collection.

Detailed studies of testicular lymph were carried out several years ago in the ram and pig (Lindner 1963, 1969; Wallace and Lascelles 1964; Setchell, Hinks, Volgmayr, and Scott 1967; Setchell 1982; Setchell *et al.* 1983). Testicular lymph in the pig does contain somewhat more testosterone than testicular venous blood plasma (Setchell *et al.* 1983) but in the ram testicular lymph contains only about 70 per cent as much testosterone as blood plasma from the internal spermatic vein (Lindner 1963, 1967). Even allowing for species differences and lymph/vascular transfer and equilibration in the cord, it is extremely difficult to reconcile these observations with the observation in the rat that testosterone concentrations in interstitial fluid are much higher than those in testicular venous blood.

We have recently employed the push-pull cannula of Gaddum (1961) to obtain interstitial fluid from the testes of anaesthetized rats (Maddocks and Setchell 1986; Maddocks, Zupp, Sowerbutts, and Setchell 1986). The push-pull cannula consists of two concentric tubes with fluid infused through the inner tube and withdrawn from the outer tube. If the infusion and withdrawal are maintained at the same constant rate, a droplet of fluid of rela-

Table 2.2

The calculated concentrations of protein, testosterone, sodium and potassium in testicular interstitial fluid collected with a push-pull cannula perfused with isotonic mannitol, with measured values for testicular venous plasma. Levels of sodium and potassium in interstitial fluid were also measured in vivo, and in collected interstitial fluid and testicular venous blood plasma using ion-sensitive electrodes. Values are presented as means ± S.E.M. Number of observations in brackets

	Interstitial fluid		Testicular venous blood plasma
	Collected with a push-pull cannula	Measured *in vivo*	
Protein (mg/ml)	75.3 ± 12.9 (4)		58.4 ± 2.8 (5)
Testosterone (ng/ml)	11.6 ± 4.2 (9)		23.4 ± 5.2 (7)
Sodium (mmol)			
—by ion-sensitive electrode	155.4 ± 21.5 (7)	149.1 ± 2.9 (8)	143.6 ± 3.7 (7)
—by flame photometry	141.7 ± 18.2 (9)		137.8 ± 1.9 (9)
Potassium (mmol)			
—by ion-sensitive electrode	4.5 ± 0.9 (7)	4.6 ± 0.3 (4)	4.7 ± 0.2 (7)
—by flame photometry	12.2 ± 2.5 (9)*		4.4 ± 0.1 (9)

* Now believed to be in error (see text).

tively constant volume is maintained at the tip of the cannula. When this cannula is introduced into tissue, substances in the extracellular fluid will mix with or diffuse into the droplet, and thus the fluid withdrawn from the cannula will contain substances found in the extracellular fluid (see reviews by Myers 1972; Szerb 1967; Yaksh and Yamamura 1974). Histological studies show that very little cell damage is caused by this procedure. While the fluid obtained is diluted by the process of collection, the amount of dilution can be corrected for by loading the animal with appropriate radioactive markers. Sampling occurs after the markers have equilibrated between blood and extracellular fluid, and by comparing levels of radioactivity in blood samples and push-pull samples, a measure of dilution can be obtained (Maddocks and Setchell 1986; 1987b). Results obtained with the push-pull cannula and ion-sensitive electrodes show comparable levels of sodium and potassium in testicular interstitial fluid and venous blood (Table 2.2; Maddocks and Setchell 1987b). The plasma-like values for sodium and potassium are in accord with the similar values obtained by introducing ion-sensitive electrodes into the interstitial tissue of anaesthetized rats (J. L. Zupp, S. Maddocks, and B. P. Setchell, unpublished observations; Table 2.2). The earlier values (Maddocks and Setchell, 1986) which suggested that potassium levels in interstitial fluid, although lower than drip-collected fluid, were still slightly higher than plasma, were much more variable and are now believed to be in error probably because the flame photometer was being used at the extreme limit of its sensitivity to potassium. Concentrations of protein in interstitial fluid can also be calculated, and these are similar to the values for blood plasma. Of greatest interest is the finding that interstitial fluid levels of testosterone are no greater than testicular venous blood, and are possibly even lower—a finding in direct contrast to previous reports (see Table 2.1).

These reults have led us to propose that Leydig cells secrete testosterone as much into blood as into extracellular fluid (Maddocks *et al.* 1986; see Fig. 2.7) which would account for the fact that the levels of testosterone measured in ram lymph have not been any higher than testicular venous levels. This hypothesis still provides for higher concentrations of testosterone around the seminiferous tubules than is seen by other target tissues, and may also explain reports that spermatogenesis can be maintained by levels 10 per cent as high as previously reported interstitial fluid testosterone levels (Cunningham and Huckins 1979; Buhl, Cornette, Kirton, and Yuan 1982; Marshall, Wickings, Ludecke, and Nieschlag 1984; Rea, Marshall, Weinbauer, and Nieschlag 1986); particularly if the control levels have been previously overestimated by a factor of 5–10 (see Sharpe 1987; see also **Section V**).

All Leydig cells, and especially those in close association with blood vessels (see Fig. 2.5) might indeed be affected by the lack of blood flow in these vessels after removal of the testis from the animal. Assuming that they can continue to secrete testosterone even to a reduced extent, the secreted product is not going to be carried away by the blood and is more likely to accumulate in the interstitial fluid. Under these conditions it is also possible that membrane function and integrity will be affected, with cells becoming "leaky". These situations would elevate the concentrations of testosterone measured in drip collected fluid to an abnormal level. It may be therefore, that the values obtained by drip collection represent post-mortem secretion and release of "stores" of Leydig cell testosterone, rather than the normal interstitial fluid levels. Further, other cells in the testis such as the macrophages may also contain quantities of testosterone that might leak into the interstitial fluid under these conditions.

Is there enough testosterone stored in the Leydig cells to produce the concentrations found if the stores were released under these conditions? Sullivan and Cooke (1984) have found that isolated, purified Leydig cells contain 0.75 ng testosterone/10^6 cells and Mori and Christensen (1980) have found that the rat testis contains 22×10^6 Leydig cells per gram. From these two values, we can estimate that Leydig cells in a 1 g testis contain 16.5 ng testosterone. Using the distribution volume of ^{51}Cr-EDTA, Setchell and Sharpe (1981) reported the interstitial fluid volume in the rat testis to be about $130\,\mu l/g$ testis. However recent evidence suggests that this value may be artifactually high due to an accumulation of the marker in the testicular capsule (B. P. Setchell, unpublished observations), and a value of about $80\,\mu l/g$ par-

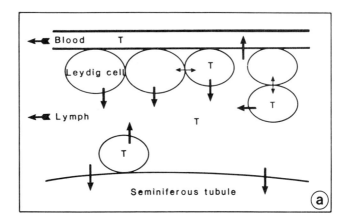

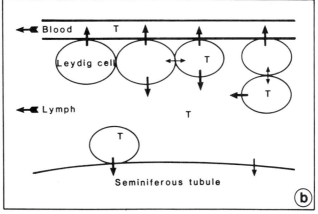

Fig. 2.7.   Diagrammatic representations of the proposed mechanisms of testosterone secretion in the rat testis. (a) The existing hypothesis proposes that the Leydig cells primarily secrete androgen into the extravascular interstitial fluid (arrows). This creates a high concentration of androgen in this fluid which supplies the seminiferous tubules with the high levels needed for spermatogenesis. Blood levels of testosterone arise from the diffusion of testosterone from the interstitial fluid into the blood capillaries. (b) The new hypothesis proposed. Some Leydig cells preferentially secrete testosterone into blood vessels (arrows). Other Leydig cells secrete into interstitial fluid, and some may supply seminiferous tubules directly. However, interstitial fluid levels of testosterone are comparable to blood levels and arise more as a consequence of blood flow rate, and capillary permeability. These factors appear to be carefully regulated by the seminiferous tubules, and cells in the interstitial tissue.

enchyma is probably more realistic. If all the testosterone in Leydig cells leaked out into this volume of fluid, levels would be increased by 206 ng/ml. As seen in Table 2.1, concentrations of this magnitude have been recorded when the interstitial fluid is obtained by drip collection. However it is likely that by isolating the Leydig cells, their testosterone stores would already be depleted and it is unlikely that all the intracellular testosterone would leak out during the fluid collection post mortem.

A more meaningful estimate may be obtained for the concentration of testosterone in the interstitial tissue by partitioning the total amount of testosterone in a rat testis. Buhl *et al.* (1982) recently reported a value of 87 ng/g, and Tapanainen *et al.* (1984) and Huang and Nieschlag (1986) have found similar values. Of this we can account for about 1 ng in the blood (blood volume 10 µl/g, Setchell and Sharpe 1981; testosterone concentrations in venous blood of recently anaesthetized animals approximately 100 ng/ml, Davies, Main, Laurie, and Setchell 1979), 12 ng in the tubular fluid (volume of luminal fluid of testis of an adult rat is 150 µl/g, Mori and Christensen 1980; testosterone concentration 78 ng/ml (mean of 65 ng/ml by the difference technique as described by Setchell (1978) from the data of Davies *et al.* (1979) and 90 ng/ml in micropuncture samples of tubular fluid, Comhaire and Vermeulen (1976)), and about 26 ng in the tubular cells (tubular concentration 35 ng/ml—about 40 per cent of total testis concentration, Cooke *et al.* 1972; de Jong *et al.* 1974; volume of tubules 750 µl/g, Mori and Christensen 1980). This leaves about 48 ng/g in the Leydig cells and the extracellular fluid of the interstitial tissue.

How this testosterone is distributed between cells and fluid depends on what figures are used for calculation. The concentration of testosterone in the extracellular interstitial fluid collected by micropuncture has been reported to be similar to that in the tubular fluid (a mean STF:IF ratio of 0.94 was reported by Comhaire and Vermeulen 1976) and if this ratio and the above mean STF concentration of 78 ng/ml is used, with a fluid volume of 80 µl/g, then this fluid would contain 6 ng of testosterone, leaving about 42 ng in the Leydig cells from 1 g of testis. Using the mean value for interstitial fluid of 150 ng/ml (Comhaire and Vermeulen 1976), the values would be 12 and 36 ng respectively. The Leydig cells comprise about 35 µl/g of testis which gives a testosterone concentration of between 1000 and 1200 ng/ml in these cells. Only one-half of this amount of testosterone would need to leak from the Leydig cells into the extracellular interstitial fluid to raise the concentration of testosterone there by 250 ng/ml, if the volume of fluid is 80 µl/g.

This value of about 40 ng testosterone in the Leydig cells in a 1 g testis is appreciably greater than the value of 16 ng in the Leydig cells in a 1 g testis obtained from the number of Leydig cells/g testis and the testosterone concentration in isolated purified cells (see page 87), but as we have already commented, the latter value for isolated Leydig cells is probably minimal; nevertheless, 16 ng/g testis in the Leydig cells is equivalent to a concentration of about 450 ng/ml inside the Leydig cells as these occupy 35 µl/g testis. Furthermore, if there are 48 ng testosterone/g testis in the Leydig cells and interstitial fluid (see above) and 16 ng of this is in the Leydig cells, then 32 ng are to be found in the extracellular interstitial fluid, giving a concentration there of about 400 ng/ml, that is comparable to that inside the Leydig cells but much greater than that measured inside the tubules or veins. Thus these calculations do not enable us to decide which concentration of testosterone in the interstitial fluid is physiological but there must be either a considerable concentration gradient between the interior of the Leydig cell and the extracellular interstitial fluid surrounding it, or between the extracellular interstitial fluid on the one hand and the venous blood plasma and tubular fluid on the other. Neither gradient fits with the idea of a freely diffusible molecule passing readily out of and through cells.

To investigate further the suggestion of leaky cells and/or continued secretion of testosterone during drip collection, we have compared the results from push-pull and drip collections of interstitial fluid from rats given testosterone-filled silastic implants for one week (Maddocks and Setchell 1987a). Testicular production of testosterone is inhibited by exogenous administration of testosterone by injection or subcutaneous implant because of reduced LH secretion (Cunningham and Huckins 1979; Rea et al. 1986). Testosterone levels in interstitial fluid obtained with the push-pull cannula or by drip collection were compared with those in testicular venous and peripheral blood samples. The results support the validity of the push-pull cannula rather than drip-collection for sampling interstitial fluid. As seen in Figure 2.8, drip-collected fluid from control animals contained significantly greater concentrations of testosterone than was found in venous blood or interstitial fluid obtained from such animals with the push-pull cannula. In testosterone implanted animals, in which the Leydig cells were non-functional, there was no significant difference between the levels of testosterone measured in the fluids obtained by drip or push-pull collection. Peripheral and testicular venous blood levels of testosterone in animals implanted with the larger implants were significantly elevated, but interstitial fluid levels were not altered. This may suggest that some saturable carrier mechanism may exist in the blood capillaries of the testis which regulates transport of testosterone from blood into the interstitial fluid.

Another interesting model is the pregnenolone-injected hypophysectomized rat, in which spermatogenesis is maintained but accessory glands regress (see Setchell 1978). Harris and Bartke (1975) suggested that in these animals, concentrations of testosterone in rete testis fluid were maintained within the normal range, whereas peripheral levels were no higher than in hypophysectomized untreated rats, and both peripheral and testicular venous plasma levels were considerably lower than in intact rats. In a later study, Turner et al. (1985) reported that in rather larger hypophysectomized rats treated with the same dose of pregnenolone, the concentration of testo-

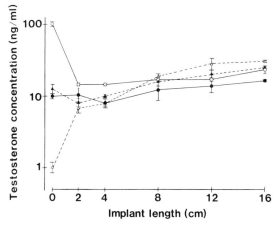

Fig. 2.8.   Levels of testosterone in fluids collected from adult male rats implanted with various sizes of testosterone-filled silastic capsules for seven days. In each animal, samples of testicular interstitial fluid were collected from one testis with a push-pull cannula (closed circles) and from the contralateral testis by drip-collection (open circles). Blood plasma was also sampled from the posterior vena cava (open triangles) and the testicular vein (closed triangles). Values presented are means $\pm$ S.E.M., n = 4 animals per group.

sterone in both peripheral blood and rete testis fluid were slightly but not significantly higher than in intact rats, but the concentrations in testicular venous blood, interstitial fluid (collected by the drip-technique) and tubular fluid were significantly lower than in intact rats by about the same proportion (40, 51, and 24 per cent respectively of levels in intact rats). In our experiments (Fig. 2.9), normal concentrations were found in peripheral and testicular venous blood, but lower than normal values in interstitial fluid (obtained by push-pull collection) with markedly regressed seminal vesicles. This again suggests that interstitial fluid collected by the drip-technique gives misleading results, and our results support the suggestion that under some circumstances the Leydig cells can secrete preferentially into the venous blood. The regressed accessory glands are a puzzle, but in the absence of values for testicular blood flow (which was unfortunately not measured in these experiments), nothing can be said about testosterone production rates. However, regressed accessory glands in the face of reasonable peripheral testosterone levels suggests that these tissues are either responding to something secreted by the pituitary or another secretory product of the testes (see **Section VII**).

   The mere presence of something in interstitial fluid is not sufficient. Knowing the *actual* composition of interstitial fluid is important if we are to understand intercellular communication and the physiological control of testicular function. It is a region of prime importance in paracrine regulation and communication (Sharpe 1986). That the majority of infertile men present with normal or near normal gonadotrophin levels and normal or varying degrees

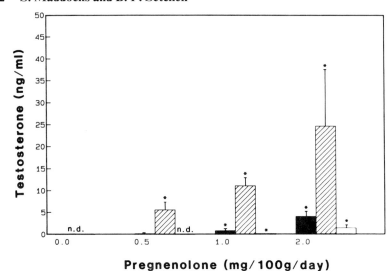

Fig. 2.9. Testosterone levels in peripheral blood plasma (open bar), testicular venous blood plasma (hatched bars) and testicular interstitial fluid (solid bars) of hypophysectomized adult rats treated with 0, 0.5, 1.0, or 2.0 mg pregnenolone/100 g bw/day for 14 days after hypophysectomy. Values are mean ± S.E.M., n = 4.*: values significantly different (P < 0.05) from untreated animals.

of subnormal numbers of spermatozoa (de Kretser and Kerr 1983) and that we remain unable to provide suitable therapy in most cases reflects our poor understanding of the control of spermatogenesis. Much of this information is probably to be found in the interstitial region, and consequently in interstitial fluid. More work is needed in this area, and techniques must be improved if any advances are to be made. Serious doubts have been raised in this review and elsewhere (see Maddocks and Setchell 1987b; Sharpe 1987) as to the usefulness of drip collection in physiological studies on interstitial fluid and the push-pull system is a novel alternative, which may itself require refinement. The process of dilution during collection makes it unsuitable for some studies. However, it has demonstrated that a re-evaluation of our understanding of the environment of the endocrine testis may be needed. Whatever approach is used, the fact remains that somewhere in the testis there exists a significant testosterone concentration gradient. Results from the push-pull collections suggest that interstitial fluid levels of testosterone are similar to blood and lymph levels, with significantly greater concentrations found within the Leydig cells, and we have discussed in some detail how this situation might occur. Alternatively, the results from drip collections suggest that the interstitial fluid should contain large concentrations of testosterone that are probably similar to levels in the Leydig cells, but significantly greater than those found in testicular blood and lymph. However, we are not aware of any attempt by those using this technique to explain how such a situation might be maintained physiologically, and it is one that we find more difficult to explain. It is important that we resolve these differences, and further investigations of the levels of steroids in interstitial fluid, and the mechanisms of secretion found in the endocrine testis, are obviously required.

VI   Control of endocrine activity in the testis

The endocrine function of the testis has historically been seen as being controlled by the anterior pituitary, since testicular development and function are so clearly dependent on the gonadotrophic hormones LH and FSH. However it is becoming increasingly evident that the day-to-day functions of the testis are also under local control (Sharpe 1983, 1986). We have discussed the role of testicular blood vessels and lymphatics in conveying substances to and from the cells in this organ. Regulation of these systems can be of immense physiological importance, and of prime consideration when discussing the mechanisms of control of endocrine function.

1   EFFECT OF PERMEABILITY OF TESTICULAR BLOOD VESSELS, BLOOD AND LYMPH FLOW

The capillaries of the testis are unusual compared with those of other endocrine tissues in the rat, in that they appear to be unfenestrated (Wolff and Merker 1966). They also show an extraordinary sensitivity to the toxic effects of cadmium salts (Setchell and Waites 1970; Aoki and Hoffer 1978). However their normal permeability would appear to be similar to that of other capillaries in the body as determined by measurements of their permeability-surface area product (PS) for Cr-EDTA, sodium, vitamin B_{12} (Bustamante and Setchell 1981) and albumin (Setchell 1987). The permeability of testicular blood vessels to albumin can be markedly increased if the rat is injected with human chorionic gonadotrophin (hCG) between 8 and 24 hours beforehand (Setchell and Sharpe 1981; Sowerbutts *et al.* 1986). The permeability returns to normal between 36 and 48 hours after a single dose of hCG; the increase is not maintained by repeated injections of hCG at 24 hourly intervals, but permeability remains at a significantly higher level if repeated injections of hCG are administered every 48 or 72 hours (Maddocks, Sowerbutts, and Setchell 1987). The increase in permeability is accompanied by an increased volume of interstitial fluid within the testis (Sharpe 1979; Widmark, Damber, and Bergh 1986) and by increased levels of testosterone in venous blood, which peak initially at 2 hours, with further peaks recorded about 16 and 72 hours after a single hCG injection (Risbridger, Kerr, Peake, and de Kretser 1981; Hodgson and de Kretser, 1982, 1984; Hodgson, Urquart, and de Kretser 1983; Sowerbutts *et al.* 1986). However testicular permeability is normal at the time of the 72 hour testosterone peak suggesting that different mechanisms may be involved in these two effects. The permeability increase does not involve androgens, prostaglandins (Veijola and Rajaniemi 1985; Sowerbutts *et al.* 1986), histamine or bradykinin, but is mediated to some extent by 5-hydroxytryptamine which may arise from the abundant mast cells found in the vicinity of the artery on the surface of the testis (Sowerbutts *et al.* 1986). Later evidence has shown that polymorphonuclear leukocytes are probably involved in the production of the increased permeability (Bergh, Widmark, Damber, and Cajander 1986) and a permeability increase 4 hours after hCG was found using colloidal carbon and *in vivo* fluorescence microscopy with fluorescein-labelled dextran, at a time when polymorphonuclear leukocytes are found adhering to the endothelium of the post-capillary venules (Bergh, Routh, Widmark, and Damber 1987). Setchell and Rommerts (1985) have reported the permeability response after hCG to be completely abolished in rats pretreated with ethane dimethane sulphonate (EDS) to eliminate Leydig cells. This suggests that either the Leydig cells are secreting some vasoactive substance as well as testosterone, or that EDS also eliminates other cells.

The increase in interstitial fluid volume is also seen after injections of LH (Damber, Widmark, and Bergh 1986) or of an LHRH-agonist (Sharpe *et al.* 1983). hCG is known to bind to LH receptors on Leydig cells and probably increases testosterone production in this way. The testicular LHRH-like substance is thought to be secreted by the Sertoli cells and may also regulate testosterone secretion as well as interstitial fluid formation. The results of Sharpe and Fraser (1980) and Sharpe *et al.* (1982) raise the possibility that the delay in action of LH and hCG on vascular permeability may in part represent the time taken for these substances to induce intratesticular secretion of LHRH. It is also interesting that changes in vascular permeability of similar magnitude to those seen in adult rats following hCG injection can also be demonstrated in young rats at about 30 days of age as they pass through puberty (Setchell *et al.* 1987; Table 2.3), although at that age the LH levels are minimal (de Jong and Sharpe 1977).

The permeability of the testicular vasculature to proteins the size of hCG is comparable to that for albumin (see Table 2.4) and high levels of hCG have been observed in testicular interstitial fluid following systemic injection of the hormone (Sharpe 1980, 1981). However the greater vascular permeability after hCG treatment will presumably also increase the access of hCG itself as well as LH and FSH to both Leydig and Sertoli cells and this may be of significance in increasing the production and secretion of testosterone (and possibly other substances).

Using the push-pull cannula to sample interstitial fluid, we have examined the levels of testosterone in this fluid and in testicular and peripheral venous blood, at various times after a single subcutaneous injection of 50 i.u. hCG (Maddocks *et al.* 1986). As seen in Figure 2.10, the levels of testosterone in interstitial fluid and testicular venous blood are significantly raised by this treatment, but the changes seen do not parallel those in peripheral blood. Of

Table 2.3

Permeability-Surface area products (PS, $\mu l\ min^{-1}\ g^{-1}$) for albumin in the testis of adult rats treated with hCG (from Setchell and Sharpe 1981) and in the testes of rats of different ages (Setchell, Pollanen, and Zupp 1987)

Hours after hCG	PS	Age (days)	PS
0	0.28	10	0.61
8	0.67	15	0.51
12	0.78	20	0.61
16	1.24	25	1.62
20	2.14	30	2.21
		33	1.70
		37	1.39
		44	0.45
		Adult	0.35

Table 2.4

Calculated vascular permeability for ^{125}I-hCG and ^{125}I-hSA in the adult rat testis. Unpublished data of B. P. Setchell and R. M. Sharpe, with V_f values from Sharpe (1981). Values are means $\pm$ S.E.M.

Time after injection (min)	Distribution volume of marker ($\mu l/g$)	
	^{125}I-hSA	^{125}I-hCG
3	11.8 ± 0.6	12.9 ± 0.5
60	43.4 ± 1.8	43.7 ± 2.2
120	53.8 ± 2.1	62.7 ± 2.2
180	69.5 ± 5.9	87.4 ± 7.9
960 (V_f)	141	273
K (min^{-1})*	0.00332	0.00186
Vascular permeability ($\mu l/g/min$)	0.468	0.508

* slope of the line of $\log_e (1 - V_t/V_f)$ against time (see Setchell, Pollanen and Zupp, 1988).

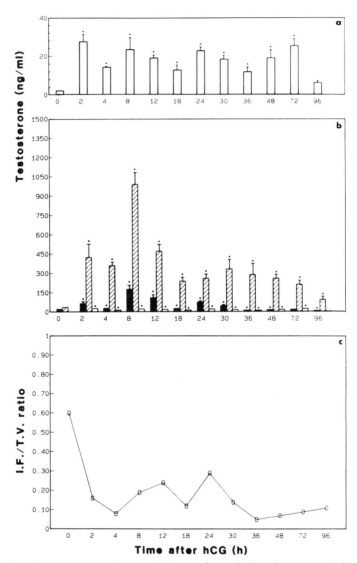

Fig. 2.10.   Testosterone levels up to 96 hours after a single subcutaneous injection of 50 i.u. hCG: (a) In peripheral blood plasma; (b) In peripheral blood plasma (open bars), testicular venous blood (hatched bars), and interstitial fluid (solid bars). Values are means ± S.E.M., n = 4.*: values significantly different (P < 0.05) from untreated animals; (c) Ratio of mean testosterone levels in interstitial fluid compared to those in testicular venous blood, reflecting the relative change in testosterone concentration in these two fluid compartments.

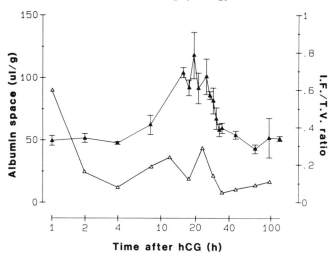

Fig. 2.11.   The ratio of the mean level of testosterone in testicular venous blood compared to that in testicular interstitial fluid (open triangles, re-presented from Fig. 2.10), and the 1-hour albumin space in the testes (solid triangles, from Sowerbutts *et al.* 1986) in adult male rats at various times after a single subcutaneous injection of 50 i.u. hCG. Values for albumin spaces are means ± S.E.M., n = 4.

particular note in this regard are the levels at 72 hours after injection of hCG. As discussed earlier, peripheral levels peak for a third time at 72 hours, but testicular vascular permeability to albumin is normal then. The results from the push-pull study show that this peak in peripheral testosterone levels also occurs without concomitant peaks in interstitial fluid or testicular venous blood. This suggests that other effects such as a change in blood flow or testosterone clearance rate contribute to the peak found in peripheral blood samples at this time. The ratio of interstitial fluid testosterone/testicular venous levels (IF/TV ratio) provides further evidence of the effect of vascular parameters on interstitial fluid testosterone levels. As can be seen in Figure 2.10, an injection of hCG leads to an immediate rise in circulating testosterone levels in peripheral and testicular venous blood. However, interstitial levels of testosterone are not increased to the same extent. The peak of testosterone in testicular venous blood at 8 hours after hCG administration is followed by increasing values of the IF/TV ratio up to 24 hours after hCG. When these data are compared with changes in vascular permeability after hCG treatment (Sowerbutts *et al.* 1986; Figure 2.11), the period of increasing vascular permeability is associated with the second major peak in peripheral blood levels of testosterone, and also coincides with a second peak in testosterone levels in interstitial fluid reflected by the increasing values for the IF/TV ratio. This evidence, and that previously published by other authors suggests that a number of factors are involved in the responses noted after an injection of hCG, and changes in testicular blood flow and capillary permeability after hCG injection are likely to be responsible for a substantial number of the changes noted. Blood flow is significantly increased between 16 and 24 hours after an injection of hCG (Setchell and Sharpe 1981; Damber, Selstam, and Wang 1981) which would reduce the levels of testosterone measured per unit volume of venous blood at this time. This corresponds with the time of increased levels of testosterone in interstitial fluid relative to blood levels, as reflected by the IF/TV ratio values presented in Figures 2.10 and 2.11. These effects are likely to be of importance in the unstimulated animal as well, and we believe these results support the suggestion that blood flow and vascular permeability are involved in regulating the endocrine activity of the physiologically normal testis.

The vascular system can control the entry or egress of substances into and out of the testis in two ways. For those substances for which there are concentration gradients across the vascular wall, movement can be regulated by varying the rate of either passive or facilitated diffusion. If this diffusion is not limiting, then transport can be regulated by changing blood flow (Setchell 1986). The finding in the aspermatogenic testis of higher than normal concentrations of testosterone in the testis and in testicular venous blood, but lower than normal levels in the peripheral circulation is probably due to a reduction in testicular blood flow in proportion to the reduction in testis weight, with continued secretion of testosterone by the Leydig cells (Setchell and Galil 1983; Galil and Setchell 1988b). Testicular blood flow is influenced by a variety of factors including posture, temperature and the common vasoactive substances such as prostaglandins and neurotransmitters (see Setchell 1970, 1978; Free 1977). It is unlikely however, that the potential ability of LH to increase testicular blood flow is of any physiological significance since the lag-time for the response is so long (Free 1977; Setchell 1978; Setchell and Sharpe 1981; Damber et al. 1986). Indeed, the enhanced secretion of testosterone by the ram testis in response to a spontaneous pulse of LH does not involve any increase in testicular blood flow (Laurie and Setchell 1978) and must be due either to a rapid change in permeability of testicular blood vessels or the walls of the Leydig cells, or an equally rapid change in the rate of testosterone synthesis by these cells (see Setchell 1986). By contrast, the lower secretion of testosterone from the abdominal testis after hCG is mainly due to reduced blood flow and not to any reduction in the ability of the Leydig cells of abdominal testes to produce testosterone (Damber, Berg, and Daehlin 1985). In fact Leydig cells isolated from cryptorchid rat testes produce more testosterone than cells from normal testes (Risbridger, Kerr, and de Kretser 1981). It remains likely that any effects of LH on testicular blood flow are indirect effects, probably mediated by testicular factors produced in response to LH (such as the LHRH-like factors described by Sharpe and co-workers), and the suggestion that some product of interstitial cells

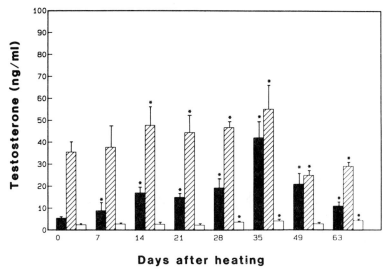

Fig. 2.12.   Levels of testosterone in peripheral blood plasma (open bars), testicular venous blood plasma (hatched bars), and testicular interstitial fluid (solid bars) of adult rats up to 63 days after heating of both testes at 43°C for 30 minutes. Values presented are means ± S.E.M., n = 4.*: values significantly different (P < 0.05) from untreated animals.

(Leydig cells or macrophages) is involved in regulating the neighbouring blood supply (Williams 1949) requires further investigation.

We have used the push-pull cannula to obtain interstitial fluid from the testes of rats in which aspects of spermatogenic and endocrine function have been altered by localized heating of the testes, or following ligation of the efferent ducts. The results obtained have provided further support for the hypothesis raised earlier that Leydig cells have an important relationship with blood vessels which is not limited to an anatomical closeness, but extends to important physiological interactions, with Leydig cell function being dramatically influenced by blood flow and capillary permeability.

Following localized heating of the testes, there is no consistent change in peripheral levels of testosterone (Main, Davies, and Setchell 1978; Main and Setchell 1980; see also Fig. 2.12). Testicular venous levels are initially unaffected by heating, but then rise between 14 and 35 days after heating. Galil and Setchell (1988a) have correlated this rise with a decrease in testis weight due to disruption of spermatogenesis. Testicular blood flow was also reduced during this time, and it was suggested that tubular mass determined testicular blood flow under these conditions (Galil and Setchell 1988b). It is therefore of interest that the levels of testosterone in interstitial fluid obtained from animals treated in the same way (Maddocks, Zupp, Sowerbutts, and Setchell

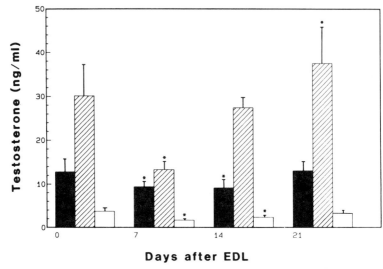

Fig. 2.13.   Levels of testosterone in peripheral blood plasma (open bars), testicular venous blood plasma (hatched bars), and testicular interstitial fluid (solid bars) of adult rats at 0, 7, 14, and 21 days after bilateral ligation of testicular efferent ducts. Values presented are means ± S.E.M., n = 4.*: values significantly different (P < 0.05) from untreated animals.

1987) increased significantly during this same time period (Fig. 2.12), and at 35 days after heat treatment were not significantly different from testicular venous blood levels.

Ligation of the efferent ducts of the testis causes fluid secreted by the testis to be retained inside the seminiferous tubules and rete testis, resulting in enlargement and eventual degeneration (Smith 1962; Setchell 1970). Following bilateral efferent duct ligation (EDL), peripheral levels of testosterone were only slightly reduced at 7 and 14 days afterwards, but not significantly different from control levels by 21 days (Fig. 2.13). Similar results have been previously reported by others (Setchell et al. 1977; Main et al. 1978; Risbridger, Kerr, Peake, and de Kretser 1981). Of interest however, is the significant reduction in testicular venous levels at 7 days after ligation, which return to normal by 14 days, and are then significantly increased by 21 days. Because interstitial fluid levels do not show a similar response, it is unlikely that the venous levels are directly attributable to changes in Leydig cell secretion. It is more likely that alterations in blood flow are of greater significance. Indeed, Wang, Galil, and Setchell (1983) concluded that during aspermatogenesis following irradiation (as with heat and efferent duct ligation) the capacity of the testis to secrete testosterone is severely limited by decreased testicular blood flow, not by the ability of the Leydig cells to release testosterone into their immediate environment. Wang, Gu, Tao, and Wu (1985) have also reported that blood flow plays a role in changes of testosterone secretion seen after EDL. At 21 days after unilateral EDL, values for testicular blood flow and organ weight of ligated testes were significantly less than the values from unligated testes. Testosterone levels in the venous blood of the ligated testes were also significantly less than in the venous blood of unligated testes, although tissue levels of testosterone were not different.

Bergh and Damber (1984) have suggested that there may be paracrine regulation of Leydig cells by seminiferous tubules, particularly at specific stages of the spermatogenic cycle (Bergh 1985a). It is possible that such effects might occur more generally in damaged testes, particularly during re-initiation of spermatogenesis and repopulation within the tubules. This is yet further cause to implicate blood flow and paracrine events in the regulation of testosterone secretion and partitioning in the testis.

Another feature of the testicular microcirculation which is of unknown significance in the control of hormone secretion is the occurrence of regular fluctuations of blood flow as measured by laser Doppler flowmetry. Variation of as much as 50 per cent of mean flow occur regularly in normal rats, with a frequency of about 10/min (Damber, Bergh, Fagrell, Lindahl, and Rooth 1986) and these disappear following treatment of the animal with hCG (Widmark *et al.* 1986) or LH (Damber *et al.* 1986). These authors have suggested that the pulsatile flow may restrict flow of fluid from the lumen of the blood vessel to the interstitium, and interstitial fluid accumulation seen after hCG or LH treatment may be a consequence of the transition from pulsatile to non-pulsatile flow. However, this is difficult to reconcile with the observation that in the rat (Setchell 1970), as in the ram (Waites and Moule 1960), dog (Blomberry, Waites, and Gow, quoted by Setchell 1970) and wallaby (Setchell and Waites 1969), the pulse pressure in the arterial blood is much reduced with only a small decrease in mean pressure during passage of the blood through the artery in the spermatic cord, although clearly the frequency of the changes in flow is lower than the heart rate.

2   EFFECTS OF BLOOD-TESTIS BARRIERS

The blood-testis barrier is probably involved in the control of endocrine activity in the testis. As emphasized by Sharpe (1983), the various barriers in the testis (vascular endothelium, myoid cells, tight-junctions of the Sertoli cells) will not only selectively prevent certain substances from entering the testis, but may also act to prevent others from leaving the testis. These various barriers allow for the creation of unique microenvironments within the tubules, and possibly in the interstitial space, and are probably essential for local interaction between the various cell types and compartments within the testis. The development of the tubular barrier formed by the Sertoli cell junctions is retarded by the absence of pituitary hormones, but not entirely prevented (Vitale-Calpe *et al.* 1973; but see also Bressler 1978), and once established seems not to be subject to hormonal modification. The only circumstance in which the barrier breaks down is following efferent duct ligation (Setchell 1986) but this is probably a result of the increased intratubular pressure and not a response to the hormonal changes which occur (Main *et al* 1978).

In addition to the major blood-testis barrier in the seminiferous tubules, Kormano (1967*b*) has demonstrated that at puberty the penetration of certain dyes into the interstitial fluid is reduced, and some may no longer penetrate at all. This implies some barrier at the level of the vascular endothelium. The barrier does not appear to restrict the passage of most proteins into interstitial fluid (Setchell and Waites 1975; Waites 1977; Waites and Gladwell 1982) unlike the blood-brain barrier, and the substances that it is intended to exclude remain to be identified. It is of interest that the establishment of this barrier also occurs at puberty suggesting some hormonal control and coincides with an increase in the activity in the blood vessels of alkaline phosphatase, an enzyme considered to be important in transporting substances across capillary walls (Kormano 1967*b*). After puberty, the testicular blood vessels are second only to the capillaries of the brain in their level of alkaline phosphatase. While the enzyme is found in the endothelial cells of all blood vessels in the testis, virtually no activity is found in other interstitial cells or the tubules (Setchell 1986). In contrast, the enzyme gamma-glutamyl transpeptidase which is also present in high concentrations in brain capillaries, is abundant in the endothelial cells of the artery on the surface of the testis and in the spermatic cord and in testicular arterioles, but not in testicular capillaries (Niemi and Setchell 1986). In other tissues, this enzyme is probably associated with amino acid transport and autoradiographic evidence for this association in the testis has been presented by Bustamante, Jarvis, and Setchell (1982). The level of this enzyme in the testis falls following hypophysectomy, but this fall can be prevented by treatment with FSH or testosterone (Sanborn, Wagle, Steinberger, and Greer-Emmert 1986). However, until more is understood of the importance of this endothelial system in regulating passage of substances into and out of the testis, further speculation on its control would be premature.

VII   Local transport of hormones to nearby tissues

It has been known for many years that the epididymis and the prostate are controlled by hormones secreted by the testis, although it is usually assumed that these hormones are secreted into the blood to be transported to the accessory organs where they have their effects. However, there have been several reports that more local routes may also be important. Suzuki, Toshimori, and Mochizuki (1955) showed that the ipsilateral epididymis and vas deferens grew much more slowly in unilaterally castrated young rats, whereas unilateral ligation of the efferent ducts was without effect. Pierrepoint *et al.* (1975) suggested that androgens carried in the deferential vein could explain the changes they found in androgen dependent activity of prostatic RNA polymerase following vasectomy. Skinner and Rowson (1968*a,b*) found that unilateral vasectomy retarded growth of the ipsilateral ampulla, and that testosterone infused into one vas deferens could produce a local stimulation of the ampulla. In humans, the volume of the peri-urethral portion of the prostate is reduced by about a quarter three months after vasectomy (Jakobsen and Juul 1985).

Albumin injected directly into a testis is carried away largely in the lymph (Galil *et al.* 1981), but when unilateral injections are made in this way in rats, higher levels of protein-bound radioactivity are found in the ipsilateral epididymis, fat pad, ductus deferens, seminal vesicle and prostate than in the contralateral tissues, suggesting that there may be local transfer from the testis in lymph to nearby tissues. Furthermore, when the albumin was infused into an actual lymphatic vessel on the surface of a ram testis, similarly increased levels of radioactivity could be found in the ipsilateral epididymis and ductus deferens (M. R. Jones and B. P. Setchell, unpublished result, quoted by Setchell 1986), supporting the idea that the larger lymphatic vessels are involved in this transfer. If albumin can be selectively transferred from one testis to the ipsilateral epididymis and other accessory glands via a lymphatic pathway, then the same pathway is likely to be involved in the transfer of peptides, particularly the larger ones. It is known that the Leydig cells of the testis contain, and presumably also secrete a number of biologically active peptides (see **Section II**). The possible actions of these peptides on accessory glands has not been investigated to any great extent, except that oxytocin is known to stimulate contractile activity of the epididymis (Knight 1974).

In addition to peptides, conjugated steroids are a class of compound which appears to be transported from the testis preferentially in lymph, but it is not yet known whether they accumulate in tissues near the testis. It is usually assumed that conjugated steroids are biologically inactive until free steroid has been liberated by enzymatic action. If the direct transfer of conjugated steroids to the epididymis and prostate is of biological significance, there should be sulphatase or glucuronidase enzymes present in the target tissues to liberate free steroid. No values for these enzymes in either epididymis or prostate appear to have been published (see Hobkirk 1985). If significant direct transfer of these conjugates from the testis to the accessory organs can be demonstrated, it will be interesting to assay tissues from various animals for the presence of these enzymes.

VIII   Conclusions

In this review, we have attempted to present a number of new concepts relating to the workings of the endocrine testis. Much of our knowledge in this field has been obtained from experiments with isolated cells and no doubt much more will come from such studies in the future. However with the recognition that cells interact and regulate each others functions *in vivo*, comes the appreciation that physiological parameters such as vascular permeability and blood and lymph flow are also of importance. The analysis of whole tissues in isolation under some conditions, must also now be questioned, and it is important that we continue to return to, and work with, the "whole" animal. Hormone transport and secretion are the result of a number of integrated and carefully regulated actions, and it is our belief that a better understanding of the regulation of the endocrine activity of the testis will come from a greater appreciation of the physiological parameters involved.

References

Aldred, L. F. and Cooke, B. A. (1982). The deleterious effect of mechanical dissociation of rat testes on the functional activity and purification of Leydig cells using Percoll gradients. *Int. J. Androl.* **5**, 191–195.

Amat, P., Paniagua, R., Nistal, M., and Martin, A. (1986). Mitosis in adult human Leydig cells. *Cell Tissue Res.* **243**, 219–221.

Aoki, A. and Hoffer, A. P. (1978). Re-examination of the lesions in rat testis caused by cadmium. *Biol. Reprod.* **18**, 579–591.

Aquilano, D. R. and Dufau, M. L. (1984). Studies on Leydig cell purification. *Ann. N.Y. Acad. Sci.* **438**, 237–258.

Armstrong, D. T., Moon, Y. S., Fritz, I. B., and Dorrington, J. H. (1975). Synthesis of estradiol-17β by Sertoli cells in culture: stimulation by FSH and dibutyryl cyclic AMP. *Curr. Top. molec. Endocr.* **2**, 85–96.

Assaykeen, T. A. and Thomas, J. A. (1965). Endogenous histamine in male organs of reproduction. *Endocrinology* **76**, 839–843.

Attramadal, A., Bardin, C. W., Gunsalus, G. L., Musto, N. A., and Hansson, V. (1981). Immunocytochemical localization of androgen binding protein in rat Sertoli and epididymal cells. *Biol. Reprod.* **25**, 983–988.

Au, C. L., Robertson, D. M., and de Kretser, D. M. (1984). An *in-vivo* method for estimating inhibin production by adult rat testes. *J. Reprod. Fert.* **71**, 259–265.

Baker, H. W. G., Bremner, W. G., Burger, H. G., de Kretser, D. M., Dulmanis, A., Eddie, L. W., Hudson, B., Keogh, E. J., Lee, V. W. K., and Rennie, G. C. (1976). Testicular control of FSH secretion. *Recent Prog. Horm. Res.* **32**, 429–469.

Bardin, C. W., Musto, N., Gunsalus, G, Kotite, N., Chang, S.-L., Larreau, F., and Becker, R. (1981). Extracellular androgen binding proteins. *A. Rev. Physiol.* **43**, 189–198.

—— Shaha, C., Mather, J., Salomon, Y., Margioris, A. N., Liotta, A. S., Gerendai, I., Chen, C. L., and Krieger, D. T. (1984). Identification and possible function of pro-opiomelanocortin-derived peptides in the testis. *Ann. N.Y. Acad. Sci.* **438**, 346–364.

Bell, J. B. G., Vinson, G. P., and Lacy, D. (1971). Studies on the ultrastructure and function of the mammalian testis III: *In vitro* steroidogenesis by the seminiferous tubules of the rat testis. *Proc. R. Soc. B.* **176**, 433–443.

Bellvé, A. R. (1979). The molecular biology of mammalian spermatogenesis. In *Oxford Reviews of Reproductive Biology* **Vol. 1** (ed. C. A. Finn) pp. 159–261. Clarendon Press, Oxford.

Benahmed, M., Grenot, C., Tabone, E., Sanchez, P., and Morera, A. M. (1985). FSH regulates cultured Leydig cell function via Sertoli cell proteins: an *in vitro study*. *Biochem. Biophys. Res. Commun.* **132**, 729–734.

Bergh, A. (1982). Local differences in Leydig cell morphology in the adult rat testis: evidence for a local control of Leydig cells by adjacent seminiferous tubules. *Int. J. Androl.* **5**, 325–330.

——(1983). Paracrine regulation of Leydig cells by the seminiferous tubules. *Int. J. Androl.* **6**, 57–65.

——(1985*a*). Development of stage-specific paracrine regulation of Leydig cells by the seminiferous tubules. *Int. J. Androl.* **8**, 80–85.

——(1985*b*). Effect of cryptorchidism on the morphology of testicular macrophages: evidence for a Leydig cell—macrophage interaction in the rat testis. *Int. J. Androl.* **8**, 86–96.

——and Damber, J. E. (1984). Local regulation of Leydig cells by the seminiferous tubules. Effect of short-term cryptorchidism. *Int. J. Androl.* **7**, 409–418.

——Rooth, P., Widmark, A., and Damber, J. E. (1987). Treatment of rats with hCG induces inflammation-like changes in the testicular microcirculation. *J. Reprod. Fert.* **79**, 135–143.

——Widmark, A., Damber, J. E., and Cajander, S. (1986). Are leukocytes involved in the human Chorionic Gonadotrophin-induced increase in testicular vascular permeability? *Endocrinology* **119**, 586–590.

Black, V. H. (1971). Gonocytes in fetal guinea-pig testes: phagocytosis of degenerating gonocytes by Sertoli cells. *Am. J. Anat.* **131**, 415–426.

Bressler, R. S. (1978). Hormonal control of postnatal maturation of the seminiferous cord. *Annls. Biol. anim. Biochim. Biophys.* **18**, 535–540.

——and Ross, M. H. (1973). On the character of the monolayer outgrowth and the fate of the peritubular myoid cells in cultured mouse testis. *Expl. Cell Res.* **78**, 295–302.

Brown, K. D., Blakeley, D. M., Henville, A., and Setchell, B. P. (1982). Rete testis fluid contains a growth factor for cultured fibroblasts. *Biochem. Biophys. Res. Commun.* **105**, 391–397.

Browning, J. Y., D'Agata, R., and Grotjan, H. E. (1981). Isolation of purified rat Leydig cells using continuous Percoll gradients. *Endocrinology* **109**, 667–669.

Buhl, A. E., Cornette, J. C., Kirton, K. T., and Yuan, Y.-D. (1982). Hypophysectomised male rats treated with polydimethylsiloxane capsules containing testosterone: Effects on spermatogenesis, fertility, and reproductive tract concentrations of androgens. *Biol. Reprod.* **27**, 183–188.

Burgos, M. H., Vitale-Calpe, R., and Aoki, A. (1970). Fine structure of the testis and its functional significance. In *The Testis* Vol. I. (ed. A. D. Johnson, W. R. Gomes, and N. L. van Demark) pp. 551–649. Academic Press, New York.

Burrow, P. V., Pholpramool, C., and Setchell, B. P. (1981). The movement of substances out of the seminiferous tubules and epididymal duct during microinfusion. *J. Physiol., Lond.* **319**, 15P–16P.

——and Setchell, B. P. (1980). The fate of radioactive substances infused into a single seminiferous tubule in anaesthetised rats. *Proc. Winter Meeting, Soc. Study Fertility* p. 22. Abstr. 32.

Bustamante, J. C., Jarvis, L. G., and Setchell, B. P. (1982). Role of the endothelium in the uptake of amino acids by the isolated perfused rat testis. *J. Physiol., Lond.* **330**, 62P–63P.

——and Setchell, B. P. (1981). Measurement of capillary permeability-surface area products in isolated perfused rat testis. *J. Physiol., Lond.* **319**, 16P–17P.

——— ——— (1982). Kinetics of the facilitated diffusion of leucine into the perfused testis of the rat. *J. Physiol., Lond.* **324**, 49P–50P.

Byers, S. W., Hadley, M. A., Djakiew, D., and Dym, M. (1986). Growth and characterization of polarized monolayers of epididymal epithelial cells and Sertoli cells in dual environment culture chambers. *J. Androl.* **7**, 59–69.

Cameron, D. F. and Merkwald, R. R. (1981). Structural responses of adult rat Sertoli cells to peritubular fibroblasts *in vitro*. *Am. J. Anat.* **160**, 343–358.

Carpenter, M. P., Manning, L. M., Robinson, R. D., and To, D. (1978). Prostaglandin synthesis by interstitial tissue of rat testis. *Prostaglandins.* **15**, 711.

Chen, G. C. C., Lin, T., Murono, E., Osterman, J., Cole, B. T., and Nankin, H. (1981). The aging Leydig cell: 2. Two distinct populations of Leydig cells and the possible site of defective steroidogenesis. *Steroids* **37**, 63–72.

Cheng, C. Y., Gunsalus, G. L., Morris, I. D., Turner, T. T., and Bardin, C. W. (1986). The heterogeneity of rat androgen binding protein (rABP) in the vascular compartment differs from that in the testicular tubular lumen. Further evidence for bidirectional secretion of rABP. *J. Androl.* **7**, 175–179.

——— ——— Musto, N. A., and Bardin, C. W. (1984). The heterogeneity of rat androgen-binding protein in serum differs from that in testis and epididymis. *Endocrinology* **114**, 1386–1394.

Christensen, A. K. (1975). Leydig cells. In *Handbook of Physiology.* Sect. **7**, Vol. **5** (ed. D. W. Hamilton and R. O. Greep) pp. 57–94. American Physiology Society, Washington.

——— and Fawcett, D. W. (1961). The normal fine structure of opposum testicular interstitial cells. *J. biophys. biochem. Cytol.* **9**, 653–670.

——— and Gillim, S. W. (1969). The correlation of fine structures and function in steroid-secreting cells, with emphasis on those of the gonads. In *The Gonads* (ed. K. W. McKerns) pp. 415–488. Appleton-Century-Crofts, New York.

——— and Mason, N. R. (1965). Comparative ability of seminiferous tubules and interstitial tissues of rat testes to synthesize androgens from progesterone-4-C14 *in vitro*. *Endocrinology* **76**, 646–656.

——— and Peacock, K. C. (1980). Increase in Leydig cell number in testes of adult rats treated chronically with an excess of human chorionic gonadotrophin. *Biol. Reprod.* **22**, 383–392.

Chubb, C. and Ewing, L. L. (1979a). Steroid secretion by *in vitro* perfused testes: secretions of rabbit and rat testes. *Am. J. Physiol.* **237**, E231–E238.

——— ——— (1979b). Steroid secretion by *in vitro* perfused testes: testosterone biosynthetic pathways. *Am. J. Physiol.* **237**, E247–E254.

Clark, R. V. (1975). Three dimensional organisation of testicular interstitial tissue and lymphatic space in the rat. *Anat. Rec.* **184**, 203–226.

Claus, R. and Hoffman, B. (1980). Oestrogens, compared to other steroids of testicular origin, in blood plasma of boars. *Acta endocr. Copenh.* **94**, 404–411.

Clegg, E. J. (1963). Studies on artificial cryptorchidism: degenerative and regenerative changes in the germinal epithelium of the rat testis. *J. Endocr.* **27**, 241–251.

——— and MacMillan, E. W. (1965a). The phagocytic nature of Schiff-positive interstitial cells in the rat testis. *J. Endocr.* **31**, 299–300.

——— ——— (1965b). The uptake of vital dyes and particulate matter by the Sertoli cells of the rat testis. *J. Anat.* **99**, 219–229.

Clermont, Y. and Perey, B. (1957). Quantitative study of the cell population of the seminiferous tubules in immature rats. *Am. J. Anat.* **100**, 241–260.

Cooke, B. A., de Jong, F. H., Van der Molen, H. J., and Rommerts, F. F. G. (1972). Endogenous testosterone concentrations in rat testis interstitial tissue and seminiferous tubules during *in vitro* incubation. *Nature, New Biol.* **237**, 255–256.

—— Magee-Brown, R., Golding, M., and Dix, C. J. (1981). The heterogeneity of Leydig cells from mouse and rat testes: evidence for a Leydig cell cycle? *Int. J. Androl.* **4**, 355–366.

Coombs, R. C., Woods, A., Ellison, J., and Jenkins, N. (1986). Molecular identification of plasminogen activators and inhibitors in the bovine Sertoli cell. In *Programme and abstracts of the annual conference of the Society for the Study of Fertility.* University of Bristol, p. 31. Abstr. 33.

Cooper, T. G. and Waites, G. M. H. (1975). Steroid entry into rete testis fluid and the blood-testis barrier. *J. Endocr.* **65**, 195–205.

Comhaire, F. H. and Vermeulen, A. (1976). Testosterone concentration in the fluid of seminiferous tubules, the interstitium and the rete testis of the rat. *J. Endocr.* **70**, 229–235.

Connell, C. J. and Christensen, A. K. (1975). The ultrastructure of the canine testicular interstitial tissue. *Biol. Reprod.* **12**, 368–382.

—— and Connell, G. M. (1977). The interstitial tissue of the testis. In *The Testis* Vol. IV (ed. A. D. Johnson and W. R. Gomes) pp. 333–369. Academic Press, New York.

Corval, P. and Bardin, C. W. (1973). Species distribution of testosterone-binding globulin. *Biol. Reprod.* **8**, 277–282.

Crone, C. (1963). The permeability of capillaries in various organs as determined by use of the "indicator diffusion method". *Acta physiol. scand.* **58**, 292–305.

Cunningham, G. R. and Huckins, C. (1979). Persistence of complete spermatogenesis in the presence of low intratesticular concentrations of testosterone. *Endocrinology* **105**, 177–186.

Damber, J. E., Bergh, A., and Daehlin, L. (1985). Testicular blood flow, vascular permeability, and testosterone production after stimulation of unilaterally cryptorchid adult rats with human Chorionic Gonadotrophin. *Endocrinology* **117**, 1906–1913.

—— —— Fagrell, B., Lindhal, O., and Rooth, P. (1986). Testicular capillary circulation in the rat studied by videophotometric capillaroscopy, fluorescence microscopy and laser doppler flowmetry. *Acta physiol. scand.* **126**, 371–376.

—— Selstam, G., and Wang, J. (1981). Inhibitory effect of estradiol-17B on human Chorionic Gonadotrophin-induced increment of testicular blood flow and plasma testosterone concentration in rats. *Biol. Reprod.* **25**, 555–559.

—— Widmark, A., and Bergh, A. (1986). The effect of LH on testicular microcirculation and vascular permeability. *Miniposters of the IV European workshop on molecular and cellular endocrinology of the testis*, 152.

Danzo, B. J., Cooper, T. G., and Orgebin-Crist, M.-C. (1977). Androgen binding protein (ABP) in fluids collected from the rete testis and cauda epididymis of sexually mature and immature rabbits and observations on morphological changes in the epididymis following ligation of the ductuli efferentes. *Biol. Reprod.* **17**, 64–77.

Davies, R. V., Main, S. J., Laurie, M. S., and Setchell, B. P. (1979). The effects of long-term administration of either a crude inhibin preparation or an antiserum to FSH on serum hormone levels, testicular function and fertility of adult male rats. *J. Reprod. Fert. Suppl.* **26**, 183–191.

—— —— and Setchell, B. P. (1978). Inhibin: evidence for its existence, methods of bioassay and nature of the active material. *Int. J. Androl. Suppl.* **2**, 102–114.

de Jong, F. H. (1987). Inhibin—its nature, site of production and function. In *Oxford Reviews of Reproductive Biology*, Vol. 9 (ed. J. R. Clarke) pp. 1–53. Clarendon Press, Oxford.

—— Hey, A. H., and Van der Molen, H. J. (1974). Oestradiol-17β and testosterone in rat testis tissue: effect of gonadotrophins, localization and production *in vitro. J. Endocr.* **60**, 409-419.

—— and Robertson, D. M. (1985). Inhibin: 1985 update on action and purification. *Molec. cell. Endocr.* **42**, 95–103.

—— and Sharpe, R. M. (1977). The onset and establishment of spermatogenesis in rats in relation to gonadotrophin levels and testosterone levels. *J. Endocr.* **75**, 197–207.

—— Welschen, R., Hermans, W. P., Smith, S. D., and Van der Molen, H. J. (1978). Effects of testicular and ovarian inhibin-like activity, using *in vitro* and *in vivo* systems. *Int. J. Androl. Suppl.* **2**, 125–138.

de Kretser, D. M. (1967). The fine structure of the testicular interstitial cells in men of normal androgen status. *Z. Zellforsch*, **80**, 594–609.

—— and Kerr, J. B. (1983). The effect of testicular damage on Sertoli and Leydig cell function. In *Monographs on endocrinology: the pituitary and the testis* (ed. D. M. de Kretser, H. G. Burger and B. Hudson) pp. 133–154. Springer-Verlag, Berlin.

Dehejia, A., Nozu, K., Catt, K. J., and Dufau, M. L. (1982). Purification of rat Leydig cells: functional and morphological evaluation. *Ann. N.Y. Acad. Sci.* **383**, 204–211.

DeMartino, C., Marcante, M. L., Floridi, A., Citro, G., Bellocci, A., Cantaforca, A., and Natali, P. G. (1977). Sertoli cells of adult rats *in vitro. Cell Tissue Res.* **176**, 69–90.

Demoulin, A., Hustin, J., Lambotte, R., and Franchimont, P. (1981). Effect of inhibin on testicular function. In *Intragonadal Regulation of Reproduction* (ed. P. Franchimont and C. P. Channing) pp. 337–342. Academic Press, New York.

Djakiew, D., Griswold, M. D., Lewis, D. M., and Dym, M. (1986). Micropuncture studies of receptor-mediated endocytosis of transferrin in the rat epididymis. *Biol. Reprod.* **34**, 691–699.

—— Hadley, M. A., Byers, S. W., and Dym, M. (1986). Transferrin-mediated transcellular transport of ^{59}Fe across confluent epithelial sheets of Sertoli cells grown in bicameral cell culture chambers. *J. Androl.* **7**, 355–366.

Dorrington, J. H. and Armstrong, D. T. (1975). Follicle stimulating hormone stimulates estradiol-17B synthesis in cultured Sertoli cells. *Proc. natn. Acad. Sci. U.S.A.* **72**, 2677–2681.

—— and Fritz, I. B. (1975). Androgen synthesis and metabolism by preparations from the seminiferous tubule of the rat testis. In *Hormonal regulation of spermatogenesis* (ed. F. S. French, V. Hansson, E. M. Ritzen, and S. N. Nayfeh) pp. 37–52. Plenum Press, New York.

—— Roller, N. F., and Fritz, I. B. (1975). Effects of follicle stimulating hormone on cultures of Sertoli cell preparations. *Molec. Cell. Endocr.* **3**, 57–70.

—— Fritz, I. B., and Armstrong, D. T. (1976). Site at which FSH regulates estradiol-17B biosynthesis in Sertoli cell preparations in culture. *Molec. Cell. Endocr.* **6**, 117–122.

Dym, M. (1973). The fine structure of the monkey Sertoli cell and its role in maintaining the blood-testis barrier. *Anat. Rec.* **175**, 639–656.

—— and Fawcett, D. W. (1970). The blood-testis barrier in the rat and the physiological compartmentation of the seminiferous epithelium. *Biol. Reprod.* **3**, 308–326.

Eik-Nes, K. B. (1970). Synthesis and secretion of androstenedione and testosterone. In *The Androgens of the Testis* (ed. K. B. Eik-Nes) pp. 1–47. Marcel-Dekker, New York.

—— Schellman, J. A., Lumry, R., and Samuels, L. T. (1954). The binding of steroids to protein I. Solubility determinations. *J. biol. Chem.* **206**, 411–419.

Eiler, H. (1986). Effect of plasma concentrations of ovarian steroids on their passage into uterine lumen in rabbits. *Am. J. Physiol.* **251**, E654–E659.

Ellis, L. C., Sorenson, D. K., and Buhrley, L. E. (1975). Mechanisms and interactions in testicular steroidogenesis and prostaglandin synthesis. *J. steroid Biochem.* **6**, 1081–1090.

Engeset, A. (1959). The route of peripheral lymph to the blood stream, an X-ray study of the barrier theory. *J. Anat.* **93**, 96–100.

Ewing, L. L., Brown, B., Irby, D. C., and Jardine, I. (1975). Testosterone and 5α-reduced androgen secretion by rabbit testes-epididymides perfused *in vitro*. *Endocrinology* **96**, 610–617.

—— Davis, J. C., and Zirkin, B. R. (1980). Regulation of testicular function: a spatial and temporal view. *Int. Rev. Physiol.* **22**, 41–115.

—— and Zirkin, B. (1983). Leydig cell structure and steroidogenic function. *Recent Prog. Horm. Res.* **39**, 599–632.

—— —— Cochran, R. C., Kromann, N., Pefers, C., and Ruiz-Bravo, N. (1979). Testosterone secretion by rat, rabbit, guinea pig, dog, and hamster testes perfused *in vitro*: correlation with Leydig cell mass. *Endocrinology* **105**, 1135–1142.

Fabbri, A., Tsai-Morris, C. H., Luna, S., Fraioli, F., and Dufau, M. L. (1985). Opiate receptors are present in the rat testis. Identification and localisation in Sertoli cells. *Endocrinology* **117**, 2544–2546.

Fawcett, D. W. (1973). Observations on the organisation of the interstitial tissue of the testis and on the occluding cell junctions in the seminiferous epithelium. *Adv. Biosci.* **10**, 83–99.

—— (1975). Ultrastructure and function of the Sertoli cell. In *Handbook of Physiology* Sect. 7, Vol. **5** (ed. D. W. Hamilton and R. O. Greep) pp. 21–55. American Physiology Society, Washington.

—— and Burgos, M. H. (1960). Studies on the fine structure of the mammalian testis II: the human interstitial tissue. *Am. J. Anat.* **107**, 245–269.

—— Leak, L. V., and Heidger, P. M. (1970). Electron microscopic observations on the structural components of the blood testis barrier. *J. Reprod. Fert. Suppl.* **10**, 105–122.

—— Neaves, W. B., and Flores, M. N. (1973). Comparative observations on intertubular lymphatics and organisation of the interstitial tissue of the mammalian testis. *Biol. Reprod.* **9**, 500–532.

Feig, L. A., Klagsbrun, M., and Bellvé, A. R. (1983). Mitogenic polypeptide of the mammalian seminiferous epithelium: biochemical characterisation and partial purification. *J. Cell. Biol.* **97**, 1435–1443.

Franchimont, P., Croze, F., Demoulin, A., Bologne, R., and Hustin, J. (1981). Effect of inhibin on rat testicular deoxyribonucleic acid (DNA) synthesis *in vivo* and *in vitro*. *Acta endocr. Copenh.* **98**, 312–320.

Francis, G. L., Triche, T. J., Brown, T. S., Brown, H. C., and Bercu, B. B. (1981). *In vivo* gonadotrophin stimulation of bovine Sertoli cell ornithine decarboxylase activity. *J. Androl.* **2**, 312–320.

Free, M. J. (1977). Blood supply to the testis and its role in local exchange and transport of hormones. In *The Testis*, Vol. IV (ed. A. D. Johnson and W. R. Gomes) pp. 39–90. Academic Press, New York.

—— and Jaffe, R. A. (1975). Dynamics of venous-arterial testosterone transfer in the pampiniform plexus of the rat. *Endocrinology* **97**, 169–177.

—— and Tillson, S. A. (1975). Local increases in concentration of steroids by venous-arterial transfer in the pampiniform plexus. In *Hormonal Regulation of Spermatogenesis* (ed. V. Hanson, E. M. Ritzen and S. N. Nayfeh) pp. 181–194. Plenum Press, New York.

Fritz, I. B., Rommerts, F. F. G., Louis, B. G., and Dorrington, J. H. (1976). Regulation by FSH and dibutyryl cyclic AMP of the formation of androgen binding protein in Sertoli cell-enriched cultures. *J. Reprod. Fert.* **46**, 17–24.

Gaddum, J. H. (1961). Push-pull cannulae. *J. Physiol., Lond.* **155**, 1–2.

Galdieri, M., Monaco, L., and Stefanini, M. (1984). Secretion of androgen binding protein by Sertoli cells is influenced by contact with germ cells. *J. Androl.* **5**, 409–415.

Galil, K. A. A., Laurie, M. S., Main, S. J., and Setchell, B. P. (1981). The measurement of the flow of lymph from the testis. *J. Physiol., Lond.* **319**, 17P.

—— and Setchell, B. P. (1988*a*). Effects of local heating of the testes of rats on the concentration of testosterone in jugular and testicular venous blood of rats and on testosterone production *in vitro*. *Int. J. Androl.* in press.

—— —— (1988*b*). Effect of local heating of the testis on testicular blood flow and testosterone secretion in the rat. *Int. J. Androl.* **11**, 73–85.

Gemmell, R. T., Stacy, B. D., and Thorburn, G. D. (1974). Ultrastructural study of secretory granules in the corpus luteum of the sheep during the estrous cycle. *Biol. Reprod.* **11**, 447–462.

Gerozissis, K. and Dray, F. (1977*a*). Prostaglandins in the isolated testicular capsule of immature and young adult rats. *Prostaglandins.* **13**, 777–783.

—— —— (1977*b*). Selective and age-dependent changes of prostaglandin E-2 in the epididymis and vas deferens of the rat. *J. Reprod. Fert.* **50**, 113–115.

—— —— (1979). Steroid and prostaglandin (PG) interrelations in the epididymis and vas deferens of prepubertal rats. *Res. Steroids* **8**, 299–301.

Gilula, N. B., Fawcett, D. W., and Aoki, A. (1976). The Sertoli cell occluding junctions and gap junctions in mature and developing mammalian testes. *Devl. Biol.* **50**, 142–168.

Giorgi, E. P. (1976). Studies on androgen transport into canine prostate *in vitro*. *J. Endocr.* **68**, 109–119.

—— (1980). The transport of steroid hormones into animal cells. *Int. Rev. Cytol.* **65**, 49–115.

—— and Stein, W. D. (1981). The transport of steroids into animal cells in culture. *Endocrinology* **108**, 688–697.

Gondos, B., Rao, A., and Ramachandran, J. (1980). Effects of antiserum to luteinizing hormone on the structure and function of rat Leydig cells. *J. Endocr.* **87**, 265–270.

Goth, A. and Johnson, A. R. (1975). Current concepts on the secretory function of mast cells. *Life Sci.* **16**, 1201–1214.

Gower, D. B., Harrison, F. A., and Heap, R. B. (1970). The identification of C19-16-unsaturated steroids and estimation of 17-oxosteroids in boar spermatic vein plasma and urine. *J. Endocr.* **47**, 357–368.

Guldenaar, S. E. F. and Pickering, B. T. (1985). Immunocytochemical evidence for the presence of oxytocin in rat testis. *Cell Tissue Res.* **240**, 485–487.

Gunsalus, G. L., Musto, N. A., and Bardin, C. W. (1980). Bidirectional release of a Sertoli cell product, androgen binding protein, into the blood and seminiferous tubule. In *Testicular Development, Structure and Function* (ed. A. Steinberger and E. Steinberger) pp. 291–297. Raven Press, New York.

Hadley, M. A., Byers, S. W., Suarez-Quian, C. A., Kleinman, H. K., and Dym, M. (1985). Extracellular matrix regulates Sertoli cell differentiation, testicular cord formation, and germ cell development *in vitro*. *J. Cell Biol.* **101**, 1511–1522.

—— Djakiew, D., Byers, S., and Dym, M. (1987). Polarised secretion of androgen-binding protein and transferrin by Sertoli cells grown in a bicameral culture system. *Endocrinology* **120**, 1097–1103.

Hagenas, L., Ritzen, E. M., and Suginami, H. (1978). Hormonal milieu of the semini-ferous tubules in the normal and cryptorchid rat. *Int. J. Androl.* **1**, 477–484.

Halley, J. B. W. (1960). Relation of Leydig cells in the human testicle to the tubules and testicular function. *Nature, Lond.* **185**, 865–866.

Hansson, V., Ritzen, E. M., French, F. S., and Nayfeh, S. N. (1975). Androgen trans-port and receptor mechanisms in testis and epididymis. In *Handbook of Physiology* **Sect. 7, Vol. V.** (ed. D. W. Hamilton and R. O. Greep) pp. 173–201. American Physiological Society, Washington.

Haour, F., Mather, J., Saez, J. M., Kouznetzova, B., and Dray, F. (1979). Role of prostaglandins in Leydig cell stimulation by hCG. *INSERM Colloq.* **91**, 75–88.

Harris, M. E. and Bartke, A. (1975). Maintenance of rete testis fluid testosterone and dihydrotestosterone levels by pregnenolone and other C_{21} steroids in hypophysec-tomised rats. *Endocrinology* **96**, 1396–1402.

Hayes, M. K. (1986). *Bovine testicular cells in vitro: establishment of primary cultures and investigations of secretory functions.* PhD thesis, University of Adelaide.

Hermo, L. and Lalli, M. (1978). Monocytes and mast cells in the limiting membrane of human seminiferous tubules. *Biol. Reprod.* **19**, 92–100.

Hilscher, W. and Makoski, H. B. (1968). Histologische und autoradiographische Untersuchungen zur "Praspermatogenese" und "Spermatogenese" der Ratte. *Z. Zellforsch. mikrosk. Anat.* **86**, 327–350.

Hinton, B. T. and Keefer, D. A. (1983). Evidence for protein absorption from the lumen of the seminiferous tubule and rete of the rat testis. *Cell Tissue Res.* **230**, 367–375.

Hirst, J. J., Rice, G. E., Jenkin, G., and Thorburn, G. D. (1986). Secretion of oxytocin and progesterone by ovine corpora lutea *in vitro*. *Biol. Reprod.* **35**, 1106–1114.

Hobkirk, R. (1985). Steroid sulfotransferases and steroid sulfate sulfatases: character-istics and biological roles. *Can. J. Biochem. Cell Biol.* **63**, 1127–1144.

Hochereau-de Reviers, M. T. and Lincoln, G. A. (1978). Seasonal variation in the his-tology of the testis of the red deer, *Cervus elaphus. J. Reprod. Fert.* **54**, 209–213.

Hodgen, G. D. and Sherins, R. J. (1973). Enzymes as markers of testicular growth and development. *Endocrinology* **93**, 585–589.

Hodgson, Y. M. and de Kretser, D. M. (1982). Serum testosterone response to single injection of hCG ovine-LH and LHRH in male rats. *Int. J. Androl.* **5**, 81–91.

—— —— (1984). The temporal response of rat testes to hCG stimulation during sexual maturation. *Int. J. Androl.* **7**, 203–214.

—— Urquart, G., and de Kretser, D. M. (1983). Effect of oestradiol and tamoxifen on the testosterone response in male rats to a single injection of hCG. *J. Reprod. Fert.* **68**, 295–304.

Hooker, C. (1970). The intertubular tissue of the testis. In *The Testis* Vol. I (ed. A. D. Johnson, W. R. Gomes, and N. L. VanDemark) pp. 483–550. Academic Press, New York.

Hovatta, O., Huhtaniemi, I., and Wahlstrom, T. (1986). Testicular gonadotrophins and their receptors in human cryptorchidism as revealed by immunohisto-chemistry and radioreceptor assay. *Acta endocr. Copenh.* **111**, 128–132.

Hsueh, A. J. W. (1980). Gonadotropin stimulation of testosterone production in primary culture of adult rat testis cells. *Biochem. Biophys. Res. Commun.* **97**, 506–512.

—— Schreiber, J. R., and Erickson, G. F. (1981). Inhibitory effect of gonadotropin releasing hormone upon cultured testicular cells. *Molec. Cell. Endocr.* **21**, 43–49.

Huang, H. F. S. and Nieschlag, E. (1986). Suppression of the intratesticular testo-sterone is associated with quantitative changes in spermatogonial populations in intact adult rats. *Endocrinology* **118**, 619–627.

Hudson, B., Baker, H. W. G., Eddie, L. W., Higginson, R. E., Burger, H. G., de Kretser, D. M., Dobos, M., and Lee, V. W. K. (1979). Bioassay for inhibin: a critical review. *J. Reprod. Fert.* Suppl. **26**, 17–29.

Huggins, C. and Moulder, P. V. (1945). Estrogen production by Sertoli cell tumours of the testis. *Cancer Res.* **5**, 510–514.

Hurkaldi, K. S., Arbatti, N. J., Mehte, S., and Sheth, A. R. (1984). Serum inhibin levels after administration of hCG. *Arch. Androl.* **12**, 45–48.

Hutson, J. C. and Stocco, D. M. (1981). Peritubular cell influence on the efficiency of androgen-binding protein secretion by Sertoli cells in culture. *Endocrinology* **108**, 1362–1368.

Jacks, F. and Setchell, B. P. (1973). A technique for studying the transfer of substances from venous to arterial blood in the spermatic cord of wallabies and rams. *J. Physiol., Lond.* **233**, 17P–18P.

Jakobsen, H. and Juul, N. (1985). Influence of vasectomy on the volume of the non-hyperplastic prostate in men. *Int. J. Androl.* **8**, 13–20.

Jamieson, J. D. and Palade, G. E. (1967a). Intracellular transport of secretory proteins in the pancreatic exocrine cell. I. Role of the peripheral element of the Golgi complex. *J. Cell Biol.* **34**, 577–596.

——— (1967b). Intracellular transport of secretory proteins in the pancreatic exocrine cell II. Transport to condensing vacuoles and zymogen granules. *J. Cell Biol.* **34**, 597–615.

Janecki, A. and Steinberger, A. (1987). Bipolar secretion of androgen-binding protein and transferrin by Sertoli cells cultured in a two-compartment culture chamber. *Endocrinology* **120**, 291–298.

Jegou, B., Dacheux, J. L., Garnier, D. H., Terqui, M., Colas, G., and Courot, M. (1979). Biochemical and physiological studies of androgen-binding protein in the reproductive tract of the ram. *J. Reprod. Fert.* **57**, 311–318.

Johnson, L. and Thompson, D. L. (1983). Age-related and seasonal variation in the Sertoli cell population, daily sperm production and serum concentrations of follicle-stimulating hormone, luteinizing hormone and testosterone in stallions. *Biol. Reprod.* **29**, 777–789.

Josso, N. and Picard, J.-Y. (1986). Anti-Mullerian hormone. *Physiol. Rev.* **66**, 1038–1090.

Kerr, J. B. and Sharpe, R. M. (1985). Stimulatory effect of FSH on rat Leydig cells. A morphologic and ultrastructural study. *Cell Tissue Res.* **239**, 405–415.

—— Donachie, K., and Rommerts, F. F. G. (1985). Selective destruction and re-generation of rat Leydig cells *in vivo*: a new method for the study of seminiferous tubular-interstitial tissue interaction. *Cell Tissue Res.* **242**, 145–156.

—— Robertson, D. M., and de Kretser, D. M. (1985). Morphological and functional characterization of interstitial cells from mouse testes fractionated on Percoll density gradients. *Endocrinology* **116**, 1030–1043.

Ketelslegers, J. M., Hetzel, W. D., Sherins, R. J., and Catt, K. J. (1978). Developmental changes in testicular gonadotropin receptors: plasma gonadotropin and plasma testosterone in the rat. *Endocrinology* **103**, 212–222.

Kissinger, C., Skinner, M. K., and Griswold, M. D. (1982). Analysis of Sertoli cell-secreted proteins by two-dimensional gel electrophoresis. *Biol. Reprod.* **27**, 233–240.

Knight, T. W. (1974). A qualitative study of factors affecting the contractions of the epididymis and ductus deferens of the ram. *J. Reprod. Fert.* **40**, 19–29.

Knorr, D. W., Vanha-Pertula, T., and Lipsett, M. B. (1970). Structure and function of the rat testis through pubescence. *Endocrinology* **86**, 1298–1304.

Kormano, M. (1967a). An angiographic study of the testicular vasculature in the postnatal rat. *Z. Anat. Entw. Gesch.* **126**, 138–153.

—— (1967b). Dye permeability and alkaline phosphatase activity of testicular capillaries in the postnatal rat. *Histochemie* **9**, 327–338.

Koskimies, A. I. and Kormano, M. (1973). The proteins in fluids from the seminiferous tubules and rete testis of the rat. *J. Reprod. Fert.* **34**, 433–444.

Krishna, A. and Terranova, P. F. (1985). Alterations in mast cell degranulation and ovarian histamine in the proestrous hamster. *Biol. Reprod.* **32**, 1211–1217.

Lacroix, M., Smith, F. E., and Fritz, I. B. (1977). Secretion of plasminogen activator by Sertoli cell-enriched cultures. *Molec. Cell. Endocr.* **9**, 227–236.

Lacy, D. (1962). Certain aspects of testis structure and function. *Br. med. Bull.* **18**, 205–208.

—— (1967). The seminiferous tubule in mammals. *Endeavour* **26**, 101–109.

—— (1973). Androgen dependency of spermatogenesis and the physiological significance of steroid metabolism *in vitro* by the seminiferous tubules. In *Endocrine Function of the Human Testis* Vol. 1 (ed. V. H. T. James, M. Serio, and L. Martini) pp. 493–532. Academic Press, New York.

Laurie, M. S. and Setchell, B. P. (1978). The continuous measurement of testicular blood flow in the ram, in relation to the pulsatile secretion of testosterone. *J. Physiol., Lond.* **287**, 10P.

Laws, A. O. (1985). *The effects of hypophysectomy and subsequent hCG treatment on Leydig cell structure and function.* PhD thesis. Monash University, Melbourne.

Le Gac, F. and de Kretser, D. M. (1982). Inhibin production by Sertoli cell cultures. *Molec. Cell. Endocr.* **28**, 487–498.

Le Lannou, D., Chambon, Y., and Le Calve, M. (1979). Role of the epididymis in reabsorption of inhibin in the rat. *J. Reprod. Fert. Suppl.* **26**, 117–121.

Lee, B. C., Pineda, J. L., Spiliotis, B. E., Brown, T. S., and Bercu, B. B. (1983). Male sexual development in the nonhuman primate. III. Sertoli cell culture and age-related differences. *Biol. Reprod.* **28**, 1207–1215.

Leinonen, P., Ruokonen, A., Kontturi, M., and Vikho, R. (1981). Effects of oestrogen treatment on human testicular unconjugated steroid and steroid sulfate production *in vivo. J. clin. Endocr. Metab.* **53**, 569–573.

Leydig, F. (1850). Zur Anatomie der männlichen Geschlechtsorgane und Analdrusen der Saugetiere. *Z. Wiss. Zool.* **2**, 1–57.

Lindner, H. R. (1963). Partition of androgen between lymph and venous blood of the testis in the ram. *J. Endocr.* **25**, 483–494.

—— (1967). Participation of lymph in the transport of gonadal hormones. *Excerpta Medica Int. Congr. Ser.* **132**, 821–827.

—— (1969). The androgenic secretion of the testis in domestic ungulates. In *The Gonads* (ed. K. W. McKerns) pp. 615–648. Appleton-Century-Crofts, New York.

Ling, N., Ying, S.-Y., Ueno, N., Shimasaki, S., Esch, F., Hotta, M., and Guilleman, R. (1986). Pituitary FSH is released by a heterodimer of the B-subunits from the two forms of inhibin. *Nature, Lond.* **321**, 779–782.

Linzell, J. L. and Peaker, M. (1971). Mechanism of milk secretion. *Physiol. Rev.* **51**, 564–597.

Loisel, G. (1903). Les graisses du testicule chez quelques mammifères. *C. r. Soc. Biol.* **55**, 1009–1012.

Lu, C. and Steinberger, A. (1977). Gamma-glutamyl transpeptidase activity in the developing rat testis. Enzyme localization in isolated cell types. *Biol. Reprod.* **17**, 84–88.

Maddocks, S. and Setchell, B. P. (1986). The composition of the interstitial fluid of the rat testis. *Miniposters of the IV European workshop on molecular and cellular endocrinology of the testis*, 154.

Maddocks, S. and Setchell, B. P. (1987*a*). Testosterone concentrations in testicular interstitial fluid collected with a push-pull cannula or by drip-collection, from adult rats with testosterone implants. *Proc. Endocr. Soc. Australia* **30**, 112.

—— —— (1987*b*). The composition of extracellular interstitial fluid in the rat testis. *J. Physiol, Lond.* in press.

—— Sowerbutts, S. F., and Setchell, B. P. (1987). Effects of repeated injections of human chorionic gonadotrophin on vascular permeability, extracellular fluid volume and the flow of lymph in the testes of rats. *Int. J. Androl.* **10**, 535–542.

—— Zupp, J. L., Sowerbutts, S. F., and Setchell, B. P. (1986). Testosterone levels in rat testicular interstitial fluid. *Proc. Aust. Soc. reprod. Biol.* **18**, 58.

—— —— —— —— (1987). Testosterone concentrations in testicular interstitial fluid and venous blood of adult rats, following local heating of the testes. *Proc. Aust. Soc. reprod. Biol.* **19**, 85.

Main, S. J. and Setchell, B. P. (1978). The facilitated diffusion of testosterone into the rete testis of the ram. *J. Physiol., Lond.* **284**, 17P–18P.

—— —— (1980). Responsiveness of the pituitary gland to androgens and of the testis to gonadotrophins following damage to spermatogenesis in rats. *J. Endocr.* **87**, 445–454.

—— Davies, R. M., and Setchell, B. P. (1978). Feedback control by the testis of gonadotrophin secretion: an examination of the inhibin hypothesis. *J. Endocr.* **79**, 255–270.

Mancini, R. E., Vilar, O., Alvarez, B., and Seiguer, A. C. (1965). Extravascular and intratubular diffusion of labelled serum proteins in the rat testis. *J. Histochem.* **13**, 376–385.

Margioris, A. N., Liotta, A. S., Vaudry, H., Bardin, C. W., and Krieger, T. (1983). Characterisation of immunoreactive proopiomelanocortin-related peptides in rat testes. *Endocrinology* **113**, 663–671.

Marshall, G. R., Wickings, E. J., Ludecke, D. K., and Nieschlag, E. (1984). Stimulation of spermatogenesis in stalk-sectioned Rhesus monkeys by testosterone alone. *J. clin. Endocr. Metab.* **57**, 152–159.

Maseki, Y., Miyake, K., Mitsuya, H., Kitamura, H., and Yamada, K. (1981). Mastocytosis occurring in the testes from patients with ideopathic male infertility. *Fert. Steril.* **36**, 814–817.

Mather, J. P. (1980). Establishment and characterization of two distinct mouse testicular epithelial cell lines. *Biol. Reprod.* **23**, 243–251.

—— Zhuang, L.-Z., Perez-Infante, V., and Phillips, D. M. (1982). Culture of testicular cells in hormone-supplemented serum-free medium. *Ann. N.Y. Acad. Sci.* **383**, 44–68.

McCullough, D. L. (1975). Experimental lymphangiography—experience with direct medium injection into parenchyma of the rat testis and prostate. *Invest. Urol.* **13**, 211–219.

McCullough, D. R. (1932). Dual endocrine activity of the testis. *Science, N.Y.* **76**, 19–20.

McIntosh, G. H. (1969). *Lymphatics of the urogenital system of the sheep.* PhD thesis. Australian National University, Canberra.

Means, A. R., Dedman, J. R., Tindall, D. J., and Welsh, M. J. (1978). Hormonal regulation of Sertoli cells. *Int. J. Androl. Suppl.* **2**, 403–421.

Mendelson, C., Dufau, M., and Catt, K. (1975). Gonadotrophin binding and stimulation of cyclic adenosine $3':5'$-monophosphate and testosterone production in isolated Leydig cells. *J. biol. Chem.* **250**, 8818–8823.

Milewich, L., Chen, G. T., Lyons, C., Tucker, Th.F., Uhr, J. W., and McDonald, P. C. (1982). Metabolism of androstenedione by guinea-pig peritoneal macrophages: synthesis of testosterone and 5α-reduced metabolites. *J. steroid Biochem.* **17**, 61–65.

Miller, S. C., Bowman, B. M., and Rowland, H. G. (1983). Structure, cytochemistry, endocytic activity and immunoglobulin (Fc) receptors of rat testicular interstitial tissue macrophages. *Am. J. Anat.* **168**, 1–13.

Molenaar, R., Rommerts, F. F. G., and Van der Molen, H. J. (1983). The steroidogenic activity of isolated Leydig cells from mature rats depends on the isolation procedure. *Int. J. Androl.* **6**, 261–274.

————— (1984). Characteristics of Leydig cells and macrophages from developing testicular cells. *Ann. N.Y. Acad. Sci.* **438**, 618–621.

Mori, H. and Christensen, A. K. (1980). Morphometric analysis of Leydig cells in the normal rat testis. *J. Cell Biol.* **84**, 349–354.

Morris, B. and McIntosh, G. H. (1971). Techniques for the collection of lymph with special reference to the testis and ovary. In *Perfusion Techniques* (ed. E. Diczfalusy) pp. 145–165. *Fourth Karolinska Symposia on Research Methods in Reproductive Endocrinology*. Karolinska Institute, Stockholm.

Muffly, K. E., Turner, T. T., Brown, M., and Hall, P. F. (1985). Content of K^+ and Na^+ in seminiferous tubule and rete testis fluids from Sertoli cell-enriched testes. *Biol. Reprod.* **33**, 1245–1251.

Murphy, H. D. (1965). Sertoli cell stimulation following intratesticular injections of FSH in the hypophysectomised rat. *Proc. Soc. exp. Biol. Med.* **118**, 1202–1205.

Myers, R. D. (1972). Methods for perfusing different structures of the brain. In *Methods in Psychobiology* (ed. R. D. Myers) pp. 169–211. Academic Press, London.

Nagendranath, N., Jose, T. M., Sheth, A. R., and Juneja, H. S. (1982). Effect of testosterone, estradiol-17β and 5α-dihydrotestosterone on inhibin production by Sertoli cells in culture. *Arch. Androl.* **9**, 217–222.

Nagy, F. (1972). Cell division kinetics and DNA synthesis in the immature Sertoli cells of the rat testis. *J. Reprod. Fert.* **38**, 389–395.

Neaves, W. B. (1973). Changes in testicular Leydig cells and in plasma testosterone levels among seasonally breeding rock hyrax. *Biol. Reprod.* **8**, 451–466.

—— (1975). Leydig cells. *Contraception* **11**, 571–606.

Nebel, B. R. and Murphy, C. J. (1960). Damage and recovery of mouse testis after 1000r acute localised x-irradiation, with reference to restitution cells, Sertoli cell increase, and type A spermatogonial recovery. *Radiat. Res.* **12**, 626–641.

Nicholson, H. D., Swann, R. W., Burford, G. D., Wathes, D. C., Porter, D. G., and Pickering, B. T. (1984). Identification of oxytocin and vasopressin in the testis and in adrenal tissue. *Regulatory Peptides* **8**, 141–146.

Niemi, M. and Ikonen, M. (1973). Histochemistry of the Leydig cells in the postnatal prepubertal testis. *Endocrinology* **72**, 443–448.

—— and Kormano, M. (1965). Contractility of the seminiferous tubule of the postnatal rat testis and its response to oxytocin. *Ann. Med. Exp. Fenn.* **43**, 40–42.

—— and Setchell, B. P. (1986). Gamma-glutamyl transpeptidase in the vasculature of the rat testis. *Biol. Reprod.* **35**, 385–391.

—— Sharpe, R. M., and Brown, W. R. (1986). Macrophages in the interstitial tissue of the rat testis. *Cell Tissue Res.* **243**, 337–344.

Nistal, M., Paniagua, R., Abaurra, M. A., and Pallardo, L. F. (1980). Kleinfelters syndrome with low FSH and LH levels and absence of Leydig cells. *Andrologia* **12**, 426–433.

—— Santamaria, L., and Paniagua, R. (1984). Mast cells in the human testis and epididymis from birth to adulthood. *Acta Anat.* **119**, 155–160.

Nykanen, M. (1980). Loose connective tissue of rat rete testis. *Cell Tissue Res.* **206**, 501–504.

Olson, G. E. and Hinton, B. T. (1985). Regional differences in luminal fluid polypeptides of the rat testis and epididymis revealed by two-dimensional gel electrophoresis. *J. Androl.* **6**, 20–34.

Orci, L. (1986). Structure-function correlates of polypeptide hormone secretion. In *Proceedings of the XXXth meeting of the International Union of Physiological Sciences*, Vancouver, BC, p. 1. Abstr. L102.01.

Orth, J. (1982). Proliferation of Sertoli cells in fetal and postnatal rats: a quantitative autoradiographic study. *Anat. Rec.* **203**, 485–492.

—— (1984). The role of follicle-stimulating hormone in controlling Sertoli cell proliferation in testes of fetal rats. *Endocrinology* **115**, 1248–1255.

Pande, J. K., Chowdhury, S. R., Dasgupta, P. R., Chowdhury, A. R., and Kar, A. B. (1966). Biochemical composition of the rat testis fluid. *Proc. Soc. exp. Biol. Med.* **121**, 899–902.

—— Dasgupta, P. R., and Kar, A. B. (1967). Biochemical composition of human testicular fluid collected post mortem. *J. clin. Endocr. Metab.* **27**, 892–894.

Pandey, K. N., Melner, M. H., Parmentier, M., and Inagami, T. (1984). Demonstration of renin activity in purified rat Leydig cells: evidence for the existence of an endogenous inactive (latent) form of enzyme. *Endocrinology* **115**, 1753–1759.

Parmentier, M., Inagami, T., Pochet, R., and Desclin, J. C. (1983). Pituitary-dependant renin-like immunoreactivity in the rat testis. *Endocrinology* **112**, 1318–1323.

Payne, A. H. and Valadares, L. E. (1980). Regulation of testicular aromatization by luteinizing hormone. In *Testicular Development, Structure and Function* (ed. A. Steinberger and E. Steinberger) pp. 185–193. Raven Press, New York.

—— Downing, J. R., and Wong, K. L. (1980). Luteinizing hormone receptors and testosterone synthesis in two distinct populations of Leydig cells. *Endocrinology* **106**, 1424–1429.

—— O'Shaugnessy, P. J., Chase, D. J., Dixon, G. E. K., and Christensen, A. K. (1982). LH receptors and steroidogenesis in distinct populations of Leydig cells. *Ann. N.Y. Acad. Sci.* **383**, 174–203.

Pelliniemi, L. J., Dym, M., Gunsalus, G. L., Musto, N. A., Bardin, C. W., and Fawcett, D. W. (1981). Immunocytochemical localization of androgen-binding protein in the male rat reproductive tract. *Endocrinology* **108**, 925–931.

—— and Niemi, M. (1969). Fine structure of the human foetal testis I: The interstitial tissue. *Z. Zellforsch*, **99**, 507–522.

Perrard, M. H., Saez, J. M., and Dazord, A. (1985). FSH stimulation of cytosolic protein synthesis in cultured pig Sertoli cells. *J. steroid Biochem*, **22**, 281–284.

Perrard-Saperi, M. H., Chatelain, P., Vallier, P., and Saez, J. M. (1986). *In vitro* interactions between Sertoli cells and steroidogenic cells. *Biochem. Biophys. Res. Commun.* **134**, 957–962.

Pierrepoint, C. G., Davies, P., Millington, D., and John, B. (1975). Evidence that the deferential vein acts as a local transport system for androgen in the rat and the dog. *J. Reprod. Fert.* **43**, 293–303.

Pollanen, P. P., Kallajocki, M., Ristelli, L., Ristelli, J., and Suominen, J. J. O. (1985). Laminen and type IV collagen in the human testis. *Int. J. Androl.* **8**, 337–347.

—— and Maddocks, S. (1988). Macrophages, lymphocytes and MHC II antigen in the ram and the rat testis. *J. Reprod. Fert.* **82**, 437–445.

Prince, F. P. (1984). Ultrastructure of immature Leydig cells in the human prepubertal testis. *Anat. Rec.* **209**, 165–176.

Rea, M. A., Marshall, G. R., Weinbauer, G. F., and Nieschlag, E. (1986). Testosterone maintains pituitary and serum FSH and spermatogenesis in gonadotrophin-releasing hormone antagonist-suppressed rats. *J. Endocr.* **108**, 101–107.

Reese, T. S. and Karnovsky, M. J. (1967). Fine structure localisation of the blood-brain barrier to exogenous peroxidase. *J. Cell. Biol.* **84**, 207–217.

Rice, G. E. and Thorburn, G. D. (1985). Subcellular localization of oxytocin in the ovine corpus luteum. *Can. J. Physiol. Pharmacol.* **63**, 309–314.

Riley, J. F. and West, G. B. (1953). The presence of histamine in the tissue mast cells. *J. Physiol., Lond.* **120**, 528–537.

Risbridger, G. P., Kerr, J. B., and de Kretser, D. M. (1981). Evaluation of Leydig cell function and gonadotrophin binding in unilateral and bilateral cryptorchidism: evidence for local control of Leydig cell function by the seminiferous tubule. *Biol. Reprod.* **24**, 534–540.

—————— Peake, R. A., and de Kretser, D. M. (1981). An assessment of Leydig cell function after bilateral or unilateral efferent duct ligation: further evidence for local control of Leydig cell function. *Endocrinology* **109**, 1234–1241.

Rommerts, F. F. G., Kruger-Sewnarain, B.Ch., Van Woerkom-Blik, A., Grootegoed, J. A., and Van der Molen, H. J. (1978). Secretion of proteins by Sertoli cell enriched cultures: effects of follicle stimulating hormone, dibutyryl cAMP and testosterone and correlation with secretion of oestradiol and androgen binding protein. *Molec. Cell Endocr.* **10**, 39–55.

Ross, M. H. (1967). The fine structure and development of the peritubular contractile cell component in the seminiferous tubules of the mouse. *Am. J. Anat.* **121**, 523–557.

Ruokonen, A., Lukkarinen, O. and Vihko, R. (1981). Secretion of steroid sulphates from human testis and their response to a single intramuscular injection of 5000 IU hCG. *J. steroid Biochem.* **14**, 1357–1360.

Russell, L. D. (1980). Sertoli-germ cell interrelations: A review. *Gamete Res.* **3**, 179–202.

Sairam, M. R. (1979). Effects of prostaglandins on the action of luteinizing hormone in dispersed rat interstitial cells. *Prostaglandins* **17**, 929–937.

Sanborne, B. M., Wagle, J. R., Steinberger, A., and Greer-Emmert, D. (1986). Maturational and hormonal influences on Sertoli cell function. *Endocrinology* **118**, 1700–1709.

Schayer, R. W. (1956). Formation and binding of histamine by free mast cells of rat peritoneal fluid. *Am. J. Physiol.* **186**, 199–206.

Schoefl, G. I. and French, J. E. (1968). Vascular permeability to particulate fat: morphological observations on vessels of lactating mammary gland and of lung. *Proc. R. Soc. B.* **169**, 153–165.

Schulze, C. (1984). Sertoli cells and Leydig cells in man. *Adv. Anat. Embryol. cell. Biol.* **88**, 1–104.

Schumacher, M., Schafer, G., Holstein, A. F., and Hilz, H. (1978). Rapid isolation of mouse Leydig cells by centrifugation in Percoll density gradients with complete retention of morphological and biochemical integrity. *FEBS Lett.* **91**, 333–338.

Sertoli, E. (1865). Dell'esistenza di particolari cellule ramificate nei canaliculi seminiferi del testicolo umano. *Morgagni* **7**, 31–40. (see Setchell 1984 for facsimile)

Setchell, B. P. (1970). Testicular blood supply, lymphatic drainage and secretion of fluid. In *The Testis* Vol. 1 (ed. A. D. Johnson, W. R. Gomes, and N. L. VanDemark) pp. 101–239. Academic Press, New York.

—— (1970*b*). The secretion of fluid by the testes of rats, rams and goats with some observations on the effect of age, cryptorchidism and hypophysectomy. *J. Reprod. Fert.* **23**, 79–85.

—— (1978). *The Mammalian Testis.* Elek Books, London.

—— (1980). Inhibin. In *Animal Models in Human Reproduction* (ed. M. Serio and L. Martini) pp. 135–147. Raven Press, New York.

—— (1982). The flow and composition of lymph from the testes of pigs with some observations on the effect of raised venous pressure. *Comp. Biochem. Physiol.* **73A**, 201–205.

—— (1984). *Male Reproduction.* Benchmark papers in human physiology series. Number 17. Van Nostrand Reinhold Company, New York.

—— (1986). The movement of fluids and substances in the testis. *Aust. J. biol. Sci.* **39**, 193–207.

—— (1987). Control of testicular function. In *Proceedings of the 1st Congress of the Asian and Oceanian Physiological Societies*, in press.

—— and Brooks, D. E. (1988). Anatomy, vasculature, innervation and fluids of the male reproductive tract. In *The Physiology of Reproduction* (ed. E. Knobil and J. D. Neill) in press. Raven Press, New York.

—— and Cox, J. E. (1982). Secretion of free and conjugated steroids by the horse testis into lymph and venous blood. *J. Reprod. Fert. Suppl.* **32**, 123–127.

—— Davies, R. V., Gladwell, R. T., Hinton, B. T., Main, S. J., Pilsworth, L., and Waites, G. M. H. (1978). The movement of fluid in the seminiferous tubules and rete testis. *Annls. Biol. anim. Biochim, Biophys.* **18**, 623–632.

—— and Gailil, K. A. A. (1983). Limitations imposed by testicular blood flow on the function of the Leydig cells *in vivo. Aust. J. Biol. Sci.* **36**, 285–293.

—— Hinks, N. T., Voglmayr, J. K., and Scott, T. W. (1967). Amino acids in ram testicular fluids and semen and their metabolism by spermatozoa. *Biochem. J.* **105**, 1061–1065.

—— and Hinton, B. T. (1981). The effects on the spermatozoa of changes in the composition of luminal fluid as it passes along the epididymis. *Prog. reprod. Biol.* **8**, 58–66.

—— Hinton, B. T., Jacks, F., and Davies, R. V. (1976). Restricted penetration of iodinated follicle-stimulating and luteinizing hormones into the seminiferous tubules of the rat testis. *Molec. Cell. Endocr.* **6**, 59–69.

—— Laurie, M. S., Flint, A. P. F., and Heap, R. B. (1983). Transport of free and conjugated steroids from the boar testis in lymph, venous blood and rete testis fluid. *J. Endocr.* **96**, 127–136.

—— —— and Jarvis, L. G. (1981). The blood-testis barrier at puberty. In *Development and Formation of Reproductive Organs* (ed. A. G. Byskov and H. Peters) pp. 186–190. Excerpta Medica Int. Cong. Ser. No. 559.

—— —— Main, S. J., and Goats, G. C. (1978). The mechanism of transport of testosterone through the walls of the seminiferous tubules of the rat testis. *Int. J. Androl. Suppl.* **2**, 506–512.

—— and Main, S. J. (1974). Bibliography (with review) on inhibin. *Bibl. Reprod.*, 245–252 and 361–367.

—— —— (1975). The blood-testis barrier and steroids. In *Hormonal regulation of spermatogenesis* (ed. F. S. French, V. Hansson, E. M. Ritzen, and S. N. Nayfeh) pp. 223–233. Plenum Press, New York.

—— —— and Davies, R. V. (1977). The effect of ligation of the efferent ducts of the testis on serum gonadotrophins and testosterone in rats. *J. Endocr.* **72**, 13P–14P.

—— Pollanen, P., and Zupp, J. L. (1987). The development of the blood-testis barrier and changes in vascular permeability at puberty in rats. *Int. J. Androl. In press.*

—— and Rommerts, F. F. G. (1985). The importance of the Leydig cells in the vascular response to hCG in the rat testis. *Int. J. Androl.* **8**, 436–440.

—— and Sharpe, R. M. (1981). Effect of injected human chorionic gonadotrophin on capillary permeability, extracellular fluid volume and the flow of lymph and blood in the testes of rats. *J. Endocr.* **91**, 245–254.

—— Voglmayr, J. K., and Waites, G. M. H. (1969). A blood-testis barrier restricting passage from blood into rete testis fluid but not into lymph. *J. Physiol., Lond.* **200**, 73–85.

—— and Waites, G. M. H. (1969). Pulse attenuation and counter-current heat exchange in the internal spermatic artery of some Australian marsupials. *J. Reprod. Fert.* **20**, 165–169.

—— —— (1970). Changes in permeability of testicular capillaries and of the "blood-testis barrier" after injection of cadmium chloride in the rat. *J. Endocr.* **41**, 81–86.

—— and Waites, G. M. H. (1975). The blood testis barrier. In *Handbook of Physiology* Sect. 7: *Endocrinology* Vol. **5** *Male Reproductive System* (ed. D. W. Hamilton and R. O. Greep) pp. 143–172. American Physiological Society, Washington.

—— and Zupp, J. L. (1988). Enhanced dissociation in the interstitial tissue of the rat testis of testosterone bound to albumin, sex-hormone-binding globulin (SHBG) or androgen-binding protein (ABP). *J. Physiol. in press.*

Shaha, C., Liotta, A. S., Krieger, D. T., and Bardin, C. W. (1984). The ontogeny of immunoreactive B-endorphin in fetal, neonatal, and pubertal testes from mouse and hamster. *Endocrinology* **114**, 1584–1591.

Sharpe, R. M. (1979). Gonadotrophin induced accumulation of "interstitial fluid" in the rat testis. *J. Reprod. Fert.* **55**, 365–371.

—— (1980). Temporal relationship between interstitial fluid accumulation and changes in gonadotrophin receptor numbers and steroidogenesis in the rat testis. *Biol. Reprod.* **22**, 851–857.

—— (1981). The importance of testicular interstitial fluid in the transport of injected hCG to the Leydig cells. *Int. J. Androl.* **4**, 64–74.

—— (1982). The hormonal regulation of the Leydig cell. In *Oxford Reviews of Reproductive Biology* Vol. 4 (ed. C. A. Finn) pp. 241–317. Clarendon Press, Oxford.

—— (1983). Local control of testicular function. *Q. Jl. exp. Physiol.* **68**, 265–287.

—— (1984). Intragonadal hormones: bibliography with review. *Bibl. Reprod.* **44**, C1–C15.

—— (1986). Paracrine control of the testis. *Clin. Endocr. Metab.* **15**, 185–207.

—— (1987). Testosterone and spermatogenesis. *J. Endocr.* **113**, 1–2.

—— and Cooper, I. (1983). Testicular interstitial fluid as a monitor for changes in the intratesticular environment in the rat. *J. Reprod. Fert.* **69**, 125–135.

—— and Fraser, H. M. (1980). Leydig cell receptors for luteinizing hormone-releasing hormone and its agonists and their modulation by administration or deprivation of the releasing hormone. *Biochem. Biophys. Res. Commun.* **95**, 256–262.

—— —— Cooper, I., and Rommerts, F. F. G. (1981). Sertoli-Leydig cell communication via a LHRH-like factor. *Nature, Lond.* **290**, 785–787.

—— —— —— —— (1982). The secretion, measurement and function of a testicular LHRH-like factor. *Ann. N.Y. Acad. Sci.* **383**, 272–294.

—— Doogan, D. G., and Cooper, I. (1983). Direct effects of a luteinizing hormone-releasing hormone agonist on intratesticular levels of testosterone and interstitial fluid formation in intact male rats. *Endocrinology* **113**, 1306–1313.

Shaw, M. J., Georgopoulos, L. E., and Payne, A. H. (1979). Synergistic effect of follicle-stimulating hormone and luteinizing hormone on testicular $\Delta 5$-3β-hydroxysteroid dehydrogenase-isomerase: application of a new method for the separation of testicular compartments. *Endocrinology* **104**, 912–918.

120   S. Maddocks and B. P. Setchell

Siiteri, P. K., Murai, J. T., Hammond, G. L., Nisker, J. A., Raymoure, W. J., and Kuhn, R. W. (1982). The serum transport of steroid hormones. *Recent Prog. Horm. Res.* **38**, 457–510.

Skinner, J. D. and Rowson, L. E. A. (1968*a*). Some effects of unilateral cryptorchidism and vasectomy on sexual development of the pubescent ram and bull. *J. Endocr.* **42**, 311–321.

—— —— (1968*b*). Effects of testosterone injected unilaterally down the vas deferens on the accessory glands of the ram. *J. Endocr.* **42**, 355–356.

Skinner, M. K. and Fritz, I. B. (1985*a*). Testicular peritubular cells secrete a protein under androgen control that modulates Sertoli cell functions. *Proc. natn. Acad. Sci. U.S.A.* **82**, 114–118.

—— —— (1985*b*). Androgen stimulation of Sertoli cell function is enhanced by peritubular cells. *Molec. Cell. Endocr.* **40**, 115–122.

—— —— (1986). Identification of a non-mitogenic paracrine factor involved in mesenchymal-epithelial cell interactions between testicular peritubular cells and Sertoli cells. *Molec. Cell. Endocr.* **44**, 85–97.

—— and Griswold, M. D. (1982). Secretion of testicular transferrin by cultured Sertoli cells is regulated by hormones and retinoids. *Biol. Reprod.* **27**, 211–221.

—— Tung, P. S., and Fritz, I. B. (1985). Co-operativity between Sertoli cells and testicular peritubular cells in the production and deposition of extracellular matrix components. *J. Cell. Biol.* **100**, 1941–1947.

Smith, G. (1962). The effects of ligation of the vasa efferentia and vasectomy on testicular function in the adult rat. *J. Endocr.* **23**, 385–399.

Sowerbutts, S. F., Jarvis, L. G., and Setchell, B. P. (1986). The increase in testicular vascular permeability induced by human chorionic gonadotrophin involves 5-hydroxytryptamine and possibly oestrogens, but not testosterone, prostaglandins, histamine or bradykinin. *Aust. J. exp. Biol. med. Sci.* **64**, 137–147.

Steinberger, A. (1983). Testicular Inhibin. *Sem. Reprod. Endocr.* **1**, 357–364.

—— Heindel, J. J., Lindsey, J. N., Elkington, J. S. H., Sanborn, B. M. and Steinberger, E. (1975). Isolation and culture of FSH responsive Sertoli cells. *Endocr. Res. Commun.* **2**, 261–272.

—— and Steinberger, E. (1971). Replication pattern of Sertoli cells in maturing rat testis *in vivo* and in organ culture. *Biol. Reprod.* **4**, 84–87.

—— —— (1977). The Sertoli cells. In *The Testis* Vol. IV (ed. A. D. Johnson and W. R. Gomes) pp. 371–399. Academic Press, New York.

Suarez-Quian, C. A., Hadley, M. A., and Dym, M. (1984). Effect of substrate on the shape of Sertoli cells *in vitro*. *Ann. N.Y. Acad. Sci.* **438**, 417–434.

Sullivan, M. H. F. and Cooke, B. A. (1984). The role of calcium in luteinizing hormone-releasing hormone agonist (ICI 118630)-stimulated steroidogenesis in rat Leydig cells. *Biochem. J.* **218**, 621–624.

Suzuki, Y., Toshimori, Y., and Mochizuki, K. (1955). Possibility of direct transportation of testicular hormone to some target organs located close to the testicles in the postnatal rat and its physiologic significance. *Endocrinology* **56**, 347–358.

Sylvester, S. R. and Griswold, M. D. (1984). Localization of transferrin and transferrin receptors in rat testes. *Biol. Reprod.* **31**, 195–203.

Szerb, J. C. (1967). Model experiments with Gaddum's push-pull cannulas. *Can. J. Physiol. Pharmacol.* **45**, 613–620.

Tähkä, K. M. (1986). Current aspects of Leydig cell functions and its regulation. *J. Reprod. Fert.* **78**, 367–380.

Tapanainen, J., Kuopio, T., Pelliniemi, L. J., and Huhtaniemi, I. (1984). Rat testicular endogenous steroids and number of Leydig cells between the fetal period and sexual maturity. *Biol. Reprod.* **31**, 1027–1035.

Theodosis, D. T., Wooding, F. B. P., Sheldrick, E. L., and Flint, A. P. F. (1986). Ultrastructural localization of oxytocin and neurophysin in the ovine corpus luteum. *Cell Tissue Res.* **243**, 129–135.

Tilney, N. L. (1971). Patterns of lymphatic drainage in the adult laboratory rat. *J. Anat.* **109**, 369–383.

Tsonis, C. G. and Sharpe, R. M. (1986). Dual gonadal control of follicle-stimulating hormone. *Nature, Lond.* **321**, 724–725.

Tuck, R. R., Setchell, B. P., Waites, G. M. H., and Young, J. A. (1970). The composition of fluid collected by micropuncture and catheterization from the seminiferous tubules and rete testis of rats. *Pflügers Arch. Europ. J. Physiol.* **318**, 225–243.

Tung, P. S. and Fritz, I. B. (1977). Isolation and culture of testicular cells: A morphological characterization. In *Techniques of Human Andrology* (ed. E. S. E. Hafez) pp. 125–146. North Holland, Amsterdam.

—— —— (1980). Interactions of Sertoli cells with myoid cells *in vitro*. *Biol. Reprod.* **23**, 207–217.

—— —— (1986). Extracellular matrix components and testicular peritubular cells influence the rate and pattern of Sertoli cell migration *in vitro*. *Dev. Biol.* **113**, 119–134.

—— Skinner, M. K., and Fritz, I. B. (1984). Fibronectin synthesis is a marker for peritubular cell contamination in Sertoli cell-enriched cultures. *Biol. Reprod.* **30**, 199–211.

—— —— —— (1985). Co-operativity between Sertoli cells and peritubular myoid cells in the formation of the basal lamina in the seminiferous tubule. *Ann. N.Y. Acad. Sci.* **438**, 435–446.

Turner, T. T., Jones, C. E., Howards, S. S., Ewing, L. L., Zegeye, B., and Gunsalus, G. L. (1984). On the androgen microenvironment of maturing spermatozoa. *Endocrinology* **115**, 1925–1932.

—— Ewing, L. L., Jones, C. E., Howards, S. S., and Zegeye, B. (1985). Androgen in male rat reproductive tract fluids: hypophysectomy and steroid replacement. *Am. J. Physiol.* **248**, E274–E280.

Vale, W., Rivier, J., Vaughan, J., McClintock, R., Corrigan, A., Woo, W., Karr, D., and Spiess, J. (1986). Purification and characterisation of an FSH releasing protein from porcine ovarian follicular fluid. *Nature, Lond.* **321**, 776–779.

Van Stratten, H. W. M. and Wensing, C. J. G. (1978). Leydig cell development in the testis of the pig. *Biol. Reprod.* **18**, 86–93.

Van Vorstenborsch, C. J. A. H. V., Colenbrander, B., and Wensing, C. J. G. (1984). Leydig cell development in the pig testis during the late fetal and early post natal period: an electron microscope study with attention to the influence of fetal decapitation. *Am. J. Anat.* **169**, 121–136.

Vazeille, M. C. and Chevalier, J. (1979). Cellules de Sertoli du porc impubère: technique d'optention de cellules isolées. *C. r. Soc. Biol.* **173**, 1064–1069.

Veijola, M. and Rajaniemi, H. (1985). The hCG-induced increase in hormone uptake and interstitial fluid volume in the rat testis is not mediated by steroids, prostaglandins or protein synthesis. *Int. J. Androl.* **8**, 69–79.

Vigieri, B., Picard, J.-Y., Campargue, J., Forest, M. G., Heyman, Y., and Josso, N. (1985). Secretion of anti-Mullerian hormone by immature bovine Sertoli cells in primary culture, studied by a competition-type radioimmunoassay: lack of modulation by either FSH or testosterone. *Molec. Cell. Endocr.* **43**, 141–150.

Vitale-Calpe, R., Fawcett, D. W., and Dym, M. (1973). The normal development of the blood-testis barrier and the effects of clomiphene and estrogen treatment. *Anat. Rec.* **176**, 333–344.

Voglmayr, J. K., Roberson, C., and Musto, N. A. (1980). Comparison of androgen

122   S. Maddocks and B. P. Setchell

levels in ram rete testis fluid, testicular lymph and spermatic venous blood plasma: evidence for a regulatory mechanism in the seminiferous tubules. *Biol. Reprod.* **23**, 29–39.

Wahlstrom, T., Huhtaniemi, I., Hovatta, O., and Seppala, M. (1983). Localisation of luteinizing hormone, follicle stimulating hormone, prolactin, and their receptors in human and rat testis using immunohistochemistry and radioreceptor assay. *J. clin. Endocr. Metab.* **57**, 825–830.

Waites, G. M. H. (1977). Fluid secretion. In *The Testis* Vol. IV (ed. A. D. Johnston and W. R. Gomes) pp. 91–123. Academic Press, New York.

—— and Gladwell, R. T. (1982). Physiological significance of fluid secretion in the testis and blood-testis barrier. *Physiol. Rev.* **62**, 624–671.

—— and Moule, G. R. (1960). Blood pressure in the internal spermatic artery of the ram. *J. Reprod. Fert.* **1**, 223–229.

Wallace, J. C. and Lascelles, A. K. (1964). Composition of testicular and epididymal lymph in the ram. *J. Reprod. Fert.* **8**, 235–242.

Walton, J. S., Evins, J. D., and Waites, G. M. H. (1978). Effect of chronic removal of testicular fluid on the release of follicle-stimulating hormone in hemicastrated rams. *J. Endocr.* **77**, 421–422.

Wang, J., Galil, K. A. A., and Setchell, B. P. (1983). Changes in testicular blood flow and testosterone production during aspermatogenesis after irradiation. *J. Endocr.* **98**, 35–46.

—— Gu, C. H., Tao, L., and Wu, X. L. (1985). Effect of surgery and efferent duct ligation on testicular blood flow and testicular steroidogenesis in the rat. *J. Reprod. Fert.* **73**, 191–196.

Wathes, D. C. (1984). Possible actions of gonadal oxytocin and vassopressin. *J. Reprod. Fert.* **71**, 315–345.

Weber, J. E., Russell, L. D., Wong, V., and Peterson, R. N. (1983). Three dimensional reconstruction of a rat stage V Sertoli cell: II. Morphometry of Sertoli-Sertoli and Sertoli-Germ cell relationships. *Am. J. Anat.* **167**, 163–179.

Welsh, M. J. and Wiebe, J. P. (1975). Rat Sertoli cells: a rapid method for obtaining viable cells. *Endocrinology* **96**, 618–624.

Widmark, A., Damber, J. E., and Bergh, A. (1986). Relationship between human chorionic gonadotrophin-induced changes in testicular microcirculation and the formation of testicular interstitial fluid. *J. Endocr.* **109**, 419–425.

Williams, R. G. (1949). Some responses of living blood vessels and connective tissue to testicular grafts in rabbits. *Anat. Rec.* **104**, 147–161.

Williamson, J. R., Chang, K., Kilzer, P., Marvel, J., and Kilo, C. (1984). Colloidal carbon given intravenously increases vascular permeation of ^{125}I-albumin. *Int. J. Microcirc.* **3**, 490.

Wilson, R. M. and Griswold, M. D. (1979). Secreted proteins from rat Sertoli cells. *Expl. Cell Res.* **123**, 127–135.

Wing, T. Y. and Lin, H. S. (1977). The fine structure of testicular interstitial cells in the adult golden hamster with special reference to seasonal changes. *Cell Tissue Res.* **183**, 385–393.

Wolff, J. and Merker, H. J. (1966). Ultrastruktur und Bildung von Poren im Endothel von porosen und geschlossenen Kapillaren. *Z. Zellforsch. mikroskop. Anat.* **73**, 174–191.

Wright, W. W., Musto, N. A., Mather, J. P., and Bardin, C. W. (1981). Sertoli cells secrete both testis-specific and serum proteins. *Proc. natn. Acad. Sci. U.S.A.* **78**, 7565–7569.

Yaksh, T. L. and Yamamura, H. I. (1974). Factors affecting performance of the push-pull cannula in brain. *J. appl. Physiol.* **37**, 428–434.

Yee, J. B. and Hutson, J. C. (1985). Effects of testicular macrophage-conditioned medium on Leydig cells in culture. *Endocrinology* **116**, 2682–2684.

Yoffey, J. M. and Courtice, F. C. (1970). *Lymphatics, Lymph and the Lymphomyeloid Complex*. Academic Press, London.

Yudilevich, D. L. and Mann, G. E. (1982). Unidirectional uptake of substrates at the blood side of secretory epithelial: stomach, salivary gland, pancreas. *Fedn. Proc.* **41**, 3045–3053.

Zirkin, B. R., Ewing, L. L., Kromann, N., and Cochran, R. C. (1980). Testosterone secretion by rat, rabbit, guinea pig, dog, and hamster testes perfused *in vitro*: correlation with Leydig cell ultrastructure. *Endocrinology* **107**, 1867–1874.

3   Molecular biology of the Sertoli cell

MICHAEL D. GRISWOLD, CARLOS MORALES AND STEVEN R. SYLVESTER

I   Introduction

Enrico Sertoli working in 1865 described a nongerminal "branched cell" in the seminiferous epithelium of human testes. The main function assigned to these cells which are now called "Sertoli cells" was that of providing support or sustenance to the various germinal cells which transit the seminiferous epithelium. According to this view, the Sertoli cells provide a number of structural and biochemical factors required by the germinal cells during spermatogenesis and spermiogenesis (for reviews of Sertoli cell structure and function, see Fawcett 1975; Steinberger and Steinberger 1977; Tindall, Rowley, Murthy, Lipshultz, and Chang 1985). Studies which have identified the Sertoli cells as the principal testicular target cell for follicle stimulating hormone (FSH) and testosterone have resulted in an expansion of the proposed role of the Sertoli cells. These studies have given rise to the concept that most or all of the endocrine regulation of spermatogenesis is manifested through the actions of hormones on Sertoli cells (for review, see Fritz 1978). While the testis, in general, and spermatogenesis, in particular, have been fertile research areas for morphologists for many years the molecular nature of the "nurse" role or the regulatory functions of the Sertoli cell products is just beginning to be defined. The molecular aspects of Sertoli cell structural interactions with other Sertoli cells and with germinal cells are totally undefined. The biochemical milieu in the adluminal compartment is probably a product of the resident cells and is determined to a large extent by the secretory activities of the Sertoli cells (for a review of testicular fluids, see Waites 1977). A number of the extragonadal regulatory roles of the testis are presumed to be a result of the production of hormone-like substances by Sertoli cells. The morphological and molecular aspects of Sertoli cell functions in spermatogenesis are intimately related: therefore the complete understanding of one requires some knowledge of the other. In this review, the overall morphology of the Sertoli cells and the progress made in understanding the molecular biological aspects will be considered as they relate to each other and subsequently to Sertoli cell functions. Most of the research on Sertoli cell functions at the molecular level which will be covered in this review has been concerned with the isolation and characterization of the proteins which are synthesized and secreted by Sertoli cells and therefore those structural aspects of the cells which relate to protein secretion and to the creation of a unique and protected compartment will be emphasized. The application of the powerful tools of recombinant DNA technology has provided a glimpse into the biochemical meaning of the term "nurse cell" and offers promise of a level of understanding of spermatogenesis at the molecular level equivalent to the level of morphological understanding engendered by the electron microscope.

II   Structural aspects of Sertoli cell functions

1   STRUCTURE OF THE SERTOLI CELL

In mammals the Sertoli cell is a columnar cell that extends from the base of the seminiferous epithelium to the lumen of the tubule. It presents numerous veil-like lateral processes that extend between spermatogonia, spermatocytes and spermatids (Russell, Tallon-Doran, Webber, Wong, and Peterson, 1983; Webber, Russell, Wong, and Peterson 1983; Wong and Russell 1983; Russell,

Gardner, and Weber 1986). The Sertoli cells interact with specific groups of germinal cells in a well defined cyclic manner which is discussed in more detail later in this review. The basal portion of the Sertoli cell is characterized by the presence of an irregularly shaped nucleus with tripartite nucleoli. The cytoplasm contains an extensive network of cisternae of the endoplasmic reticulum (ER) which show relatively few ribosomes attached to their surface. Part of this ER network is associated with junctional complexes or tight junctions existing at the base of these cells (Brokelman 1963, Flickinger and Fawcett 1967; Dym and Fawcett 1970) and to subsurface specializations, also called ectoplasmic specializations, found facing pachytene spermatocytes, round and elongated spermatids (Russell and Clermont 1976; Russell 1977c). The Golgi apparatus is a large supranuclear organelle formed by separate stacks of membranes, each consisting of a few short parallel saccules and associated small vesicles (Fawcett 1975). In a more recent work, Rambourg, Clermont and Hermo (1979) analyzed the tridimensional structure of the Golgi apparatus, and by using metallic impregnation on semi-thin sections demonstrated the existence of communication networks between adjacent Golgi stacks via intersaccular connecting tubules. The Sertoli cells also contain clusters of membrane-bound granules of different sizes and densities. These membrane-bound bodies showed acid phosphatase and arylsulfatase activity and were considered as part of the lysosomal system of the Sertoli cells (Dietert 1966; Fawcett 1975; Lalli, Tang, and Clermont 1984; Morales, Clermont, and Hermo 1985; Morales, Clermont, and Nadler 1986). The remaining cytoplasmic components include cytoskeletal elements (filaments and microtubules) free ribosomes, glycogen granules and numerous spherical and elongated mitochondria (Fawcett 1975; 1977). In general, the large Golgi and other cellular components indicate that the Sertoli cell is very actively involved in the secretion of proteins but no obvious secretory granules or vesicles have been identified by electron microscopy.

2   STRUCTURAL INTERACTIONS BETWEEN SERTOLI CELLS

In 1963 Brokelman described a structure between adjacent Sertoli cells located near the base of the seminiferous epithelium characterized by the presence of tight and gap junctions. These structures were later called junctional specializations (Flickinger and Fawcett 1967; Nicander 1967). Generally, the membranes of adjacent Sertoli cells are separated by a space of 15 nm–20 nm. However, in the region of the junctional specializations they approach to within 2 nm of each other and form gap junctions. In addition to the gap junctions, the plasma membranes of adjacent Sertoli cells form linear tight junctions (Dym and Fawcett 1970). Running parallel to the Sertoli cell plasma membrane in the areas of junctional specializations are also flat saccules or cisternae of endoplasmic reticulum (ER) (Flickinger and Fawcett

1967). Between these flat cisternae of the ER and plasma membrane of adjoining Sertoli cells are bundles of actin filaments. The filaments when viewed in cross sections at high magnification exhibit a hexagonal arrangement (Flickinger and Fawcett 1969; Dym and Fawcett 1970).

The concept of the blood-testis barrier was first presented in early physiological studies which showed that when vital dyes were introduced into the blood stream they rapidly appeared in the testicular lymph but not in the fluid collected from the rete testis. Further insight was provided by the studies of Setchell (1967; 1969; 1977) and Tuck, Setchell, Waites, and Young (1970) which demonstrated that the concentrations of proteins, ions and amino acids in the seminiferous tubular fluid, the plasma, and the testicular lymph were all very different. As a result of these data, it was postulated that a blood-testis barrier associated with the seminiferous tubules was capable of excluding from the tubular lumen many substances normally present in the blood and lymph (Setchell 1969). Fawcett and his collaborators (Dym and Fawcett 1970; Dym 1972; Aoki and Fawcett 1975) demonstrated that the tight junctions formed an effective barrier that prevented the penetration of molecules of relatively large size from the interstitial space to the lumen of the seminiferous tubules.

The tight junctions also constitute barriers which demarcate distinct compartments, that is, a latero-basal compartment which faces the interstitial space and houses the spermatogonia and preleptotene spermatocytes, and an adluminal compartment which houses the primary and secondary spermatocytes, and the spermatids. Obviously this inter-Sertoli cell barrier cannot be maintained indefinitely, as clones of preleptotene spermatocytes must traverse it at regular intervals. However, Sertoli-Sertoli cell contacts are established both above and below the clone of spermatocytes as they move from the basal into the adluminal compartment. Therefore, the formation of an "intermediate compartment" takes place for a short period of time (Russell 1977b; Russell 1978a; Russell et al. 1983). As in several other epithelia, the tight junctions between Sertoli cells also demarcate plasma membrane domains which may have different properties.

In physiological terms the blood-testis barrier has a more complex function than the simple exclusion of different types of macromolecules. It is known that Sertoli cells are capable of secreting fluids, proteins and ions, and that the composition of the seminiferous tubular fluid is different from that of the plasma and testicular lymph. Thus, one of the main functions of the blood-testis barrier is to allow the creation of a special environment within the seminiferous tubules in which the germinal cells undergo meiosis and develop into spermatids (Setchell 1980). The barrier also regulates the concentration of substances of endocrinological importance such as androgen binding protein (ABP) which appears highly concentrated in the seminiferous tubular fluid (Setchell 1980). Another possible function of the blood-testis barrier is to prevent the immunological system of the body from entering into contact with proteins produced by the haploid spermatids and spermatozoa (Alexander 1977; Silber 1978; Setchell 1980).

3   STRUCTURAL INTERACTIONS BETWEEN SERTOLI AND GERM CELLS

Sertoli cells are intimately related to the germ cells in a very complex manner. As described above, the inter-Sertoli cell tight junctions result in the formation of a basal compartment where these cells interact with the spermatogonia and preleptotene spermatocytes and an adluminal compartment where these cells interact with the primary and secondary spermatocytes and the differentiating spermatids. One of the most interesting structural interactions between Sertoli cells and germ cells are the ectoplasmic specializations (Ross 1976; Russell and Clermont 1976; Russell 1977c).These structures are present in the Sertoli cell cytoplasm adjacent to the plasma membrane facing the spermatocytes and the acrosomes of the spermatids. Each consists of a single layer of flattened cisternae of the endoplasmic reticulum lying parallel to the Sertoli cell surface and bundles of actin filaments which occupy the layer of cytoplasm between the subsurface ER cisternae and the Sertoli plasma membrane (Brokelmann 1963; Flickinger and Fawcett 1967; Nicander 1967; Franke, Grund, Fink, Weber, Jockush, Zentgraph, and Osborn 1978; Russell et al. 1983; Vogl and Soucey 1985; Grove and Vogl 1986). These junctional specializations have generally been considered as attachment devices serving to bind the germinal cells in the adluminal compartment (Nicander 1963; Flickinger and Fawcett 1967; Fawcett and Phillips 1969; Ross 1976).

Other structures referred to as the "tubulobulbar complexes" have been described at the interface of Sertoli cells and the heads of maturation-phase spermatids in the rat and several other mammalian species (Russell and Clermont 1976; Russell 1979a; Malone 1979). In the rat these complexes consist of long, narrow tubular projections of the spermatid's plasma membrane, located on the ventral aspect of the sickle-shaped nucleus, which invaginate the subjacent plasma membrane of Sertoli cells and terminate as distended, bulbous extremities. Although flattened cup-shaped cisternae of endoplasmic reticulum are seen in close apposition to the bulbar portion of this complex, these cisternae, in contrast to those forming the ectoplasmic specializations, are not associated with bundles of cytoplasmic filaments (Russell and Clermont 1976). These structures have been implicated in the transport of fluids from the late spermatids to the Sertoli cells. This speculation arose from the observations made by Russell (1978b) on the testes of hypophysectomized rats receiving testosterone, and by Gravis (1979) on the testes of hamsters injected with dibutyryl cyclic AMP. These authors noted that such treatments prevented the formation of the tubulobulbar complexes and as a result the perinuclear cytoplasm of the late spermatids remained dilated and did not show the expected condensation. More recently Russell (1979*b*) presented some morphometric evidence suggesting that the condensation of the cytoplasm of the late spermatids in the rat coincided with the development of the tubulobulbar complexes. While the function of these structures remains to be clarified the various structural features of the tubulobulbar complexes stress the intimate functional relationship which exists between spermatids and Sertoli cells (Clermont, McCoshen, and Hermo 1980).

Another structural interaction between Sertoli cell and germ cells is the desmosome-gap junction (Russell *et al.* 1983). These structures were initially described as desmosome-like junctions (Nicander 1967; Russell 1977*a*) in association with small gap junctions (McGinley, Pozalaky, Porvaznik, and Russell 1979). Consequently, as the name depicts, desmosomes and gap junctions exist together as a complex in the seminiferous epithelium (McGinley *et al.* 1979; Russell *et al.* 1983). The desmosome-gap junctions have been shown to be abundant between Sertoli cells and pachytene spermatocytes and less frequent between the former and early round spermatids (Russell *et al.* 1983). Desmosome-gap junctions have been shown to possess strong adherent properties (Russell 1977*a*). Cell to cell communications may be one of the functions of the gap junctional component of the desmosome-gap junction. Ziparo, Siracusa, Palombi, Russo, and Stefanini (1982) demonstrated that Sertoli cells in culture are capable of transferring ^{3}H-choline to germ cells. Transfer of information between germ cells could also seemingly be influenced by cytoplasmic bridges, thus appearing to reduce the requirement that all germ cells need to be coupled electrically or chemically to the Sertoli cells (Russell *et al.* 1983).

Finally, it should be noted that Sertoli cells possess a phagocytic activity for germinal cell components. A number of features indicate that phagocytosis of residual bodies is possibly of the receptor-mediated variety. For example, the presence of thin cytoplasmic processes of Sertoli cells can be seen closely apposed to and rapidly enclosing the newly formed residual bodies (Morales *et al.* 1985; Morales *et al.* 1986). Even more important is the fact that at the time of sperm release, while Sertoli cell processes adhere firmly to the membrane of residual bodies and prevent their loss from the surface of the seminiferous epithelium (Fawcett and Phillips 1969), the heads and proximal part of the tail of the spermatozoa smoothly disengage from the large processes which encapsulate or surround them and are freed into the tubular lumen (Russell and Clermont 1976; Russell 1984). It appears that the selective retention and phagocytosis of residual bodies is specific and results from the Sertoli cell recognition of the plasma membrane which delimits these bodies (Clermont, Morales, and Hermo in press). Finally, the plasma membrane delimiting the residual bodies has antigenic determinants which differ from those present in the membrane of the rest of the spermatozoon (Millette 1979; Millette and Bellvé 1980; Bellvé and Moss 1983). Thus, this apparent differential partitioning of molecules in the plasma membrane of the spermatid may well contribute to the recognition of these distinct domains by Sertoli cells (Clermont *et al.* in press).

4 MORPHOLOGICAL EVIDENCE FOR A CYCLIC FUNCTIONAL MODE FOR SERTOLI CELLS

The seminiferous epithelium of sexually mature animals is composed of Sertoli cells, one or two generations of spermatogonia along the basement membrane, one or two generations of spermatocytes, and one or two genera- tions of spermatids (Clermont 1972). Each generation of germ cells is com- prised of a group of cells at approximately the same stage of development which evolve synchronously throughout the spermatogenic process. These various generations of germ cells are not associated at random but form very specific cellular associations (Leblond and Clermont 1952). The cellular asso- ciations have also been referred to as stages of the cycle, designated by Leblond and Clermont (1952) I through XIV in the rat, and they follow each other in a specific and fixed sequence in a given area of seminiferous tubules. With time, the succession of a complete series of stages in an area of the semi- niferous tubule constitutes a cycle of the seminiferous epithelium (Leblond and Clermont 1952; Clermont 1972).

Morphological, histochemical and biochemical observations strongly sug- gest the existence of a cycle of Sertoli cells corresponding to different func- tional modes the cells may enter during the cyclic development of the seminiferous epithelium (Leblond and Clermont 1952; Niemi and Kormano 1965; Kerr and deKretser 1975; Hilscher, Passia, and Hilscher 1979; Parvi- nen 1982; Morales *et al.* 1986). The first well documented suggestion that Sertoli cells have a functional cycle was made by Leblond and Clermont (1952). These authors described changes in nuclear morphology of Sertoli cells in relation to the stages of the cycle of the seminiferous epithelium of the rat. Some recent morphometric analyses have demonstrated variations in the relative volume of the Sertoli cells in the monkey and the rat (Cavicchia and Dym 1977; Bugge and Ploen 1986). Also, the amount of lipid in Sertoli cells changes during the cycle. For example, shortly after phagocytosis of the residual bodies the lipid content of the Sertoli cells dramatically increases during stages IX–X (Niemi and Kormano 1965; Posalaki, Szabo, Bacsi, and Okros 1968; Kerr and deKretser 1975).

The distribution and number of secondary lysosomes have been shown to change during the cycle of the seminiferous epithelium. At most stages the secondary lysosomes form clusters which are located in the supranuclear region of the Sertoli cells near the apex of the elongated nuclei of late sperma- tids. However, at stage VI, VII and VIII, the lysosomes accumulate near the plasma membrane at the base of the Sertoli cell (Morales *et al.* 1985). At stage IX the lysosomes migrate towards the supranuclear region and merge with the released residual bodies (Morales *et al.* 1985; Morales *et al.* 1986).

Analysis of the turnover kinetics and life span of secondary lysosomes in Sertoli cells of the rat have provided evidence that they are disposed of most rapidly between stages IX and X (Morales *et al.* 1986). This data substantiated the notion that secondary lysosomes of Sertoli cells fuse with phagocytosed residual bodies and contribute directly to their lysis (Morales *et al.* 1985). In this process, residual bodies and the lysosomes are eliminated from the Sertoli cell cytoplasm.

Sertoli cells contain receptors for both FSH and testosterone: however, the adult seminiferous tubules respond minimally to FSH stimulation (Fritz 1978; Ritzen, Hansson and French 1981). There is some variation in the binding of FSH and in FSH-stimulated cAMP production in relation to the stages of the cycle in adult seminiferous tubule (Parvinen 1982). FSH appears maximally bound at stage I and minimally bound at stage VII (Parvinen 1982). On the other hand, endogenous testosterone appears to reach its maximal concentration at stage VIII (Parvinen and Ruokonen 1982).

III   Molecular studies on Sertoli cells

1   SERTOLI CELLS IN CULTURE

The use of Sertoli cells in primary cell culture has been vital for establishing the Sertoli cell origin of secreted components. Techniques for the effective culture of Sertoli cells from immature rats (20 days of age) were developed in several laboratories (Dorrington and Fritz 1975; Steinberger, Heindel, Lindsey, Elkington, Sanborn, and Steinberger 1975a; Steinberger, Elkington, Sanborn, Steinberger, Heindel, and Lindsey 1975b; Welsh and Wiebe 1975). Sertoli cells can be cultured in serum-free medium for 1–2 weeks and during this time they continue to respond to hormones and to secrete proteins into the medium. Sertoli cells are usually plated as aggregates of cells or as small pieces of the seminiferous tubule from which the peritubular and interstitial cells have been removed by differential enzymatic digestion (Tung, Dorrington, and Fritz 1975). Under ideal conditions the cultured cells form a nondividing monolayer which consists of approximately 90 per cent Sertoli cells (Fritz, Louis, Tung, Griswold, Rommerts, and Dorrington 1975). This is a significant enrichment of Sertoli cells compared to the whole testis but even a 10 per cent or less contamination by germinal cells or peritubular cells could lead to difficulties in the interpretation of results. In dense cultures of Sertoli cells, the extent of contamination by peritubular or germinal cells is very difficult to determine. However, in sparse cultures where the Sertoli cells have the opportunity to spread out on the surface of the culture dish, the peritubular cells can be readily distinguished and the germinal cells are released after 2 to 3 days of culture. Residual germinal cells can be removed from the monolayer by a brief hypotonic treatment (Galdieri, Ziparo, Palombi, Russo, and Stefanini 1981; Wagle, Heindel, Sanborn, and Steinberger 1986). After 6–8 days in culture the response to hormones and the protein secretory activity of the Sertoli cells begins to decline (Skinner and Griswold 1982; Karl and Griswold 1980). The peritubular cells can easily be cultured virtually free of Sertoli cells and these cultures can and should be tested for the presence of any putative Sertoli cell product. A definitive measure of the purity of Sertoli cell cultures from 20-day old rats is the absence in the culture medium of radiolabelled fibronectin. This protein has been shown to be a major product of peritubular myoid cells (Tung, Skinner, and Fritz 1984). The addition of serum or other medium supplements will encourage the proliferation of the peritubular cells and eventually their secretory activities will supplant those of the Sertoli cells (Tung *et al.* 1984). The presence of peritubular myoid cells in the cultures has been shown to have strong positive effects on the morphology and functions of the Sertoli cells. The secretion of androgen binding protein (ABP) was shown to be increased if the Sertoli cell cultures contained some peritubular cells (Hutson and Stocco 1981; Hutson 1983; Skinner and Fritz 1985*a,b*). The presence of peritubular cells greatly increased the plating efficiency and viability and altered the morphology of Sertoli cells in culture (Tung and Fritz 1980; Cameron and Markwald 1981; Cameron and Snydle 1985). Skinner and Fritz (1985*a,b*) described a factor obtained from the medium of cultured peritubular cells which modulated the action of androgens on the Sertoli cells. This factor which was termed P-Mod-S for "peritubular modifies Sertoli" stimulated the synthesis of transferrin and ABP by the cultured Sertoli cells.

While the culture techniques for the Sertoli cells from the 20-day rat have been well established, the culture of Sertoli cells from younger or older rats is feasible but it is not as well defined. The culture of Sertoli cells from adult rats has been done successfully although only a small percentage of the total cells in the adult rat testis are Sertoli cells (Steinberger *et al.* 1975*b*). These cell cultures often contain a significant number of germinal cells which can be removed by hypotonic treatment of the cell monolayer (Galdieri *et al.* 1981). The culture of Sertoli cells from rats younger than 20 days of age is complicated by the vigorous mitotic activity of contaminant peritubular cells. Proliferation of peritubular cells in cultures of Sertoli cells from 20 day old rats or older can be prevented by the inclusion of cytosine arabinoside in the culture medium (Tung, LaCroix, and Fritz 1980). However, Sertoli cells prepared from younger rats are also capable of significant mitotic activity so the cytosine arabinoside cannot be used in these cultures (Griswold, Solari, Tung, and Fritz 1977). The determination of optimal conditions for the culture of Sertoli cells from other than 20 day old rats would be of use to many investigators. New techniques for the culture of Sertoli cells include the use of extracellular matrix on plastic and on suspended filters (Tung and Fritz 1984; Hadley, Byers, Suarez-Quien, Kleinman, and Dym 1985; Byers, Hadley, Djakiew, and Dym 1986). These techniques yield cultures of Sertoli cells which maintain a morphology which more closely resembles that of cells *in* *vivo.* Cells cultured on matrix may also be more active in protein secretion (Mather, Wolpe, Gunsalus, Bardin, and Phillips 1984) than their counterparts cultured on plastic but the matrix material may contribute greatly to the total protein content of the culture medium and could complicate efforts to purify and characterize a particular component.

The original protocols for the preparation and culture of Sertoli cells were established for rat tissue (Dorrington and Fritz 1975; Steinberger *et al.* 1975*a,b*; Welsh and Wiebe 1975). Rat Sertoli cells continue to be utilized by most researchers in studies where cell culture is necessary, but in addition the techniques have been modified and applied to Sertoli cells from bovine (Smith and Griswold 1981), porcine (Tabone, Benahmed, and Saez 1984; Perrard-Sapori, Saez, and Dazord 1985), Cynomolgus monkey (Keeping, Winters, and Troen 1985), and human tissue (Lipshultz, Murthy, and Tindall 1982).

The availability of continuous cell lines which retain the characteristics of Sertoli cells would simplify many of the problems encountered in the use of primary cell cultures. Two cell lines have been described which are of putative Sertoli cell origin. One line termed "TM-4" was obtained from murine Sertoli cell cultures and the other line termed "TR-ST" was obtained from a spontaneously arising rat testicular tumor (Mather 1980; Mather, Zhuang, Perez-Infante, and Philips 1982). One recent report about the action of vitamin A on TM-4 cells was attributed to Sertoli cells (Carson and Lennarz 1983). Another report drew biological inferences about Sertoli cell functions from data on a unique vitamin A binding protein which was found to be secreted by TM-4 cells (Carson, Rosenberg, Blaner, Kato, and Lennarz 1984). This vitamin A binding protein could not be found in the medium of cultured Sertoli cells (Blaner, Galdieri and Goodman 1986). While both of the cell lines listed above appear to have some Sertoli cell-like characteristics they have not been completely characterized and both are clearly different in a number of ways from primary cultures of Sertoli cells. Data on the growth regulation of these cell lines or the biological significance of the synthesis of specific proteins by these cells should not be extrapolated directly to Sertoli cells. However, these cell lines may be potentially useful if they could be shown to have a product similar to that of cultured primary Sertoli cells.

2 SECRETION OF GLYCOPROTEINS BY SERTOLI CELLS

The morphological data described above suggests that the Sertoli cell is active in the secretion of protein and may do so in very specialized regions of the tubule. From molecular studies it has been shown that Sertoli cells secrete macromolecules presumably either into the adluminal or the basal compartments of the seminiferous tubule. From 4 to 14 per cent of the total newly synthesized protein from cultured Sertoli cells from 20 day old rats was found to be in the form of secreted glycoproteins (Wilson and Griswold

1979). It seems likely that the glycoproteins secreted by the Sertoli cells *in vivo* have direct regulatory roles in germinal cell development and in other biological processes related to reproduction. The determination of the biochemical composition and biological activities of these secretion products has been a major goal for investigators.

Primarily as a result of studies using cell culture and gel electrophoresis a number of different putative Sertoli cell secreted glycoproteins have been described (for a recent review see Griswold 1987). Proteins which are identified by investigators as Sertoli cell secretion products should accumulate in the medium of cultured Sertoli cells but not in the medium of cultured peritubular or germinal cells and should be actively synthesized during culture as shown by radiolabelling experiments. In general, the proteins secreted by Sertoli cells can be placed in two categories based on their relative abundance. Proteins which function as hormones or growth factors or have enzymatic activities are secreted at very low levels and yet may have important biological activities. Examples of this type of secretion product include anti-Mullerian hormone (Josso, Picard, and Tran 1977; Josso and Picard 1986), and inhibin (deJong and Robertson 1985, review). Proteins which function primarily as transport proteins such as transferrin (Skinner and Griswold 1980), and ceruloplasmin (Skinner and Griswold 1983) are secreted at relatively high levels. Other proteins which are secreted at high levels such as sulfated glycoproteins 1 and 2 are characterized structurally but as yet have unknown functions (Sylvester, Skinner, and Griswold, 1984). A summary of the putative secretion products of Sertoli cells and their proposed functions or characteristics is given in Table 3.1. As seen in Table 3.1, different investigators may have reported on the same protein using a different nomenclature.

Most of the more abundant Sertoli cell secreted proteins were first detected in the medium of cultured cells by the use of polyacrylamide gel electrophoresis followed by fluorography (Kissinger, Skinner, and Griswold 1982). Many secreted polypeptides can be detected by this technique but only transferrin, ceruloplasmin and sulfated glycoproteins 1 and 2 (SGP-1 and SGP-2) have been characterized or identified (Skinner and Griswold 1980; 1983; Skinner, Cosand, and Griswold 1984; Sylvester *et al.* 1984). Transferrin, ceruloplasmin and SGP-1 and 2 comprise more than 80 per cent of the total mass of proteins secreted by cultured Sertoli cells and SGP-2 alone represents as much as 50 per cent (Griswold, Huggenvik, Skinner, and Sylvester 1984). Anti-Müllerian hormone and androgen binding protein are relatively low abundance products and cannot be detected by these methods without selectively concentrating the protein by immunoprecipitation. The profile of secretion products revealed by 2-dimensional gel electrophoresis and fluorography is shown in Figure 3.1 Note that SGP-2 comprises two major families of spots which demonstrate pronounced charge heterogeneity. SGP-2 is composed of 2 subunits (47 and 34 kD) which are linked under nonreducing

Table 3.1
Major Sertoli cell secreted proteins

Only those proteins that have been characterized to some extent and are clearly established as Sertoli cell secretion products are included in this Table. Other proteins which have been described by gel electrophoresis but which have not been further characterized can be found in Kissinger et al., 1982; Sanborn et al., 1986; Kierszenbaum et al., 1986; Cheng et al., 1986. "Possible correlates" represents the authors' analysis of the literature and indicates which proteins with different names might be the same.

Name (Reference)	Molecular Weight/pI	Characteristics and/or Function	Possible Correlates
Ceruloplasmin (1)	130 kD/5.5–6.0	Copper transport; ferroxidase	------
Transferrin (2)	75 kD/7.2–7.4	Iron transport to germinal cells	------
Cyclic Protein-2 (3)	22 kD/4.9–6.2	Differential appearance with stages	Prot. 5b (12)
SGP-1 (4)	70 kD/4.2–4.6	Sperm coating	S70 (13) Band 4 (12)
SGP-2 (5)	47;34 kD/4.2–4.6	Sperm coating; made in epididymis	S45;S35 (13) DAG-protein (12)
Clusterin (6)	45;35 kD	Isolated from ram Sertoli cells	CMB-21 (14) SGP-2 (5,12) SGP-1 (5,12)
Testibumin (7)	68 kD	Characterized by 1-D gel only	
ABP (8)	45;41 kD/4.7–5.5	Androgen transport to epididymis	
Plasminogen activator (9)	38–40 kD;75 kD	Remodelling of tubule; tight junctions	
Anti-Müllerian (10)	72 kD	Synthesized in pre-natal Sertoli cells	
Inhibin (11)	32 kD (ovarian)	Sertoli cell origin is probable in males	

References: (1) Skinner and Griswold, 1983; (2) Skinner and Griswold, 1982; (3) Wright and Luzarraga, 1986; (4) Kissinger et al., 1982; Griswold et al., 1986; (5) Griswold et al., 1986a; Sylvester and Griswold, 1984. (6) Blaschuk and Fritz, 1984; (7) Cheng and Bardin, 1986; (8) Tindall et al., 1985 (review); (9) LaCroix et al., 1977; (10) Josso and Picard, 1986 (review) (11) deJong and Robertson, 1985 (review); Ling et al., 1985; Mason et al., 1985 (12) Kissinger et al., 1982; (13) Kierszenbaum et al., 1986; (14) Cheng et al., 1986.

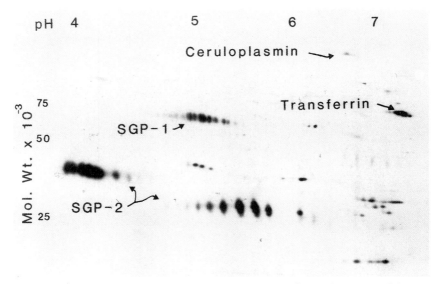

Fig. 3.1   Two-dimensional gel electrophoresis map of proteins secreted by rat Sertoli cells in culture. Sertoli cell cultures were incubated with ^{35}S-methionine for 24 h and the media concentrated and desalted. After two-dimensional electrophoresis, the gel was fluorgraphed onto X-ray film and subsequently developed. The spots indicated are those which have been identified and discussed in the text. Under non-reducing conditions, SGP-2 would appear as a single row of spots at approximately 72,000 MW.

conditions by disulfide bonds. This overall pattern of secretion products of Sertoli cells from 20 day old rats is very characteristic and is qualitatively similar to that obtained from Sertoli cells from adults rats (Kissinger *et al.* 1982). However, the pattern of secretion products is very different from that obtained from cultured peritubular cells (Kissinger *et al.* 1982; Kierszenbaum, Crowell, Shabinowitz, DePhilip, and Tres 1986). The appearance of significant amounts of high molecular weight polypeptides (greater than 130 kD) in medium obtained from Sertoli cell cultures is an indication that the cultures contain significant numbers of peritubular cells. It has been demonstrated that some of these peritubular cell-specific, high-molecular weight polypeptides are components of fibronectin (Tung, Skinner, and Fritz 1984). Quantitative determinations of some of the proteins such as ABP and transferrin have been used as markers for the endocrine regulation of Sertoli cells in cell culture (Skinner and Griswold, 1982; Perez-Infant, Bardin, Gunsalus, Musto, Rich, and Mather, 1986). The details of endocrine regulation of the secretion products has not been emphasized in this review but has very recently been reviewed by Sanborn, Caston, Buzak, and Ussef, 1987.

While the overall profile of rat Sertoli cell secretion products shown in Figure 3.1 has been duplicated in several laboratories (Kissinger *et al.* 1982;

Sanborn, Wagle, Steinberger, and Greer-Emmert 1986; Shabinowitz and Kierszenbaum 1986) analysis by electrophoresis of Sertoli cell secretions from any animal other than the rat is limited. When the products secreted from cultured mouse Sertoli cells are analyzed by gel electrophoresis and fluorography the results are very similar to those shown for the rat except the relative amount of radioactivity incorporated into SGP-2 is increased and the amount into transferrin is decreased (unpublished observations of M. D. Griswold). Profiles of secretion products from cultured porcine Sertoli cells have also been described and large amounts of radioactivity were associated with 2 polypeptides which have molecular weights and isoelectric points similar to those of rat SGP-1 and SGP-2 (Perrard-Sapori, *et al.* 1985).

3 TESTICULAR TRANSFERRIN—STRUCTURE AND FUNCTION

Testicular transferrin was the first protein to be identified as one of the major secretion products of rat Sertoli cells (Skinner and Griswold, 1980). The protein was identified in cell culture media by its ability to bind iron and by its cross-reaction with antibodies to rat sero-transferrin. Testicular transferrin co-migrated with sero-transferrin in sodium dodecylsulphate (SDS) poly-acrylamide gel electrophoresis. Further analysis of transferrins derived from Sertoli cells and serum revealed that the peptide backbone was identical but that the proteins exhibited different glycosylation patterns (Skinner *et al.* 1984). Studies of Sertoli cells in culture revealed that transferrin secretion could be stimulated by treating cells with insulin, follicle stimulating hormone and retinol (Skinner and Griswold 1982). Another study found that transferrin secretion could be increased by epidermal growth factor and by plating at higher cell densities but was unaffected by follicle stimulating hormone (Perez-Infante *et al.* 1986). It is not apparent that follicle stimulating hormone, epidermal growth factor, insulin and retinol act on Sertoli cells through any common mechanism. It seems that transferrin synthesis in Sertoli cells cultured in serum—free medium can be affected by many growth factors normally present in serum, as well as by very selective hormones such as follicle stimulating hormone, and by culture conditions.

The search for the role of transferrin in spermatogenesis began with experiments designed to determine the fate of the secreted protein. The secretion of transferrin by Sertoli cells *in vivo* could be multidirectional or unidirectional to fluid spaces or to neighbouring cells. By radioimmunoassay, rat testicular lymph was found to contain 3.7 mg/ml transferrin, the same level observed in serum. If Sertoli cells were secreting the protein basally, one might expect higher levels in testicular lymph. Seminiferous tubule fluid had 141 μg/ml and rete testis fluid collected from the efferent ducts contained 47 μg/ml transferrin (Sylvester and Griswold 1984). The blood-testis barrier should preclude

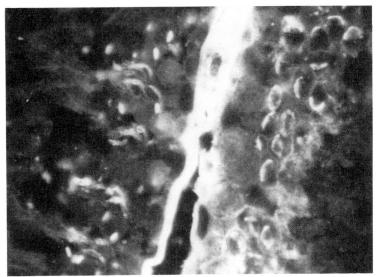

Fig. 3.2   Indirect immunofluorescence of transferrin in paraffin sections of rat testes. Rehydrated tissue sections were treated sequentially with rabbit anti-rat transferrin, biotinylated goat anti-rabbit IgG, and avidin-fluoresciein isothiocyanate with appropriate washes interspersed. The interstitium stains brightly while acrosomal regions of spermatids are stained somewhat less bright. Approximately 1000X.

entry of sero-transferrin into the lumen of the seminiferous tubule, so that the protein present in tubule fluid might be largely derived from Sertoli cells.

Indirect immunofluorescence was used to determine the localization of transferrin in paraffin sections of testis (Sylvester and Griswold 1984). The most intense staining is in the interstitium, the area bathed by lymph (Fig. 3.2) In the seminiferous epithelium, the acrosomal region of round and elongating spermatids exhibit staining. Staining was less intense on spermatids in the final stages of testicular maturation and no staining was observed on sperm isolated from epididymal fluid. These findings suggested that spermatids might receive the protein from Sertoli cells as they lie adluminal to the occlusive tight junctions between Sertoli cells and would not be exposed to lymph.

Most mammalian cells take up transferrin via specific receptor proteins at the membrane. Isolated rat pachytene spermatocytes were shown to have specific binding sites for transferrin as were isolated Sertoli cells (Holmes, Bucci, Lipshultz, and Smith 1983). An indirect immunofluorescence technique was used to localize transferrin receptors in paraffin sections of rat testes (Sylvester and Griswold 1984). In this study, receptors were localized in the juxtanuclear regions of spermatocytes and round spermatids, and were qualitatively strongest in germinal cells undergoing division. This staining was consistent with findings from studies of other cell types that dividing and differentiating cells express transferrin receptor (Hu, Gardner, Aisen, and Skoultchi 1977; Trowbridge and Omary 1981) and that internalized receptor-ligand complexes reside in the juxtanuclear region (Hopkins 1983). Brown (1985) used a monoclonal antibody raised against rat transferrin receptor and localized the receptor protein in frozen sections of testis. His findings corroborated those discussed above.

Transferrin is ascribed the role of transporting iron in a useable and soluble form (for reviews see Aisen and Listowsky 1980; Chung 1984). At physiological pH and in the presence of an anion such as bicarbonate, transferrin binds two ferric ions. If pH is lowered, iron dissociates from the protein. Transferrin receptors have a high affinity for differic-transferrin at pH 7.2 and have a high affinity for apotransferrin at pH 5. The current concept of iron transport (Octave, Schneider, Trouet, and Chrichton 1983) is that differic transferrin binds to transferrin receptors at the surface of a cell and then the receptor-ligand complex is internalized. Rather than entering a lysosome, transferrin-receptor complexes sort to another sub-cellular compartment referred to as the compartment of uncoupling and recycling of ligand (CURL) (Ciechanover, Schwartz, Dautry-Varsat, and Lodish 1983). Acidification in the CURL causes the release of iron but maintains the apotransferrin-receptor complex. By some unknown means, iron moves into the cell and apotransferrin is recycled with receptor back to the cell surface.

A model for iron transport in the rat testis involving Sertoli cell-secreted transferrin was proposed from this knowledge and the findings discussed above (Huggenvik, Sylvester, and Griswold 1984). In this model (Fig. 3.3), differic sero-transferrin interacts with Sertoli cell transferrin receptors on the basal surface and is internalized. Iron is released to the Sertoli cell and then serum apotransferrin and receptor are recycled to the basal surface. The iron remaining in the cell is incorporated into testicular transferrin which has been synthesized in the Sertoli cell. Differric testicular transferrin is then secreted into the fluid space surrounding the germinal cells where it can bind germinal cell receptors. The process results in a vectorial movement of iron from lymph to germinal cells where it can be used in heme and non-heme iron proteins and presents a path by which iron transport can circumvent the blood-testis barrier.

Many recent studies support this model of iron transport. Morales and Clermont (1986) and Wauben-Penris and coworkers (Wauben-Penris, Strous, and Donk 1986) have studied transferrin receptors in seminiferous tubules and give evidence for receptor mediated endocytosis. Sertoli cells plated on reconstituted basement membrane in dual environment culture chambers preferentially secrete transferrin into the apical compartment which should correspond to the luminal compartment of seminiferous tubules (Hadley, Djakiew, Byers, and Dym 1987). Djakiew and coworkers (Djakiew, Hadley, Byers, and Dym 1986) used similar culture chambers to follow the transport of radioactive iron across confluent cultures of Sertoli

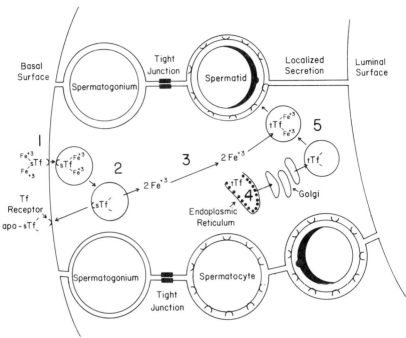

Fig. 3.3   Model for iron transport in the rat testis. Diferric sero-transferrin (sTf) binds transferrin receptors at the basal surface of Sertoli cells (1) and is internalized in endocytic vessicles. The endocytic vesicles are acidified causing the release of iron into the cytoplasm (2) and then apo-transferrin and receptor are returned to the basal surface. Testicular transferrin (tTf) is synthesized *de novo* in the Sertoli cell (4), and prepared for secretion (5). Cytoplasmic iron is then delivered to testicular transferrin (3) and the diferric testicular transferrin is then secreted to the fluid space surrounding germinal cells and binds germinal cell receptors or enters the luminal fluid.

cells. After human transferrin with bound ^{59}Fe was added to the basal chamber, immunopurified rat transferrin from the apical chamber was shown to contain radioactive iron.

Shabanowitz and Kierszenbaum (1986) presented evidence that transferrin secretion by rat Sertoli cells may be an artifact of tissue culture and that transferrin in the testis might arise from other cell types. However, Morales, Hugly and Griswold (1987) showed by *in situ* hybridization with a specific cRNA probe that the Sertoli cells was the only type in the testis which contained substantial amounts of transferrin message. Further, Morales, Sylvester, and Griswold (1987) demonstrated *de novo* synthesis of transferrin in the rat testes *in vivo* and traced the route of iron transport by radioautography. After giving ^{59}Fe-transferrin intratesticularly, the radioautographs showed a movement of iron from the basal aspect of Sertoli cells into the area of germinal cells and into the lumen of the seminiferous tubule. Additionally, experiments using ^{59}Fe-^{125}I-transferrin showed movement of iron into the seminiferous tubule while transferrin recycled to the exterior. Finally, transferrin was found in purified germinal cells by immunodetection on Western blots of polyacrylamide gel electrophoresis of cell homogenates (Morales, Sylvester, and Griswold unpublished results). In these same cells, Northern blots failed to detect any transferrin message. Taken as a whole, there is considerable data to support the model of iron transport shown in Fig. 3.3.

Testicular transferrin has been extensively studied in the rat but only in a preliminary manner in humans and even less in other species. Iron binding proteins in seminal plasma of cattle, sheep, chicken, possum, rabbit and man were examined by Roberts and Boettcher (1971). They found proteins which bound radioactive iron in the seminal plasmas of all the animals except sheep. In cattle, possum and man, two electrophoretically distinct iron binding proteins were found, transferrin and lactoferrin (the iron binding protein found in milk). Transferrin has also been found to be present in human seminal plasma by immunological techniques (Quinlivan and Sullivan 1972) but its cellular origin was unknown until recently. Holmes, Lipshultz, and Smith (1982) determined transferrin levels in seminal plasma from vasectomized men and found only 20 per cent of the level observed in normal intact men thus suggesting that the balance of the protein was of testicular origin. They subsequently studied human Sertoli cells in culture and found transferrin to represent 1–4 per cent of the total proteins secreted into culture media (Holmes et al. 1984). A correlation between sperm count and seminal plasma transferrin levels has led to the suggestion that the protein might be a good marker of seminiferous tubule function (Holmes, et al. 1982, Orlando, Caldini, Barni, Wood, Strasburger, Natali, Maver, Forti, and Serio 1985).

Tissue cultures of whole testis from rhesus monkey and rabbit were shown to synthesize a protein which cross reacted with monospecific antibodies to serum transferrins of human and rabbit, respectively (Thorbecke, Liem, Knight, Cox, and Muller-Eberhard 1973). Secreted proteins from porcine Sertoli cells have been examined by two-dimensional chromatography and have a protein of appropriate molecular weight and pI to be transferrin, but the identity has not been confirmed (Perrard-Sapori et al. 1985). Each animal must have some mechanism for transporting iron around the blood testis barrier. It remains to be seen if these mechanisms are similar to the model proposed for rat.

4 SGP-1 AND SGP-2

The sulfated glycoproteins SGP-1 and SGP-2 constitute nearly 10 and 50 per cent, respectively, of the total protein secreted by Sertoli cells. Sulfated glycoprotein-1 is a monomer of 70,000 molecular weight and exhibits extensive charge heterogeneity in two-dimensional electrophoresis gels with isoelectric points in the range of 4.0 to 4.8 (Kissinger et al. 1982). The protein elutes from reversed phase HPLC very late in a gradient of increasing acetonitrile

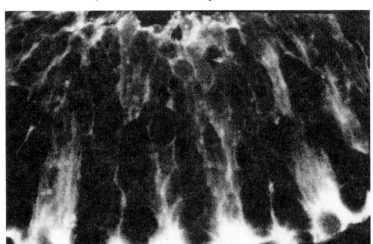

Fig. 3.4   Indirect immunofluorescence of SGP-1 in paraffin sections of rat testis. Rehydrated tissue sections were treated with rabbit anti-rat SGP-1 and as described in Fig. 3.2. The fluorescence shown here coincides with the cytoplasm of Sertoli cells. Approximately 1100X.

suggesting the protein is quite hydrophobic in nature. Purified protein has been used to produce antibodies in rabbits. Immunohistochemical studies show staining of Sertoli cell cytoplasm and weak staining on the head and tail of spermatozoa (Fig. 3.4). Immunodetection on Western blots of electropheretograms of testicular fluids reveal the presence of the protein in seminiferous tubule fluid and epididymal fluid (Sylvester and Griswold unpublished results). The secretion of SGP-1 by Sertoli cells in dual chamber cell culture systems has been shown to be polarized with most of the protein secreted apically (Danahey, Letscher, and DePhilip 1986). The function of SGP-1 in spermatogenesis is unknown.

Sulphated glycoprotein-2 is the major secretion product of Sertoli cells in culture. The protein exhibits a molecular weight of 70,000–73,000 by SDS polyacrylamide gel electrophoresis and is composed of two disulfide linked subunits of 34,000 and 47,000 molecular weight (Kissinger et al. 1982). The protein is heavily glycosylated (24 per cent of the molecular weight) and the sulfation appears to be associated with a triantenary oligosaccharide conformation (Griswold, Roberts, and Bishop 1986b). Pulse chase studies show that the protein is synthesized as a large single chain precursor and then is cleaved proteolytically to the two subunits (Collard and Griswold 1987). The protein does not stain well in polyacrylamide gels with conventional protein stains, nor does it iodinate well. Therefore, metabolic labelling and immunodetection present the best ways to study the protein.

Antibodies to the purified protein have been used in Western blot analysis to show that the protein is present in serum, seminiferous tubule fluid, rete testis fluid and in a lower molecular weight form, in epididymal fluid (Sylvester *et al.* 1984). Indirect immunofluorescence was used to locate the protein in paraffin sections of rat testes (Fig. 3.5). Sertoli cell cytoplasm was stained and spermatids exhibited staining in their final stages of development. Spermatozoa isolated from the testis, epididymis or vas deferens were stained on the acrosome, at the neck, and on the distal tail. Studies in dual chamber culture systems suggest that SGP-2 is secreted apically and is capable of preferentially binding to spermatozoa *in vitro* (Danahey *et al.* 1986). *In situ* hybridization with a cRNA to SGP-2 message shows Sertoli cells to be the only testicular cell with significant amounts of SGP-2 messenger RNA (Morales *et al.* 1987). These findings suggest that SGP-2 is synthesized by Sertoli cells and then binds to spermatozoa. The function of this major Sertoli cell product in the process of spermatogenesis is unknown at this time.

IV   Application of recombinant DNA technology to the study of Sertoli cell functions

1   CLONED cDNA PROBES TO SERTOLI CELL SECRETED PROTEINS

The utility of cDNA probes for investigations of Sertoli cell functions is apparent. If the cDNA is derived from a mRNA sequence which is present only in the Sertoli cells of the testis then measurements of the levels of that specific mRNA may provide insight into the functional state of Sertoli cells. Quantitation of a specific mRNA by hybridization techniques utilizing either cDNA or cRNA are sensitive and can be done on whole testis tissue despite the presence of germinal cells or other cell types (Griswold *et al.* 1986a). These determinations can be done by dot blot or Northern blotting techniques as well as DNA-RNA or RNA-RNA solution hybridization. In addition to the quantitative data, knowledge of the cDNA sequence and subsequently, the derived amino acid sequence may lead to direct identification or may give some clue to the function of an otherwise unknown protein.

At this time well-characterized and completely sequenced cDNA probes have been constructed for several Sertoli cell secretion products. The cDNA for rat ABP was obtained by immunological screening of a testis cDNA library cloned in lambda phage (Joseph, Hall, and French 1985; 1987). The sequence of this cDNA showed that rat ABP is composed of 2 identical subunits of 41,183 Daltons. The mRNA codes for a single subunit which has 2 potential N-glycosylation sites, and only 1 gene for ABP was detected. Previously observed heterogeneity in the size of the two subunits was attributed to differential glycosylation. Of particular interest was the observation that rat ABP and human steroid hormone binding globulin (SHBG) share considerable homology. Both rat ABP and SHBG monomers have 373 amino acid residues in which there is a 68 per cent homology in the corresponding residues (Joseph *et al.* 1987) (see also Moore and Bulbrook, and Pardridge: this volume).

A transferrin cDNA corresponding to the 3' half of the mRNA was cloned from rat liver and has been used to monitor mRNA synthesis in cultured Sertoli cells and in whole testis (Griswold *et al.* 1986*b*; Huggenvik, Idzerda, Haywood, Lee, McKnight, and Griswold 1987; Hugly and Griswold 1987; Morales *et al.* 1987). Both sero-transferrin and testicular transferrin are apparently transcribed from the same gene (Idzerda, Huebers, Finch, and McKnight 1986).

A cDNA comprising the complete mRNA transcript for SGP-2 has been cloned and sequenced (Griswold 1986*a*; Collard and Griswold 1987). It was determined from the sequence information that both subunits of the SGP-2 are transcribed from a single mRNA into a glycosylated precursor protein of approximately 62 kD. This cytoplasmic precursor is subsequently sulfated, cleaved by proteolysis into 2 dissimilar disulfide-linked subunits and secreted into the lumen of the seminiferous tubules. Northern blot analysis of cellular RNA revealed very high concentrations of SGP-2 mRNA in Sertoli cells and epididymides with detectable amounts of the sequence in RNA from liver, brain, kidney, and spleen (Collard and Griswold 1987). Comparison of the SGP-2 sequence with the sequences in release 7.0 of the Protein Identification Resource (PIR) database did not identify any proteins with direct homology to SGP-2. However, using a program called "RELATE" it was determined that the SGP-2 sequence was related to the apolipoprotein A-1 family. This relationship may be a distant one but the probability that it resulted from chance is less than 2.1×10^{-5} (Dayhoff, Barber, and Hunt 1983). This purported relationship of SGP-2 with lipoproteins is consistent with some of the previously reported properties of the protein. Even though SGP-2 is acidic, highly charged and glycosylated, it has a number of hydrophobic properties (Griswold *et al.* 1986*b*; Griswold 1987). During purification procedures both SGP-2 and SGP-1 have a tendency to aggregate and fall out of solution, and both proteins elute from reverse phase HPLC columns as if they were hydrophobic (Griswold *et al.* 1986*b*). It was also shown that both of these proteins ultimately end up as components of the sperm membrane (Sylvester *et al.* 1984). A cDNA probe for SGP-1 has also been constructed but the sequence has not yet been reported (Griswold *et al.* 1986*a*).

Combined gas-phase sequence analysis of the protein and synthetic oligo-nucleotide screening of testis cDNA libraries resulted in the successful isolation of cloned cDNA and gene sequences for bovine and human anti-Müllerian hormone or Müllerian Inhibiting Substance (MIS) (Cate *et al.* 1986). The MIS protein was found to be a primary transcript of 58,000 which apparently undergoes post-translational modification to attain a monomeric molecular weight of 70 to 74 kD. Inhibin, which has been purported to be a

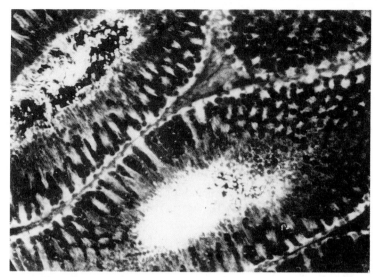

Fig. 3.5 Indirect immunofluorescence of SGP-2 in paraffin sections of rat testis. Rehydrated tissues sections were treated with rabbit anti-rat SGP-2 and as described in Fig. 3.2. The fluorescence coincides with Sertoli cell cytoplasm and the tails of spermatids. Approximately 400X.

Sertoli cell product (Chowdhury, Steinberger, and Steinberger 1978; Demoulin, Koulischer, Hustin, Hazee-Hagelstein, Lambotte, and Franchimont 1979; LcGac and deKretser 1982) has been isolated and characterized from ovarian follicular fluid (Mason *et al.* 1985). A cDNA complementary to the inhibin mRNA was constructed, cloned and sequenced. The ovarian inhibin was shown to be a protein consisting of 2 dissimilar subunits of 18 Kd and 14 kD cross-linked by disulfide bonds (Mason *et al.* 1985). At this time there is no information on the similarity of the ovarian inhibin to Sertoli cell produced inhibin. Of considerable interest in the studies on both MIS and inhibin are the extensive homologies found between both of these proteins and transforming growth factor β (TGF-β), with the homology between TGF-β and inhibin subunit β being particularly striking (Mason *et al.* 1985; Cate *et al.* 1986).

2 USE OF cDNA PROBES TO ASSAY FOR SERTOLI CELL FUNCTIONS

The cDNAs for both transferrin and SGP-2 have been used to examine Sertoli cell mRNA levels in cultured cells and in testicular tissue. Biotinylated cDNA probes were hybridized to paraffin sections of testis and visualized by combination with avidin-peroxidase. The cDNAs for both transferrin and for SGP-2 hybridized to mRNA that was located only in Sertoli cells (Morales *et al.* 1987) (Fig. 3.5). The amount of transferrin mRNA present in rat Sertoli cells cultured in the presence of hormones and vitamins was determined by dot-blot hybridization to ^{32}P-cDNA. High levels of transferrin mRNA in cultured cells was shown to be dependent on the presence of insulin, FSH and retinol in the culture medium (Huggenvik *et al.* 1987). The amount of transferrin mRNA in Sertoli cells in whole testis was shown to be decreased dramatically in rats which were vitamin A deficient (Griswold *et al.* 1986*a*; Hugly and Griswold 1987), and in rats which were hypophysectomized and maintained for 20 days (Griswold *et al.* 1986*a*). SGP-2 mRNA levels in Sertoli cells were also shown to be dramatically decreased after hypophysectomy but vitamin A deficiency produced a pronounced increase in SGP-2 mRNA levels per Sertoli cell so at least in this latter case the regulation of the mRNA's for transferrin and for SGP-2 is different (Griswold *et al.* 1986*a*; Hugly and Griswold 1987).

An interesting aspect of the studies described above is that a direct mathematical relationship was observed between the testicular levels of transferrin mRNA, but not SGP-2 mRNA, and the testicular weight. If the weight of the testis was decreased as a result of hypophysectomy or vitamin A deprivation or cryptorchidism, or increased as a result of restoration of FSH and testosterone or vitamin A, this direct relationship was observed. Testicular weight changes in these studies were the result of the loss or gain in the number of germ cells in the testis (Hugly and Griswold 1897). Other investigators using DNA synthesis inhibitors have demonstrated that the number of Sertoli cells per testis is limiting to the size of the testis and therefore to the daily sperm production (Amann 1970; Orth 1985). Thus, the amount of transferrin produced by the testis may be a limiting factor in the number of germ cells a Sertoli cell can support. It is not surprising that this relationship does not hold for SGP-2 mRNA levels since localization studies suggest that SGP-2 is involved with spermatozoa following spermiation and not necessarily during spermatogenesis or spermiogenesis (Sylvester *et al.* 1984).

V   Molecular studies of Sertoli cell cyclicity

1   CYCLIC PATTERN OF PROTEIN SECRETION

Several Sertoli cell specific proteins were detected in the medium of cultured seminiferous tubules of defined stages. The proteins secreted from these tubules exhibited a pronounced stage-dependent cyclicity of secretion. These experiments were possible because in the rat, stage specific segments can be manually dissected from tubules which are viewed by transillumination (Parvinen and Vanha-Pertulla 1982). The known proteins secreted in a cyclic manner are: plasminogen activator, androgen binding protein, sulfated glycoproteins-1 and 2, testicular transferrin and cyclic protein-2 (Parvinen

1982; Wright, Parvinen, Musto, Gunsalus, Phillips, Mather, and Bardin 1983; Shabanowitz, DePhilip, Crowell, Tres, and Kierszenbaum 1986; Vihko, Topperi, and Parvinen 1987). Plasminogen activator, for example, is principally secreted during stages VII and VIII (Lacroix *et al.* 1981; Vihko *et al.* 1987). It has been speculated that the transfer of the preleptotene spermatocytes through the tight junctions of the neighbouring Sertoli cells during stages VII-VIII of the cycle involves the secretion of plasminogen activator by the Sertoli cells which may be instrumental in the alteration of the junctional specializations (Lacroix, Smith, and Fritz 1977). It has also been speculated that plasminogen activator may play a role in the release of maturing spermatids from the seminiferous epithelium at stage VIII (Parvinen 1982). Nevertheless, until now this hypothesis remains to be proven and further investigation is needed to clarify the function of this protein. Androgen binding protein showed maximal secretion at stages VII and VIII, remaining relatively high from stages IX through XIV. Low detectable levels were found between stages I and VI (Ritzen, Boitani, Parvinen, French, and Feldman 1982). Cyclic protein-2 (CP-2), another Sertoli cell product, exhibited a pronounced stage dependent cyclicity of secretion (Wright *et al.* 1983). This protein was detected at 30-fold higher levels from tubules at stage VI than from tubules at stages XII-XIV. The role of CP-2 in spermatogenesis has not yet been determined (Wright *et al.* 1983; Wright, and Luzzaraga 1986).

Sulfated glycoproteins 1 and 2 (also designated S70 and S45 and S35 by Shabanowitz *et al.* 1986) apparently are expressed in sexually mature rats in a stage dependent manner as shown by two dimensional electrophoresis of secreted proteins. Both proteins were seen at stages VII–VIII but not at stage IV of the cycle. Unlike sulfated glycoprotein 1 (S70) sulfated glycoprotein 2 (S45-S35) was also reported at stages X and XIV (Shabanowitz *et al.* 1986). Preliminary evidence using immunocytochemistry indicated, however, that SGP-1 is expressed throughout the 14 stages of the cycle (unpublished observations of the authors).

Thus, the majority of these proteins showed an apparent maximal secretion during stages VI through VIII of the cycle. This part of the cycle coincides with the movement of preleptotene spermatocytes from the basal to the adluminal compartment which results in the temporary formation of an intermediate compartment and with spermiation which is characterized by the release of the late spermatids and the retention of the residual bodies at the apex of Sertoli cell processes.

Testicular transferrin was also shown to be expressed in a stage dependent manner. However, unlike the preceding proteins, testicular transferrin is maximally secreted at stages XIII–XIV and minimally at stages VII–VIII (Wright *et al.* 1983). All of the available data shows that transferrin is made in Sertoli cells throughout all stages of the cycle but it also suggests a special role for this protein during meiotic division. Transferrin has been shown to be important because of its actions in stimulating cell proliferation. It is possible that the increased synthesis of transferrin in stages XIII–XIV of the cycle is stimulatory to the meiotically dividing cells and thus transferrin could be considered a major regulatory factor in spermatogenesis (Morales *et al.* 1987, in press).

2 QUANTITATIVE *IN SITU* HYBRIDIZATION

The combination of *in situ* hybridization and quantitative radioautography for the examination of mRNA expression in Sertoli cells in relation to the different stages of the cycle of the seminiferous epithelium offers a powerful alternative to other methods currently in use. The dissection procedure followed by transillumination is laborious, subject to dissection errors, limited by the amount of tissue available, and the results can be difficult to interpret due to the many cell types present in the testis (Parvinen 1982). As discussed above, the use of *in situ* hybridization in combination with biotinylated cRNA allowed the specific localization of transferrin and SPG-2 mRNAs in the main body of the Sertoli cells throughout the tissue section (Fig. 3.6). Thus, based on these results the localization and stage dependent level of transferrin and SGP-2 was also examined in rat testis by *in situ* hybridization with specific cRNA probes. For this study, the cDNAs for transferrin and SGP-2 was transcribed into ^{3}H-cRNA and subsequently hybridized *in situ* to paraffin sections of adult rat testis (Morales *et al.* 1987, in press). Meaningful results obtained by this technique were dependent on the observation of Bustos-Obregon (1970) and Wing and Christensen (1982) that the number of Sertoli cells per unit length of seminiferous tubules remains constant throughout the 14 stages of the cycle. The number of Sertoli cells per cross section and the proportion of their total volume in perfect transverse sections of seminiferous tubules also remains constant. Thus, the relative levels of mRNA per Sertoli cell could be quantified by simply counting silver grains per cross section of seminiferous tubules. When this was done it became apparent that transferrin mRNA levels were maximal at stages XIII and XIV. Conversely, SGP-2 mRNA levels were maximal at stages VII–VIII (Morales *et al.* 1987, in press).

All the evidence clearly suggests that Sertoli cells assume cyclic functional modes which correlate with the different stages of the cycle of the seminiferous epithelium. Sertoli cells associated with specific cellular associations may respond to different hormonal stimuli and produce different levels of secretion products. Similarly, the complex interaction between stage specific germ cells and Sertoli cells may be of major importance in the regulation of gene expression in the Sertoli cells. Therefore, further investigation of the heterogeneity of Sertoli cells and the factors which control it, is of major importance.

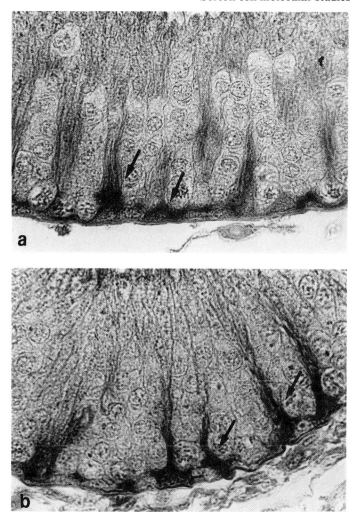

Fig. 3.6   Cross sections of seminiferous tubules hybridized *in situ* with biotinylated cRNA probes transcribed from SP65 plasmids, and visualized with the avidin-glucose oxidase/tetranitroblue reaction. Black deposits of insoluble formazan indicate that both transferrin mRNA (a) and SGP-2 mRNA (b) are localized in Sertoli cells but not in germ cells (arrows).

3   SYNCHRONIZATION OF THE STAGES OF THE CYCLE: A NEW APPROACH FOR THE STUDY STAGE REGULATION OF SERTOLI CELLS

Vitamin A deficiency (VAD) causes progressive germ cell depletion and cessation of spermatogenesis resulting in seminiferous tubules which contain Sertoli cells, spermatogonia and a small number of spermatocytes (Mason

1933; Thompson, Howell, and Pitt 1964; Huang and Hembree 1979; Huang, Dysenfurth, and Hembree 1983). The spermatogenic arrest appears to be at the preleptotene stage of spermatocytes (Huang and Hembree 1979; Morales *et al.* 1987) and can be rapidly reversed by vitamin A replacement (Mason 1933; Huang *et al* 1983).

In experiments initially designed to study the influence of vitamin A *in vivo* on the transcriptional activity and secretion of various specific Sertoli cell products we observed a stage synchronization of most seminiferous tubules in VAD male rats subsequently treated with retinol (Morales and Griswold 1986). Since the duration of the spermatogenic cycle is constant in a given species and is known in this particular strain (Clermont 1972), the stages of the seminiferous tubules of the retinol treated VAD rats could be predicted with reasonable accuracy by calculating the evolution of the most advanced cell (preleptotene spermatocyte) during the period between the first administration of retinol and the day of sacrifice. In the majority of the cases synchronous development of seminiferous tubules evolved in the predicted way particularly during the shorter times (Fig. 3.7). With longer periods of time (62 78 days), synchronization persisted, but the synchronized stages drifted beyond the predicted range, indicating either that the duration of the cycle was longer in our Sprague-Dawley strain or that the initial estimations were shorter than the real values. Nevertheless, synchronization could be maintained at least up to 78 days after the initiation of retinol administration, through 3 rounds of spermiation. This evidence would suggest a type A1 spermatogonial arrest in addition to the already depicted spermatogenic arrest at the preleptotene spermatocyte stage. Furthermore, this system presents the advantage that synchronization takes place simultaneously in both testes. Thus, while one testis may be fixed and submitted to morphological and histochemical analyses, the second testis can be utilized for other types of studies. Using this experimental approach testes enriched in seminiferous tubules at specific stages can be obtained. Therefore, this model should be important in any future studies which will require the isolation of large amounts of stage-specific seminiferous tubules such as the isolation of stage-specific and cycling proteins, the isolation of enriched populations of germ cells and of Sertoli cells from specific stages of the cycle.

VI   Summary and conclusions

The "nurse cell' concept developed as a result of the morphological relationships between germ cells and Sertoli cells, and because of the junctional barrier defined by Sertoli cells which allowed the creation of a compartment in which the constituents could be regulated by Sertoli cells. The molecular approach to studies of the role of Sertoli cells in spermatogenesis has already led to a better understanding of the "nurse cell" concept. Molecular studies on the function of Sertoli cells have for the most part been dependent on the

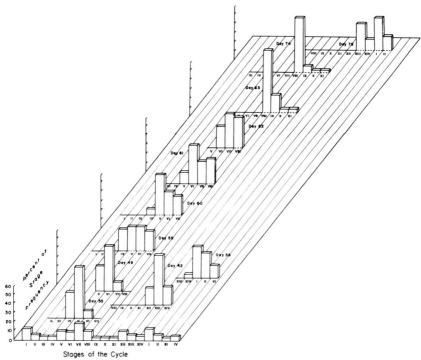

Fig. 3.7 Schematic representation of a survey of testicular sections from rats recovering from vitamin-A deficiency. Vitamin-A deficient rats were treated with two subcutaneous injections of retinol followed by daily oral administration of retinol and then sacrificed between 35 and 78 days after the initiation of recovery. The Roman numerals indicate the stages of the cycle according to the classification of Leblond and Clermont (1952). The columns in the first row represent the frequency of stage appearance in cross sections of normal Sprague-Dawley rat testes. The rest of the grouped columns represent the frequency of stage appearance observed in the experimental testes whereas the solid lines under these columns indicate the predicted stages. The stages were predicted on the basis of the amount of time following retinol treatment and the relative duration of each stage of the cycle.

Sertoli cell culture techniques which were first published in 1975 (Dorrington and Fritz 1972; Steinberger *et al.* 1975*a*; Welsh and Wiebe 1975). As a result of this technical innovation Sertoli cells which remained responsive to hormones and which continued to carry on secretory activities could be obtained relatively free of other cell types. In the spent medium from cultured Sertoli cells specific secretion products could be detected such as ABP, transferrin, and SGP-2; components with inhibin-like activity; and metabolic products such as lactate. Thus, for the first time some molecular correlates to the "nurse cell" role of Sertoli cells have been described. In addition to the close morphological relationships between germ cells and Sertoli cells it has now been demonstrated that protein products of the Sertoli cells directly interact with germ cells. The Sertoli cell-mediated iron transport to germ cells via transferrin is the clearest demonstration of this interaction. Both SGP-1 and SGP-2 interact with spermatozoa and can be found tightly associated with the plasma membrane of these cells. As more of the secretion products of Sertoli cells are characterized the important role of these cells in spermatogenesis and spermiogenesis will be underscored.

The identification and characterization of the Sertoli cell secretion products is important information to obtain, but an understanding of the function of these products also requires temporal knowledge about their synthesis. The interdependence of the morphological observations of the testis and the new technology of molecular biology is most clearly illustrated by the studies involving quantitative *in situ* hybridization. This technique has been utilized to quantify the amount of transferrin and SGP-2 mRNA present in Sertoli cells associated with different stages of the cycle of the seminiferous epithelium and to establish positively the Sertoli cell location of a specific mRNA (Morales *et al.* 1987). The technique is potentially applicable to any Sertoli cell specific protein product for which a cDNA probe becomes available. The discovery of a method to synchronize the testis of rats to a given stage or group of stages should further stimulate progress in determining the molecular changes which occur in Sertoli cells during the cycle of the seminiferous epithelium.

References

Alexander, N. T. (1977). Immunological aspects of vasectomy. In *Immunological Influence in Human Fertility*. (ed. B. Boettcher) pp. 25–39. Academic Press, Sydney.

Amann, R. P. (1970). Sperm production rates. In *The Testis* (eds A. D. Johnson, W. R. Gomes and N. L. Vandermark.) Vol. I, pp. 107–239. Academic Press, New York.

Aoki, A. and Fawcett, D. W. (1975). Impermeability of Sertoli cell junctions to prolonged exposure to peroxidase. *Andrologia.* 7, 63–76.

Aisen, P. and Listowsky, L., (1980). Iron transport and storage proteins. *A. Rev. Biochem.* **49**, 357–393.

Bellvé, A. R. and Moss, S. (1983). Monoclonal antibodies as probes of reproductive mechanism. *Biol. Reprod.* **28**, 1–26.

Blaner, W. S., Galdieri, M., and Goodman, D. S. (1986). Distribution and levels of cellular retinol and cellular retinoic acid-binding protein in various types of rat testis cells. *Biol. Reprod.* **36**, 130–137.

Blaschuk, O. and Fritz, I. B. (1984). Isoelectric forms of clusterin isolated from ram rete testis fluid and from secretions of primary cultures of ram and rat Sertoli-cell enriched cultures. *Can. J. Biochem. Cell Biol.* **62**, 456–461.

Brokelman, J. (1963). Fine structure of germ cells and Sertoli cells during the cycle of the seminiferous epithelium in the rat. *Z. Zellforsch.* **59**, 820–850.

Brown, W. R. A. (1985). Immunohistochemical localization of the transferrin receptor in the seminiferous epithelium of the rat. *Gamete Res.* **12**, 317–326.

Bugge, H. P. and Ploen, L. (1986). Changes in the volume of Sertoli cells during the cycle of the seminiferous epithelium in the rat. *J. Reprod. Fert.* **76**, 39–42.

Bustos-Obregon, E. (1970). On Sertoli cell number and distribution in rat testis. *Arch. Biol.* (Liege). **81**, 99–108.

Byers, S. W., Hadley, M. A., Djakiew, D., and Dym, M. (1986). Growth and characterization of polarized monolayers of epididymal epithelial cells and Sertoli cells in dual environment culture chambers. *J. Androl.* **7**, 59–69.

Cameron, D. F. and Markwald, R. R. (1981). Structural response of adult rat Sertoli cells to peritubular fibroblasts in vitro. *Am. J. Anat.* **160**, 343–358.

Cameron, D. F. and Snydle, E. (1985). Selected histochemistry of Sertoli cells 2. Adult rat Sertoli cells in co-culture with peritubular fibroblasts. *Andrologia* **17**, 185–193.

Carson, D. D. and Lennarz, W. J. (1983). Vitamin A deprivation selectively lowers uridine nucleotide pools in cultured Sertoli cells. *J. biol. Chem.* **258**, 1632–1636.

Carson, D. D., Rosenberg, L. L., Blaner, W. S., Kato, M., and Lennarz, W. (1984). Synthesis and secretion of a novel binding protein for retinol by a cell line derived from Sertoli cells. *J. biol. Chem.* **259**, 3117–3123.

Cate, R. L., Mattaliano, R. J., Hession, S., Tizard, R., Farber, N. M., Cheung, A. Ninfa, E. G., Ramachandran, K. L., Ragin, R. C., Manganaro, T. F., MacLaughlin, and Donahoe, P. K. (1986). Isolation of the bovine and human genes for Müllerian inhibiting substance and expression of the human gene in animal cells. *Cell* **45**, 685–698.

Cavicchia, J. C. and Dym, M. (1977). Relative volume changes of Sertoli cells in monkey seminiferous epithelium: A stereological analysis. *Am. J. Anat.* **150**, 501–507.

Cheng, C. Y. and Bardin, C. W. (1986). Rat testicular testibumin is a protein responsive to follicle stimulating hormone and testosterone that shares immunodeterminants with albumin. *Biochemistry* **25**, 5276–5288.

Chowdhury, M., Steinberger, A. and Steinberger, E. (1978). Inhibition of *de novo* synthesis of FSH by the Sertoli cell factor (SCF). *Endocrinology* **103**, 644–647.

Chung, M. C-M. (1984). Structure and function of transferrin. *Biochem. Educ.* **12**, 146–154.

Ciechanover, A., Schwartz, A. L., Dautry-Varsat, A., and Lodish, H. F. (1983). Kinetics of internalization and recycling of transferrin and the transferrin receptor in a human hepatoma cell line. *J. biol. Chem.* **258**, 9681–9689.

Clermont, Y. (1972). Kinetics of spermatogenesis in mammals: Seminiferous epithelium cycle and spermatogonial renewal. *Physiol. Rev.* **52**, 198–235.

Clermont, Y., McCoshen, J., and Hermo, L. (1980). Evolution of the endoplasmic reticulum in Sertoli cell cytoplasm encapsulating the head of late spermatids in the rat. *Anat. Rec.* **196**, 83–99.

Clermont, Y., Morales, C., and Hermo, L. (1987) Endocytic activities of Sertoli cells in the rat. *Ann. N.Y. Acad. Sci.* (in press).

Collard, M. and Griswold, M. D. (1987). Biosynthesis and molecular cloning of sulfated glycoprotein-2 secreted by rat Sertoli cells. *Biochemistry* **26**, 3297–3303.

Danahey, D. G., Letscher, R. L., and DePhillip, R. M. (1986). Evidence for the establishment of functional polarity by Sertoli cells in vitro. *J. Cell Biol.* **183**, 484a.

Dayhoff, M. O., Barker, W. C., and Hunt, L. T. (1983). Establishing homologies in protein sequences. *Methods Enzym.* **91**, 524–545.

deJong, F. H. and Robertson, D. M. (1985). Inhibin: 1985 update on action and purification. *Mol. Cell. Endocr.* **42**, 95–103.

Demoulin, A., Koulischer, L., Hustin, J., Hazee-Hagelstein, M. T., Lambotte, R., and Franchimont, P., (1979). Organ culture of mammalian testis. III. Inhibin secretion. *Hormone Res.* **10**, 177–190.

Dietert, S. F. (1966). Fine structure of the formation and fate of the residual bodies of mouse spermatozoa with evidence for the partition of lysosomes. *J. Morphol.* **120**, 317–346.

Djakiew, D., Hadley, M. A. Byers, S. W., and Dym, M. (1986). Transferrin-mediated transcellular transport of ^{59}Fe across confluent epithelial sheets of Sertoli cells grown in bicameral cell culture chambers. *J Androl.* **7**, 355–366.

Dorrington, J. H. and Fritz, I. B. (1975). Cellular localization of 5α-reductase and 3α-hydroxysteroid dehydrogenase in the seminiferous tubule of the rat testis. *Endocrinology* **96**, 879–889.

Dym, M. (1972). The fine structure of the monkey Sertoli cell and its role in establishing the blood-testis barrier. *Biol. Reprod.* **7**, 129a.

Dym, M. and Fawcett, D. (1970). The blood-testis barrier in the rat and the physiological compartmentalization of the seminiferous epithelium. *Biol. Reprod.*, **3**, 308–326.

Fawcett, D. W. (1975). Ultrastructure and function of the Sertoli cell. In *Handbook of Physiology*, **7**. *Endocrinology*. **V**. (ed. E. B. Aswood and R. O. Greep) pp. 21–55. American Physiological Society, Bethesda, Maryland.

Fawcett, D. W. (1977). The ultrastructure and functions of the Sertoli cell. In *Frontiers in contraceptive development*. (ed. R. O. Greep and M. A. Koblinsky) pp. 302–320. MIT Press, Cambridge, Massachusetts.

Fawcett, D. W. and Phillips, D. M. (1969). Observations on the release of spermatozoa and on changes in the head during passage through epididymis. *J. Reprod. Fert.* **6**, 405–418.

Flickinger, C. and Fawcett, D. W. (1967). The junctional specializations of Sertoli cells in the seminiferous epithelium. *Anat. Rec.* **158**, 207–222.

Franke, W. D., Grund, C., Fink, A., Weber, K. Jockush, B. M., Zentgraf, H., and Osborn, M. (1978). Location of actin in the microfilament bundles associated with the junctional specializations between Sertoli cells and spermatids. *Biol. Cell.* **31**, 7–14.

Fritz, I. (1978). Sites of actions of androgens and follicle stimulating hormone on cells of the seminiferous tubule. In *Biochemical Actions of Hormones*, Vol. V, (ed. G. Litwack), pp. 249–278. New York, Academic Press.

Fritz, I. B., Louis, B. G., Tung, P. S., Griswold, M. D., Rommerts, F. G., and Dorrington, J. H. (1975). In *Hormonal Regulation of Spermatogenesis* (eds F. S. French, V. Hansson, E. M. Ritzen and S. Nayfeh) pp. 367–382, Plenum Press, New York.

Galdieri, M., Ziparo, E., Palombi, F., Russo, M. A., and Stefanini, M. J., (1981). Pure Sertoli cell cultures: a new model for the study of somatic-germ cell interactions. *J. Androl.* **2**, 249–254.

Gravis, C. (1979). Inhibition of spermiation using dibutyryl cAMP. *Anat. Rec.*, **193**, 553a.

Griswold, M. D. (1987). Protein secretions of Sertoli cells. *Int. Rev. Cytol.*, *in press*.

Griswold, M. D., Collard, M., Hugly, S., and Huggenvik, J. (1986*a*). The use of specific cDNA probes to assay Sertoli cell functions. In *Molecular Aspects of Reproduction*. (ed. D. Dhindsa) pp. 301–317. Plenum Press, New York.

Griswold, M. D., Roberts, K., and Bishop, P. (1986*b*). Purification and characterization of a sulfated glycoprotein secreted by Sertoli cells. *Biochemistry* **25**, 7265–7270.

Griswold, M. D., Solari, A., Tung, P. S., and Fritz, I. B. (1977). Stimulation by FSH of DNA synthesis and of mitosis in cultured Sertoli cells prepared from the testes of immature rats. *Mol. Cell. Endocr.* **7**, 151–165.

Griswold, M. D., Huggenvik, J., Skinner, M. K., and Sylvester, S. R. (1984). The interactions of Sertoli cells glycoproteins with germinal cells in the testis. In *Endo-*

crinology. (eds. F. Labrie and L. Prouix) pp. 619–622. Elsevier Science Publishers, New York.

Grove, B. D. and Vogl, A. W. (1986). Actin-binding proteins in ectoplasmic specializations of rat Sertoli cells. *J. Cell Biol.* **103**, 538a.

Hadley, M. A., Byers, S. W., Suarez-Quian, C. A., Klienman, H. K., and Dym, M., (1985). Extracellular matrix regulates Sertoli cell differentiation, testicular cord formation, and germinal cell development. *J. Cell Biol.* **101**, 1511–1522.

Hadley, M. A., Djakiew, D., Byers, S. W., and Dym, M. (1987). Polarized secretion of androgen-binding protein and transferrin by Sertoli cells grown in a bicameral culture system. *Endocrinology* **120**, 1097–1103.

Hilscher, B., Passia, D., and Hilscher, W. (1979). Kinetics of the enzymatic pattern in the testis. I. Stage dependence of enzymatic activity and its relation to cellular interactions in the testis of the Wistar rat. *Andrologia.* **11**, 169–181.

Holmes, S. D., Bucci, L. R., Lipshultz, L. I., and Smith, R. G. (1983). Transferrin binds specifically to pachytene spermatocytes. *Endocrinology* **113**, 1916–1918.

Holmes, S. D., Lipshultz, M. D., and Smith, R. G. (1982). Transferrin and gonadal dysfunction in man. *Fertil. Steril.* **38**, 600–604.

Holmes, S. D., Lipshultz, M. D., and Smith, R. G. (1984). Regulation of transferrin secretion by human Sertoli cells cultured in the presence or absence of human peritubular fibroblasts. *J. clin. Endocr. Metab.* **59**, 1058–1062.

Hopkins, C. R., (1983). Intracellular routing of transferrin and transferrin receptors in epidermoid carcinoma A431 cells. *Cell* **35**, 321–330.

Hu, H. Y., Gardner, J., Aisen, P., and Skoultchi, A. I. (1977). Inducibility of transferrin receptors of Friend erythroleukemic cells. *Science, N.Y.* **197**, 559–561.

Huang, H. F. S., Dyrenfurth, I., and Hembree, W. C. (1983). Endocrine changes associated with germ cell loss during vitamin A deficiency and vitamin A induced recovery of spermatogenesis. *Endocrinology.* **12**, 1163–1171.

Huang, H. F. S. and Hembree, W. C. (1979). Spermatogenic response to vitamin A in vitamin A deficient rats. *Biol. Reprod.* **21**, 891–904.

Huggenvik, J., Idzerda, R. L., Haywood, L., Lee, D. C., McKnight, G. S., and Griswold, M. D. (1987). Transferrin messenger ribonucleic acid: Molecular cloning and hormonal regulation in rat Sertoli cells. *Endocrinology* **120**, 332–340.

Huggenvik, J., Sylvester, S. R., and Griswold, M. D. (1985). Control of transferrin mRNA synthesis in Sertoli cells. *Ann. N.Y. Acad. Sci.* **438** 1–7.

Hugly, S. and Griswold, M. D. (1987). Regulation of levels of specific Sertoli cell mRNAs by vitamin A. *Devl. Biol.* **121**, 316–324.

Hutson, J. C. (1983). Metabolic cooperation between Sertoli cells and peritubular cells in culture. *Endocrinology* **112**, 1375–1381.

Hutson, J. C. and Stocco, D. M. (1981). Peritubular cell influence on the efficiency of androgen binding protein secretion by Sertoli cells in culture. *Endocrinology* **108**, 1362–1368.

Idzerda, R. L., Huebers, H., Finch, C. A., and McKnight, G. S. (1986). Rat transferrin gene expression: Tissue specific regulation by iron deficiency. *Proc. natn. Acad. Sci. U.S.A.* **83**, 3723–3727.

Joseph, D. R., Hall, S. H., and French, F. S. (1985). Identification of complementary DNA clones that encode rat androgen binding protein. *J. Androl.* **6**, 291–298.

Joseph, D. R., Hall, S. H., and French, F. S. (1987). Rat androgen-binding protein: Evidence for identical subunits and amino acid sequence homology with human sex hormone-binding globulin. *Proc. natn. Acad. Sci. U.S.A.* **84**, 339–343.

Josso, N. and Picard, J. (1986). Anti-Müllerian Hormone, *Physiol. Rev.* **66**, 1038–1090.

Josso, N., Picard, J., and Tran, D. (1977). The anti-Müllerian Hormone, *Recent. Prog. Horm. Res.* **33**, 117–167.

Karl, A. F. and Griswold, M. D. (1980). Prolonged ABP synthesis by Sertoli cells cultured in defined medium. *Cell Biol. Int. Rep.* **4**, 669–674.

Keeping, H. S., Winters, S. J., and Troen, P. (1985). Identification of androgen-binding protein from testis cytosol and Sertoli cell culture medium of the Cyno-molgus monkey, *Macaca fascicularis. Endocrinology* **117**, 1521–1529.

Kerr, J. B. and de Kretser, D. M. (1975). Cyclic variations in Sertoli cell lipid content throughout the spermatogenic cycle of the rat. *J. Reprod. Fert.* **43**, 1–8.

Kierszenbaum, A. L., Crowell, J. A., Shabinowitz, R. B., DePhilip, R. M., and Tres, L. (1986). Protein secretory patterns of rat Sertoli and peritubular cells are influenced by culture conditions. *Biol. Reprod.* **35**, 239–251.

Kissinger, C., Skinner, M. K., and Griswold, M. D. (1982). Analysis of Sertoli cell secreted proteins by two-dimensional gel electrophoresis. *Biol. Reprod.* **27**, 233–240.

Lacroix, M., Parvinen, M., and Fritz, I. B. (1981). Localization of plasminogen activator in discrete portions (stages VII and VIII) of the seminiferous tubule. *Biol. Reprod.* **25**, 143–146.

Lacroix, M., Smith, F. E., and Fritz, I. B. (1977). Secretion of plasminogen activator by Sertoli cell enriched cultures. *Molec. Cell. Endocr.* **9**, 227–236.

Lalli, M. F., Tang, X. M., and Clermont (1984). Glycoprotein synthesis in Sertoli cells during the cycle of the seminiferous epithelium of adult rats: a radioautographic study. *Biol. Repod.* **30**, 493–505.

Leblond, C. P. and Clermont, Y. (1952). Definition of the stages of the cycle of the seminiferous epithelium in the rat. *Ann. N.Y. Acad. Sci.* **55**, 548–573.

LeGac, F. and deKretser, D. M. (1982). Inhibin production by Sertoli cell cultures. *Molec. Cell. Endocr.* **28**, 487–498.

Ling, N., Ying, S-Y., Ueno, N., Esch, F., Denoroy, L., and Guillemin, R. (1985). Isolation and partial characterization of a M_r 32 000 protein with inhibin activity from porcine follicular fluid. *Proc. natn. Acad. Sci. U.S.A.* **82**, 7217–7221.

Lipshultz, L. I., Murthy, L., and Tindall, D. J. (1982). Characterization of human Sertoli cells *in vitro. J. clin. Endocr. Metab.* **82**, 228–237.

Malone, J. (1979). A study of Sertoli-spermatid tubulobulbar complexes in selected mammals. *Anat. Rec.* **193**, 610a.

Mason, K. E. (1933). Differences in testes injury and repair after vitamin A deficiency, vitamin E deficiency and inanition. *Am. J. Anat.* **52**, 153–239.

Mason, A. J., Hayflick, J. S., Ling, N., Esch, F., Ueno, N., Ying, S-Y., Guilliman, R., Niall, H., and Seeburg, P. H. (1985). Complementary DNA sequences of ovarian follicular fluid inhibin show precurser structure and homology with transforming growth factor-β. *Nature, Lond.* **318**, 659–663.

Mather, J. P. (1980). Establishment and characterization of two distinct mouse testicular epithelial cell lines. *Biol. Reprod.* **23**, 243–252.

Mather, J. P., Zhuang, L., Perez-Infante, V., and Philips, D. M. (1982). Culture of testicular cells in hormone-supplemented serum-free medium. *Ann. N.Y. Acad. Sci.* **383**, 44–68.

Mather, J. P., Wolpe, S. D., Gunsalus, G. L., Bardin, C. W., and Phillips, D. M. (1984). Effect of purified and cell-produced extracellular matrix components on Sertoli cell function. *Ann. N.Y. Acad. Sci.* **438**, 572–575.

McGinley, D. M., Pozalaky, Z., Porvaznik, M. and Russell, L. D. (1979). Gap junctions between Sertoli cells and germ cells of rat seminiferous tubules. *Tissue and Cell.* **11**, 741–754.

Millete, C. F. (1979). Appearance and partitioning of plasma membrane antigens during mouse spermatogenesis. In *The spermatozoon* (eds D. W. Fawcett and J. M. Bedford) pp. 177–186. Urban and Schwarzenberg Inc., Baltimore-Munich.

Millette, C. F. and Bellvé, A. R. (1980). Selective partitioning of membrane antigens during mouse spermatogenesis, *Devl. Biol.* **79**, 309–324.

Morales, C., Clermont, Y., and Hermo, L. (1985). Nature and function of endocytosis in Sertoli cells of the rat. *Am. J. Anat.* **173**, 203–217.

Morales, C., Clermont, Y., and Nadler, N. J. (1986). Cyclic endocytic activity and kinetics of lysosomes in Sertoli cells of the rat: A morphometric analysis. *Biol. Reprod.* **34**, 297–218.

Morales, C. and Griswold, M. D. (1987). Retinol-induced stage synchronization in seminiferous tubules of the rat. *Endocrinology*, **121**, 432–434.

Morales, C., Hugly, S., and Griswold, M. D. (1987). Stage dependent levels of specific mRNA transcripts in Sertoli cells. *Biol Reprod.* **36**, 1035–1046.

Morales, C., Sylvester, S. R., and Griswold, M. D. Transport of iron and transferrin synthesis by the seminiferous epithelium of the rat *in vivo*. *Biol. Reprod.* (in press).

Nicander, L. (1963). Some ultrastructural features of mammalian Sertoli cells. *J. Ultrastruct. Res.* **8**, 190–191.

Nicander, L. (1967). An electron microscopical study of cell contacts in the seminiferous tubules of some mammals. *Z. Zellforsch.*, **83**, 375–397.

Niemi, M. and Kormano, M. (1965). Cyclical changes in and significance of lipids and acid phosphatase activity in the seminiferous tubules of the rat testis. *Anat. Rec.* **151**, 159–170.

Octave, J-N., Schneider, Y-J., Trouet, A., and Crichton, R. R. (1983). Iron uptake and utilization by mammalian cells. I: Cellular uptake of transferrin and iron. *Trends Bioch. Sci.* **8**, 217–220.

Orlando, C., Caldini, T., Wood, W. G. Strasburger, C. J. Natali, A., Maver, A., Forti, G., and Serio, M. (1985). Ceruloplasmin and transferrin in human seminal plasma: are they an index of seminiferous tubular function? *Fertil. Steril.* **43**, 290–294.

Orth, J. (1985). Effect of Sertoli cell depletion during neonatal life on spermatid production in adults. *Anat. Rec.* **211**, 144a.

Parvinen, M. (1982). Regulation of the seminiferous epithelium. *Endocr. Rev.* **3**, 404–417.

Parvinen, M. and Ruokonen (1982). Endogenous steroids in rat seminiferous tubules. Comparison of different spermatogenic stages isolated by transillumination-assisted microdisection. *J. Androl.* **3**, 211–220.

Parvinen, M. and Vanha-Pertula, T. (1972). Identification and enzyme quantitation of the stages of the seminiferous epithelial wave in the rat. *Anat. Rec.* **174**, 435–450.

Perez-Infante, V., Bardin, C. W., Gunsalus, G. L., Musto, N. A., Rich, K. A., and Mather, J. P. (1986). Differential regulation of testicular transferrin and androgen-binding protein secretion in primary cultures of rat Sertoli cells. *Endocrinology* **118**, 383–392.

Perrard-Sapori, M. H., Saez, J. M., and Dazord, A. (1985). Hormonal regulation of proteins secreted by cultured pig Sertoli cells: characterization by two-dimensional polyacrylamide gel electrophoresis. *Molec. Cell. Endocr.* **43**, 189–197.

Posalaki, Z., Szabo, D., Basci, E., and Okros, I. (1968). Hydrolitic enzymes during spermatogenesis in rat. An electron microscopic study. *J. Histochem. Cytochem.* **16**, 249–262.

Quinlivan, W. L. G. and Sullivan, H. (1972). The identity and origin of antigens in human semen. *Fertil. Steril.* **23**, 873–878.

Rambourg, A., Clermont, Y., and Hermo, L. (1979). Three-dimensional architecture of the Golgi apparatus in Sertoli cells of the rat. *Am. J. Anat.* **154**, 455–475.

Ritzen, E. M., Boitani, C., Parvinen, M., French, F. C., and Feldman, M. (1982). Stage-dependent secretion of ABP by rat seminiferous tubules. *Molec. Cell. Endocr.* **25**, 25–33.

Ritzen, E. M., Hansson, V., and French, F. S. (1981). The Sertoli cell. In *The Testis. Comprehensive Endocrinology*. (ed. H. Burger and D. deKretser) pp. 171–194. Raven Press, New York.

Ross, M. H. (1976). The Sertoli cell junctional specialization during spermiogenesis and at spermiation. *Anat. Rec.* **186**, 79–104.

Roberts, T. K. and Boettcher, B. (1971). Iron-binding proteins in the reproductive tract. *J. Reprod. Fert.* **24**, 129.

Russell, L. D. (1977*a*). Observations on rat Sertoli ectoplasmic ("junctional") specializations in their association with germ cells of the rat testis. *Tissue and Cell.* **9**, 475–498.

Russell, L. D. (1977*b*). Movement of spermatocytes from the basal to the adluminal compartment of the rat testis. *Am. J. Anat.* **148**, 313–328.

Russell, L. D. (1978*a*). The blood-testis barrier and its formation relative to spermatocyte maturation in the adult rat: A lanthanum tracer study. *Anat. Rec.* **190**, 99–112.

Russell, L. D. (1978*b*). Testosterone induced deformities in rat spermiogenesis: Failure of tubulobulbar complexes to form in late spermatids. *Anat. Rec.* **190**, 527a.

Russell, L. D. (1977*c*). Desmosome-like junctions between Sertoli cells and germ cells in the rat testis. *Am. J. Anat.* **148**, 313–328.

Russell, L. D. (1979*a*). Further observations on tubulobulbar complexes formed by late spermatids and Sertoli cells in the rat testis. *Anat. Rec.* **194**, 213–232.

Russell, L. D. (1979*b*). Spermatid-Sertoli tubulobulbar complexes as devices for elimination of cytoplasm from the head region of late spermatids of the rat. *Anat. Rec.* **194**, 233–246.

Russell, L. D. (1984). Spermiation—the sperm release process: Ultrastructural observations and unresolved problems. In *Electron microscopy in biology and medicine. Ultrastructure of Reproduction* (eds J. van Blerkon and P. M. Motta) pp. 46–65. Plenum Press, New York.

Russell, L. D. and Clermont, Y. (1976). Anchoring device between Sertoli cells and later spermatids in rat seminiferous tubules. *Anat. Rec.*, **185**, 259–278.

Russell, L. D., Gardner, R. L., and Webber, J. (1986). Reconstruction of a type-B configuration monkey Sertoli cell: Size, shape, and configurational and specialized cell-to-cell relationships. *Am. J. Anat.* **175**, 73–90.

Russell, L. D., Tallon-Doran, M., Webber, J. E., Wong, V., and Peterson, R. N. (1983). Three-dimensional reconstruction of rat stage V Sertoli cell: III. A study of specific cellular associations. *Am. J. Anat.* **167**, 181–192.

Sanborn, B. M., Wagle, J. R., Steinberger, A., and Greer-Emmert, D. (1986). Maturational and hormonal influences on Sertoli cell function. *Endocrinology* **118**, 1700–1709.

Sanborn, B. M., Caston, L. A., Buzak, S. W., and Ussuf, K. K. (1987). Hormonal Regulation of Sertoli cell function. In *Regulation of Ovarian and Testicular Function* (ed. V. B. Mahesh) Plenum Press (in press).

Setchell, B. P. (1967). The blood-testis barrier in sheep. *J. Physiol., Lond.* **189**, 63–65.

Setchell, B. P. (1969). Do the Sertoli cells secrete the rete testis fluid? *J. Reprod. Fert.* **19**, 391–392.

Setchell, B. P. (1980). The functional significance of the blood-testis barrier. *Andrology*. 1, 3–10.

Setchell, B. P. (1977). The blood-testis barrier. In *Frontiers in reproduction and fertility control. A review of the reproductive sciences and contraceptive development* (eds R. O. Greep and M. A. Koblinsky) Part 2, pp. 338–352. MIT Press, Cambridge, Massachusetts.

Shabanowitz, R. B., DePhillip, R. M., Crowell, J. A., Tres, L., and Kierszenbaum, A. L. (1986). Temporal appearance and cyclic behavior of Sertoli cell-specific secretory proteins during the development of the rat seminiferous tubule. *Biol. Reprod.* 35, 745–760.

Shabinowitz, R. B. and Kierszenbaum, A. L. (1986). Newly synthesized proteins in seminiferous intertubular and intratubular compartments of rat testis. *Biol. Reprod.* 35, 179–190.

Silber, S. J. (1978). Vasectomy and vasectomy reversal. *Fertil. Steril.* 29, 125–140.

Skinner, M. K., Cosand, W. L., and Griswold, M. D. (1984). Purification and characterization of testicular transferrin secreted by rat Sertoli cells. *Biochem. J.* 218, 313–320.

Skinner, M. K. and Fritz, I. B., (1985a). Androgen stimulation of Sertoli cell function is enhanced by peritubular cells. *Molc. cell. Endocr.* 40, 115–122.

Skinner, M. K. and Fritz, I. B. (1985b). Testicular peritubular cells secrete a protein under androgen control that modulates Sertoli cell functions. *Proc. natn. Acad. Sci. U.S.A* 82, 114–118.

Skinner, M. K. and Griswold, M. D. (1980). Sertoli cells synthesize and secrete a transferrin-like protein. *J. biol. Chem.* 255, 9523–9525.

Skinner, M. K. and Griswold, M. D. (1983). Sertoli cells synthesize and secrete a ceruloplasmin-like protein. *Biol. Reprod.* 28, 1225–1229.

Skinner, M. K. and Griswold, M. D. (1982). Secretion of testicular transferrin by cultured Sertoli cells is regulated by hormones and retinoids. *Biol. Reprod.* 27, 211–221.

Smith, B. C. and Griswold, M. D. (1981). Supporting cells from bovine testes-characterization of the cell culture. *In Vitro* 17, (7), 612–618.

Steinberger, A., Heindel, J. J., Lindsey, J. N., Elkington, J. S. H., Sanborn, B. M., and Steinberger, E. (1975a). Isolation and culture of FSH responsive Sertoli cells. *Endocr. Res. Commun.* 2, 261–272.

Steinberger, A., Elkington, J. S. H., Sanborn, B. M., Steinberger, E., Heindel, J. J., Lindsey, J. N. (1975b). In *Hormonal regulation of spermatogenesis* (eds F. S. French, V. Hansson, E. M. Ritzen and S. Nayfeh) pp. 399–411, Plenum Press, New York.

Steinberger, A. and Steinberger, E. (1977). The Sertoli cells. In *The testis* Vol. IV (eds. A. Johnson and W. Gomes) pp. 371–394. Academic Press, New York.

Sylvester, S. R. and Griswold, M. D. (1984). Localization of transferrin and transferrin receptors in rat testes. *Biol. Reprod.* 31, 195–203.

Sylvester, S. R., Skinner, M. K., and Griswold, M. D. (1984). A sulfated glycoprotein synthesized by Sertoli cells and by epididymal cells is a component of the sperm membrane. *Biol. Reprod.* 31, 1087–1101.

Tabone, E., Benahmed, M., and Saez, J. M. (1984). Interactions between immature porcine Leydig cells and Sertoli cells *in vitro*. *Cell Tissue Res.* 237, 357–362.

Thompson, J. N., Howell, J. McC., and Pitt, G. A. J. (1964). Vitamin A and reproduction in rats. *Proc. R. Soc.* B 159, 510–535.

Thorbecke, G. J., Liem, H. H., Knight, S., Cox, K., and Muller-Eberhard, U. (1973). Sites of formation of the serum proteins transferrin and hemopexin. *J. clin. Invest.* 52, 725–731.

Tindall, D., Rowley, D., Murthy, L., Lipshultz, L., and Chang, C., (1985). Structure and biochemistry of the Sertoli cell. *Int. Rev. Cytol.* **94**, 127–149.

Trowbridge, I. S. and Omary, M. B. (1981). Human cell surface glycoprotein related to cell proliferation is the receptor for transferrin. *Proc. natn. Acad. Sci. U.S.A.* **5**, 3039–3043.

Tuck, R. R., Setchell, B. P., Waites, G. H., and Young, J. A. (1970). The composition of fluids collected by micropuncture and catheterization from seminiferous tubules and rete testis of rats. *Pflügers Arch.* **318**, 225–243.

Tung, P. S. and Fritz, I. B. (1980). Interactions of Sertoli cells with myoid cells *in vitro. Biol. Reprod.* **23**, 207–217.

Tung, P. S., Dorrington, J. H., and Fritz, I. B., (1975). Structural changes induced by follicle-stimulating hormone or dibutyryl cyclic AMP on presumptive Sertoli cells in culture. *Proc. natn. Acad. Sci. U.S.A.* **72**, 1838–1842.

Tung, P. S., LaCroix, M., and Fritz, I. B., (1980). Effects of cytosine arabinoside on properties of testicular preparations in culture. *Biol. Reprod.* **22**, 1255–1261.

Tung, P. S., Skinner, M. K., and Fritz, I. B., (1984). Fibronectin synthesis is a marker for peritubular contaminants in Sertoli cell-enriched cultures. *Biol. Reprod.* **30**, 199–211.

Tung, P. S. and Fritz, I. B. (1984). Extracellular matrix promotes histotypic expression *in vitro. Biol. Reprod.* **30**, 213–229.

Vernon, R. J., Kopec, B., and Fritz, I. B. (1984). Observations on the binding of androgens by rat testis seminiferous tubules and testis extracts. *Mol. Cell. Endocr.* **1**, 167.

Vihko, K. K., Toppari, J., and Parvinen, M. (1987). Stage-specific regulation of plasminogen activator in the rat seminiferous epithelium. *Endocrinology.* **120**, 142–145.

Vogl, A. W. and Soucey, L. J. (1985). Arrangement and possible function of actin filament bundles in ectoplasmic specializations of ground squirrel Sertoli cells. *J. Cell. Biol.* **100**, 814–825.

Wagle, J. R., Heindel, J. J., Sanborn, B. M., and Steinberger, A., (1986). The effect of hypotonic treatment on Sertoli cell purity and cell function in culture. *In Vitro* **22**, 325–331.

Waites, G. M. H. (1977) Fluid secretion. In *The Testis.* Vol. IV (eds. A. Johnson and W. Gomes,) Academic Press, N.Y., pp. 91–123.

Wauben-Penris, P. J., Strous, G. J., and van der Donk, H. A. (1986). Transferrin receptors of isolated rat seminiferous tubules bind both rat and human transferrin. *Biol. Reprod.* **35**, 1227–1234.

Webber, J., Russell, L. D., Wong, V., and Peterson, R. N. (1983). Three dimensional reconstruction of a rat stage V Sertoli cell: II. Morphometry of Sertoli-Sertoli and Sertoli-germ-cell relationships. *Am. J. Anat.* **167**, 163–179.

Welsh, M. J. and Wiebe, J. P. (1975). Rat Sertoli cells: a rapid method for obtaining viable cells. *Endocrinology* **96**, 618–624.

Wilson, R. M. and Griswold, M. D. (1979). Secreted proteins from rat Sertoli cells. *Expl. Cell. Res.* **123**, 127–135.

Wing, T. Y. and Christensen, A. K. (1982). Morphometric studies on rat seminiferous tubules. *Am. J. Anat.* **165**, 13–25.

Wong, V. and Russell, L. D. (1983). Three dimensional reconstruction of rat stage V Sertoli cell: I. Methods, basic configuration, and dimensions. *Am. J. Anat.* **167**, 143–161.

Wright, W. W. and Luzzarga, M. L. (1986). Isolation of cyclic protein-2 from rat seminiferous tubule fluid and Sertoli cell culture medium. *Biol. Reprod.* **35**, 761–772.

Wright, W. W., Parvinen, M., Musto, N. A., Gunsalus, G. L., Phillips, D. M., Mather, J. P., and Bardin, C. W. (1983). Identification of stage-specific proteins synthesized by rat seminiferous tubules. *Biol. Reprod.* **29**, 257–270.

Ziparo, E., Siracusa, G., Palombi, F., Russo, M. A., and Stefanini, M. (1982). Formation in vitro of intercellular junctions between isolated germ cells and Sertoli cells in the rat. *Ann. N.Y. Acad. Sci.*, **383**, 511–512.

4   The developmental history of female germ cells in mammals

ANNE McLAREN

I   Introduction

More people are more interested in female germ cells today than ever before. Twenty five years ago no more than a dozen or so live human oocytes had ever been seen. By 1978, the year Louise Brown was born, this number must have risen to several hundred, almost all of them recovered by Patrick Steptoe and his colleagues in Oldham. But today there are hundreds of centres practising oocyte recovery, distributed throughout all the developed countries of the world. The number of babies born as a result of *in vitro* fertilization (IVF) must now be around ten thousand, and the number of live human oocytes examined is probably closer to 1,000,000 than to 100,000.

Although a number of descriptive (including ultrastructural and histochemical) studies have been made, we still have limited understanding of the biology of early stages of germ cell development in women or indeed in any other mammal. Figure 4.1 shows the cycle of female germ cell development: this cycle is similar for all mammals, but the duration of the different stages varies from one species to another. In this paper I shall attempt to summarize very briefly what we know of the development of female germ cells, from fertilization through to oocyte maturation. My account will be based mainly on studies in mice, the species in which most experimental work has been done, supplementing where appropriate with information from other species.

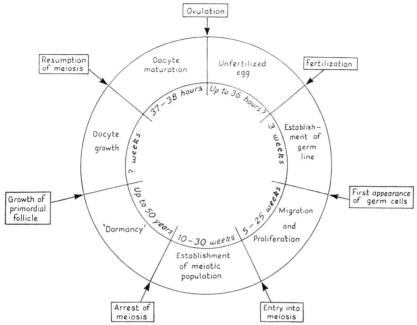

Fig. 4.1   Female germ cell cycle in mammals. The timings given are for the human. 36 hours is thought to be the maximum time that can elapse after ovulation if fertilization is to occur successfully. The duration of oocyte growth in the human ovary is not known precisely, but is likely to be several weeks.

II   Establishment of the germ line

It seems unlikely that a determined germ line exists in mice, and perhaps not in any mammal. By a determined germ line (McLaren 1983), we mean that some cytoplasmic determinant of germ cell development is present in the fertilized egg and segregates in early cleavage, so that any cell lacking it is incapable of giving rise to germ cells. This is the situation in *Drosophila*, and probably also in frogs, though it has recently been shown that primordial germ cells of *Xenopus* are not irreversibly determined even in their migratory phase (Wylie, Heasman, Snape, O'Driscoll, and Holwill 1985). In mice, we know that no segregation of the germ line occurs in the first two cell divisions after fertilization, since reaggregation studies show that descendants of each cell at the 4-cell stage can be found in both the germinal and the somatic cell population (Kelly 1977). Cleavage results in a blastocyst, with an inner cell mass and a surrounding layer of trophectoderm; after implantation, the mouse inner cell mass differentiates into an egg cylinder, with an inner core of epiblast (embryonic ectoderm), and an outer layer of primary endoderm. Even at the late blastocyst and early egg cylinder stage, most if not all of the cells of the inner cell mass and epiblast respectively have the potential to give rise to both germ cells and somatic cells after transplantation to a host blastocyst (Gardner 1977; Table 4 in McLaren 1981a; Gardner, Lyon, Evans, and Burtenshaw 1985). Although such findings render a *determined* germ line very unlikely, they have of course no bearing on whether or not an *allocated* germ line exists, in the sense of a consistent cell lineage followed by the germ line in the normal, undisturbed conceptus.

We know that primordial germ cells, at least in mice, arise from the epiblast and not, as was at one time believed, from the yolk sac endoderm (a derivative of the primary endoderm). When epiblast cells are transplanted to a host blastocyst, their descendants include germ cells as well as somatic cells, but similar transplantation experiments using cells from the primary endoderm have shown that these never give rise to germ cells (Gardner and Rossant 1979). In addition, enzyme studies (McMahon, Fosten, and Monk 1981) have shown that X-chromosome inactivation in primordial germ cells is random, as in other epiblast derivatives, while in all tissues formed from primary endoderm the paternal X chromosome is preferentially inactivated. Endodermal derivatives in the mouse embryo characteristically carry receptors for the lectin *Dolichos biflorus agglutinin* (Noguchi, Noguchi, Watanabe, and Muramatsu 1982), but mouse primordial germ cells do not react with this lectin (Hahnel and Eddy 1986).

III Migration and proliferation

Mouse primordial germ cells can first be identified, by virtue of their high alkaline phosphatase activity, in the mesoderm at the base of the allantois and at the extreme posterior end of the primitive streak (Ozdzenski 1967), about 8 days *post coitum*. Studies in which segments of the egg cylinder were removed 7 days *post coitum* and cultured *in vitro* for 24 hours, then histochemically tested for alkaline phosphatase activity, showed that the progenitors of these cells were concentrated at the 7-day stage in a small region located somewhat more anteriorly (Snow 1981). At this stage the germ cell lineage may already be determined; alternatively, some inductive influence from the tissue environment may be acting on a pluripotent cell population. Copp, Roberts, and Polani (1986) have shown that germ cells can develop from primitive streak cells grafted orthotopically into this region, but not from adjacent endoderm cells or ectoderm/mesoderm cells from the anterior part of the embryo, grafted into the same region. This suggests that by $7\frac{1}{2}$ days *post coitum* the competence to form primordial germ cells may be restricted to epiblast derivatives in the posterior primitive streak.

Between 7 and 8 days *post coitum* the future allantois and tail bud region is extruded backwards, taking with it the nascent germ cell population. As the yolk sac endoderm and mesoderm invaginate to form the hindgut, the primordial germ cells are again carried passively along (Snow and Monk 1983). However, their migration out of the wall of the hind gut, up the dorsal mesentery and round the coelomic angle into the urogenital ridge appears to involve active cell locomotion. Indeed, germ cells isolated at this stage are actively motile *in vitro*, and behave as invasive cells do in culture (Donovan, Stott, Cairns, Heasman, and Wylie 1986). We do not know to what extent fibronectin or other extracellular matrix substances are important as substrates for germ cell locomotion, nor whether contact guidance plays the same role as has been shown in Amphibia (Wylie, Heasman, Swan, and Anderton 1979).

Primordial germ cells can first be identified in the human embryo about 3 weeks after fertilization, and they begin to enter the genital ridges at about 8 weeks. Their migration was first described in the classic paper of Witschi (1948). As in the mouse, human primordial germ cells can be identified histochemically by their high alkaline phosphatase activity (McKay, Hertig, Adams, and Danziger 1953), but this seems not to be so for all mammals. It has been claimed that, at least in larger mammals such as the cow, where the migration route is longer, some of the germ cells may travel to the genital ridges via the blood stream (as occurs in birds) rather than through the solid tissues of the hind gut (Wartenberg 1983); but no direct evidence for this mode of migration has as yet been reported.

Primordial germ cells proliferate at a rather uniform rate during the period of migration (Tam and Snow 1981), reaching a maximum number in the mouse of about 25 000 per gonad 13–14 days *post coitum* (about 3 days after entering the urogenital ridge). In the human ovary, the maximum number of germ cells (over 3 million per ovary) is reached 5–6 months *post coitum* (Baker 1963); the comparable figures for the tammar wallaby are about 450,000 50 days after birth (Alcorn and Robinson 1983). In the mouse, most of the germ cells enter the region of the future gonad, but a few remain in the mesonephric region of the ridge, and a few others enter the primordium of the adrenal cortex. Two mutants are known in the mouse, Steel (*Sl*) and White-spotting (*W*), which in homozygous condition cause sterility due to absence of germ cells from the gonads. The primordial germ cells migrate normally towards the urogenital ridges, at least in *Sl/Sl* embryos, but either die or fail to proliferate, so that few or even none reach the ridges (McCoshen and McCallion 1975). Aggregates of granulofibrillar material termed nuage have been observed in the cytoplasm of migrating germ cells in the mouse (Spiegelman and Bennett 1973) and rat (Eddy 1974), as well as in some later stages of germ cell development in the human (Kellokumpu-Lehtinen and Söderström 1978) and rat (Eddy 1974): the function of nuage is unknown, but if it is of general occurrence it may turn out to be a useful marker for the germ cell lineage.

IV   Establishment of meiotic population

From the first moment at which they can be identified, throughout the period of migration and the initial proliferative phase within the genital ridge, primordial germ cells and gonia (oogonia, pro-spermatogenia), as they are termed once they have entered the genital ridge, are indistinguishable in male and female embryos, both in appearance and in behaviour. In the mouse, the testis can be distinguished from the ovary by the formation of testis cords at $12\frac{1}{2}$ days *post coitum*, about 2 days after the first germ cells have entered the genital ridge; and within the next day or two the germ cells in the testis and in the ovary become very different from one another.

The prospermatogenia of the mouse, surrounded by the pre-Sertoli cells that form the testis cords, undergo a last prenatal mitotic division and then arrest in the G_1 (or G_0) phase of the cell cycle (McLaren 1984). DNA replication and mitosis are resumed shortly after birth, spermatogonia type A and B are formed, and by about 10 days after birth the first meiotic germ cells are seen (Hilscher and Hilscher 1976).

In contrast, the germ cells in the ovary (oogonia) enter meiosis prenatally, at about the same time as the prospermatogonia are undergoing mitotic arrest. After their final mitotic division, the oogonia undergo one further round of DNA replication before entering the leptotene stage of meiotic prophase. Within a few days they pass through zygotene, when homologous chromosomes pair and synaptonemal complexes are formed, and through pachytene, when crossing-over takes place, so that by the time of birth or shortly after most oocytes in the mouse have reached the diplotene stage of meiotic prophase (Borum 1961). The process of meiosis is thought to be similar in all mammals, including marsupials (Alcorn and Robinson 1983), but the timing and duration of the various stages differ from one species to another (Peters 1970). In some (for example, rabbit), the onset of meiosis is delayed until after birth, but still occurs much earlier in the ovary than in the testis (Byskov 1979). The first germ cells to enter meiosis in the human ovary are seen about 13 weeks *post coitum* (Baker and Franchi 1967), and in the tammar wallaby ovary about 4 weeks *post partum* (that is, 8 weeks *post coitum*) (Alcorn and Robinson 1983).

Many germs cells degenerate throughout oogonial and oocyte development. This loss (atresia) reaches a peak at about the time of birth in mice, during the pachytene to diplotene transition. The biological basis for such a massive loss of germ cells is not understood.

Although primordial germ cells in male and female embryos are indistinguishable at a gross morphological level, they are of course chromosomally different, XY and XX respectively. Since the ancestors of the germ cell population undergo random X chromosome inactivation at the same time as the other cells in the mouse epiblast (McMahon, Fosten, and Monk 1983), XY and XX germ cells will be "dosage compensated" during the period of migration. During oogenesis, on the other hand, both X chromosomes in XX oocytes are expressed. This was first established by Gartler, Liskay, and Gant (1973), who observed that the oocytes of women heterozygous for two isozymes of the X-linked enzyme glucose-6-phosphate dehydro-genase (G6PD) showed a hybrid band, indicating that both alleles were being transcribed in the same oocyte. A number of lines of evidence (for example, Monk and McLaren 1981) have since shown that reactivation of the silent X chromosome, at least in the mouse, occurs in the fetal ovary, at about the time of onset of meiosis. A more general restoration of totipotency must take place at some stage prior to the development of the mature oocyte, since the epiblast which gives rise to the primordial germ cells is a tissue of restricted potency, no longer able to form trophoblast or primary endoderm; but whether this restoration step is also associated with meiosis is not known.

Prenatal entry into meiosis characterizes not only germ cells in the ovary, but also some of those that remain in the mesonephric region (Francavilla and Zamboni 1985; McLaren unpublished observations), and even the small population of germ cells that misguidedly enter the adrenal, whether in the female or in the male fetus (Upadhyay and Zamboni 1982). In the testis itself, a few germ cells may enter meiosis before birth if genetic or environmental circumstances are abnormal, for example in Sex-reversed (*Sxr*) XX mice (McLaren 1981*b*), male XX ↔ XY chimeras (Mystkowska and Tarkowski 1970; McLaren, Chandley, and Kofman-Alfaro 1972), and normal fetal testes whose cord structure has been disrupted by transplantation under the kidney capsule (Ozdzenski 1972).

It seems that wherever a germ cell enters meiotic prophase before birth, whether in the ovary or in the testis or in an extra-gonadal site, whether in a female or a male fetus, whether the germ cell concerned is XX or XY in chromosome constitution, in all cases the germ cell continues in the female pathway of development and undergoes oogenesis. Thus growing oocytes have been found postnatally in the male adrenal (Upadhyay and Zamboni 1982), in Sex-reversed and chimeric testes (McLaren 1980; Mystkowska and Tarkowski 1970), and in testes transplanted to the kidney (Ozdzenski 1972). We may conclude that the female germ cell, the oocyte, is female not because of its chromosome constitution or its location in an ovary or in a female indi-vidual, but because it entered meiosis before birth rather than after.

So what determines time of entry into meiosis? There is general agreement that germ cells enclosed in normal testis cords are inhibited from entering meiosis before birth. This inhibition seems more likely to be due to some dif-fusible substance, released perhaps by pre-Sertoli cells, than to physical sequestration, since it extends to some germ cells in the mesonephric region and some within the testis but outside the cords (Francavilla and Zamboni 1985; McLaren, unpublished observations). Zamboni and Upadhyay (1983) take the view that all germ cells not subject to this inhibition enter spon-taneously into meiosis before birth; Byskov and her colleagues on the other hand believe that a meiosis-inducing substance is synthesized in the fetus, which induces any germ cell to enter meiotic prophase unless it is protected by the testicular inhibiting factor (for example, Byskov, Anderson, and

Westergaard 1983). This question has not yet been resolved (for a summary of the evidence to date, see McLaren 1984).

In the fetal mouse ovary, variation in the time at which germ cells enter the leptotene stage of meiotic prophase is no more than 2 days ($13\frac{1}{2}$–$15\frac{1}{2}$ days *post coitum*), but in the human ovary this variation may be as much as 4–5 months. Henderson and Edwards (1968) suggested that the substantial increase in frequency of aneuploid embryos that occurs with increasing maternal age both in women and in some strains of mice could be explained if the later germ cells to enter meiosis had a higher incidence of chromosomal pairing anomalies, and if these later forming oocytes were also those that were ovulated later in life (the "production line" hypothesis). Speed and Chandley (1983) have recently used a surface-spreading technique to analyze pairing defects in zygotene and pachytene oocytes in fetal mouse ovaries. Their results show that few errors in synapsis occurred and oocytes containing univalents were rare and did not increase in frequency with gestational age. This finding lends no support to the "production line" hypothesis, but does not refute it.

V   "Dormancy"

In the mouse ovary, flattened follicle cells can already be seen adjacent to some of the oocytes shortly before birth. Within a day or two of birth, all the surviving oocytes are in the diplotene stage of meiotic prophase, and are contained within primordial follicles. In all mammals, the single-layered primordial follicles, which constitute by far the most numerous class of ovarian follicle, are located mainly towards the periphery of the ovary. The plasma membranes of oocyte and follicle cells are closely apposed and may be connected by desmosomes and even by gap junctions, but the investment of the oocyte by the follicle cells in eutherian mammals seems to be incomplete. Zamboni (1974) showed for the rhesus monkey, and confirmed on mouse, rabbit and human material, that some parts of the quiescent oocyte are not covered by even a thin layer of follicle cell cytoplasm, but are exposed directly to all the elements of the ovarian stroma, including blood capillaries. In primordial follicles of the bandicoot, on the other hand, Ullmann (1978) reports that the oocytes are completely invested by long pseudopod-like interdigitating projections from the follicle cells. The bandicoot primordial oocyte contains a number of cytoplasmic organelles lacking from its eutherian counterpart, some of which may serve a yolk-like nutritive role.

Most female germ cells spend by far the greatest part of their existence contained within primordial follicles. In mice this stage lasts from a few days to more than a year, in humans from 10 years to 50 years. The maternal age effect referred to above implies that it is the oocytes that have spent longest in primordial follicles that are most likely to give rise to abnormal embryos. It is therefore tempting to seek some environmental influence that might have a cumulative adverse effect on these primordial oocytes, but none has yet been identified.

The description of primordial oocytes as dormant or quiescent is perhaps misleading, as they continue to be metabolically active, though the level of RNA and protein synthesis is low. In mice and rats the chromatin in the nucleus is in a very diffuse, decondensed state ("dictyate"), similar to that seen in interphase, but in most mammals the chromosomes remain partially condensed, in the diplotene condition.

As a female gets older and the number of primordial follicles decreases, the proportion of growing follicles increases. Baker, Challoner, and Burgoyne (1980) established that it is the size of the primordial pool in the mouse as a whole that determines the number of growing follicles, not the size of the pool in an individual ovary. We still do not know whether there is any pre-determined order in which the follicles and the oocytes within them start to grow (as suggested for example by the "production line" hypothesis), or whether the time of onset of growth for any particular follicle is entirely random, like radioactive decay.

VI Oocyte growth

The growth phase is believed to start in the follicle cells rather than in the oocyte itself. The first sign that a follicle is leaving the primordial pool is an increase in the number of cells investing the oocyte, so that while still forming a single layer they cease to be flattened and attenuated, and become cuboidal or columnar, surrounding the oocyte completely. Lintern-Moore and Moore (1979) reckon that any follicle in the mouse ovary containing more than 8 cells may be reckoned to have entered the growth phase. With 9 follicle cells an increase in oocyte nucleolar RNA polymerase activity was detected, and with 10 follicle cells an increase in oocyte area. As the follicle cells continue to divide and the follicle becomes first two-layered, then multi-layered, the oocyte too shows a steady increase in size. This continues in the mouse until the follicles have reached stage 5 (Pedersen and Peters 1968), just before antrum formation; oocyte growth then stops, but antrum and follicle growth continue until the mature Graafian follicle is formed. During all this period the oocyte nucleus remains arrested in the diplotene or dictyate stage of meiotic prophase, and meiosis is only resumed with the onset of oocyte maturation, which *in vivo* is initiated hormonally, by the LH surge. The total increase in volume of the oocyte during the growth phase is over 300-fold.

The onset of mitotic activity of the follicle cells coincides with the appearance of patches of zona pellucida material, at least in the rhesus monkey ovary (Zamboni, 1974). Since all three of the principal zona pellucida proteins in the mouse are now known to be synthesized in the oocyte rather than in the follicle cells (Haddad and Nagai 1977; Bleil and Wassarman 1980), this implies that the period of oocyte dormancy ends as soon as the follicle cells start to proliferate. Oocyte growth is accompanied by other changes in gene expression: for example the oocyte-specific pattern of expression of the enzyme glucose phosphate isomerase (GPI) is not seen in fetal oocytes nor in oocytes from primordial follicles, but only in oocytes that have begun their growth phase, whether *in vivo* or *in vitro* (McLaren and Buehr 1981; Buehr and McLaren 1985). The activities of G6PD and lactate dehydrogenase also begin to rise sharply once oocyte growth has begun (Mangia and Epstein 1975).

RNA polymerase activity of oocytes is low during the period of dormancy, increases during the growth phase, reaching a maximum as the oocytes approach their maximum size, and is low again in oocytes from large antral follicles (Moore and Lintern-Moore 1978). The rate of transcription is never high, and light and electron microscopic studies of mouse oocytes from pachytene through the period of dormancy and the growth phase (Bachvarova, Burns, Spiegelman, Choy, and Chaganti 1982) confirm that mouse oocytes never develop the true lampbrush chromosomes responsible for the very rapid transcription that occurs in amphibian oocytes. mRNA is synthesized primarily in the growth phase rather than during the period of dormancy or in fully grown oocytes (Bachvarova and de Leon 1980).

The desmosomes formed between follicle cells and oocyte in the primordial follicle persist during the growth phase, so that as the zona pellucida thickens it becomes traversed by long cytoplasmic follicle-cell processes, terminating in bulbous projections anchored on the oocyte surface or embedded in the vitellus (Zamboni 1972, 1974; Anderson 1974). Gap junctions between follicle-cell processes and the oocyte also persist. The growing oocyte is therefore in intimate contact with the surrounding follicle cells, while being at the same time effectively isolated from the extrafollicular environment.

When oocyte growth is initiated in the absence of follicle cells, as in the male or female adrenal (Upadhyay and Zamboni 1982) or in the testis of XX sex-reversed males (McLaren 1980), the oocytes always degenerate within a week or two. It seems likely that degeneration sets in as soon as the secreted zona pellucida material forms a complete investment around the oocyte, since without follicle cell processes penetrating the zona, no further nutrients would then be able to reach the oocyte.

The profound influence that the follicle cells exert on oocyte metabolism has been reviewed by Moor (1983). For most metabolites present in follicle-enclosed oocytes, more than 85 per cent was initially taken up by follicle cells and transferred to oocytes via gap junctions (Heller, Cahill, and Schultz 1981). The follicle cells not only provide energy substrates and RNA precursors for the growing oocyte by this type of metabolic co-operation, but they also regulate the rate of carrier-mediated uptake across the oolemma. Further, Schultz, Letourneau, and Wassarman (1979) calculated, on the basis of estimates of the rate of protein synthesis in growing oocytes, that the oocyte's own protein-synthetic machinery is capable of making less than half the total amount of protein that the mature oocyte contains. If this is so, the rest must be taken in from outside the cell in a macromolecular form, probably by pinocytosis of degenerating follicle cell cytoplasm (Zamboni 1974). Signals from the follicle cells even influence the pattern of proteins synthesized within the oocyte, at least in sheep (Moor 1983).

VII   Culture of germ cells and oocytes

Germ cells developing *in situ*, in their normal undisturbed somatic environment, are largely inaccessible to experimental investigation. We therefore need to explore the possibilities of *in vitro* culture systems if we are to approach such tantalizing problems as the mechanism of germ cell migration, the regulation of entry into meiosis, X-chromosome reactivation, and the initiation and control of oocyte growth.

In a pioneering study, Blandau, White, and Rumery (1963) established that mouse germ cells in squash preparations of the hind gut or early genital ridges, and oogonia from 11–12 week human ovaries, were capable of pseudopodial activity, but they did not attempt to maintain these cells in culture for prolonged periods. Some preliminary observations on mouse and rat primordial germ cell locomotion *in vitro* have been made by Merchant-Larios and Alvarez-Buylla (1984), but we still have little understanding of the mechanism of locomotion and the means of navigation of migrating germ cells. Fetal macrophages may be a source of confusion, as these cells show superficial similarities to germ cells, and are highly motile (De Felici, Heasman, Wylie, and McLaren 1985).

Using mouse germ cells identified immunologically as well as histochemically, Donovan *et al.* (1986) compared the behaviour *in vitro* of germ cells isolated before and after entry into the genital ridges. Cultured on feeder layers, the migrating germ cells were actively motile and showed behaviour characteristic of invasive cells, but those that had already entered the genital ridges showed little locomotory activity and were not invasive.

Germ cells isolated from mouse gonads $13\frac{1}{2}$–$16\frac{1}{2}$ days *post coitum* with minimal somatic cell contamination can be maintained in culture on a plastic surface at 37°C for up to a week (De Felici and McLaren 1983). The cells adhere to one another to form large or small clumps, but do not attach to the plastic surface. At $13\frac{1}{2}$ days, mitotic proliferation occurred in germ cells from both male and female embryos, but its rate declined during culture, as it would *in vivo*. Germ cells from female embryos progressed through meiotic prophase *in vitro*: more were detected in meiosis on each day of culture, and progressively later stages of prophase were seen each day, up to pachytene. In contrast, germ cells from male embryos were never observed in meiotic prophase *in vitro*.

Unfortunately, germ cells isolated from genital ridges $11\frac{1}{2}$ or $12\frac{1}{2}$ days *post coitum*, prior to the onset of meiosis or mitotic arrest, did not survive in this culture system. Viability could only be maintained if the temperature was reduced to below 30°C. We have so far been unsuccessful in devising any system in which isolated germ cells from these younger embryos can be kept alive at 37°C (Wabik-Sliz and McLaren 1984). At lower temperatures DNA replication and mitotic activity is maintained, though at a slow rate. Since meiotic progression, even in older germ cells, appears to require a higher temperature, we have been unable to investigate the factors required to induce the onset of meiosis in our isolated germ cells.

At the other extreme, when entire genital ridges are cultured in an organ culture system, the onset of meiosis and the initiation of oocyte growth take place *in vitro* just as they do *in vivo*. Meiotic germ cells are not seen in male genital ridges, unless the ridge area is explanted at the very beginning of ridge formation, $10\frac{1}{2}$ days *post coitum* (McLaren, unpublished observations). Testis cords do not form in these very early explants, though some differentiation of pre-Sertoli cells is seen. When male and female gonads are co-cultured under various conditions, meiotic germ cells only appear in the male gonads if testis cord formation has been disrupted: these studies are reviewed in McLaren (1984).

When oocytes are dissociated from either female genital ridges $13\frac{1}{2}$ days *post coitum* or neonatal ovaries, and cultured in the presence of ovarian somatic cells of the same age, the somatic cells attach to the plastic culture dish as a discontinuous monolayer, with unattached germ cells rolling freely over their surface. Under these culture conditions oocyte growth and zona pellucida formation is initiated, and within 10–16 days oocytes up to 60 μm in diameter can be found (Buehr and McLaren 1985). The oocyte-specific pattern of GPI expression can be demonstrated in the growing oocytes, but not in those of 20 μm diameter or less, which have not yet begun their growth phase. This experiment shows that oocyte growth and the associated changes in gene expression do not require the presence of a normal follicle, nor indeed any stable cellular investment around the oocyte. It is possible however that transient gap junctions form between the oocytes and the underlying layer of somatic cells (Heller *et al.* 1981), allowing the passage of small molecules that may exert either a signalling or a nutrient function. Somatic cells of other than ovarian origin will not support oocyte growth, while in the presence of an intervening layer of agar growth still occurs but is markedly retarded (Bachvarova, Baran, and Tejblum 1980).

Although a fibroblast monolayer is unable to support oocyte growth, it will maintain for several days the viability of growing oocytes isolated from their follicles. Viability is maintained equally well if the oocytes are cultured in conditioned media, but again no growth occurs. The spontaneous resumption of meiosis in these explanted oocytes was studied by Canipari, Palombi, Riminucci and Mangia (1984). Fully grown oocytes (65 μm or more in diameter) resumed meiosis as soon as they were freed of their follicular investment, but smaller oocytes required a period of culture before they were competent to resume meiosis. The smaller the oocyte, the longer the period of culture that was required. This intriguing result implies that, at least in the mouse, the development in an oocyte of competence to resume meiosis is not dependent on oocyte growth, or the presence of follicle cells, or contact with any feeder somatic cells. It appears to depend on a definite time program, operating autonomously within the oocyte itself.

Thus while studies *in vitro* have so far been unsuccessful in shedding much light on earlier aspects of female germ cell development, including the initiation of meiosis, they are yielding valuable information on oocyte growth and on the development of the capacity to undergo maturation.

VIII Oocyte maturation

The final burst of follicular growth and the formation of the mature Graafian follicle is stimulated by follicle-stimulating hormone (FSH). At the same time FSH induces the formation of luteinizing hormone (LH) receptors in follicle cells. Inhibin, present in high concentration in pre-ovulatory follicles in sheep, may play a role in suppressing FSH during the late follicular phase of the cycle (Tsonis, Quigg, Lee, Leversha, Trounson, and Findlay 1983). *In vivo*, the oocyte resumes meiosis in response to the LH surge, the germinal vesicle (oocyte nucleus) breaks down, the chromosomes complete the first meiotic division, and the first polar body is extruded. The follicle cells respond to LH by interruption of the cytoplasmic processes that pass through the zona pellucida, so that oocyte and follicle cells are no longer electrically and metabolically coupled. Oestrogen is replaced by progesterone as the dominant steroid in the follicular fluid, the follicle cells luteinize, and the egg is ovulated. The chromosomes now once again enter meiotic arrest, this time in second meiotic metaphase. This final period of differentiation, from the resumption of meiosis onwards, is termed oocyte maturation. In addition to the nuclear events, it involves cytoplasmic maturation, and renders the oocyte competent to undergo fertilization and embryonic development.

The literature on the regulation of oocyte maturation is extensive, especially since the advent of IVF. Only a brief review will be given here; a critical review of the subject up to 1983 has already been published in this series (McGaughey 1983).

The molecular events underlying germinal vesicle breakdown vary from one species to another. In the mouse, the resumption of meiosis and the breakdown of the germinal vesicle requires neither transcription (Crozet and Szöllösi 1980) nor new protein synthesis (Golbus and Stein 1976). Genetic differences in the rate of oocyte maturation must be determined before meiosis is resumed, as they are expressed cell-autonomously, even when cumulus-free oocytes are matured *in vitro* (Polanski 1986). The proteins involved in the transition to metaphase I are already present in the prophase oocyte in an inactive form, and are activated through the agency of maturation promoting factor (MPF), which is itself thought to be rendered active (as in *Xenopus*) by a progesterone-dependent decrease in intracellular cyclic AMP levels (for review, see Masui and Clarke 1979). In the sheep, where oocyte maturation has been studied in some detail (for reviews, see Moor and Warnes 1978; Moor and Osborn 1984), inhibition of either transcription or protein synthesis within 1–2 hours after the onset of maturation prevents breakdown of the germinal vesicle. Unlike in the mouse, no decrease in intracellular cyclic AMP levels occurs at this time, but a new protein is synthesized shortly before germinal vesicle breakdown which could form part of the MPF activation system (Moor and Crosby 1986).

The interruption of intercellular coupling that occurs in response to LH must exert a profound effect on oocyte function. It is unlikely, however, that the actual resumption of meiosis is due to disruption of gap junctions, since for some hours after meiosis has resumed the junctional complexes are still morphologically intact (Szöllösi, Gérard, Ménézo, and Thibault 1978) and capable of transferring choline from the follicle cells to the oocyte (Moor, Smith, and Dawson 1980). Disruption of the gap junctions may however trigger cytoplasmic maturation: in mouse, rabbit, pig and cow oocytes, the disappearance of gap junctions is followed immediately by a reduction in charge on the oocyte surface and migration of the cortical granules to their definitive position beneath the cell membrane (Szöllösi *et al.* 1978).

The spontaneous maturation that occurs *in vitro* when fully grown oocytes are removed from their follicles and cultured, indicates that within the follicle the oocyte is maintained in the immature germinal vesicle stage by maturation-inhibiting factors. Recent work suggests that one such factor may be hypoxanthine, an inhibitor of phosphodiesterase activity. Hypoxanthine is present in high concentration in both pig and mouse follicular fluid, and is able to maintain isolated oocytes in meiotic arrest *in vitro* (Down, Coleman, Ward-Bailey, and Eppig 1985).

Although removal from the follicle allows the oocyte to resume meiosis and to progress through to second meiotic metaphase *in vitro*, developmental competence is not always achieved, probably because cytoplasmic maturation is deficient (Moor and Osborn 1984). In sheep, gonadotrophins, steroids and other factors from the follicle cells all interact to provide essential support for the maturing oocyte *in vivo*. Staigmiller and Moor (1984) showed that sheep oocytes totally denuded or surrounded only by corona cells resumed meiosis *in vitro* but remained developmentally incompetent, even when steroids and gonadotrophins were present in the medium. Although the addition of follicle cells to the culture system did not help the denuded oocytes, it successfully rescued those surrounded by corona cells, so that fertilization and subsequent embryonic development resembled that achieved by oocytes matured in follicles, and normal lambs were born. It seems that full oocyte maturation requires both direct cell contact, and the presence of a minimum number of follicle cells.

In mice, immature oocytes recovered from follicles before the LH surge have been successfully matured *in vitro*, whether surrounded by granulosa cells or denuded. The presence of serum in the culture medium was essential. The fertilization rate was increased by the presence of granulosa cells, but some live young were produced even from the denuded oocytes (Schroeder and Eppig, 1984). This suggests that the conditions required for successful oocyte maturation are less stringent in the mouse than in the sheep, perhaps because transcription and protein synthesis are not required for mouse oocyte maturation.

Culture within the follicle is not feasible for human oocyte maturation. Oocyte recovery in clinical IVF programmes is therefore timed so that most of the maturation period takes place *in vivo*. Recently, however, Templeton and his colleagues recovered immature human oocytes from women receiving no hCG and with no endogenous LH surge, and compared them with oocytes recovered 12, 24 or 36 hours after the administration of hCG (Templeton 1985). The oocytes were cultured *in vitro* for 42, 30, 18 or 6 hours respectively, before insemination. More than half the oocytes were fertilized and cleaved normally, even those that had never been exposed to hCG. A live birth has also been reported from an immature oocyte that had apparently failed to extrude the first polar body after hCG administration, and was subsequently matured *in vitro* (Veeck, Wortham, Witmyer, Sandow, Acosta, Garcia, Jones, G. S., and Jones, H. W. 1983). Thus it may be that human oocytes resemble those of the mouse more than those of the sheep in the conditions that they require for successful maturation.

There is no doubt that a reliable, safe and effective method of inducing maturation of immature oocytes *in vitro* would revolutionize human IVF, and would also be of considerable economic importance in animal breeding. But caution is required. Inappropriate hormonal conditions during critical stages of oogenesis and oocyte maturation could produce small defects that would not affect fertilization or early development, but would lead to abnormalities later in development. As the great embryologist E. B. Wilson wrote in 1925 "Embryogenesis begins in oogenesis".

References

Alcorn, G. T. and Robinson, E. S. (1983). Germ cell development in female pouch young of the tammar wallaby (*Macropus eugenii*). *J. Reprod. Fert.* **67**, 319–325.

Anderson, E. (1974). Comparative aspects of the ultrastructure of the female gamete. *Int. Rev. Cytol.* **4**, 1–70.

Bachvarova, R., Baran, M. M., and Tejblum, A. (1980). Development of naked growing mouse oocytes *in vitro*. *J. exp. Zool.* **211**, 159–169.

—— Burns, J. P., Spiegelman, I., Choy, J., and Chaganti, R. S. K. (1982). Morphology and transcriptional activity of mouse oocyte chromosomes. *Chromosoma* **86**, 181–196.

—— and de Leon, V. (1980). Polyadenylated RNA of mouse ova and loss of maternal RNA in early development. *Devl. Biol.* **74**, 1–8.

Baker, T. G. (1963). A quantitative and cytological study of germ cells in human ovaries. *Proc. R. Soc. B.* **158**, 417–433.

—— Challoner, S., and Burgoyne, P. S. (1980). The number of oocytes and the rate of atresia in hemiovariectomized mice up to eight months after surgery. *J. Reprod. Fert.* **60**, 449–456.

—— and Franchi, L. L. (1967). The fine structure of oogonia and oocytes in human ovaries. *J. Cell Sci.* **2**, 213–224.

Blandau, R. J., White, B. J., and Rumery, R. E. (1963). Observations on the movement of the living primordial germ cells in the mouse. *Fert. Steril.* **14**, 482–489.

Bleil, J. D. and Wasserman, P. M. (1980). Synthesis of zona pellucida proteins by denuded and follicle-enclosed mouse oocytes during culture *in vitro. Proc. natn. Acad. Sci. U.S.A.* **77**, 1029–1033.

Borum, K. (1961). Oogenesis in the mouse. A study of the meiotic prophase. *Exp. Cell Res.* **24**, 495–507.

Buehr, M. and McLaren, A. (1985). Expression of glucose-phosphate isomerase in relation to growth of the mouse oocyte *in vivo* and *in vitro. Gamete Res.* **11**, 271–281.

Byskov, A. G. (1979). Regulation of meiosis in mammals. *Annls. Biol. anim. Biochim. Biophys.* **19**, 1251–1261.

—— Anderson, C. Y., and Westergaard, L. (1983). Dependence of the onset of meiosis on the internal organization of the gonad. In *Current Problems in Germ Cell Differentiation.* Brit. Soc. Devl. Biol. symposium 7 (eds A. McLaren and C. C. Wylie) pp. 215–224. Cambridge University Press, Cambridge.

Canipari, R., Palombi, F., Riminucci, M., and Mangia, F. (1984). Early programming of maturation competence in mouse oogenesis. *Devl. Biol.* **102**, 519–524.

Copp, A. J., Roberts, H. M., and Polani, P. E. (1986). Chimaerism of primordial germ cells in the early postimplantation mouse embryo following microsurgical grafting of posterior primitive streak cells *in vitro. J. Embryol. exp. Morph.* **95**, 94–115.

Crozet, N. and Szöllösi, D. (1980). Effects of actinomycin D and α-amanitin on the nuclear ultrastructure of mouse oocytes. *Biol. cellulaire* **38**, 163–170.

De Felici, M. (1984). Binding of fluorescent lectins to the surface of germ cells from fetal and early postnatal mouse gonads. *Gamete Res.* **10**, 423–432.

—— Heasman, J., Wylie, C. C., and McLaren, A. (1985). Macrophages in the urogenital ridge of the mid-gestation mouse fetus. *Cell Differentiation* **18**, 119–129.

—— and McLaren, A. (1983). *In vitro* culture of mouse primordial germ cells. *Exp. Cell Res.* **144**, 417–427.

Donovan, P. J., Stott, D., Cairns, L. A., Heasman, J., and Wylie, C. C. (1986). Migratory and postmigratory primoridial germ cells behave differently in culture. *Cell* **44**, 831–838.

Downs, S. M., Coleman, D. L., Ward-Bailey, P. F., and Eppig, J. J. (1985). *Proc. natn. Acad. Sci. U.S.A.* **82**, 454–458.

Eddy, E. M. (1974). Fine structural observations on the form and distribution of nuage in germ cells of the rat. *Anat. Rec.* **178**, 731–758.

Francavilla, S. and Zamboni, L. (1985). Differentiation of mouse ectopic germ cells in intra- and perigonadal locations. *J. exp. Zool.* **233**, 101–109.

Gardner, R. L. (1977). Developmental potency of normal and neoplastic cells of the early mouse embryo. In *Birth Defects* (eds J. W. Littlefield and J. de Grouchy). *Excerpta Medica Int. Congr. Ser.* **432**, 154–166.

—— Lyon, M. F., Evans, E. P., and Burtenshaw, M. D. (1985). Clonal analysis of X-chromosome inactivation and the origin of the germ line in the mouse embryo. *J. Embryol. exp. Morph.* **88**, 349–363.

—— and Rossant, J. (1979). Investigation of the fate of 4.5 day post-coitum mouse inner cell mass cells by blastocyst injection. *J. Embryol. exp. Morph.* **52**, 141–152.

Gartler, S. M., Liskay, R. M., and Gant, N. (1973). Two functional X-chromosomes in human fetal oocytes. *Expl. Cell Res.* **82**, 464–466.

Golbus, M. S. and Stein, M. P. (1976). Qualitative patterns of protein synthesis in the mouse oocyte. *J. exp. Zool.* **198**, 337–342.

Haddad, A. and Nagai, M. E. T. (1977). Radioautographic study of glycoprotein bio-synthesis and renewal in the ovarian follicles of mice and the origin of the zona pellucida. *Cell Tissue Res.* **177**, 347–369.

Hahnel, A. C. and Eddy, E. M. (1986). Cell surface markers of mouse primordial germ cells defined by two monoclonal antibodies. *Gamete Res.* **15**, 25–34.

Heller, D. T., Cahill, D. M., and Schultz, R. M. (1981). Biochemical studies of mam-malian oogenesis: metabolic cooperativity between granulosa cells and growing mouse oocytes. *Devl. Biol.* **84**, 455–464.

Henderson, S. A. and Edwards, R. G. (1968). Chiasma frequency and maternal age in mammals. *Nature, Lond.* **218**, 22–28.

Hilscher, W. and Hilscher, B. (1976). Kinetics of the male gametogenesis. *Andrologia* **8**, 105–116.

Kellokumpu-Lehtinen, P. and Söderström, K. (1978). Occurrence of nuage in fetal human germ cells. *Cell Tissue Res.* **194**, 171–177.

Kelly, S. J. (1977). Studies of the developmental potential of 4- and 8-cell mouse blastomeres. *J. exp. Zool.* **200**, 365–376.

Lintern-Moore, S. and Moore, G. P. M. (1979). The initiation of follicle and oocyte growth in the mouse ovary. *Biol. Reprod.* **20**, 773–778.

Mangia, F. and Epstein, C. J. (1975). Biochemical studies of growing mouse oocytes: preparation of oocytes and analysis of glucose-6-phosphate dehydrogenase and lactate dehydrogenase activities. *Devl. Biol.* **45**, 211–220.

Masui, Y. and Clarke, H. J. (1979). Oocyte maturation. *Int. Rev. Cytol.* **57**, 186–282.

McCoshen, J. A. and McCallion, D. J. (1975). A study of the primordial germ cells during their migratory phase in steel mutant mice. *Experientia* **31**, 589–590.

McGaughey, R. W. (1983). Regulation of oocyte maturation. In *Oxford Reviews of Re-productive Biology* Vol. 5 (ed. C. A. Finn) pp. 108–130. Clarendon Press, Oxford.

McKay, D. G., Hertig, A. T., Adams, E. C., and Danziger, S. (1953). Histochemical observations on the germ cells of human embryos. *Anat. Rec.* **117**, 201–209.

McLaren, A. (1980). Oocytes in the testis. *Nature, Lond.* **283**, 688–689.

—— (1981*a*). *Germ cells and soma: a new look at an old problem.* Yale University Press, New Haven and London.

—— (1981*b*). The fate of germ cells in the testis of fetal *Sex-reversed* mice. *J. Reprod. Fert.* **61**, 461–467.

—— (1983). Primordial germ cells in mice. *Biblthca. anat.* **24**, 59–66.

—— (1984). Meiosis and differentiation of mouse germ cells. In *Controlling Events in Meiosis*, (ed. H. G. Dickinson). Soc. Exp. Biol. Symp. 38, pp. 7–23. Cambridge: Cambridge University Press.

—— and Buehr, M. (1981). GPI expression in female germ cells of the mouse. *Genet. Res.* **37**, 303–309.

—— Chandley, A. C., and Kofman-Alfaro, S. (1972). A study of meiotic germ cells in the gonads of foetal mouse chimaeras. *J. Embryol. exp. Morph.* **27**, 515–524.

McMahon, A., Fosten, M., and Monk, M. (1981). Random X-chromosome inactiva-tion in female primordial germ cells in the mouse. *J. Embryol. exp. Morph.* **64**, 251–258.

—— —— —— (1983). X-chromosome mosaicism in the three germ layers and the germ line of the mouse embryo. *J. Embryol. exp. Morph.* **74**, 207–220.

Merchant-Larios, H. and Alvarez-Buylla, A. (1986). The role of extracellular matrix and tissue topographic arrangement in mouse primordial germ cell migration. In

Development and Function of the Reproductive Organs. Serono Symp. Rev. **11** (eds A. Eshkol, B. Eckstein, N. Dekel, H. Peters, and A. Tsafriri) pp. 1–11. Ares-Serono Symposia, Rome.

Monk, M. and McLaren, A. (1981). X-chromosome activity in fetal germ cells of the mouse. *J. Embryol. exp. Morph.* **63**, 75–84.

Moor, R. M. (1983). Contact, signalling and cooperation between follicle cell and dictyate oocytes in mammals. In *Current Problems in Germ Cell Differentiation.* (eds A. McLaren and C. C. Wylie) Brit. Soc. Devl. Biol. Symp. 7, pp. 307–326. Cambridge University Press, Cambridge.

—— and Crosby, I. M. (1986). Protein requirements for germinal vesicle breakdown in ovine oocytes. *J. Embryol. exp. Morph.* **94**, 207–220.

—— and Osborn, J. C. (1984). Somatic control of protein synthesis in mammalian oocytes during maturation. In *Molecular Biology of Egg Maturation.* Ciba Fdn Symp. **98**, (eds R. Porter and J. Whelan) pp. 178–191. Pitman, London.

—— Smith, M. W. and Dawson, R. M. C. (1980). Measurement of intercellular coupling between oocytes and cumulus cells using intracellular markers. *Expl Cell Res.* **126**, 15–29.

—— and Warnes, G. M. (1978). Regulation of oocyte maturation in mammals. In *Control of Ovulation* (eds D. G. Crighton, G. R. Foxcroft, N. B. Haynes, and G. E. Lamming) pp. 159–176. Butterworths, London.

Moore, G. P. M. and Lintern-Moore, S. (1978). Transcription of the mouse oocyte genome. *Biol. Reprod.* **17**, 865–870.

Mystkowska, E. T. and Tarkowski, A. K. (1970). Behaviour of germ cells and sexual differentiation in late embryonic and early postnatal mouse chimaeras. *J. Embryol. exp. Morph.* **23**, 395–405.

Noguchi, M., Noguchi, T., Watanabe, M., and Muramatsu, T. (1982). Localization of receptors for *Dolichos biflorus* agglutinin in early post-implantation embryos in mice. *J. Embryol. exp. Morph.* **72**, 39–52.

Ozdzenski, W. (1967). Observations on the origin of primordial germ cells in the mouse. *Zool. Pol.* **17**, 367–379.

—— (1972). Differentiation of the genital ridges of mouse embryos in the kidney of adult mice. *Arch. Anat. Micr. Morph. Exp.* **61**, 267–278.

Pedersen, T. and Peters, H. (1968). Proposal for a classification of oocytes and follicles in the ovary of the mouse. *J. Reprod. Fert.* **17**, 555–557.

Peters, H. (1970). Migration of gonocytes into the mammalian gonad and their differentiation. *Phil. Trans. R. Soc. Lond.* B, **259**, 91–101.

Polanski, Z. (1986). *In-vivo* and *in-vitro* maturation rate of oocytes from two strains of mice. *J. Reprod. Fert.* **78**, 103–109.

Schroeder, A. C. and Eppig, J. J. (1984). The developmental capacity of mouse oocytes that matured spontaneously *in vitro* is normal. *Devl. Biol.* **102**, 493–497.

Schultz, R. M., Letourneau, G. E., and Wassarman, P. M. (1979). Program of early development in the mammal: Changes in the pattern and absolute rates of tubulin and total protein synthesis during oocyte growth in the mouse. *Devl. Biol.* **73**, 120–133.

Snow, M. H. L. (1981). Autonomous development of parts isolated from primitive streak-stage mouse embryos. Is development clonal? *J. Embryol. exp. Morph.* **65**, suppl. 269–287.

—— and Monk, M. (1983). Emergence and migration of mouse primordial germ cells. In *Current Problems in Germ Cell Differentiation* (eds A. McLaren and C. C. Wylie), Brit. Soc. Devl. Biol. Symp. 7, 115–135. Cambridge University Press, Cambridge.

Speed, R. M. and Chandley, A. C. (1983). Meiosis in the foetal mouse ovary.

II. Oocyte development and age-related aneuploidy. Does a production line exist? *Chromosoma* **88**, 184–189.

Spiegelman, M. and Bennett, D. (1973). A light- and electron-microscopy study of primordial germ cells in the early mouse embryo. *J. Embryol. exp. Morph.* **30**, 97–118.

Staigmiller, R. B. and Moor, R. M. (1984). Effect of follicle cells on the maturation and developmental competence of ovine oocytes matured outside the follicle. *Gamete Res.* **9**, 221–229.

Szöllösi, D., Gérard, M., Ménézo, Y., and Thibault, C. (1978). Permeability of ovarian follicle; corona cell-oocyte relationship in mammals. *Annls. Biol. anim. Biochim. Biophys.* **18**, 511–521.

Tam, P. P. L. and Snow, M. H. L. (1981). Proliferation and migration of primordial germ cells during compensatory growth in the mouse embryo. *J. Embryol. exp. Morph.* **64**, 133–147.

Templeton, A. A. (1985). Ovulation timing and IVF. In *In Vitro Fertilisation and Donor Insemination.* (eds W. Thompson, D. N. Joyce and J. R. Newton). Proc. 12th Study Group RCOG, pp. 45–59. Royal College of Obstetricians and Gynaecologists, London.

Tsonis, C. G., Quigg, H., Lee, V. W. K., Leversha, L., Trounson, A. O., and Findlay, J. K. (1983). Inhibin in individual ovine follicles in relation to diameter and atresia. *J. Reprod. Fert.* **67**, 83–90.

Ullmann, S. L. (1978). Observations on the primordial oocyte of the bandicoot *Isoodon macrourus* (Peramelidae, Marsupialia). *J. Anat.* **128**, 619–631.

Upadhyay, S. and Zamboni, L. (1982). Ectopic germ cells. A natural model for the study of germ cell sexual differentiation. *Proc. natn. Acad. Sci. U.S.A.*, **79**, 6584–6588.

Veeck, L. L., Wortham, J. W. C., Witmyer, J., Sandow, B. A., Acosta, A. A., Garcia, J. E., Jones, G. S., and Jones, H. W. (1983). Maturation and fertilization of morphologically immature human oocytes in a program of *in vitro* fertilization. *Fertil. Steril.* **39**, 594–602.

Wabik-Sliz, B. and McLaren, A. (1984). Culture of mouse germ cells isolated from fetal gonads. *Expl. Cell Res.* **154**, 530–536.

Wartenburg, H. (1983). Germ cell migration induced and guided by somatic cell interaction. *Bibl. Anat.* **24**, 93–110.

Wilson, E. B. (1925). *The Cell in Development and Heredity.* Macmillan, New York.

Witschi, E. (1948). Migration of the germ cells of human embryos from the yolk sac to the primitive gonadal folds. *Carnegie Inst. Contr. Embryol.* **32**, 67–80.

Wylie, C. C., Heasman, J., Snape, A., O'Driscoll, M., and Holwill, S. (1985). Primordial germ cells of *Xenopus laevis* are not irreversibly determined early in development. *Devl. Biol.* **112**, 66–72.

—————— Swan, A. P., and Anderton, B. H. (1979). Evidence for substrate guidance of primordial germ cells. *Expl. Cell Res.* **121**, 315–324.

Zamboni, L. (1972). Comparative studies on the ultrastructure of mammalian oocytes. In *Oogenesis* (eds J. D. Biggers and A. W. Schuetz) pp. 5–45. University Park Press, Baltimore.

—— (1974). Fine morphology of the follicle wall and follicle cell-oocyte association. *Biol. Reprod.* **10**, 125–149.

—— and Upadhyay, S. (1983). Germ cell differentiation in mouse adrenal glands. *J. exp. Zool.* **228**, 173–193.

5   The epidemiology and function of sex hormone-binding globulin

JOHN W. MOORE AND RICHARD D. BULBROOK

I   Introduction

Our interest in sex-hormone-binding globulin[1] (SHBG) arose because we were unable to demonstrate a convincing relationship between plasma oestradiol concentrations and any aspect of the biology of human breast cancer. This was puzzling because indirect evidence from epidemiological studies strongly implicated ovarian function as a prime determinant of risk of the disease. In 1981, Siiteri, Hammond and Nisker claimed that women with breast cancer had abnormally high proportions of non-protein-bound oestradiol in their blood. Their concept was that an increase in the fraction of the blood oestradiol available for biological activity was of importance in the aetiology of the disease and since this fraction was largely controlled by the concentration of SHBG, the epidemiology of this protein became important.

It rapidly became apparent that the variation of SHBG concentrations between women was very large and it was obviously desirable to attempt to identify some of the causes of this variability. This was partly because of an

1. It was decided, by vote, at the First International Symposium on Steroid Hormone Binding Proteins, Lyons, France, April 1986, that this protein should be called either sex hormone-binding globulin (SHBG) or sex steroid-binding protein (SBP). It was also agreed that other names such as testosterone-estradiol binding globulin (TeBG) or steroid binding β-globulin (SBβG) should no longer be used.

interest in the biology of the protein but mainly because it had been suggested that assays of SHBG might be a useful means of identifying women at risk of breast cancer who could then be treated with anti-oestrogenic drugs in a trial of prevention (Cuzick, Wang, and Bulbrook 1986). These considerations led to a massive epidemiological study in which serum SHBG was measured in a population of nearly 5000 women.

In order to interpret data from such a study, many factors had to be taken into account including the effects of drug administration, disease and physiological conditions on SHBG concentration as well as fundamental questions concerning the role of blood steroid-binding proteins and the availability of sex-steroids to target organs. Some of these questions have been considered in the two extensive monographs by Westphal (1971; 1986) and in reviews by Anderson (1974), Heyns (1977), Wagner (1978) and Lobl (1981). From our reading of the recent literature it became clear that there continues to be a keen interest in sex-steroid binding proteins and exciting developments have occurred in many areas. For this reason, in our review, we have tried to encompass a fairly broad spectrum of the literature on SHBG. Areas to be covered include an examination of current concepts of how blood steroid-binding proteins determine cellular availability and metabolic clearance of sex steroids, recent advances in the characterization and measurement of SHBG, physiological and pharmacological factors affecting SHBG activity and finally an examination of SHBG in disease states in the human.

The study of binding of sex steroids to blood proteins began in the early 1950s with the extensive investigation of the interactions of steroids with albumin. In 1958, Daughaday and Kozak discovered an α-globulin, subsequently called corticosteroid-binding-globulin (CBG), with a high affinity for steroids bearing the C-21 side chain such as cortisol and progesterone and a lower affinity for testosterone. Direct evidence for a testosterone-binding protein, distinct from CBG and albumin was obtained by Mercier, Alfsen, and Baulieu (1966) and Pearlman and Crepy (1967). It was shown shortly after this that oestradiol binds to the protein at the same site as testosterone (van Baelen, Heyns, Schonne, and De Moor 1968; Murphy 1968). For extensive reviews of these early studies the reader is referred to Westphal (1971), Anderson (1974) and King and Mainwaring (1974).

The site of synthesis of SHBG is assumed to be the liver and there is indirect evidence to support this. Thus hepatoma cells (Hep G2) in culture secrete a protein which is indistinguishable from SHBG (Kahn, Knowles, Aden, and Rosner 1981) and immunofluorescent studies have shown that monkey and human hepatocytes are specifically labelled after incubation with anti-human SHBG antiserum (Bordin and Petra 1980; Mercier-Bodard and Baulieu 1986). Plasma proteins to which sex steroids bind with high affinity have been found in most species. In human and other primates, SHBG binds naturally occurring 17β-hydroxy-androgens and oestradiol whereas in other animals, for example, the cat, dog and rabbit, they are mainly androgen binding proteins. Adult rats, mice and guinea pigs do not have SHBG. For a summary of the literature on SHBG in non-human mammals and non-mammalian vertebrates see Westphal (1986).

II   Blood binding proteins and compartmentalization of sex steroids

1   SHBG AND ALBUMIN

In human blood, the most biologically active sex steroids, 5α-dihydrotesto-sterone, testosterone and oestradiol circulate in the non-protein-bound form or in association with SHBG, albumin and possibly other proteins (see II, 2). Albumin binding of steroids is of low affinity, with association constants (Ka) in the region of $10^4 \ M^{-1}$, but of enormous capacity because of the high concentration of the protein (0.6–0.8 mmol/l) and the large number of binding sites on the molecule for some steroids. Steroid binding to albumin usually follows the polarity rule with less polar steroids, such as oestradiol, binding with higher affinity than those of greater polarity such as dihydro-testosterone and testosterone.

SHBG, on the other hand, is present in the blood at much lower concentrations than albumin and has one binding site of high affinity and specificity for 17β-hydroxy steroids (section III). The association constants for the binding of sex steroids to SHBG are in the region of $10^9 M^{-1}$ and there is general agreement that the Ka for dihydrotestosterone binding to SHBG is approximately three times greater than that of testosterone which, in turn, is about twice that of oestradiol. For extensive details of steroid-protein binding see the monographs by Westphal (1971, 1986).

The factors which determine the blood distribution of steroids between the free compartment and the various binding proteins are very complex and involve the association and dissociation constants for steroid-protein binding as well as the concentrations of interacting molecules. From some of these variables, Dunn, Nisula, and Rodbard (1981) have calculated the theoretical distribution of most of the naturally occurring steroids between the free and protein-bound compartments.

Table 5.1 shows the distribution of oestradiol, testosterone and 5α-dihydrotestosterone, experimentally determined in our laboratory. It can be seen that testosterone and 5α-dihydrotestosterone are bound more to SHBG than to albumin whereas in the case of oestradiol, there is greater binding to albumin. This reflects the greater affinity of oestradiol for albumin and its lower affinity for SHBG compared with testosterone and 5α-dihydrotesto-sterone.

The percentages of free (non-protein-bound) oestradiol shown in Table 5.1 are lower than those calculated by Dunn et al. (1981) for men and women. They are also lower than the experimentally determined results of Burke and Anderson (1972) and our own previous data for normal women which were

Table 5.1

Mean (± SD) percentages of free, SHBG-bound and albumin-bound 17β-OH steroids in normal male and female sera

Steroid	n	Free (%)[1]	SHBG bound (%)[2]	Albumin bound (%)[2]	n	Free (%)[1]	SHBG bound (%)[2]	Albumin bound (%)[2]
		Male				Female		
Testosterone	10	1.73 ± 0.32	59.9 ± 9.3	38.3 ± 9.0	10	1.18 ± 0.20	71.9 ± 4.6	27.0 ± 4.4
Oestradiol	10	1.21 ± 0.11	33.6 ± 5.6	65.2 ± 5.5	10	0.96 ± 0.14	39.8 ± 9.7	59.2 ± 9.5
Dihydro-testosterone	10	1.42 ± 0.26	74.3 ± 5.2	24.3 ± 4.9	10	1.18 ± 0.21	80.2 ± 4.7	18.6 ± 4.5

SHBG concentrations: Male = 28.4 ± 9.3 nmol/l,   Female = 50.4 ± 21.6 nmol/l.
[1] Measured by centrifugal-ultrafiltration-dialysis (Hammond, Nisker, Siiteri, and Jones 1980) as modified (Moore, Hoare, Quinlan, Clark, and Wang 1987). [2] Calculated from the percentage of free steroid in serum, heat-treated (60°C/60 min) to denature SHBG (see Siiteri, Murai, Hammond, Nisker, Raymoure, and Kuhn 1982; Hammond, Lähteenmäki, Lähteenmäki and Lukkainen 1982).

obtained using centrifugal-ultrafiltration-dialysis as originally described (Moore, Clark, Bulbrook, Hayward, Murai, Hammond, and Siiteri 1982). The disagreement with our previous data is because we have now improved the methodology and the interested reader is referred to our recent publication for technical details (Moore, Hoare, Quinlan, Clark, and Wang 1987).

Our results in serum from non-pregnant females are very similar to those of Dowsett, Mansfield, Griggs, and Jeffcoate (1984) who also found percentages of free oestradiol in the region of 1 per cent. The very low levels of steroid hormones in saliva are thought to be equal in concentration to the non-protein-bound steroid levels in serum (Riad-Fahmy, Read, Walker, and Griffiths 1982) and it has recently been shown, by direct radioimmunoassay, that oestradiol concentrations in samples of saliva from premenopausal women are about 1 per cent of those of serum (Wang, Fantl, Habibollahi, Clark, Fentiman, Hayward, and Bulbrook 1986) which lends support to our findings.

Within the physiological range of SHBG values found in male, female and pregnancy sera, highly significant negative correlations between the percentages of free testosterone and oestradiol and the concentration of SHBG are observed. Our non-protein-bound testosterone and oestradiol data from Table 5.1 are plotted against the concentrations of SHBG (see Fig. 5.1). As shown by Burke and Anderson (1972) changes in the concentration of SHBG have a much greater effect on the percentage of free testosterone than they have on free oestradiol. When the relationship between SHBG and the percentages of albumin-bound oestradiol and testosterone are considered, however, it is apparent that the calculated regression lines, though different in elevation, are similar in shape over the SHBG concentration range studied

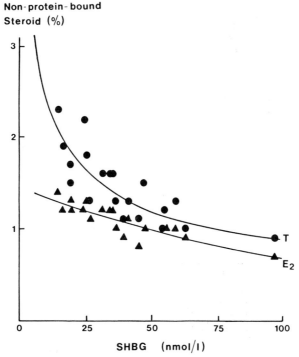

Fig. 5.1.   Relationship between SHBG and the percentages of non-protein bound oestradiol (E2) and testosterone (T) in human serum. Calculated linear regression equations: log non-protein bound T $(\%) = 1.89$–0.436 log SHBG (nmol/l); log non-protein bound E2 $(\%) = 0.358$–0.00746 SHBG (nmol/l).

(Fig. 5.2). Changes in SHBG, therefore, result in similar percentage changes in the albumin-bound fractions of oestradiol and testosterone. Thus, in this study group, the percentages of albumin-bound oestradiol and testosterone are about 25 per cent higher at the low end of the SHBG range compared with those at the high end.

The relationship between the percentage of SHBG-bound steroids and the amount of SHBG present in the sample is the converse of that seen for albumin-bound steroids and SHBG concentration. The small changes in concentration of albumin, within the physiological range, do not affect the distribution of steroids because of the low affinities of the steroid-albumin interactions (Moore *et al.* 1982).

2   OTHER SEX STEROID-BINDING PROTEINS IN HUMAN BLOOD

SHBG and albumin are not the only binding proteins for sex-steroids in the

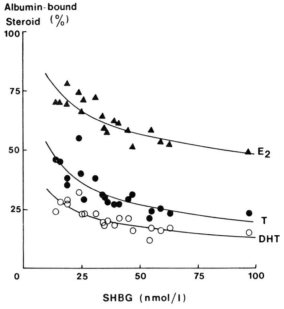

Fig. 5.2.   Relationship between SHBG and the percentages of albumin-bound (Alb-Bound) oestradiol (E2), testosterone (T) and 5α-dihydrotestosterone (DHT) in human serum. Calculated linear regression equations: log Alb-Bound E2 (%) = 4.96–0.24 log SHBG (nmol/l); log Alb-Bound T (%) = 4.99–0.44 log SHBG (nmol/l); log Alb-Bound DHT (%) = 4.42–0.41 log SHBG (nmol/l).

human circulation. Others include orosomucoid (α_1-acid glycoprotein), oestradiol-binding protein, corticosteroid-binding globulin (CBG), fetal steroid-binding protein and fetal steroid-binding glycoprotein.

Orosomucoid binds progesterone, testosterone and androstenedione with affinities about ten times higher than albumin, whereas oestradiol binds with an affinity similar to that of its binding to albumin. The protein has only one binding site for most of the steroids studied. The concentration of orosomucoid in serum, less than 25 μmol/l, is very much lower than that of albumin and therefore its importance in determining the distribution of steroids in the blood is small (see Englebienne 1984; Westphal 1986).

Oestradiol-binding protein, another protein distinct from SHBG, which has a high affinity and low capacity for oestradiol but which does not bind 5α-dihydrotestosterone or testosterone significantly, was identified in human serum by O'Brien, Higashi, Kanasugi, Gibbons and Morrow (1982). It is suggested that there is a relationship between oestrogen status and synthesis of this protein since concentrations range from 0.26 nM in male blood to 2 nM in blood from females and it is reported that in pregnancy the levels are higher. Concentrations of oestradiol-binding protein are very much less than those of SHBG (see III, 3) and its importance remains to be determined.

CBG in the human and several other species shows some affinity for testosterone ($Ka = 1.4 \times 10^6$ M^{-1} in the human) but no significant binding to oestradiol. Although in the adult mouse and rat, where SHBG is absent, it may be an important testosterone-binding protein, it is unlikely that binding of testosterone to CBG is particularly extensive in the human (see Dunn et al. 1981; Westphal 1986).

Fetal steroid-binding protein (FSBP) was recently discovered in cytosolic preparations of human fetal liver because, unlike SHBG, it is immobilized by the cibachrome blue-agarose (blue gel) portion of a two-tier column described by Iqbal and Johnson (1977). 5α-Dihydrotestosterone, testosterone and oestradiol bind to FSBP with high affinity and in this regard the protein shows similarities to SHBG. However, unlike SHBG, the protein does not bind 5α-androstane-3β, 17β-diol specifically and appears to differ from SHBG immunologically and in molecular weight (Wilkinson, Iqbal, and Williams 1983; Iqbal, Wilkinson, and Williams 1983; Iqbal, Forbes, Wilkinson, Moore, Williams, and Bulbrook 1987).

Another binding protein distinct from SHBG, with high affinities for sex-steroids was identified in cord blood and amniotic fluid by Dalton (1984) and termed fetal steroid binding globulin (FiSBOG). It shares some steroid binding characteristics with FSBP and is found in human blood only in late pregnancy where the reported mean level was 11.15 nmol 5α-dihydrotestosterone bound/l. Further investigation of FSBP and FiSBOG is required. It is possible that they may be related proteins and could also be variants of SHBG.

III   Purification, characterization and measurement of SHBG

1   PURIFICATION OF SHBG

Attempts at purification of SHBG were first undertaken by Mercier-Bodard, Alfsen, and Baulieu (1970) who were able to partially purify the protein. SHBG was first purified to homogeneity by Mickelson and Petra (1975) using an affinity matrix in which 5α-dihydrotestosterone was linked to agarose through the 17β position. Yields were very low, however, and the introduction of affinity matrices in which the steroidal 17β-hydroxyl group were exposed (Rosner and Smith 1975; Iqbal and Johnson 1979) did not much alter the situation.

The most dramatic improvement in yield came with the introduction of affinity adsorbents in which derivatives of 5α-dihydrotestosterone were coupled to the supporting media through the 17α position leaving both the 3-oxo function and the 17β-hydroxyl group free to interact with SHBG (Suzuki, Itagaki, Mori, and Hosoya 1977; Petra and Lewis 1980; Cheng, Musto, Gunsalus, and Bardin 1983). Further refinements in technique have made it possible to isolate milligram quantities of SHBG from serum within 48 hours on a routine basis by incorporating these types of chromatographic methods into the purification procedures (Khan, Ehrlich, Birken, and Rosner 1985; Hammond, Robinson, Sugino, Ward, and Finne 1986; Petra, Namkung, Senear, McCrae, Rousslang, Teller, and Ross 1986*a*).

2   CHARACTERIZATION AND STRUCTURE

Native SHBG is considered to be a homodimeric glycoprotein with one steroid binding site (Petra, Stanczyk, Senear, Namkung, Novy, Ross, Turner, and Brown 1983; Hammond *et al.* 1986; Petra, Kumar, Hayes, Ericsson, and Titani 1986). A comparison of the physical characteristics of SHBG from the rabbit, baboon, macaque and human reveal that the protein in each species has similar molecular weights (ca 85,000 Daltons) and the same dimeric molecular organization although they differ in carbohydrate content (Petra *et al.* 1986*a*; Petra, Namkung, Titani, and Walsh 1986*b*).

The amino acid sequence of the purified protein has been largely elucidated (Walsh, Titani, Takio, Kumar, Hayes, and Petra, 1986). Hammond *et al.* (1986) have sequenced the amino (N)-terminal of the protein and obtained the same sequence as Walsh and colleagues but observed an additional N-terminal leucine residue. Walsh *et al.* (1986) suggest that their failure to detect this residue was due to its proteolytic removal during storage of serum prior to purification. This probably also accounts for previously revealed heterogeneity at the N-terminus (Fernlund and Laurell 1981; Petra *et al.* 1983). Although affinity labelling and X-ray diffraction analysis will be needed to provide proof, Petra *et al.* (1986*b*) have identified an internal amino acid sequence repeat within a hydrophobic region which they suggest may be the steroid binding site of the protein.

There is no sequence similarity between human SHBG and either the human oestradiol and glucocorticoid receptors or with the oncogene product of the erythroblastosis virus (v-*erb*A) with which the receptor proteins share some homology (Hollenberg, Weinberger, Ong, Cerelli, Oro, Lebo, Thompson, Rosenfeld, and Evans 1985; Weinberger, Hollenberg, Rosenfeld, and Evans 1985; Green, Gilna, Waterfield, Baker, Hort, and Shine 1986). The amino acid sequences of the rat androgen-binding protein (Joseph, Hall, and French, 1986; 1987) show many regions of homology including the putative binding site (Petra, Titani, Walsh, Joseph, Hall, and French 1986).

3   MEASUREMENT OF SHBG

Until the late 1970s, the measurement of SHBG in serum or plasma has relied upon the determination of the binding capacity of the protein by saturation of high affinity binding sites with tritium or ^{14}C labelled 17β-hydroxy steroids followed by the separation of specifically bound from non-specifically bound and free label. Many methods of separation have been employed and these include equilibrium dialysis (Vermeulen, Stoica, and Verdonck 1971), ammonium sulphate precipitation (Rosner, 1972; Anderson, Lasley, Fisher,

Shepherd, Newman, and Hendrickx 1976), DEAE filter paper adsorption (Mickelson and Petra 1974), agar gel electrophoresis (Wagner 1978), charcoal separation (Hammond and Lähteenmäki 1983) and affinity chromatography (Iqbal and Johnson 1977; Nisula, Loriaux, and Wilson 1978; Bruning, Bonfrer and Nooyen 1985). This methodology has recently been reviewed (Englebienne 1984; Rosner 1986).

Following the isolation of purified preparations of SHBG and the production of specific antibodies (Bordin, Lewis, and Petra 1978), several antibody-based assays have been established which allow quantitation of the concentration (in contrast to the binding capacity) of the protein. These include "rocket" immunoelectrophoresis (Laurell and Rannevik 1979), radioimmunoassay (Mercier-Bodard, Renoir, and Baulieu 1979; Kahn, Ewan, and Rosner 1982; Cheng, Bardin, Musto, Gunsalus, Cheng, and Ganguly 1983; Maruyama, Aoki, Suzuki, Sinohara, and Yamamoto 1984; Lapidus, Lindstedt, Lunberg, Bengtsson, and Gredmark 1986), enzyme-linked immunoassay (Bordin, Torres, and Petra 1982), immunoprecipitation (Degrelle 1986) and immunoradiometric assay (Hammond, Langley, and Robinson 1985). Since it is established that there is one steroid binding site per mole of SHBG, binding capacity measurements are equivalent to concentrations of the protein measured by immunological techniques so for the sake of simplicity, in this review, we have not distinguished between binding capacities and amounts of SHBG except where such a distinction would be instructive.

With so many methods of analysis available it is inevitable that some variation in the ranges of SHBG in comparable physiological situations occurs when different methods are used or when the same method is employed in different laboratories. As discussed by Rosner (1986) this disagreement has many reasons, a major one being the lack of a standard preparation of SHBG with which to calibrate individual assays. Table 5.2 shows "normal ranges" obtained with some of the more widely used methods.

One major advantage of the immuno-type of assay compared with the binding capacity assay is that sensitivity is much greater and this has allowed the detection of SHBG, with ease, in various body fluids where concentrations of the protein are much lower than those found in serum or plasma. SHBG has been found in amniotic fluid, ovarian follicular fluid, breast cyst fluid and saliva (Hammond, Leinonen, Bolton, and Vihko 1983; Rosner, Kahn, Breed, Fleisher, and Bradlow 1985; Ben-Raphael, Mastroianni, Meloni, Lee, and Flickinger 1986; Hammond and Langley 1986). Another advantage is that larger numbers of samples can be assayed easily.

IV Significance of SHBG

Although SHBG has been called a transport protein, sex steroids, at the concentration found in the blood are fully soluble and do not require a high

Table 5.2

SHBG binding capacities and concentrations ($nmol/l \pm S.E.M.$) in male, female, and pregnant female blood

Reference	Male	Female	Pregnancy	Assay types	Separation method
Rosner (1972)	32±2 (27)	64±4 (16)	427±23 (19) (stage not specified)	Binding capacity (^{3}H-DHT)	Ammonium sulphate precipitation
Anderson, Lasley, Fisher, Shepherd, Newman, and Hendrickx (1976)	35±2 (23)	74±9 (40)	367±21 (24) (at term)	Binding capacity (^{3}H-DHT)	
Rudd, Duignan, and London (1974)	50±4 (10)	78±12 (12)	290±38 (10) (third trimester)	Binding capacity (^{14}C-T)	
Iqbal and Johnson (1977)	26±6 (13)	60±6 (18)	—	Binding capacity (^{3}H-DHT)	Affinity chromatography
Hammond, Langley, and Robinson (1985)	23±2 (20)	53±4 (32)	402±77 (5) (third trimester)	Immunoradiometric assay	—
Cheng, Bardin, Musto, Gunsalus, Cheng, and Ganguly (1983)	18±3 (12)	54±5 (8)	115±24 (4) (weeks 10–15) 212±17 (4) (weeks 20–25) 374±22 (6) (weeks 35–40)	Radioimmunoassay	—

^{3}H-DHT = [^{3}H]-5α-dihydrotestosterone, ^{14}C-T = [^{14}C]-testosterone, Numbers of subjects shown in parenthesis.

affinity binding protein for this purpose (Anderson 1974). The conventional view of steroid uptake by cells enunciated by Tait and Burstein (1964) is that only the free fraction of steroid is transported across the cell membrane. Protein-bound steroid in a capillary must first dissociate from the protein before it becomes available for uptake. In recent years, Pardridge and his colleagues have intensively investigated factors such as cell membrane permeability, capillary transit time and steroid-protein dissociation rates which affect the availability of steroids to organs such as the brain, liver and uterus. They have concluded that, in general, the free and albumin-bound fractions of testosterone and oestradiol are available for uptake by most tissues and that in the liver even some of the SHBG bound fractions may be available (see Pardridge 1981 and also this volume).

It is apparent that since the amount of SHBG in the blood undoubtedly controls the proportion of free oestradiol and testosterone and also the proportions which are bound to albumin, it is of profound importance in controlling the hormonal stimulus to responsive cells, if the conventional dogma of steroid uptake is correct. Evidence from studies in the human and non-human primates that the SHBG-bound steroid is generally not available for uptake comes from metabolic and *in vitro* studies which will be briefly reviewed.

1   INFLUENCE OF SHBG BINDING ON THE METABOLIC CLEARANCE RATE OF SEX STEROIDS

Indirect evidence that specific binding to proteins protects steroid hormones from metabolism came from early studies on the metabolic clearance rate (MCR) of testosterone. Serum SHBG is higher in adult females than in adult males and the MCR of testosterone in normal women is lower than in normal men. In pregnancy, when SHBG levels are high, the MCR of testosterone is diminished compared to that of non-pregnant women. Hirsute women, who often have low SHBG capacities, tend to have higher MCR for testosterone than do normal women (Bardin and Lipsett, 1967; Vermeulen, Verdonck, Van der Straeten, and Orlie 1969; Vermeulen and Ando 1979).

Highly significant correlations between the percentages of free and non-SHBG bound testosterone and also 5α-dihydrotestosterone and the metabolic clearance rates of these steroids were demonstrated in normal males and postmenopausal females (Vermeulen and Ando 1979). Since the percentages of free and non-SHBG bound steroids in the blood are dependent on the SHBG activity, negative correlations between the MCRs of sex-steroids and SHBG binding capacity would be expected. This has been confirmed for testosterone and oestradiol in postmenopausal and perimenopausal women (Siiteri, Murai, Hammond, Nisker, Raymoure, and Kuhn 1982; Longcope, Hui, and Johnston 1987).

In hyperthyroid patients, where SHBG capacities are usually elevated

(Anderson 1974; Ridgway, Longcope, and Maloof 1975), the metabolic clearance rates of testosterone and oestradiol are diminished (Gordon, Southren, Tochimoto, Rand, and Olivo 1969; Gordon, Olivo, Rafii, and Southren 1975; Ridgway *et al.* 1975) whereas those of oestrone and andro-stenedione, which do not bind to SHBG, are normal (Ridgway, Maloof, and Longcope 1982). This implies an important role for SHBG in controlling the metabolic clearance of the two former steroids.

Direct experimental evidence demonstrating a role for SHBG in regulating the metabolic clearance rate of testosterone in non-human primates was provided by Petra and his colleagues (Petra, Stanczyk, Namkung, Fritz, and Novy 1985; Stanczyk, Namkumg, Fritz, Novy, and Petra 1986) who infused pure human or rhesus SHBG into female rhesus macaques and observed increases of between 150 and 300 per cent in the levels of SHBG which was accompanied by a decrease of about 10 per cent in the metabolic clearance rates. In further experiments they infused purified antibodies to human SHBG, which cross-reacted with macaque SHBG, and observed a steady increase in metabolic clearance with decreasing SHBG levels. Using equations derived from the law of mass action they calculated the distribution of testosterone between SHBG and albumin and were able to show that as long as the percent of testosterone bound to SHBG was equal to or higher than that bound to albumin, the effect on the clearance rate was small. A dramatic increase in metabolic clearance of testosterone occurred, however, if SHBG levels were reduced to the extent that testosterone was mostly bound to albumin.

It must be said that the small fall in MCR of testosterone which accompanied quite a large rise in SHBG in the monkey does not accord with the situation in the human where in males the SHBG capacity is about half but the MCR of testosterone is twice that of females (Vermeulen and Ando 1979). Neither does the further finding that the effect on the clearance rate of testosterone is small if the percentage bound to SHBG is greater or equal to 50 per cent since, as shown in Table 5.1, the percentage of testosterone bound to SHBG in the male and female are both above 50 per cent and yet the clearance rates of testosterone are very different. This may reflect differences between the calculated and experimentally determined binding of testosterone to SHGB, or that the regulation of the MCR of testosterone in monkeys by SHBG may be slightly different from that in humans.

In contrast to the unanimity in the literature that SHBG and MCR of oestradiol and testosterone are negatively related, Hotchkiss (1985) found that in the prepubertal rhesus monkey (*Macacca mulatta*) both the MCR of oestradiol and SHBG binding capacities were high when compared to the adult. The author points out that, in addition to SHBG levels, the mature and immature animals differ in a great many respects and speculates that other factors which affect clearance (see Pardridge 1981) such as hepatic blood flow, may change during development.

2   INFLUENCE OF SHBG ON THE PERIPHERAL CONVERSION OF STEROIDS

Interconversions between androstenedione, testosterone and 5α-dihydro-testosterone occur widely in peripheral tissues (see Vermeulen and Ando 1979); indeed, as shown by Horton and Tait (1966), approximately 60 per cent of testosterone in female plasma arises from extraglandular conversion from androstenedione. Vermeulen and Ando (1979) observed that, in addition to metabolic clearance rates, the blood conversion ratios of testosterone to androstenedione and to a lesser extent those of testosterone to 5α-dihydro-testosterone were significantly correlated with both the free and the percentage of non-SHBG bound testosterone but not with the total amount of the hormone. Decreased testosterone—androstenedione conversion ratios in hyperthyroid patients are fully consistent with elevated SHBG causing a decrease in the fraction of circulating testosterone available for extraction by peripheral tissues (Ridgway *et al.* 1982). *In vitro* studies in various cell systems have arrived at similar conclusions (Anderson 1974; Egloff, Savoure, Tardival-Lacombe, Massart, Nicol, and Degrelle 1981).

3   POSSIBLE BIOLOGICAL AVAILABILITY OF SHBG-BOUND SEX STEROIDS

Bordin and Petra (1980) have advanced the hypothesis that steroid hormones bound to their specific binding proteins may be available for transfer across cell membranes in some circumstances, possibly by a receptor mediated process (Stanczyk *et al.* 1986). Subsequently, others have also proposed the same idea (Siiteri *et al.* 1982; Siiteri 1986). Evidence in support of this concept include the immunocytochemical localisation of SHBG in the prostate, testis and epididymis of the monkey (Bordin and Petra, 1980) and human (Egloff, Vendrely, Tardival-Lacombe, Dadoune, and Degrelle 1982), and in human breast cancer tissue (Tardival-Lacombe, Egloff, Mazabraud, and Degrelle 1984). Using [125]Iodine labelled SHBG, Kahn, and Rosner (1985) have found two binding sites on human prostate membranes, one of high affinity and one of low affinity. Strel'chyonok, Avvakumov, and Survilo (1984) showed that [125]Iodine labelled SHBG binds to decidual endometrial membranes only when complexed with oestradiol. SHBG devoid of steroid or complexed with testosterone does not bind. They propose that in plasma membranes of oestrogen target tissues, there is a recognition system for SHBG-oestradiol which may allow these cells to take up from the blood, not only free oestradiol but also oestradiol complexed with the binding proteins.

Baulieu (1986) defines three categories of steroid binding proteins; those originating in the liver and circulating in the blood; those found in a restricted system such as seminiferous tubules; and those which are found within target cells. In the latter case, he suggests that the binding proteins are synthesized *in situ*. The possible function of SHBG (or SHBG like molecules) is discussed extensively by Baulieu (1986). He advances the delightful hypothesis that "we are seeing the situation upside down", that is to say that the binding proteins were originally intracellular molecules which controlled the access to various cellular compartments. The binding proteins in the blood would then be secretory products playing a secondary role in controlling the availability of steroids to other cells.

4 TISSUE CONCENTRATIONS OF STEROIDS

During the last few years, some extraordinary data have appeared concerning the concentrations of steroids within tissues. If oestradiol is considered briefly as an example, amounts of this steroid within normal and malignant breast tissue and fat are some twenty fold greater than those in the peripheral blood (Santen 1986; Vermeulen 1986). What is more, no correlation is found between blood and tissue oestradiol concentrations.

How this concentration gradient is achieved is not clear. Some of the oestrogen is synthesized locally via the aromatization of androstenedione to oestrone. Some comes from the action of sulphatase on oestrone sulphate but it seems doubtful if the activity of the enzymes involved would account for extreme cases where the concentration gradient may be 200-fold. Santen (1986) points out that while the plasma levels of oestradiol in postmenopausal women are 4 to 40 per cent of those found in premenopausal women, the tumour tissue concentrations are similar. He goes on to argue that the blood/tissue gradient is not wholly dependent on the classical oestrogen receptor proteins because a substantial gradient is still found in receptor-negative breast tumours. Also, the receptor binds oestrone with about one third of the affinity of oestradiol but the two steroids are present in equal concentrations within the tissue. If tissue concentrations are determined by receptor protein, one would expect lower concentrations of oestrone. Finally, no correlation has been found between receptor levels and tissue concentrations of oestradiol (for review see Vermeulen 1986).

The question has to be asked whether extracellular (or the hypothesized intracellular) SHBG or other blood binding proteins play any part in the determination of tissue concentrations? This is an area which has been pursued vigorously by Pardridge and is fully discussed in his chapter (this volume).

V SHBG in physiological conditions

In vivo evidence suggests that SHBG synthesis by the liver is stimulated by oestrogens and thyroid hormones and inhibited by androgens (see Anderson 1974 and Section **VI**). Although, in some physiological situations the impact of the net androgen/oestrogen/thyroid balance on SHBG synthesis is apparent, in others it is not so obvious. There follows, therefore, a review of rele-

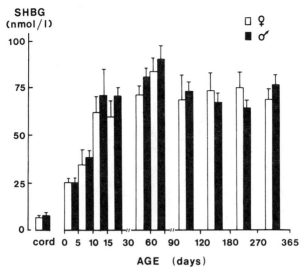

Fig. 5.3.   Change in plasma levels of SHBG (mean + S.E.M.) in infants. Reproduced with permission from Forest, Bonneton, Lecoq, Brebant, and Pugeat (1986).

vant literature on changes in SHBG in various physiological situations and an attempt has been made to relate these to the prevailing endocrine environment.

1   SHBG LEVELS IN CHILDREN

i   SHBG in children and during puberty

At birth, SHBG levels in cord plasma and in the plasma of new-born infants of both sexes are less than 10 per cent of the mother at term and similar to those of the adult male (August, Tkachuk, and Grumbach 1969; Forest, Ances, Tapper, and Migeon 1971; Anderson et al. 1976). This gradient is independent of fetal sex and weight (Forest et al. 1971; Anderson et al. 1976). After delivery SHBG rises rapidly within 30 days to achieve levels approximating to those found in non-pregnant adult females (August et al. 1969; Wenn, Kamberi, Vossough, Kariminejad, Torabee, Ayoughi, Keyvanjah, and Sarberi 1977; Forest, Bonneton, Lecoq, Brebant, and Pugeat 1986). The data from Forest et al. (1986) are shown in Fig. 5.3.

   Most studies suggest that the high levels of SHBG achieved during the weeks after birth are maintained until 8 to 10 years of age (August et al. 1969; Horst, Bartsch, and Dirksen-Thedens 1977; Wenn et al. 1977; Bartsch, Horst, and Derwahl 1980; Lee, Lawder, Townend, Wetherall, and Hahnel 1985; Forest et al. 1986). The literature is not entirely unanimous on this point, however, since Belgorosky and Rivarola (1986) in their study of 91 boys demonstrated a steady decline in SHBG from the age of 3 months

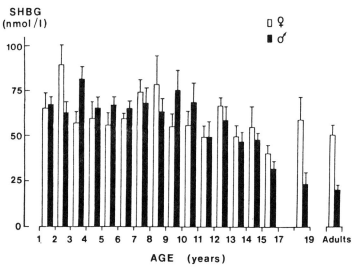

Fig. 5.4. Changes in SHBG (mean + S.E.M.) in males and females from childhood to adulthood. Reproduced with permission from Forest, Bonneton, Lecoq, Brebant, and Pugeat (1986).

onward. SHBG levels were not found to differ with the sex of prepubertal children (Wenn *et al.* 1977; Lee *et al.* 1985) although an earlier report suggested there was a difference (August *et al.* 1969).

There is general agreement that from around the ages of 8 to 10 in boys, blood levels of SHBG decline steadily and stabilize by about the age of 18 at levels approximately four-fold lower than those found before puberty (Horst *et al.* 1977; Blank, Attanasio, and Gupta 1978; Forest *et al.* 1986; Belgorosky and Rivarola 1986).

In a cross sectional examination of puberty in girls, Bartsch *et al.* (1980) demonstrated that by the age of about 15 (approximately mid-puberty), SHBG levels were about half the prepubertal levels and in a longitudinal study (Apter, Bolton, Hammond, and Vihko 1984), the concentration of SHBG, measured by an immunoradiometric assay, declined gradually (30 per cent) during puberty. This study also revealed that girls with menarche before 13 years of age had lower SHBG concentrations at 10–15.9 years than girls with later menarche.

In late female puberty, SHBG levels higher than those found at mid-puberty have been reported (Odlind, Carlstrom, Michaelsson, Vahlquist, Victor, and Mellbin 1982; Forest *et al.* 1986), and normal young adult females have only slightly lower SHBG capacities than those found before puberty (Bartsch *et al.* 1980; Forest *et al.* 1986). The changes in SHBG levels before and during puberty in boys and girls found by Forest *et al.* (1986) are shown in Fig. 5.4.

In a small study of pubertal changes in boys and girls, Gaidano, Berta,

Rovero, Valenzano, and Rosatti (1980) showed that the binding capacities for testosterone and 5α-dihydrotestosterone diminished. They also observed relative changes in capacity according to the ligand used in the assay and suggested that the apparent changes in SHBG were in fact changes in the binding characteristics of the protein. This appears to be unlikely, however, since Belgorosky and Rivarola (1986) found no differences in association constant (Ka) of SHBG binding to 5α-dihydrotestosterone between boys of different ages.

ii   Hormonal control of SHBG before and during puberty

The hormonal control of SHBG synthesis in the fetus and neonate is not clear. Levels of oestrogen in fetal blood at term are high, falling abruptly after birth (Tulchinsky and Chopra 1973) so the oestrogens do not appear to induce synthesis of SHBG by the neonatal liver. A stimulatory role for thyroid hormones in the neonatal rise in SHBG synthesis has been suggested (Anderson *et al.* 1976). Forest, Cathiard, and Bertrand (1973) and Wenn *et al.* (1977) showed that although newborn males have on average five times more testosterone than newborn females there were no differences in SHBG.

In children before puberty, no correlations were observed between the major sex steroids (testosterone, 5α-dihydrotestosterone or oestradiol) and SHBG (Horst *et al.* 1977; Bartsch *et al.* 1980; Belgarosky and Rivarola 1986).

During male puberty, the decline in SHBG is thought to be due to increasing adrenal and gonadal androgen production (but see below). The increase in testosterone is about 10 fold during puberty (Blank *et al.* 1978).

In girls, the 50 per cent fall in SHBG described during puberty by Bartsch *et al.* (1980) was accompanied by a 10-fold increase in oestradiol and a 5-fold increase in testosterone. Multivariate analysis showed that only the androgens were significantly related (negatively) with SHBG, suggesting a predominant role of these hormones in the control of the synthesis of the protein in these subjects. Apter *et al.* (1984) observed weakly significant negative correlations between SHBG and blood androgens (testosterone and androstenedione). The closest relationships shown, however, were between SHBG and bodyweight (and body fat percentage) suggesting that factors other than steroids have to be considered in the regulation of SHBG during puberty in girls. Among factors known in the human to regulate SHBG synthesis, thyroid hormones are unlikely candidates since they do not appear to change during the course of puberty (Lamberg, Kantero, Saarinen, and Widholm 1973; Bartsch *et al.* 1980). Growth hormone is a possibility since it decreases SHBG levels (De Moor, Heyns, and Bouillon 1972) and an increase in secretion at mid-puberty has been reported (Kantero, Wide, and Widholm 1975; Bierich 1983).

Some doubt has been cast on the importance of androgens in the pubertal decline in SHBG values in both boys and girls by the finding that in patients

with untreated isolated gonadotrophin deficiency there was a highly significant inverse correlation between SHBG levels and age during the second
decade of life even though testosterone levels did not rise (Cunningham,
Loughlin, Culliton, and McKenna 1984). The numbers of patients in this
study were small (n = 4) and the maximum period over which individuals
were studied was 24 months so confirmation of these findings in a larger
study would be helpful to our understanding. Cunningham *et al.* (1984) also
observed a decline in SHBG during puberty in two 46 XY siblings, pheno-
typically female with complete androgen insensitivity.

In summary, factors controlling SHBG synthesis in children before and
during puberty are poorly understood. The high levels in prepubertal children may stem from a lack of endocrine suppression of synthesis. This
becomes manifest as androgens increase towards puberty. The marked fall
during male puberty is probably due to increasing androgens, and the
gradual decline in SHBG in girls may be due to increasing androgens
opposed by oestrogens. The greater increase in oestrogens compared with
that of androgens would indicate a higher potency for the androgens in controlling SHBG synthesis during puberty in the female. It is evident, however,
that the role of androgens is becoming increasingly controversial and bodyweight appears to be much more important than steroid secretion (cf. Adams
and Steiner, this volume).

2   EFFECT OF AGEING ON SHBG IN THE ADULT

i   Men

Many studies have demonstrated that SHBG capacities in men older than
60–65 are higher than those of younger men (Vermeulen, Rubens, and Ver-
donk 1972; Pirke and Doerr 1973; Bartsch 1980; Winters and Troen, 1982).
Recently, significantly positive correlations between age and SHBG capacities (Purifoy, Koopmans and Mayes 1981) and concentrations (Maruyama *et
al.* 1984) have been observed between the third and ninth decades of life.
Using multivariate analysis a similarly significant relationship was observed
between the third and fifth decades when corrections were made for obesity
and endurance fitness measured on a bicycle ergometer (Semmens, Rouse,
Beilin, and Masarei 1983*a*). These increases in SHBG activity with age probably reflect the androgen-oestrogen balance brought about by the decreasing
androgen output by the Leydig cells (Vermeulen *et al.* 1972; Pirke and Doerr
1973; Bartsch 1980) and increasing levels of oestradiol found by many but
not all workers in the field (see Winters and Troen 1982).

ii   Women

The situation in women is less clear. We have measured SHBG concentrations, by the immunoradiometric method of Hammond *et al.* 1985, in serum samples from 1200 women between the ages of 35 and 75 who had never used oral contraceptives or hormone replacement therapy, had no history of endocrine diseases or cancer, and were not taking drugs likely to affect SHBG. The data were analysed by multivariate analysis. There was little variation in weight-adjusted SHBG levels with age during the reproductive years as also found by Semmens *et al.* (1983*a*). In postmenopausal subjects the mean SHBG concentration were very significantly lower than those of premenopausal women, a finding which is consistent with the cessation of ovarian function. Curiously, when the postmenopausal SHBG levels were considered there was a statistically significant, approximately linear relationship with the number years past the menopause (Moore, Key, Bulbrook, Clark, Allen, Wang, and Pike 1987). Reasons for this increase are not clear and require further investigation.

The diminution in SHBG following the menopause has previously been observed (Murayama Sakuma, Udagawa, Utsonomiya, Okamoto, and Asano 1978; Moore *et al.* 1982; Moore, Clark, Takatani, Wakabayashi, Hayward, and Bulbrook 1983).

Longcope *et al.* (1987) studied SHBG, free oestradiol and free testosterone in 78 perimenopausal women and concluded that the menopause is not associated with changes in SHBG or percent free steroids. Compared to our population, however, the age range of women in this study (42–58 years) was considerably narrower and the numbers were very much smaller and this may have obscured an effect of age on SHBG.

In a study of 168 normal non-pregnant Japanese females between the ages of 20 and 90 years, a highly significant correlation between age and SHBG concentration was observed (Maruyama *et al.* 1984), the postmenopausal group clearly showed higher concentrations than the younger women. No account was taken of body weight in this study, however.

3   THE RELATIONSHIP BETWEEN SHBG AND BODYWEIGHT

Of all the variables affecting SHBG levels in the blood, the inverse relationship between bodyweight or some index of body size is the most consistently reported. Low SHBG levels have been found in obese men, women, and children. Higher than normal levels were found in male and female patients with anorexia nervosa (see Table 5.3 for references). Diminished SHBG in obese women is reversible with weight loss (Enriori, Orsini, Cremona, Etkin, Cardillo, and Reforzo-Membrives 1986) and in female anorectics weight gain is associated with a decline in SHBG binding capacity (Estour *et al.* 1986). In the study by Wheeler *et al.* (1983) weight gain in male anorectics was not associated with a decrease in SHBG despite an increase in levels of testosterone. Among possible reasons, they suggest that this could be due to compensatory changes in oestradiol production which was not measured.

Table 5.3

References to SHBG in obese men, women and children and in patients with anorexia nervosa

Conditions studied	Blood SHBG	Reference
Obesity in men	Low	Glass, Swerdloff, Bray, Dahms, and Atkinson (1977) Amatruda, Harman, Pourmo, and Lockwood (1978) Schneider, Kirschner, Berkowitz, and Ertel (1979)
Obesity in women	Low	Hosseinan, Kim, and Rosenfield (1976) Kopelman, Pilkington, White, and Jeffcoate (1980) Plymate, Fariss, Bassett, and Matej (1981) Cunningham, Loughlin, Culliton, and McKenna (1985) Lapidus, Lindstedt, Lunberg, Bengtsson, and Gredmark (1986)
Obesity in children	Low	Dunkel, Sorva, and Voutilainen (1984) Apter, Bolton, Hammond, and Vihko (1984)
Anorexia nervosa	High	Wheeler, Crisp, Hsu, and Chen (1983) Estour, Pugeat, Lang, Dechaud, Pellet, and Rousset (1986)

In men and women within the normal weight range, statistically significant negative correlations between SHBG activity and weight have usually been observed (Moore *et al.* 1982; Semmens *et al.* 1983a; Moore, Clark, Hoare, Millis, Hayward, Quinlan, Wang, and Bulbrook 1986). In our study of women in Guernsey (Moore *et al.* 1987), we observed a highly significant, inverse correlation between SHBG and Quetelet's index $\left(\dfrac{\text{weight, kg}}{\text{height, m}^2}\right)$ in both premenopausal and postmenopausal women. The average SHBG concentration in the serum of women with Quetelet's index of less than 20 are almost twice that of those with the index of greater than 32 (see Fig. 5.5). It was noted that in premenopausal women there was little decrease in SHBG until a Quetelet's index of greater than 26 was reached.

Obesity is often associated with anovulation and amenorrhoea (Rogers and Mitchell 1952) and most of the studies which have attempted to explain why obese women have diminished SHBG have been part of wider investigations and have often included only small numbers of obese but otherwise normal women (Hosseinian *et al.* 1976; Plymate *et al.* 1981). In obese, oligomenorrhoeic patients with or without hirsutism, strikingly lower SHBG capacities were associated with significantly increased serum testosterone

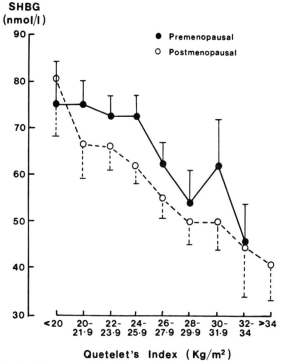

Fig. 5.5. Relationship between Quetelet's index and SHBG in a population of normal women. Calculated from data of Moore, Key, Bulbrook, Clark, Allen, Wang, and Pike (1987).

(Hosseinian *et al.* 1976). However, in a small group (n = 6) who were obese but without signs of hirsutism or ovulatory disturbances, no significant differences were observed in either SHBG or testosterone compared with women of normal weight. Other workers, though consistently reporting diminished SHBG, have failed to find any abnormalities in testosterone concentrations in obese women (Kopelman *et al.* 1980; Plymate *et al.* 1981; Cunningham *et al.* 1985). The same applies to oestradiol and FSH (Plymate *et al.* 1981; Cunningham *et al.* 1985). In this latter study, oestrone was elevated and Kopelman *et al.* (1980) found a non-significant increase in the oestrone/oestradiol ratio and androstenedione levels.

A simple hormonal mechanism, therefore, does not explain the changes in SHBG associated with altered weight in women. This is even more apparent in men where morbid obesity is associated with low SHBG, low levels of serum testosterone (Glass, Swerdloff, Bray, Dahms, and Atkinson 1977; Amatruda, Harman, Pourmo, and Lockwood, 1978; Schneider, Kirschner, Berkowitz and Ertel 1979) and increased oestradiol (Schneider *et al.* 1979). In children no differences in testosterone or oestradiol were observed in obese boys and girls compared to children of normal weight (Dunkel, Sorva, and Voutilainen 1984). In their study, however, Apter *et al.* (1984) found that in premenarchial girls, low SHBG correlated with high oestradiol levels although androgens (testosterone and DHEA) were normal.

A consequence of lower SHBG and normal or elevated testosterone levels found in obesity is, of course, that the calculated free testosterone index is usually higher than normal (Hosseinian *et al.* 1976; Schneider *et al.* 1979; Cunningham *et al.* 1985). Cunningham *et al.* (1985) have suggested that the increased free testosterone resulting from diminished SHBG contributes to further diminution of SHBG production by the liver. Although this is a possibility, it is probable that other factors cause the observed changes in SHBG with weight.

Obesity is associated with increased conversion of androgens to oestrogens and a two-fold increase in serum oestradiol levels compared to controls was found in obese men by Schneider *et al.* (1979). They suggested, but no evidence was presented, that the failure of the liver to produce more SHBG under this greater oestrogenic stimulus was due to weight-related alterations in the activity of oestrogen receptors on the hepatocyte.

In summary, there is no question that SHBG concentrations are negatively related to weight and while this is a strong correlation, it should be borne in mind that in our population study, weight only accounts for 20 per cent of the variance in SHBG levels. The hormone related changes in obesity remain to be clarified and do not appear to be closely associated with variations in SHBG.

4   CHANGES IN SHBG DURING THE MENSTRUAL CYCLE

The postulated relationship between SHBG synthesis and oestrogen secretion suggest that changes in blood levels of the protein ought to be seen over the menstrual cycle. Some workers found no differences in the SHBG binding capacities during the cycle in normally ovulating women (Wu, Motohasi, Abdel-Rahman, Flickinger, and Mikhail 1976; Motohashi, Wu, Abdel-Ramen, Marymor, and Mikhail 1979; Odlind, Elamsson, Englund, Victor, and Johansson 1982; Bolufer, Antonio, Garcia, Munoz, Rodriguez, and Romeu 1983; Cerutti, Gibin, Fede, Mozzanega, and Marchesoni 1984). In contrast, Solomon, Iqbal, Dalton, Jeffcoate, and Ginsburg (1979) demonstrated a marginally significant increase in binding capacity in the luteal phase compared to the follicular phase in a small group of spontaneously ovulating women. In a more extensive investigation, Dowsett, Attree, Virdee, and Jeffcoate (1985) observed significant increases of 15 per cent in the mean SHBG binding capacity two days after ovulation which were maintained for the first 10 days of the luteal phase. The initial increase was found to correlate with levels of oestradiol in the follicular phase. It was presumed that the elevated levels in the luteal phase were maintained by the secondary rise in oestradiol.

Several other groups who have examined this question have reported that, compared to the follicular phase, SHBG does increase significantly during the luteal phase (Mattsson, Silfverstolphe, and Samsioe 1984; Plymate, Moore, Cheng, Bardin, Southworth, and Levinski 1985; Apter *et al*. 1984) and this does not occur in non-ovulatory cycles (Apter *et al*. 1984). The ability to detect the small changes which occur over the cycle is undoubtedly due to the high precision of the methods used (Dowsett *et al*. 1985; Plymate *et al*. 1985) and also to the study of large populations (for example, Apter *et al*. 1984, studied 145 adolescent cycles). In our cross-sectional study of over 600 premenopausal women SHBG was higher in the first 12 days of the luteal phase compared with the rest of the cycle (Moore *et al*. 1987).

In their study, Apter *et al*. (1984) found a positive correlation between SHBG and progesterone concentration in late luteal phase blood. This was not confirmed in the mid-luteal phase by Dowsett *et al*. (1985): however, Dalton (1984*a*) reported increases in SHBG binding capacity in women with premenstrual syndrome given therapeutic dosages of oral progesterone (see Section **VI, 2**).

In summary, the bulk of recent evidence now supports an increase in SHBG in the luteal phase of the cycle and this is probably due to an effect of oestradiol but the role of progesterone is not certain.

5   EFFECT OF AGE AT MENARCHE ON SHBG

In our population study in premenopausal women there was an approximately linear relationship between SHBG and age at menarche. This relationship was reduced in magnitude by adjusting for Quetelet's index because there is a marked inverse relationship between Quetelets index and age at menarche. In postmenopausal women, the unadjusted values showed a tendency for SHBG to be higher in women who had a late menarche, but this relationship disappeared after adjustment for Quetelet's index (Moore *et al*. 1987).

6   SHBG IN PREGNANCY

i   Changes during pregnancy

Between weeks five and seven of pregnancy, SHBG in the maternal circulation starts to rise, achieving levels five to ten times higher than those found in the non-pregnant female by week 30. The most dramatic rise occurs during the first half of pregnancy, presumably under the influence of the rapidly increasing oestrogen concentration (de Hertogh, Thomas, and Vanderheyden 1976; Uriel, Dupiers, Rimbout, and Buffe 1981; Hertz and Johnsen

1983). During the second half of pregnancy, despite a further three- to four-fold rise in oestradiol concentrations, there is only a small (30 per cent) further increase in SHBG (de Hertogh *et al.* 1976; Uriel *et al.* 1981). It is not known whether this indicates maximal synthesis by the end of the first half of pregnancy or antagonism by other hormones during the later stages of gestation. The maternal levels of the binding protein fall after delivery with a half life of about 7 days (Anderson *et al.* 1976).

Although the increase in SHBG is thought to be stimulated by the increase in oestradiol from the non-pregnant luteal phase levels of 0.2–0.5 ng/ml to 20–30 ng/ml in late pregnancy (see Reed and Murray 1979), there is no evidence for this, apart from what is inferred from the administration of exogenous oestrogens. Indeed in the pregnant rhesus macaque, although there is a slight rise in SHBG in early gestation, the levels later actually fall in spite of increasing oestradiol concentrations, so that before delivery the SHBG activity is lower than that of the non-pregnant animal, whereas the total oestradiol levels are increased 4–5 fold (Anderson *et al.* 1976; Stanczyk, Hess, Namkung, Senner, Petra, and Novy 1986).

Bernstein and her colleagues reported that the SHBG binding capacity was approximately 10 per cent higher, and the percentage free and total oestradiol levels lower, in the early part of a woman's second pregnancy compared to that found early in her first pregnancy (after adjustments were made for stage of pregnancy and weight at the beginning of pregnancy). They argue that this may result in increased exposure of the germ cells to oestrogen and account for the increased risk of cryptorchidism and testicular cancer in first-born males (Bernstein, Depue, Ross, Judd, Pike, and Henderson 1986).

ii   SHBG in abnormal pregnancies

In a study of SHBG capacities in women who were admitted for vaginal bleeding between the sixth and eighteenth gestational week, those whose pregnancies ended in spontaneous abortion had lower SHBG capacities compared to those whose pregnancies went to term. This was especially true after the 13th week. There were, however, many women with high SHBG who subsequently aborted and it was concluded that measurements of SHBG would not be as reliable a marker for threatened abortion as other hormonal parameters such as chorionic gonadotrophin, placental lactogen, progesterone and pregnancy specific β_1-glycoprotein (Hertz and Johnsen 1983).

Uriel *et al.* (1981) measured SHBG concentration by an immunodiffusion technique and also the binding capacity for 5α-dihydrotestosterone using the filter disc method of Mickelson and Petra (1974). The correlation between these methods in non-pregnant and healthy pregnant women was highly significant ($r = 0.85$). Six patients who had abnormal pregnancies had normal SHBG concentrations, but in five of these the ratios of the binding capacity to the SHBG concentrations were much lower than normal. This would appear to indicate defective binding by SHBG and needs to be confirmed in a larger study.

iii   SHBG in amniotic fluid

SHBG has been detected in amniotic fluid and levels fifty fold lower than in the maternal circulation do not appear to vary much between early (weeks 13–20) and late pregnancy (weeks 36–37) (Caputo and Hosty 1972; Hammond et al. 1983; Dalton 1984; Forest et al. 1986). The physico-chemical characteristics of SHBG derived from the amniotic fluid do not differ from those of pregnancy serum and it is probable that the source of SHBG and other serum proteins is largely the maternal circulation via the amniochorion (Hammond et al. 1983).

iv   Possible role for SHBG in pregnancy

It is often suggested that the purpose of SHBG in the maternal circulation is to protect the mother from the high levels of testosterone and other 17β-hydroxy androgens found in pregnancy by reducing the free, biologically active steroid (see Anderson 1974). While it is true that in pregnancy the free testosterone concentrations are below those of non-pregnant women in the presence of an approximate doubling in concentration of total testosterone (Vermeulen 1979), the testosterone levels are elevated because of the diminished clearance due to increased peripheral SHBG. Therefore, if SHBG did not rise, since there is no evidence that the production rate of testosterone is increased in pregnancy, it is probable that total and free testosterone concentrations would remain at the levels found before pregnancy which in the normal female do not cause virilization.

SHBG may act by sequestering androgens from the fetal into the maternal circulation, preventing virilization of the female fetus (Anderson 1974). Hammond et al. (1983) suggest that SHBG in amniotic fluid may also play a role in buffering any virilizing effect of androgens on the fetus since the small amounts of SHBG in amniotic fluid are well in excess of the testosterone concentrations. As pointed out by Forest et al. (1986), however, the percentage of free testosterone is higher in amniotic fluid than it is in plasma. This is presumably a consequence of the lower albumin and total protein concentrations in amniotic fluid compared with plasma (cf. Hammond et al. 1983).

vi   Long-term effects of pregnancy on SHBG

As part of a wider study aimed at elucidating the hormonal basis for the association between reproductive factors and risk of breast cancer (see Yu, Gerkins, Henderson, Brown, and Pike 1981), it was shown that young parous women had a 12 per cent greater mean SHBG binding capacity and lower urinary and plasma oestrogen levels than nulliparous women of similar age and cycle length (Bernstein, Pike, Ross, Judd, Brown, and Henderson 1985). This was confirmed in our study. However, further examination of our data showed that this was due solely to the lower concentration of SHBG in the unmarried, nulliparous group. We found a similar pattern postmeno-pausally, in that unmarried nulliparous women had lower SHBG levels than married nulliparous women. The reason for this is not clear (Moore *et al.* 1987).

7   NYCHTHEMERAL VARIATION

Plasma SHBG binding capacity for testosterone shows a significant nych-themeral rhythm in healthy adult men with the lowest capacities between midnight and 6 a.m. (Clair, Claustrat, Jordan, Dechaud, and Sassolas 1985). We have recently investigated nychthemeral variation of SHBG concentra-tion in blood samples taken every 2 hours in a group of 28 normal women and found essentially the same variation over the 24 hours (Clark, Moore, Fentiman, and Wang unpublished).

These findings are consistent with previous investigations which have shown that total serum protein is lower by 5 to 10 per cent in supine subjects compared with subjects who are erect. The minimum is obtained within 2 to 4 hours after assuming the horizontal position and the maximum within 2 hours after rising. This phenomenon is usually ascribed to haemoconcentra-tion in the erect position (Fawcett and Wynn 1956; Henry 1968).

To summarize, while we are prepared to accept that a wide variety of fac-tors such as the stage of the menstrual cycle, pregnancy, menopausal status and nychthemeral rhythms affect SHBG concentrations, it has to be admit-ted that the evidence for a precise hormonal control of this protein is equivo-cal and unconvincing.

VI   The effect of drugs on SHBG

1   OESTROGENS

The oestrogenic components of oral contraceptives (ethinyloestradiol or its 3-methyl ether, mestranol) are potent inducers of SHBG synthesis in men (van Look and Frolich 1981) and women (van Kammen, Thijssen, Rademaker, and Schwartz 1975; Briggs 1975; Helgason, Damber, Damber, von Schoultz, Selstam, and Sodergard 1982). Significant dose dependent in-creases in SHBG are also observed after the oral administration of conju-gated equine oestrogens to postmenopausal women (Pogmore and Jequier 1979; Geola, Frumar, Tataryn, Lu, Hershman, Eggena, Sambhi, and Judd 1980; Mathur, Landgrebe, Moody, Semmens, and Williamson 1985). Other oestrogen inducible liver proteins include CBG, TBG and pregnancy zone protein. SHBG synthesis is more sensitive to oestrogen administration than either CBG or TBG (Geola *et al.* 1980) but less sensitive than pregnancy zone protein (Ottosson, Damber, Damber, Selstam, Solheim, Stigbrand, Soder-gard, and von Schoultz 1981). In terms of potency, ethinyloestradiol is much more oestrogenic as an inducer of SHBG synthesis than oestradiol or oestrone sulphate when administered orally (see Helgason *et al.* 1982).

2 PROGESTATIONAL AGENTS

The progestational agents commonly used in contraceptive medications are usually derivatives of 19-nortestosterone, 17αOH-progesterone or 19-norprogesterone. In addition to their progestational activity the 19-nortestosterone derivatives also have anti-oestrogenic activity and this is evidenced by the depressive effect on hepatic synthesis of oestrogen-sensitive proteins such as SHBG. They appear to operate through androgen receptors rather than the oestrogen receptors in the liver (Bergink, Hamburger, de Jager, and van der Vies, 1981; Hammond, Langley, Robinson, Numi, and Lund 1984).

Levonorgestrel administered orally (Crona, Silfverstolpe, and Samsioe 1984; Ruokonen and Kaar 1985) or slowly released from a vaginal ring (Cekan, Jia, Landgren, and Diczfalusy 1985) reduced blood levels of SHBG significantly. Oral desogestrel and lynestrenol (Ruokonen and Kaar 1985) and dl-norgestrel (El Makhzangy, Wynn, and Lawrence 1979) had similar effects. In this latter study, women using norethisterone followed the trend towards lower SHBG although the decrease was not significant (see Table 5.4).

Of the 17α-hydroxy-progesterone derivatives, medroxyprogesterone acetate (MPA) (250 mg i.m. weekly) significantly decreased serum binding of tritiated 5α-dihydrotestosterone in girls with precocious puberty, the changes being related to the duration of treatment (Forest and Bertrand 1972). Reductions in the SHBG binding capacity were also observed in women given MPA (150 mg every third month) by van Kammen *et al.* 1975. Victor and Johansson (1977) argue that a direct effect of MPA on liver synthesis of SHBG is unlikely because of its close similarity to megestrol acetate which actually increased SHBG slightly as shown by van Kammen *et al.* (1975) and subsequently by El Makhzangy *et al.* (1979). Rather they suggest that SHBG diminution may be associated with the significant reduction in oestradiol levels induced by MPA (Jeppsson and Johansson 1976).

No change in SHBG binding capacity occurred when MPA was adminis-tered by means of impregnated silastic intravaginal rings (Victor and Johans-son 1977). The reason is probably dose related since the intramuscular dosage used by van Kammen *et al.* (1974) gives rise to 3–8 times higher

Table 5.4

Effect of some progestagen only and progestogen/ethinyloestradiol (EO) combination oral contraceptives (OC) on blood SHBG (mean ± S.E.M.) in women

OC composition	Duration of treatment[1]	Baseline SHBG (nmol/l)	Treatment SHBG (nmol/l)	P	Reference
dl-Norgestrel (75 µg)	3–24 months	44 ± 5	24 ± 3	<0.05	El Makhzangy Wynn and Lawrence (1979)
Norethisterone (350 µg)	3–16 months	38 ± 6	27 ± 4	NS	
Levonorgestrel (125 µg)	30 days	65 ± 5	43 ± 4[2]	<0.001	Ruokonen and Kaar (1985)
Desogestrel (125 µg)	30 days	65 ± 5	31 ± 3[2]	<0.001	
Lynestrenol (5 mg)	30 days	65 ± 5	38 ± 3[2]	<0.001	
EO (50 µg) Megestrol (4 mg)	3–18 months	48 ± 4	205 ± 8	<0.001	El Makhzangy Wynn and Lawrence (1979)
EO (50 µg) Norethisterone (1 mg)	3–24 months	39 ± 4	81 ± 9	<0.001	
EO (30 µg) Desonorgestrel (150 µg)	80 days[3]	48 ± 3	150 ± 10	<0.001	Hammond, Robinson, Nummi, and Lund (1984)
EO (30 µg) Levonorgestrel (150 µg)	80 days[3]	50 ± 4	60 ± 4	NS	

[1] Indicates duration of treatment before blood samples taken; [2] Estimated from graphical data; [3] Approximation: samples were taken between days 18 and 21 of the third cycle during treatment.

plasma MPA concentrations than the vaginal rings (Jeppsson and Johansson 1976).

The 19-norprogesterone derivative ST-1435 (Merck) did not affect SHBG or CBG when administered subcutaneously (Lahteenmaki, Hammond, and Luukkainen 1983). As previously mentioned, there is one report that dose dependent increases in SHBG binding capacity were observed with therapeutic administration of progesterone to women with premenstrual syndrome (Dalton 1984*a*).

3   COMBINED OESTROGEN-PROGESTOGEN PREPARATIONS

The relative effect of the synthetic progestogens in relation to liver stimulation of SHBG can be seen when they are combined with synthetic oestrogens in oral contraceptives. As shown in Table 5.4, when ethinyloestradiol was accompanied by megestrol acetate, there was no inhibition of SHBG and levels rose dramatically to a mean of 205 nmol/l during treatment from a basal level of 48 nmol/l. Norethisterone acetate combined with ethinyloestradiol produced a far less significant surge in SHBG, indicating some antagonism by the progestogen (El Makhzangy *et al.* 1979). The reduction induced by norethisterone was shown to be dose dependent (Ottosson *et al.* 1981).

Similar diminutions in the ethinyloestradiol-stimulated synthesis of SHBG were observed with combinations containing norethisterone (Ottosson *et al.* 1981; Granger, Roy, and Mishell 1982) and desogestrel (Crona *et al.* 1984; Hammond *et al.* 1984; Cullberg, Hamberger, Mattsson, Mobacken, and Samsioe 1985). Levonorgestrel (150 μg) in combination with ethinyloestradiol (30 μg) resulted in no significant changes in SHBG capacity or concentration (El Makhzangy *et al.* 1979; Crona *et al.* 1984; Hammond *et al.* 1984).

4 BINDING OF CONTRACEPTIVE STEROIDS TO SHBG

In contrast to the naturally occurring oestrogens, ethinyloestradiol in human serum is bound only to serum albumin with an apparent affinity constant about three times that of oestradiol-17β (Akpoviroro and Fotherby 1980). This has recently been confirmed for undiluted serum at physiological temperatures and it has been further shown that the percentage of free steroid is between 1 and 2 per cent of the total (Hammond, Lähteenmäki, Lähteenmäki, and Luukkainen 1982).

Progestogens synthesized from either 17-hydroxyprogesterone or 19-norprogesterone do not bind specifically to SHBG (Lähteenmäki *et al.* 1983). In contrast 19-nortestosterone derivatives display significant but varying affinities for the binding protein. Levonorgestrel has the highest affinity for SHBG, the association constant behind midway between those of testosterone and oestradiol (Victor, Weiner, and Johansson 1976; Jenkins and Fotherby 1980; Bergink *et al.* 1981; Pugeat, Dunn, and Nisula 1981). Cekan, *et al.* (1985) have shown that blood levels of levonorgestrel in young women who had undergone vaginal implantation with silastic rings containing the steroid, were positively and significantly correlated with SHBG levels obtained before and during treatment. This is of course entirely consistent with the role of SHBG as an important binding protein for levonorgestrel. Pre-treatment levels of SHBG were also related to their potency as anti-ovulatory agents.

The serum distribution between SHBG and albumin of several commonly used contraceptive steroids has recently been determined in undiluted serum at 37°C (Lähteenmäki *et al.* 1983; Hammond *et al.* 1984). For more detailed discussion of the biological importance of the interactions between oral contraceptives and their metabolites and blood and cellular binding proteins the interested reader is referred to these papers and also to those of Bergink *et al.* (1981), Bergink, Holma and Pyorola (1981) and Fotherby (1984).

5   ANDROGENS

i   Anabolic steroid and testosterone administration

The antagonistic effect of the administration of androgens on SHBG synthesis in the normal androgen-responsive human is indisputable. With self-administration, by athletes, of extreme pharmacological doses of testosterone and other anabolic steroids, a 90 per cent fall in basal SHBG levels was observed and they remained low for 16 weeks after withdrawal of the drug (Ruokonen, Alen, Bolton, and Vihko 1985). Men with testicular insufficiency often show increased levels of SHBG (Anderson 1974) and administration of testosterone (200 mg i.m. every two weeks) to 5 normal men and 5 men with Klinefelter's syndrome resulted in significant increases in total and free testosterone concentrations and decreased SHBG binding capacity after 3 months (Plymate, Leonard, Paulsen, Fariss, and Karpas 1983).

The administration of depot dehydroepiandrosterone enanthate to oophorectomized women resulted in small but significant decreases in SHBG and high density lipoprotein cholesterol (HDL-C) (Mattson, Cullberg, Tangkeo, Zador, and Samsioe 1980). In terms of clinical usefulness, Belgorosky and Rivarola (1985) were able to distinguish abnormalities in androgen secretion from abnormal target cell responses in prepubertal patients with male pseudohermaphroditism by measuring SHBG after administration of testosterone.

ii   Danazol and Gestrinone

Danazol is a derivative of ethisterone widely used in the treatment of endometriosis presumably acting by reducing peripheral oestrogen levels (Meldrum, Pardridge, Karow, Rivier, Vale, and Judd 1983). Androgenic side-effects (Potts 1977; Wynn 1977) are probably due to the rise in levels of free, biologically active, testosterone (Nilsson, Sodergard, Damber, Damber, and von Schoultz 1982; Dowsett, Forbes, Rose, Mudge, and Jeffcoate 1986). Part of this increase is due to the rapid fall in the SHBG binding capacity (Schwarz, Tappeiner, and Hintner 1981; Nilsson *et al.* 1982; Gershagen, Doberl, and Rannevik 1984; Dowsett *et al.* 1986) and also to competitive displacement of testosterone from SHBG by danazol and its metabolites (Nilsson *et al.* 1982; Dowsett *et al.* 1986).

Gestrinone is another effective agent in the treatment of endometriosis which also suppresses SHBG synthesis. In a recent study, Dowsett *et al.* (1986) showed that in patients with endometriosis both danazol and gestrinone reduced blood levels of SHBG to similar extents after one week of treatment. There were, however, significantly greater increases in the percentage of free testosterone in plasma samples of patients treated with danazol. Further, *in vitro* studies suggested that this additional increase in free testosterone was largely due to the competitive effect of ethisterone which is a major metabolite of danazol with a greater affinity for SHBG (Pugeat *et al.* 1981; Pugeat, Nicolas, Tourniaire and Forest 1984).

6 GONADOTROPHINS

In adult men, Leydig cell stimulation by long or short term administration of human chorionic gonadotrophin (hCG) does not result in a decrease in SHBG (Plymate *et al.* 1983; Willemse, Sleijfer, Pratt, Sluiter, and Doorenbos 1984). This is contrary to the situation which obtains in prepubertal boys where significant decreases have been observed (Belgorosky and Rivarola 1982; Dunkel 1985). The probable reasons for this difference are twofold. First, in boys, the increase in testosterone following a single dose of hCG can be 100 fold (Dunkel 1985) whereas in adults the increase is only 2.4 times base line levels (Saez and Forest 1979). Second, as well as testosterone stimulation, there is significant oestradiol stimulation after acute administration of hCG in adult males which probably counteracts the rise in testosterone with regard to SHBG induction (Forest, Lecoq, and Saez 1979; Plymate *et al.* 1983). In the pre-pubertal male, oestradiol concentrations remain low following hCG (Dunkel 1985).

Odlind *et al.* (1982) studied eight amenorrhoeic and previously anovulatory women who were undergoing treatment with gonadotrophins and ovulation induction with hCG. They observed a pronounced increase in SHBG capacity, from a mean of 31 nmol/l at the beginning of treatment to a mean of 64 nmol/l 14 days after a very large, hCG induced, ovulatory peak of serum oestradiol (a mean of three times the normal concentration). Similar findings were reported by Dowsett *et al.* (1985), and also by Clair, Claustrat, Brun, Dechaud and Thoulon (1985).

7 GLUCOCORTICOIDS

The literature is conflicting. In a variety of clinical conditions, glucocorticoid administration has been reported to have no effect on SHBG (Anderson 1974; Kim, Rosenfield, and Dupon 1976; Lachelin, Judd, Swanson, Hanck, Parker, and Yen 1982; Lobo, Paul, March, Granger, and Kletzky 1982; Pugeat, Forest, Nisula, Corniau, de Peretti, and Tourniaire 1982; Darley, Moore, Besser, Munro, and Kirby 1983). Others found that small doses of dexamethasone (0.5 mg at night) for three months increased SHBG (Cunningham, Loughlin, Culliton, and McKenna 1983) and some evidence for a direct agonistic effect of dexamethasone on SHBG synthesis comes from *in vitro* studies in a hepatoma-derived cell line (Mercier-Bodard and Baulieu 1986).

Decreased SHBG capacities in men with chronic asthmatic bronchitis taking large doses of dexamathasone were found by Vermeulen *et al.* (1969).

More recently, Wu (1982) found that the mean percentage of testosterone bound to SHBG in hirsute women treated with dexamethasone (2 mg/day) was significantly decreased from pretreatment levels. Further support for the lowering of SHBG by dexamethasone comes from studies in cycling rhesus macaques (Stanczyck, Petra, Senner, and Novy 1985). In children and adolescents, long term (3 day) ACTH stimulation which resulted in massive increases in blood cortisol levels, was accompanied by significant reductions in SHBG binding capacities (Forest *et al.* 1986). The reasons for the conflicting literature are not clear but dosage and length of treatment are probably important.

8 ANTI-OESTROGENS

i Tamoxifen

Tamoxifen has been widely used in the treatment of breast cancer. The mode of action of the drug is thought to be that of an anti-oestrogen blocking oestradiol at the oestrogen receptor sites in the tumour. However the biological activity of the compound varies with the dose from species to species and even from organ to organ and it can act as an oestrogen (Nicholson, Walker, and Davies 1986). As an anti-oestrogen, it might have been expected to inhibit SHBG synthesis, but this is not the case. Tamoxifen caused significant increases, not only in SHBG but also in other oestrogen-sensitive proteins such as CBG and pregnancy-zone protein in patients under treatment for breast cancer (Sakai, Cheix, Clavell, Colon, Mayer, Pommatau, and Saez 1978; Fex, Adielsson, and Mattson 1981; Caleffi, Fentiman, Clark, Wang, Needham, Clark, La Ville, and Lewis 1988). In human hepatoma cells in culture tamoxifen significantly induced SHBG synthesis (Mercier-Bodard and Baulieu 1986).

ii Clomiphene

Clomiphene citrate is structurally related to tamoxifen and has a similar oestrogenic effect on SHBG synthesis. Increases in this protein have been observed in women undergoing treatment for infertility (Lobo *et al.* 1982) and in men being treated for oligozoospermia (Adamopoulos, Vassilopoulos, Kapolla, and Kontogeorgos 1981). As pointed out by the latter group, this is an unwanted side effect of the drug in the treatment of this type of male infertility where the need is usually to increase testicular androgen production. Although gonadotrophins increase and testicular androgen production improves, testicular oestrogen production also increases and the rise in SHBG upsets the endocrine balance by diminishing the bioavailability of testosterone to a greater extent than that of oestradiol.

9 ANTI-CONVULSANT DRUGS

SHBG binding capacity was shown to be increased in male and female epileptic patients being treated with phenytoin (Victor, Lundberg, and Johansson 1977; Dana-Haeir, Oxley, and Richens 1982; Beastall, Cowan, Gray, and Fogelman 1985). In the study carried out by Beastall *et al.* (1985), binding capacities for CBG, as well as SHBG, were increased during treatment and there was a strong positive association between the capacities of these proteins and serum levels of phenytoin. Levels of TBG or the binding capacity of vitamin D binding protein were not affected, but there were reductions in levels of cortisol and thyroid hormones which suggests that anticonvulsant therapy causes widespread disturbances in hormonal homeostatsis.

In new-born infants, far from inducing synthesis of SHBG, a single intramuscular injection of phenobarbital was associated with a marked delay in the post-natal rise in SHBG, which suggested that the drug may retard synthesis in the neonate (Forest, Lecoq, Salle, and Bertrand 1981).

In summary, administration of oestrogenic compounds increase whereas androgenic medications and some progestogens decrease the concentration of SHBG. The combined oestrogen/progestogen contraceptives have a variety of effects depending on the type and dose of progestogen used. The situation regarding the effect of glucocorticoids is still controversial but the weight of evidence suggests that large doses inhibit synthesis. Anticonvulsants appear to induce SHBG synthesis except in neonates.

VII SHBG and disease

1 HYPERANDROGENISM AND THYROID DISEASES

One of the few areas where SHBG measurements are widely used clinically is in the diagnosis and management of diseases associated with hyperandrogenization in women such as hirsutism, polycystic ovary syndrome and acne vulgaris. In these conditions blood levels of testosterone, other 17βOH-steroids and adrenal androgens may not be consistently elevated but SHBG is usually significantly depressed. This results in increased levels of biologically available testosterone (Anderson 1974; Lawrence, Katz, Robinson, Newman, McGarrigle, Shaw, and Lachelin 1981; Odlind *et al.* 1982; Carter, Holland, Alaghband-Zadeh, Rayman, Dorrington-Ward and Wise 1983; Cunningham *et al.* 1983; Darley *et al.* 1983).

Hyperthyroidism and the administration of large doses of thyroid hormones is associated with increased circulating SHBG activity, and hypothyroidism with decreased SHBG levels (see Yosha, Fay, Longcope, and Braverman 1984 for references). Although SHBG measurements are of minor importance in the diagnosis of common thyroid diseases (Lindstedt, Lundberg, Hammond, and Vihko 1985), they are reported as being valuable in the diagnosis of thyroid hormone resistance where SHBG levels are normal, despite markedly elevated thyroid hormone levels (Cooper, Ladenson, Nisula, Dunn, Chapman, and Ridgway 1982; de Nayer, Lambot, Desmons, Rennotte, Malvaux, and Beckers 1986).

2 BENIGN PROSTATIC HYPERPLASIA AND PROSTATIC CANCER

Growth and function of the prostate gland are regulated predominantly by testosterone and its metabolites, particularly 5α-dihydrotestosterone, although other hormones such as oestradiol and prolactin are also involved (Pasqualini 1982; Peeling and Griffiths 1986; Griffiths, Davies, Eaton, Harper, Peeling, Turkes, Wilson, and Pierrepoint 1987).

SHBG inhibits the uptake, metabolism and action of testosterone in rat prostate glands in organ culture (Lasnitzki and Franklin 1975) and in excised human hypertrophic tissue (Pachman 1984). The possibility that peripheral SHBG capacities might be different in men with benign prostatic hyperplasia compared with normal has been examined by several authors. No differences were observed by some (Dennis, Horst, Kreig, and Voight, 1977; Bartsch, Becker, Pinkenburg, and Krieg 1979). Pachman (1984a), however, showed that the mean SHBG binding capacity was significantly higher and testosterone levels lower in patients and suggested that the increase in SHBG might represent a defence mechanism against further androgenic stimulation of the gland. Curiously, excision of the hypertrophied tissue resulted in a significant fall in SHBG without any alteration in testosterone concentration.

Prostatic tissue contains oestradiol receptors (Chaisiri and Pierrepoint 1980). Although the importance of endogenous oestrogens in prostatic cancer is uncertain there is some evidence that they are protective. Thus, autopsy studies of patients with cirrhosis of the liver, where testosterone levels are low (Gordon, Altman, Southren, Rubin, and Lieber 1976) and oestradiol levels are high (Siiteri and MacDonald 1973), indicate a lower incidence of prostatic cancer compared with controls of the same age (Glantz 1964). Blood oestradiol concentrations and SHBG binding capacity measurements in 116 patients with prostatic cancer supports this view. Significantly higher SHBG capacities and lower oestradiol levels were associated with poorly differentiated tumours while the mean amounts of free oestradiol were 30 per cent higher in those with the most differentiated tumours (Haapiainen, Rannikko, Adlercreutz, and Alfthan 1986).

3 BREAST CANCER

Early work on mammary cancer in murine species (Noble 1964), epidemiological studies in the human (Kelsey 1979; Pike and Ross 1984) and clinical experience with endocrine treatment of breast cancer (Hayward and Rubens

1987) has led to the conclusion that ovarian hormones, especially the oestrogens, are intimately involved in the induction of breast cancer in the human. Despite this, however, there is very little convincing evidence that there are consistent abnormalities in blood levels of ovarian, adrenal, pituitary or hypothalamic hormones in patients either with breast cancer or at high risk of developing breast cancer (see Moore, Thomas, and Wang 1986, for review).

Recently, attention has focused on the biologically available fraction of oestradiol in the blood of patients with breast cancer and several laboratories have found significantly increased percentages of non-protein-bound oestradiol in patients with breast cancer compared with matched controls (Siiteri et al. 1981; Moore et al. 1982; Reed, Cheng, Noel, Dudley, and James 1983; Langley, Hammond, Bardsley, Sellwood, and Anderson 1985; Ota, Jones, Jackson, Jackson, Kemp, and Bauman 1986). Significantly increased percentages of the albumin-bound fraction of oestradiol were also found in two of these studies (Langley et al. 1985; Ota et al. 1986). In all of these investigations, the available fractions of oestradiol (free and albumin-bound) were correlated with the SHBG binding capacity. In two studies (Moore et al. 1982 and Ota et al. 1986) but not in the others (Siiteri et al. 1981; Reed et al. 1983; Langley et al. 1985) where no differences were observed, the increase in the available oestradiol fraction was partially dependent on diminished SHBG binding. Adami, Johansson, Vegelius, and Victor (1979) also found marginally but significantly lower capacities in 122 patients with newly diagnosed breast cancer compared to the same number of age matched controls. On the other hand Sulkes, Fuks, Gordon, and Gross (1984) and Meyer, Brown, Morrison, and MacMahon (1986) found that women with breast cancer had higher SHBG binding capacities than normal women of the same nutritional status.

The reason for the increased percentage of free oestradiol in the presence of normal or only slightly diminished SHBG capacities is not clear and has been ascribed to interference of oestradiol binding by free fatty acids (see Moore et al. 1982). Credence to this hypothesis is given by Bruning and Bonfrer (1986) who, though unable to confirm that the percentage of free oestradiol is elevated in patients with breast cancer (Bruning, Bonfrer, and Hart 1985) have shown *in vitro* that the polyunsaturated free fatty acids (arachidonic, linoleic and linolenic) tend to displace oestradiol from both SHBG and albumin. They also found, in a study involving 56 women with breast and other forms of cancer, that an increase in free fatty acids during an overnight fast was accompanied by an increase in the percentage free oestradiol. A similar increase was observed after lipase activation by intravenous injection of heparin (500 i.u.). Further evidence that fatty acids can displace steroids from blood binding proteins is presented by Umstot and Andersen (1986) and by Reed and his colleagues (Reed, Baranek, Cheng, and James 1986; Reed, Cheng, Beranek, Few, Franks, Gilchik, and James 1986).

A prospective study involving 5000 women living on the island of Guernsey (see below) has shown that those who developed breast cancer usually had a higher percentage of free oestradiol than matched controls and that, again, this was at least in part explained by diminished SHBG levels (Moore *et al.* 1986).

Siiteri *et al.* (1981) were the first to demonstrate the increase in percent free oestradiol in patients with breast cancer but they have subsequently been unable to confirm their original findings. They suggest that this was due to the age of the serum samples used in the first study (Siiteri, Simberg, and Murai, 1984). Certainly, Langley *et al.* (1985), whose study was confirmatory of the free oestradiol hypothesis, showed a change in affinity of SHBG for 5α-dihydrotestosterone during prolonged storage at $-20°C$ and matched their samples from cases and controls appropriately.

Studies in Japan and the USA have led to the claim that there are positive correlations between tumour oestradiol receptor levels and SHBG binding capacities indicating that SHBG may be a strong indicator of response to hormone therapy (Murayama *et al.* 1978; Murayama, Utsonomiya, Takahashi, Kitamura, Tominga 1979; Plymate, Stutz, and Farris 1984). High SHBG binding capacities were also associated with a long disease free interval (Muryama, Utsonomiya, Asano, and Bulbrook 1979). At least two studies have failed to confirm the relationship between SHBG and receptor status (Mason, Miller, Hawkins, and Forrest 1981; Sulkes *et al.* 1984) and Harris, Smith, Dowsett, Jeffcoate, Coombes, Powles, and Neville (1981) were unable to confirm the relationship between SHBG capacity and recurrence of breast cancer in British women. The differences may be ascribed to differing study populations or to the methods used (Harris *et al.* 1981).

The evidence that raised levels of plasma oestradiol are important in the aetiology and clinical course of breast cancer is weak and results on bioavailable oestrogens are still not definitive. Whether risk of breast cancer, free fatty acid concentrations and bioavailable oestradiol are interrelated remains to be determined.

4   ENDOMETRIAL AND CERVICAL CANCER

Ratajczak, Twaddle and Hahnel (1980) found significantly higher mean SHBG binding capacity in postmenopausal patients with oestrogen-receptor positive endometrial or cervical cancer than in those with oestrogen-receptor negative tumours.

Gambone, Pardridge, Lagasse, and Judd (1982) used the blood/brain assay (which measures the transport of labelled steroid through the brain capillary wall and, hence, the proportion of steroid bound to SHBG) to compare the availability of oestradiol in patients with endometrial cancer and weight-matched controls. They found no significant differences in the two groups but brain uptake was related to body size. Thus obese women show a

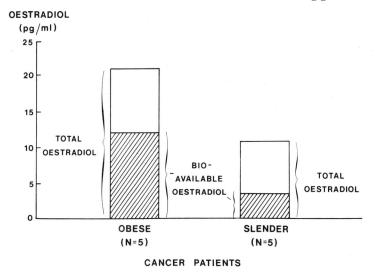

Fig. 5.6.   Available oestradiol in patients with endometrial cancer. (Reprinted with permission from Gambone, Pardridge, Lagasse, and Judd (1982).

dual effect: their total plasma oestradiol is twice as high as that of slender controls and, because of diminished binding, their available oestradiol levels are four times higher (see Figure 5.6).

Weight is a known risk factor for endometrial cancer and the use of weight-matched controls is questionable. In a subsequent publication, Laufer and his colleagues showed that patients with endometrial cancer had higher levels of oestradiol and lower levels of SHBG than women with osteoporotic hip fractures who were used as, somewhat unsatisfactory, controls (Laufer, Davidson, Ross, Lagasse, Siiteri, and Judd 1983).

5   OTHER DISEASES

i   Osteoporosis

It is widely accepted that endocrine function is of primary importance in osteoporosis, especially the oestrogens, but attempts to implicate particular hormonal abnormalities with the disease have not been strikingly successful (see, for example Davidson, Riggs, Coulam, and Toft 1980). This situation is similar to that found for breast cancer.

Osteoporosis may be a disease in which SHBG is important. Davidson, Ross, Paganini-Hill, Hammond, Siiteri, and Judd (1982) compared elderly women who had a hip fracture with controls and, while they found no differences in the total amounts of plasma oestrogens, SHBG levels were higher and unbound oestradiol was lower in the fracture group.

There has been a proposal that the anti-oestrogen Tamoxifen, might be used for the prevention of breast cancer and that SHBG assays could be used for identifying women at high risk since there is some evidence that low levels of this protein are related to enhanced risk (Cuzick *et al.* 1985). One of the drawbacks to this proposal is that women might become osteoporotic and it would be a matter of importance to monitor bone densities if preventitive experiments were carried out.

ii   Multiple myeloma

De Moor, Louwagie, Faict, and Vanham (1986) showed that male patients with multiple myeloma (Kahler's disease) have higher serum SHBG and lower CBG binding activity than normal, age-matched blood donor controls. Serum CBG in patients whose myeloma was secreting immunoglobulin light chains of the lambda variety were significantly lower whereas those patients secreting kappa light chains had a significantly higher mean SHBG activity. De Moor *et al.* (1986) suggest that low CBG levels are more likely to arise from the lambda repertoire of clones and the high SHBG from the kappa variety, and suggest that immunoglobulin light chain *V* genes may be genetic determinants for low CBG and high SHBG binding capacity in serum.

iii   Cardiovascular disease

SHBG has been measured in several studies aimed at clarifying the relationships between the endocrine environment and risk factors for cardiovascular disease. Lapidus *et al.* (1986) have shown an intriguing relationship between SHBG concentration and risk of myocardial infarction. The plot of the 12-year incidence against SHBG was in fact U-shaped, indicating that high or low values are associated with higher risk.

Although Heller, Wheeler, Micallef, Miller, and Lewis (1983) found a negative association between SHBG and HDL-C, positive relationships have usually been observed (Semmens, Rouse, Beilin, and Maserei 1983; Hämäläinen, Adlercreutz, Enholm, and Puska 1986). In our population study (Moore *et al.* 1987) we observed that premenopausal women who smoked had, on average, serum levels of SHBG (corrected for weight and other covariates) which were 14 per cent higher than non-smokers. Postmenopausally, the trend was the same although not statistically significant. Lapidus *et al.* (1986) made similar observations in postmenopausal women, although in men SHBG was not related to smoking habits (Lindholm *et al.* 1982).

VIII   Concluding comments

In this review, we have examined some of the recent advances in knowledge of SHBG over a fairly broad front. At the fundamental level, the elucidation of the structure of human SHBG represents a highly significant landmark and the culmination of years of intensive effort. These advances will result in the introduction of many new and exciting techniques such as those of immunocytochemistry and molecular biology which could, in a short time, clarify questions concerning synthesis and localization of SHBG within tissues.

New methods of analysis where the actual concentration of SHBG can be measured with high precision, accuracy, sensitivity and speed, has already enabled the assay of SHBG in large populations which have helped to identify some of the factors which affect the considerable variation in SHBG. Our own study in normal women in Guernsey has shown that in addition to weight, which has a strong effect on SHBG, other factors which influence peripheral levels of the protein include age at menarche and parity in pre-menopausal women, age in postmenopausal women and possibly smoking history. The mechanisms by which these factors affect SHBG are totally unknown.

The unconfirmed report (Uriel *et al.* 1982) that SHBG variants exist which show defective binding during pregnancy raises the question about the re-lationship between assays which measure the binding capacity and those which measure concentration. Hammond *et al.* (1985) in a limited study could not demonstrate any discrepancies between SHBG concentration measured by their immunoradiometric assay and the 5α-dihydrotestosterone binding capacity. It would be helpful to know, in a large population, if such variants exist and if so, their frequency.

With respect to the effects of drugs on SHBG, although stimulation or sup-pression of synthesis by the liver is widely assumed, it remains a possibility that the changes in peripheral SHBG activity brought about by drug ad-ministration could be the result of alterations in degradation of the protein. The use of the hepatoma-derived cell line (Hep G2) has already confirmed that oestradiol and thyroid hormones stimulate SHBG synthesis (Mercier-Bodard and Baulieu 1986; Lee, Dawson, Wetherall, and Hahnel 1987). The surprising finding that androgens also stimulate SHBG synthesis in these cells (Lee *et al.* 1987) is contrary to the traditional concept of hormonal con-trol and obviously needs to be investigated further.

The prevalent concepts about transport of steroidal hormones into cells has been thoroughly covered by Pardridge in this volume and it would be superfluous to add significantly to his discussion. However, the finding of very high intra-cellular concentrations of several steroid hormones (200-fold for oestradiol in extreme cases), and the lack of correlation between these levels and the amount of oestrogen receptor protein, leads us to the supposi-tion that some form of intra-cellular control mechanism exists. The cellular levels of steroids are well within the range at which they would have a pro-nounced effect on several key enzymes (for example glucose-6-phosphate dehydrogenase). It would be illogical to agonize over 2- or 3-fold differences in availability of steroids in the blood brought about by changes in SHBG concentration, while ignoring the fact that tissue levels of oestradiol (in the breast for example) average 20 times the blood levels. If the key events in the final biological action of steroids is mainly affected by intra-cellular availability, it would not be surprising that so many of the results that we have reviewed are indecisive. This is exactly what would be expected if blood levels of a binding protein are only poorly correlated to intra-cellular events.

Present indications point to the fact that concentrations of the protein may be related to the aetiology of diseases which account for a substantial proportion of the early deaths in man (hormone-related cancers, cardiovascular disease). Indeed Lapidus and his colleagues (1986) have found an inverse correlation between SHBG levels and the 12 year mortality rates in women. The importance of elucidating the precise role of SHBG cannot be underestimated.

Acknowledgements

We wish to thank Professors R. V. Brooks, M. G. Forest, and K. Fotherby, and Drs M. Dowsett and D. Y. Wang for their helpful comments during the preparation of this review and also to Mrs Maureen Cobbing and Mrs Audrey Symons for their enduring patience and assistance.

References

Adami, H. O., Johansson, E. D. B., Vegelius, J., and Victor, A. (1979). Serum concentrations of estrone, androstenedione, testosterone and sex-hormone-binding-globulin in postmenopausal women. *Uppsala J. med. Sci.* **84**, 259–274.

Adamopoulos, D. A., Vassilopoulos, P., Kapolla, N., and Kontogeorgos, L. (1981). The effect of clomiphene citrate on sex hormone binding globulin in normospermic and oligozoospermic men. *Int. J. Androl.* **4**, 639–645.

Akpoviroro, J. and Fotherby, K. (1980). Assay of ethinyloestradiol in human serum and its binding to plasma proteins. *J. steroid Biochem.* **13**, 773–779.

Amatruda, J. M., Harman, S. M., Pourmo, G., and Lockwood, D. H. (1978). Depressed plasma testosterone and fractional binding of testosterone in obese males. *J. clin. Endocr. Metab.* **47**, 268–271.

Anderson, D. C. (1974). Sex-hormone-binding globulin. *Clin. Endocr.* **3**, 69–96.

—— Lasley, B. L., Fisher, R. A., Shepherd, J. H., Newman, L., and Hendrickx, A. G. (1976). Transplacental gradients of sex-hormone-binding globulin in human and simian pregnancy. *Clin. Endocr.* **5**, 657–669.

Apter, D., Bolton, N. J., Hammond, G. L., and Vihko, R. (1984). Serum sex hormone binding globulin during puberty in girls and in different types of adolescent menstrual cycles. *Acta endocr. Copenh.* **107**, 413–419.

August, G. P., Tkachuk, M., and Grumbach, M. M. (1969). Plasma testosterone-binding affinity and testosterone in umbilical cord plasma, late pregnancy, pre-pubertal children, and adults. *J. clin. Endocr. Metab.* **29**, 891–899.

Bardin, C. W. and Lipsett, M. B. (1967). Testosterone and androstenedione blood production rates in normal women and women with idiopathic hirsutism or polycystic ovaries. *J. clin. Invest.* **46**, 891–902.

Bartsch, W. (1980). Interrelationships between sex hormone-binding globulin and tes-tosterone, 5α-dihydrostestosterone and oestradiol-17β in blood of normal men. *Maturitas* **2**, 109–118.

—— Becker, H., Pinkenburg, F. A., and Krieg, M. (1979). Hormone blood levels and their inter-relationships in normal men and men with benign prostatic hyperplasia (BPH). *Acta endocr. Copenh.* **90**, 727–736.

—— Horst, H. J., and Derwahl, K. M. (1980). Interrelationships between sex hormone-binding globulin and 17β-estradiol, testosterone, 5α-dihydrotesto-sterone, thyroxine, and triiodothyronine in prepubertal and pubertal girls. *J. clin. Endocr. Metab.* **50**, 1053–1056.

Baulieu, E. E. (1986). Steroid hormone binding plasma proteins and their intra and extra-cellular congeners. In *Binding Proteins of Steroid Hormones*, Colloque INSERM Vol. **149** (eds M. G. Forest and M. Pugeat), pp. 1–11, John Libbey Eurotext Ltd., London and Paris.

Beastall, G. H., Cowan, R. A., Gray, J. M. B., and Fogelman, I. (1985). Hormone binding globulins and anticonvulsant therapy. *Scot. med. J.* **30**, 101–105.

Belgorosky, A. and Rivarola, M. A. (1982). Sex hormone-binding globulin response to human chorionic gonadotrophin stimulation in children with cryptorchidism, anorchia, male pseudohermaphroditism, and micropenis. *J. clin. Endocr. Metab.* **54**, 698–704.

—— —— (1985). Sex hormone binding globulin response to testosterone. An andro-gen sensitivity test. *Acta endocr. Copenh.* **109**, 130–138.

—— —— (1986). Progressive decrease in serum sex hormone-binding globulin from infancy to late prepuberty in boys. *J. clin. Endocr. Metab.* **63**, 510–512.

Ben-Rafael, Z., Mastroianni, L., Meloni, F., Lee, M. S., and Flickinger, G. L. (1986). Total estradiol, free estradiol, sex hormone-binding globulin, and the fraction of estradiol bound to sex hormone-binding globulin in human follicular fluid. *J. clin. Endocr. Metab.* **63**, 1106–1111.

Bergink, E. W., Hamburger, A. D., de Jager, E., and van der Vies, J. (1981). Binding of a contraceptive progestogen ORG 2969 and its metabolites to receptor proteins and human sex hormone binding globulin. *J. steroid Biochem.* **14**, 175–183.

—— Holma, P., and Pyorala, T. (1981). Effects of oral contraceptive combinations levonorgestrel or desogestrel on serum proteins and androgen binding. *Scand. J. clin, Lab. Invest.* **41**, 663–668.

Bernstein, L., Depue, R. H., Ross, R. K., Judd, H. L., Pike, M. C., and Henderson, B. E. (1986). Higher maternal levels of free estradiol in first compared to second pregnancy: early gestational differences. *J. natl. Cancer Inst.* **76**, 1035–1039.

—— Pike, M. C., Ross, R. K., Judd, H. L., Brown, J. B., and Henderson, B. E. (1985). Estrogen and sex hormone binding globulin levels in nulliparous and parous women. *JNCI* **74**, 741–745.

Bierich, J. R. (1983). Treatment of constitutional delay of growth and adolescence with human growth hormone. *Klin. Paediat.* **195**, 309–316.

Blank, A., Attanasio, A., Rager, K., and Gupta, D. (1978). Determination of serum sex hormone binding globulin (SHBG) in pre-adolescent and adolescent boys. *J. steroid Biochem.* **9**, 121–125.

Bolufer, P., Antonio, P., Garcia, R., Munoz, J, Rodriguez, A., and Romeu, A. (1983). Role of the ovary in the regulation of sex hormone binding globulin and its contri-bution to peripheral levels of androstenedione. *Exp. clin. Endocr.* **82**, 29–34.

Bordin, S., Lewis, J., and Petra, P. H. (1978). Monospecific antibodies to the sex steroid-binding protein (SBP) of human and rabbit sera: Cross-reactivity with other species. *Biochem. Biophys. Res. Commun.* **85**, 381–576.

—— and Petra, P. H. (1980). Immunocytochemical localisation of sex steroid-binding protein of plasma in tissues of the adult monkey *Macaca nemestrina. Proc. natn. Acad. Sci. U.S.A.* **77**, 5678–5682.

—— Torres, R., and Petra, P. H. (1982). An enzyme-immunoassay (ELISA) for the sex steroid-binding protein (SBP) of human serum. *J. steroid Biochem.* **17**, 453–457.

Briggs, M. H. (1975). Hormonal contraceptives and plasma sex hormone binding globulin. *Contraception* **12**, 149–153.

Bruning, P. F. and Bonfrer, J. M. G. (1986). Free fatty acid concentrations correlated with the available fraction of estradiol in human plasma. *Cancer Res.* **46**, 2606–2609.

—————— and Hart, A. A. M. (1985). Non protein-bound estradiol, sex-hormone-binding-globulin, breast cancer and breast cancer risk. *Br. J. Cancer* **51**, 479–484.

—————— and Nooyen, W. J. (1985). A routine solid-phase assay for the binding capacity of sex hormone binding globulin. *Clin. Chim. Acta* **148**, 69–76.

Burke, C. W. and Anderson, D. C. (1972). Sex-hormone-binding globulin is an oestrogen amplifier. *Nature, Lond.* **240**, 38–40.

Caleffi, M., Fentiman, I. S., Clark, G. M. G., Wang, D. Y., Needham, J., Clark, K., La Ville, A., and Lewis, B (1988). The effect of tamoxifen on oestrogen binding, lipid and lipoprotein concentrations and blood clotting parameters in premenopausal women with breast pain. *J. Endocr. submitted.*

Caputo, M. J. and Hosty, T. A. (1972). The presence of sex binding globulin in amniotic fluid. *Am. J. Obstet. Gynec.* **113**, 803–811.

Carter, G. D., Holland, S. M., Alaghband-Zadeh, J., Rayman, G., Dorrington-Ward, P., and Wise, P. H. (1983). Investigation of hirsutism: testosterone is not enough. *Annls. clin. Biochem.* **20**, 262–263.

Cekan, S. Z., Jia, M., Landgren, B.-M., and Diczfalusy, E. (1985). The interaction between sex hormone binding globulin and levonorgestrel released from vaginal rings in women. *Contraception* **31**, 431–439.

Cerutti, R., Gibin, P., Fecde, T., Mozzanega, B., and Marchesoni, D. (1984). SHBG pattern during the menstrual cycle. *Clin. exp. Obstet. Gynec.* **11**, 60–63.

Chaisiri, N. and Pierrepoint, C. G. (1980). Examination of the distribution of oestrogen receptor between the stromal and epithelial compartments of the canine prostate. *Prostate* **1**, 357–366.

Cheng, C. Y., Bardin, C. W., Musto, N. A., Gunsalus, G. L., Cheng, S. L., and Ganguly, M. (1983). Radioimmunoassay of testosterone-estradiol-binding globulin in humans: a reassessment of normal values. *J. clin. Endocr. Metab.* **56**, 68–75.

—— Musto, N. A., Gunsalus, G. L., and Bardin, C. W. (1983). Demonstration of heavy and light protomers of human testosterone-estradiol-binding globulin. *J. steroid Biochem.* **19**, 1379–1389.

Clair, P., Claustrat, B., Brun, J., Dechaud, H., and Thoulon, J. M. (1985). Evolution des taux plasmatiques de la SHBG (sex-hormone binding globulin) au cours de l'induction de l'ovulation par les gonadotrophines. *J. Obstet. Gynaec. reprod. Biol.* **14**, 19–25.

—————— Jordan, D., Dechaud, H., and Sassolas, G. (1985). Daily variations of plasma sex hormone-binding globulin binding capacity, testosterone and luteinizing hormone concentrations in healthy rested adult males. *Horm. Res.* **21**, 220–223.

Cooper, D. S., Ladenson, P. W., Nisula, B. C., Dunn, J. F., Chapman, E. M., and Ridgway, E. C. (1982). Familial thyroid hormone resistance. *Metabolism* **31**, 504–509.

Crona, N., Silfverstolpe, G., and Samsioe, G. (1984). Changes in serum apolipoprotein AI and sex-hormone-binding globulin levels after treatment with two different progestins administered alone and in combination with ethinyl estradiol. *Contraception* **29**, 261–270.

Cullberg, G., Hamberger, L., Mattsson, L., Mobacken, H., and Samsioe, G. (1985). Effects of a low-dose desogestrel-ethinylestradiol combination on hirsutism, androgens and sex hormone binding globulin in women with a polycystic ovary syndrome. *Acta obstet. gynec. scand.* **64**, 195–202.

Cunningham, S. K., Loughlin, T., Culliton, M., and McKenna, T. J. (1983). Plasma sex hormone-binding globulin and androgen levels in the management of hirsute patients. *Acta endocr. Copenh.* **104**, 365–371.

———— (1984). Plasma hormone-binding globulin levels decrease during the second decade of life irrespective of pubertal status. *J. clin. Endocr. Metab.* **58**, 915–918.

———— (1985). The relationship between sex steroids and sex hormone-binding globulin in plasma in physiological and pathological conditions. *Annals. clin, Biochem.* **22**, 489–497.

Cuzick, J., Wang, D. Y., and Bulbrook, R. D. (1986). The prevention of breast cancer. *Lancet* **i**, 83–86.

Dalton, M. E. (1984). Another foetal steroid binding glycoprotein in cord blood and amniotic fluid distinct from sex hormone binding globulin. *IRCS Med. Sci.* **12**, 743–744.

—— (1984a). The effect of progesterone administration on sex hormone binding globulin binding capacity in women with severe premenstrual syndrome. *J. steroid Biochem.* **20**, 437–439.

Dana-Haeri, J., Oxley, J., and Richens, A. (1982). Reduction of free testosterone by antiepileptic drugs. *Br. med. J.* **284**, 85–86.

Darley, C. R., Moore, J. W., Besser, G. M., Munro, D. D., and Kirby, J. D. (1983). Low dose prednisolone or oestrogen in the treatment of women with late onset or persistent acne vulgaris. *Br. J. Derm.* **108**, 345–353.

Daughaday, W. H. and Kozak, I. (1958). Binding of corticosteroids by plasma proteins. 3. The binding of corticosteroids and related hormones by equilibrium dialysis. *J. clin. Invest.* **37**, 511–518.

Davidson, B. J., Riggs, B. L., Coulam, C. B., and Toft, D. O. (1980). Concentration of cytosolic estrogen receptors in patients with postmenopausal osteoporosis. *Am. J. Obstet. Gynec.* **136**, 430–433.

—— Ross, R. K., Paganini-Hill, A., Hammond, G. D., Siiteri, P. K., and Judd, H. L. (1982). Total and free estrogens and androgens in postmenopausal women with hip fractures. *J. clin. Endocr. Metab.* **54**, 115–120.

De Hertogh, R., Thomas, K., and Vanderheyden, I. (1976). Quantitative determination of sex hormone-binding globulin capacity in the plasma of normal and diabetic pregnancies. *J. clin. Endocr. Metab.* **42**, 773–777.

De Moor, P., Heyns, W., and Bouillon, R. (1972). Growth hormone and the steroid binding β-globulin of human plasma. *J. steroid Biochem.* **3**, 593–600.

—— Louwagie, A., Faict, D., and Vanham, G. (1986). Steroid-carrier proteins in patients with multiple myeloma. *J. steroid Biochem.* **25**, 261–264.

De Nayer, P., Lambot, M. P., Desmous, M. C., Rennotte, B., Malvaux, P., and Beckers, C. (1986). Sex hormone-binding protein in hyperthyroxinaemic patients: a discriminator for thyroid status in thyroid hormone resistance (TRH) and familial dysalbuminemic hyperthyroxinemia (FDH). *J. clin. Endocr. Metab.* **62**, 1309–1312.

Degrelle, H. (1986). Dosage immunologique de la "sex steroid-binding protein" (SBP) dans le serum humain. In *Binding Proteins of Steroid Hormones*, Colloque

INSERM Vol. **149** (eds. M. G. Forest and M. Pugeat), pp. 221–226, John Libbey Eurotext Ltd., London and Paris.

Dennis, M., Horst, H. J., Krieg, M., and Voigt, K. D. (1977). Plasma sex hormone-binding globulin binding capacity in benign prostatic hypertrophy and prostatic carcinoma: comparison with age dependent rise in normal human males. *Acta endocr. Copenh.* **84**, 207–214.

Dowsett, M., Attree, S. L., Virdee, S. S., and Jeffcoate, S. L. (1985). Oestrogen-related changes in sex hormone binding globulin levels during normal and gonado-trophin-stimulated menstrual cycles. *Clin. Endocr.* **23**, 303–312.

—— Forbes, K. E., Rose, G. L., Mudge, J. E., and Jeffcoate, S. L. (1986). A comparison of the effects of danazol and gestrinone on testosterone binding to sex hormone binding globulin *in vitro* and *in vivo*. *Clin. Endocr.* **24**, 555–563.

—— Mansfield, M. D., Griggs, D. J., and Jeffcoate, S. L. (1984). Validation and use of centrifugal ultrafiltration-dialysis in the measurement of percent non-protein-bound oestradiol in serum. *J. steroid Biochem.* **21**, 343–345.

Dunkel, L. (1985). Decrease in serum sex hormone binding globulin during human chorionic gonadotrophin stimulation in prepubertal boys. *Acta endocr. Copenh.* **109**, 423–427.

—— Sorva, R., and Voutilainen, R. (1984). Low levels of sex hormone-binding globulin in obese children. *J. Pediat.* **107**, 95–97.

Dunn, J. F., Nisula, B. C., and Rodbard, D. (1981). Transport of steroid hormones: binding of 21 endogenous steroids to both testosterone-binding globulin and corti-costeroid-binding globulin in human plasma. *J. clin. Endocr. Metab.* **53**, 58–68.

Egloff, M., Savoure, N., Tardival-Lacombe, J., Massart, C., Nichol, M., and Degrelle, H. (1981). Influence of sex hormone-binding globulin and serum albumin on the conversion of androstenedione to testosterone by human erythrocytes. *Acta endocr. Copenh.* **96**, 136–140.

—— Vendrely, E., Tardivel-Lacombe, J., Dadoune, J. P., and Degrelle, H. (1982). Immunohistochemical study of human testis and epididymis with a monospecific antiserum against the sex-steroid-binding-plasma protein. *C. r. Acad. Sc. Paris* **111**, 107–111.

El Makhzangy, M. N., Wynn, V., and Lawrence, D. M. (1979). Sex hormone binding globulin capacity as an index of oestrogenicity or androgenicity in women on oral contraceptive steroids. *Clin. Endocr.* **10**, 39–45.

Englebienne, P. (1984). The serum steroid transport proteins: biochemistry and clinical significance. *Molec. Aspects Med.* **7**, 313–396.

Enriori, C. L., Orsini, W., Cremona, M. C., Etkin, A. E., Cardillo, L. R., and Reforozo-Membrives, J. (1986). Decrease of circulating level of SHBG in post-menopausal obese women as a risk factor in breast cancer: reversible effect of weight loss. *Gynecol. Oncol.* **23**, 77–86.

Estour, B., Pugeat, M., Lang, F., Dechaud, H., Pellet, J., and Rousset (1986). Sex hormone binding globulin in women with anorexia nervosa. *Clin. Endocr.* **24**, 571–576.

Fawcett, J. K. and Wynn, V. (1956). Variation of plasma electrolyte and total protein levels in the individual. *Br. med. J.* **ii**, 582–585.

Fernlund, P. and Laurell, C. B. (1981). A simple two-step procedure for the simultaneous isolation of corticosteroid binding globulin and sex hormone binding globulin from human serum by chromatography on cortisol-Sepharose and phenyl-Sepharose. *J. steroid Biochem.* **14**, 545–552.

Fex, G., Adielsson, G., and Mattson, W. (1981). Oestrogen-like effects of tamoxifen on the concentration of proteins in plasma. *Acta endocr. Copenh.* **97**, 109–113.

Forest, M. G., Ances, I. G., Tapper, A. J., and Migeon, C. J. (1971). Percentage binding of testosterone, androstenedione and dehydroisoandrosterone in plasma at the time of delivery. *J. clin. Endocr. Metab.* **32**, 417–425.

—— and Bertrand, J. (1972). Studies of the protein binding of dihydrotestosterone in human plasma. *Steroids* **19**, 197–214.

—— Bonneton, A., Lecoq, C., Brebant, C., and Pugeat, M. (1986). Ontogénèse de la proteine de liaison des hormones sexuelles (SBP) et de la transcortine (CBG) chez les primates; variations physiologiques et études dans différents milieux biologiques. In *Binding Proteins of Steroid Hormones*, Colloque INSERM Vol. **149** (eds M. G. Forest and M. Pugeat), pp. 263–291, John Libbey Eurotext Ltd., London and Paris.

—— Cathiard, A. M., and Bertrand, J. A. (1973). Total and unbound testosterone levels in the newborn and in normal and hypogonadal children: use of a sensitive radioimmunoassay for testosterone. *J. clin. Endocr. Metab.* **36**, 1132–1142.

Forest, M. G., David, M., and Pugeat, M. (1987). Sex steroid binding protein (SBP), androgens and tropic hormones (hCG, ACTH, prolactin). *Ann. N.Y. Acad. Sci.* in press.

—— Lecoq, A., and Saez, J. M. (1979). Kinetics of human chorionic gonadotrophin-induced steroidogenic response of the human testis. II. Plasma 17α-hydroxyprogesterone, Δ4-androstenedione, estrone and 17β-estradiol: evidence for the action of human chorionic gonadotrophin on intermediate enzymes implicated in steroid biosynthesis. *J. clin. Endocr. Metab.* **49**, 284–291.

—— —— Salle, B., and Bertrand, J. (1981). Does neonatal phenobarbital treatment affect testicular and adrenal function and steroid binding in plasma in infancy? *J. clin. Endocr. Metab.* **52**, 103–110.

Fotherby, K. (1984). A new look at progestogens. *Clin. Obstet. Gynec.* **11**, 701–721.

Furman, R., Howard, R., Norcia, L., and Keaty, E. (1958). The influence of androgens, estrogens and related steroids on serum lipids and lipoproteins. Observations in hypogonadal and normal human subjects. *Am. J. Med.* **59**, 80–97.

Gaidano, G., Berta, L., Rovero, E., Valenzano, C., and Rosatti, P. (1980). Dynamics of the binding capacity of plasma sex hormone binding globulin (SHBG) for testosterone and dihydrotestosterone during puberty. *Clin. Chim. Acta* **100**, 91–97.

Gambone, J. C., Pardridge, W. M., Lagasse, L. D., and Judd, H. L. (1982). In vivo availability of circulating estradiol in postmenopausal women with and without endometrial cancer. *Obstet. Gynec.* **59**, 416–420.

Geola, F. L., Frumar, A. M., Tataryn, I. V., Lu, K. H., Hershman, J. M., Eggena, P., Sambhi, M. P., and Judd, H. L. (1980). Biological effects of various doses of conjugated equine estrogens in postmenopausal women. *J. clin. Endocr. Metab.* **51**, 620–625.

Gershagen, S., Doberl, A., and Rannevik, G. (1984). Changes in the SHBG concentration during danazol treatment. *Acta obstet. gynec. scand.* Suppl. **123**, 117–123.

Glantz, G. M. (1964). Cirrhosis and carcinoma of the prostate gland. *J. Urol.* **91**, 291–293.

Glass, A. R., Swerdloff, R. S., Bray, G. A., Dahms, W. T., and Atkinson, R. L. (1977). Low serum testosterone and sex-hormone-binding-globulin in massively obese men. *J. clin. Endocr. Metab.* **45**, 1211–1219.

Gordon, G. G., Altman, K., Southren, A. L., Rubin, E., and Lieber, C. S. (1976). Effect of alcohol (ethanol) administration on sex hormone metabolism in normal men. *New Engl. J. Med.* **295**, 793–797.

—— Olivo, J., Rafii, F., and Southren, A. L. (1975). Conversion of androgens to oestrogens in cirrhosis of the liver. *J. clin. Endocr. Metab.* **40**, 1018–1026.

—— Southren, A. L., Tochimoto, S., Rand, J. J., and Olivo, J. (1969). Effect of hyperthyroidism and hypothyroidism on the metabolism of testosterone and androstenedione in man. *J. clin. Endocr. Metab.* **29**, 164–170.

Granger, L. R., Roy, S., and Mishell, D. R. (1982). Changes in unbound sex steroids and sex hormone binding globulin capacity during oral and vaginal progestogen administration. *Am. J. Obstet. Gynec.* **144**, 578–584.

Green, G. L., Gilna, P., Waterfield, M., Baker, A., Hort, Y., and Shine, J. (1986). Sequence and expression of human estrogen receptor complementary DNA. *Science N.Y.* **321**, 1150–1154.

Griffiths, K., Davies, P., Eaton, C. L., Harper, M. E., Peeling, W. B., Turkes, A. O., Turkes, A., Wilson, W. D., and Pierrepoint, C. G. (1987). Cancer of the prostate: endocrine factors. *Oxford Reviews of Reproductive Biology* (ed. J. R. Clarke) Vol. **9**, 192–259. Clarendon Press, Oxford.

Haapiainen, R., Ranniko, S., Adlercreutz, H., and Alfthan, O. (1986). Correlation of pretreatment plasma levels of estradiol and sex hormone-binding globulin-binding capacity with clinical stage and survival of patients with prostatic cancer. *The Prostate* **8**, 127–137.

Hämäläinen, E., Adlercreutz, H., Ehnholm, C., and Puska, P. (1986). Relationships of serum lipoproteins and apoproteins to sex hormones and to the binding capacity of sex hormone binding globulin in healthy Finnish men. *Metabolism* **35**, 535–541.

Hammond, G. L. and Lähteenmäki, P. L. A. (1983). A versatile method for the determination of serum cortisol binding globulin and sex hormone binding globulin binding capacities. *Clin. Chim. Acta* **132**, 101–110.

—— —— Lähteenmäki, P., and Luukkainen, T. (1982). Distribution and percentages of non-protein bound contraceptive steroids in human serum. *J. steroid Biochem.* **17**, 375–380.

—— Langley, M. S. (1986). Identification and measurement of sex hormone binding globulin (SHBG) and corticosteroid binding globulin (CBG) in human saliva. *Acta endocr. Copenh.* **112**, 603–608.

—— —— Robinson, P. A. (1985). A liquid phase immunoradiometric assay for human sex-hormone-binding-globulin. *J. steroid Biochem.* **23**, 451–460.

—— —— —— Nummi, S., and Lund, L. (1984). Serum steroid binding protein concentrations, distribution of progestogens, and bioavailability of testosterone during treatment with contraceptives containing desogestrel or levonorgestrel. *Fert. Steril.* **42**, 44–51.

—— Leinonen, P., Bolton, N. J., and Vihko, R. (1983). Measurement of sex hormone binding globulin in human amniotic fluid: its relationship to protein and testosterone concentrations, and fetal sex. *Clin. Endocr.* **18**, 377–384.

—— Nisker, J. A., Jones, L. A., and Siiteri, P. K. (1980). Estimation of percentage of free steroid in undiluted serum by centrifugal ultrafiltration-dialysis. *J. biol. Chem.* **255**, 5023–5026.

—— Robinson, P. A., Sugino, H., Ward, D. N., and Finne, J. (1986). Physicochemical characteristics of human sex hormone binding globulin: evidence for two identical subunits. *J. steroid Biochem.* **24**, 815–824.

Harris, A. L., Smith, I. E., Dowsett, M., Jeffcoate, S. L., Coombes, R. C., Powles, T. J., and Neville, A. M. (1981). Sex hormone binding globulin level and prognosis in early breast cancer. *Lancet* **i**, 279.

Hayward, J. L. and Rubens, R. D. (1987). UICC multidisciplinary project on breast cancer. Management of early advanced breast cancer. *Int. J. Cancer* **39**, 1–5.

Helgason, S., Damber, J. E., Damber, M. G., von Schoultz, B., Selstam, and G. Sodergard, R. (1982). A comparative longitudinal study on sex hormone binding globulin capacity during estrogen replacement therapy. *Acta obstet. gynec. scand.* **61**, 97–100.

Heller, R. F., Wheeler, M. J., Micallef, J., Miller, N. E., and Lewis, B. (1983). Relationship between high density lipoprotein cholesterol with total and free testosterone and sex hormone binding globulin. *Acta endocr. Copenh.* **104**, 253–256.

Henry, R. J. (1968). *Clinical Chemistry: Principles and Technics*, Harper and Row, New York, N.Y.

Hertz, J. B. and Johnsen, S. G. (1983). Sex-hormone-binding globulin (SHBG) in serum in threatened abortion. *Acta endocr. Copenh.* **104**, 381–384.

Heyns, W. (1977). The steroid-binding β-globulin of human plasma. *Adv. steroid Biochem. Pharmac.* **6**, 59–79.

Hollenberg, S. M., Weinberger, C., Ong, E. S., Cerelli, G., Oro, A., Lebo, R., Thompson, E. B., Rosenfeld, M. G., and Evans, R. M. (1985). Primary structure and expression of a functional human glucocorticoid receptor cDNA. *Nature, Lond.* **318**, 635–641.

Horst, H. J., Bartsch, W., and Dirksen-Thedens, I. (1977). Plasma testosterone, sex hormone binding globulin binding capacity and per cent binding of testosterone and 5α-dihydrotestosterone in prepubertal, pubertal and adult males. *J. clin. Endocr. Metab.* **45**, 522–527.

Horton, R. and Tait, J. F. (1966). Androstenedione production and interconversion rates, measured in peripheral blood and studies on the possible site of its conversion to testosterone. *J. clin. Invest.* **45**, 301–313.

Hosseinian, A. H., Kim, M. H., and Rosenfield, R. L. (1976). Obesity and oligomenorrhea are associated with hyperandrogenism independent of hirsutism. *J. clin. Endocr. Metab.* **42**, 765–769.

Hotchkiss, J. (1985). Changes in sex hormone-binding globulin binding capacity and percent free estradiol during development in the female rhesus monkey (*Macaca mulatta*): relation to the metabolic clearance rate of estradiol. *J. clin. Endocr. Metab.* **60**, 786–792.

Hryb, D. J., Khan, M. S., and Rosner, W. (1985). Testosterone-estradiol-binding globulin binds to human prostatic cell membranes. *Biochem. Biophys. Acta* **128**, 432–440.

Iqbal, M. J., Forbes, A., Wilkinson, M. L., Moore, J. W., Williams, R., and Bulbrook, R. D. (1987). Foetal steroid binding protein in British and Japanese women. *Acta endocr. Copenh.* **114**, 584–588.

—— and Johnson, M. W. (1977). Study of steroid-binding by a novel "Two-Tier" column employing Cibachrom Blue F3 G-A-Sepharose 4B. 1. Sex-hormone-binding-globulin. *J. steroid. Biochem.* **8**, 977–983.

———— (1979). Purification and characterization of human sex hormone binding globulin. *J. steroid. Biochem.* **10**, 535–540.

—— Wilkinson, M. L., and Williams, R. (1983). A new circulatory sex-steroid binding protein in benign and malignant liver disease and other malignancies associated with sex-steroids. *IRCS Med. Sci.* **11**, 1125–1126.

Jenkins, N. and Fotherby, K. (1980). Binding of the contraceptive steroids norgestrel and norethisterone in human plasma. *J. steroid Biochem.* **13**, 521–527.

Jeppsson, S. and Johansson, E. D. B. (1976). Medroxyprogesterone acetate, estradiol, FSH and LH in peripheral blood after intramuscular administration of depoprovera to women. *Contraception* **14**, 461–469.

Joseph, D. R., Hall, S. H., and French, F. S. (1986). Rat androgen binding protein: structure of the gene, mRNA and protein. In *Binding Proteins of Steroid Hormones*. Colloque INSERM Vol. **149** (eds M. G. Forest and M. Pugeat), pp. 123–135, John Libbey Eurotext Ltd., London and Paris.

—— —— —— (1987). Rat androgen-binding protein: evidence for identical subunits and amino acid sequence homology with human sex hormone-binding globulin. *Proc. natn. Acad. Sci. U.S.A.* **84**, 339–343.

Kantero, R. L., Wide, L., and Widholm, O. (1975). Serum growth hormone and gonadotrophins and urinary steroids in adolescent girls. *Acta endocr. Copenh.* **78**, 11–21.

Kelsey, J. L. (1979). A review of the epidemiology of human breast cancer. *Epidem. Rev.* **1**, 74–109.

Khan, M. S., Ehrlich, P., Birken, S., and Rosner, W. (1985). Size isomers of testosterone-estradiol-binding globulin exist in the plasma of individual men and women. *Steroids* **45**, 463–472.

—— Ewen, E., and Rosner, W. (1982). Radioimmunoassay for human testosterone-estradiol-binding globulin. *J. clin. Endocr. Metab.* **54**, 705–710.

—— Knowles, B. B., Aden, D. P., and Rosner, W. (1981). Secretion of testosterone-estradiol-binding globulin by a human hepatoma-derived cell line. *J. clin. Endocr. Metab.* **53**, 448–449.

Kim, M. H., Rosenfield, R. L., and Dupon, C. (1976). The effects of dexamethasone on plasma free androgens during the normal menstrual cycle. *Am. J. Obstet. Gynecol.* **126**, 982–986.

King, R. J. B. and Mainwaring, W. I. P. (1974). *Steroid-cell interactions.* Butterworth & Co., London.

Kopelman, P. G., Pilkington, T. R. E., White, N., and Jeffcoate, S. L. (1980). Abnormal sex steroid secretion and binding in massively obese women. *Clin. Endocr.* **12**, 363–369.

Lachelin, G. C. L., Judd, H. L., Swanson, S. C., Hanck, M. E., Parker, D. C., and Yen, S. S. C. (1982). Long term effects of dexamethasone administration in patients with polycystic ovarian disease. *J. clin. Endocr. Metab.* **55**, 768.

Lähteenmäki, P. L. A., Hammond, G. L., and Luukkainen, T. (1983). Serum non-protein bound percentage and distribution of progestin ST-1435: no effect of ST-1435 treatment on plasma SHBG and CBG binding capacities. *Acta endocr. Copenh.* **102**, 307–313.

Lamberg, B. A., Kantero, R. L., Saarinen, P., and Widholm, O. (1973). Endocrine changes before and after the menarche. *Acta endocr. Copenh.* **74**, 685–694.

Langley, M. S., Hammond, G. L., Bardsley, A., Sellwood, R. A., and Anderson, D. C. (1985). Serum steroid binding proteins and the bioavailability of estradiol in relation to breast diseases. *J. natl. Cancer Inst.* **75**, 823–829.

Lapidus, L., Lindstedt, G., Lundberg, P. A., Bengtsson, C., and Gredmark, T. (1986). Concentrations of sex-hormone binding globulin and corticosteroid binding globulin in serum in relation to cardiovascular disease and overall mortality in postmenopausal women. *Clin. Chem.* **32**, 146–152.

Lasnitzki, I. and Franklin, H. R. (1975). The influence of serum on the uptake, conversion and action of dihydrotestosterone in rat prostate glands in organ culture. *J. Endocr.* **64**, 289–297.

Laufer, L. R., Davidson, B. J., Ross, R. K., Lagasse, L. D., Siiteri, P. K., and Judd, H. L. (1983). Physical characteristics and sex hormone levels in patients with osteoporotic hip fractures or endometrial cancer. *Am. J. Obstet. Gynecol.* **145**, 585–590.

Laurell, C. B. and Rannevik, G. (1979). A comparison of plasma protein changes induced by danazol, pregnancy and estrogens. *J. clin. Endocr. Metab.* **49**, 719–725.

Lawrence, D. M., Katz, M., Robinson, T. W. E., Newman, M. C., McGarrigle, H. G., Shaw, M., and Lachelin, G. C. L. (1981). Reduced sex hormone binding globulin and derived free testosterone levels in women with severe acne. *Clin. Endocr.* **15**, 87–91.

Lee, I. R., Lawder, L. E., Townend, D. C., Wetherall, J. D., and Hahnel, R. (1985). Plasma sex hormone binding globulin concentration and binding capacity in children before and during puberty. *Acta endocr. Copenh.* **109**, 276–280.

——Dawson, S. A., Wetherall, J. D., and Hahnel, R. (1987). Sex hormone-binding globulin secretion by human hepatocarcinoma cells is increased by both estrogen and androgens. *J. clin. Endocr. Metab.* **64**, 825–831.

Lindholm, J., Winkel, P., Brodthagen, U., and Gyntelberg, F. (1982). Coronary risk factors and plasma sex hormones. *Am. J. Med.* **73**, 648–651.

Lindstet, G., Lundberg, P. A., Hammond, G. L., and Vihko, R. (1985). Sex hormone-binding globulin—still many questions. *Scand. J. clin. lab. Invest.* **45**, 1–6.

Lobl, T. J. (1981). Androgen transport proteins: physical properties, hormonal regulation, and possible mechanism of TeBG and ABP action. *Arch. Androl.* **7**, 133–151.

Lobo, R. A., Paul, W., March, C. M., Granger, L., and Kletzky, O. A. (1982). Clomiphene and dexamethasone in women unresponsive to clomiphene alone. *Obstet. Gynec.* **60**, 497–501.

Longcope, C., Gorbach, S., Goldin, B., Woods, M., Dwyer, J., and Warram, J. (1985). The metabolism of estradiol; oral compared to intravenous administration. *J. steroid Biochem.* **23**, 1065–1070.

——Hui, S. L., and Johnston Jr., C. C. (1987). Free estradiol, free testosterone and sex hormone-binding globulin in perimenopausal women. *J. clin. Endocr. Metab.* **64**, 513–517.

Lyrenas, S., Carlstrom, K., Backstrom, T., and von Schoultz, B. (1981). A comparison of serum oestrogen levels after percutaneous and oral administration of oestradiol-17β. *Br. J. Obstet. Gynaec.* **88**, 181–187.

Maruyama, Y., Aoki, N., Suzuki, Y., Sinohara, H., and Yamamoto, T. (1984). Variation with age in the levels of sex-steroid-binding plasma protein as determined by radioimmunoassay. *Acta endocr. Copenh.* **106**, 428–432.

Mason, R. C., Miller, W. R., Hawkins, R. A., and Forrest, A. P. M. (1981). Plasma sex hormone binding globulin and tumour oestrogen receptor status in breast cancer patients. *Lancet* **i**, 617.

Mathur, R. S., Landgrebe, S. C., Moody, L. O., Semmens, J. P., and Williamson, H. O. (1985). The effect of estrogen treatment on plasma concentrations of steroid hormones, gonadotropins, prolactin and sex-hormone-binding globulin in postmenopausal women. *Maturitas* **7**, 129–133.

Mattson, L.-A., Cullberg, G., Tangkeo, P., Zador, G., and Samsioe, G. (1980). Administration of dehydroepiandrosterone enanthate to oophorectomized women—effects on sex hormones and lipid metabolism. *Maturitas* **2**, 301–309.

——Silfverstolpe, G., and Samsioe, G. (1984). Lipid composition of serum lipoproteins in relation to gonadal hormones during the normal menstrual cycle. *Eur. J. Obstet. Gynec. reprod. Biol.* **17**, 327–335.

Meldrum, D. R., Pardridge, W. M., Karow, W. G., Rivier, J., Vale, W., and Judd, H. L. (1983). Hormonal effects of danazol and medical oophorectomy in endometriosis. *Obstet. Gynecol.* **62**, 480–485.

Mercier, C., Alfsen, A., and Baulieu, E. E. (1966). A testosterone binding globulin. In *Androgens in Normal and Pathological Conditions. Int. Congr. Ser.* No. **101** (eds A. Vermeulen and D. Exley) P. 212. Excerpta Medica Foundation.

Mercier-Bodard, C., Alfsen, A., and Baulieu, E. E. (1970). Sex steroid binding plasma protein (SBP). *Acta endocr. Copenh.* **64**, 204–224.

—— and Baulieu, E. E. (1986). Hormonal control of SBP in Human Hepatoma Cells. *J. steroid Biochem.* **24**, 443–448.

Mercier-Bodard, C., Renoir, M., and Baulieu, E. E. (1979). Further characterization and immunological studies of human sex steroid-binding protein. *J. steroid Biochem.* **11**, 253–259.

Meyer, F., Brown, J. B., Morrison, A. S., and MacMahon, B. (1986). Endogenous sex hormones, prolactin, and breast cancer in premenopausal women. *J. natl Cancer Inst.* **77**, 613–615.

Mickelson, K. E. and Petra, P. H. (1974). Purification of the sex steroid binding protein from human serum. *Biochemistry N.Y.* **14**, 957–963.

Moore, J. W., Clark, G. M. G., Bulbrook, R. D., Hayward, J. L., Murai, J. T., Hammond, G. L., and Siiteri, P. K. (1982). Serum concentrations of total and non-protein-bound oestradiol in patients with breast cancer and in normal controls. *Int. J. Cancer* **29**, 17–21.

—— —— Takatani, O., Wakabayashi, Y., Hayward, J. L., and Bulbrook, R. D. (1983). Distribution of 17β-estradiol in the sera of normal British and Japanese women. *J. natl Cancer Inst.* **71**, 749–753.

—— —— Hoare, S. A., Millis, R. R., Hayward, J. L., Quinlan, M. K., Wang, D. Y., and Bulbrook, R. D. (1986). Binding of oestradiol to blood proteins and the aetiology of breast cancer. *Int. J. Cancer* **38**, 625–630.

—— Hoare, S. A., Quinlan, M. K., Clark, G. M. G., and Wang, D. Y. (1987). Centrifugal ultrafiltration dialysis for non-protein-bound oestradiol in blood: Importance of the support. *J. steroid Biochem.* **28**, 677–681.

—— Key, T. J. A., Bulbrook, R. D., Clark, G. M. G., Allen, D. S., Wang, D. Y., and Pike, M. C. (1987). Sex hormone binding globulin concentrations in a population of normal women who had never used exogenous hormones. *Br. J. Cancer*, in press.

—— Thomas, B. S., and Wang, D. Y. (1986). Endocrine status and the epidemiology and clinical course of breast cancer. *Cancer Surveys* **5**, 537–559.

Motohashi, T., Wu, C. H., Abdel-Rahman, H. A., Marymor, N., and Mikhail, G. (1979). Estrogen/androgen balance in health and disease. *Am. J. Obstet. Gynec.* **135**, 89–95.

Murayama, Y., Sakuma, T., Udagawa, H., Utsunomiya, J., Okamoto, R., and Asano, K. (1978). Sex hormone-binding globulin and estrogen receptor in breast cancer: technique and preliminary clinical results. *J. clin. Endocr. Metab.* **46**, 998–1006.

—— Utsonomiya, J., Asano, K., and Bulbrook, R. D. (1979). Sex hormone-binding globulin and recurrence after mastectomy. *Gan* (Tokyo) **70**, 715–716.

—— —— Takahashi, I., Kitamura, M., and Tominaga, T. (1979). Sex hormone binding globulin as a reliable indicator of hormone dependence in human breast cancer. *Ann. Surg.* **190**, 133–137.

Murphy, B. E. P. (1968). Binding of testosterone and estradiol in plasma. *Can. J. Biochem.* **46**, 299–302.

Nicholson, R. I., Walker, K. J., and Davies, P. (1986). Hormone agonists and antagonists in the treatment of hormone sensitive breast and prostate cancer. *Cancer Surveys* **5**, 463–486.

Nilsson, B., Sodergard, R., Damber, M.-G., Damber, J.-E., and von Schoultz, B. (1982). Free testosterone levels during danazol therapy. *Fert. Steril.* **39**, 505–509.

Nisula, B. C., Loriaux, D. L., and Wilson, Y. A. (1978). Solid phase method for measurement of the binding capacity of testosterone—estradiol binding globulins in human serum. *Steroids* **31**, 681–690.

Noble, R. R. (1964). In *The Hormones* (eds. G. Pincus, K. V. Thiman, and E. B. Astwood) pp. 559–695. Academic Press, London and New York.

O'Brien, T. J., Higashi, M., Kanasugi, H., Gibbons, W. E., and Morrow, C. P. (1982). A plasma/serum estrogen-binding protein distinct from testosterone-estradiol-binding globulin. *J. clin. Endocr. Metab.* **54**, 793–797.

Odlind, V., Carlstrom, K., Michaelsson, G., Vahlquist, A., Victor, A., and Mellbin, T. (1982). Plasma androgenic activity in women with acne vulgaris and in healthy girls before, during and after puberty. *Clin. Endocr.* **16**, 243–249.

——Elamsson, K., Englund, D. E., Victor, A., and Johansson, E. D. B. (1982). Effects of oestradiol on sex hormone binding globulin. *Acta endocr. Copenh.* **101**, 248–253.

Ota, D. M., Jones, L. A., Jackson, G. L., Jackson, P. M., Kemp, K., and Bauman, D. (1986). Obesity, non-protein-bound estradiol levels and distribution of estradiol in the sera of breast cancer patients. *Cancer* **57**, 558–562.

Ottosson, U. B., Damber, J. E., Damber, M. G., Selstam, G., Solheim, F., Stigbrand, T., Sodergard, R., and von Schoultz, B. (1981). Effects on sex hormone binding globulin capacity and pregnancy zone protein of treatment with combination of ethinyloestradiol and norethisterone. *Maturitas* **3**, 295–300.

Pachman, A. (1984). Studies on the role of the sex-hormone-binding globulin (SHBG) in prostatic gland hypertrophy in men. II *In vitro* research *Int. Urol. Nephrol.* **16**, 211–217.

——(1984a). Studies on the role of sex-hormone-binding globulin (SHBG) in benign prostatic hypertrophy in men. I Clinical research *Int. Urol. Nephrol.* **16**, 141–147.

Pardridge, W. M. (1981). Transport of protein-bound hormones into tissues *in vivo*. *Endocr. Rev.* **2**, 103–125.

Pasqualini, J. R. (1982). Hormonal regulation of the prostatic gland: physiology and pathology. *Triangle* **21**, 21–26.

Pearlman, W. H. and Crepy, O. (1967). Steroid-protein interaction with particular reference to testosterone binding by human serum. *J. biol. Chem.* **242**, 182–189.

Peeling, W. B., and Griffiths, K. (1986). Endocrine treatment of prostatic cancer. In *The Prostate* (eds J. P. Blandy and B. Lytton) pp. 188–207. Butterworth, London.

Petra, P. H., Kumar, S., Hayes, R., Ericsson, L. H., and Titani, K. (1986). Molecular organization of the sex steroid-binding protein (SBP) of human plasma. *J. steroid Biochem.* **24**, 45–49.

——and Lewis, J. (1980). Modification in the purification of the sex steroid-binding protein of human serum by affinity chromatography. *Analyt. Biochem.* **105**, 165–169.

——Namkung, P. C., Senear, D. F., McCrae, D. A., Rousslang, K. W., Teller, D. C., and Ross, J. B. A. (1986a). Molecular characterization of the sex steroid binding protein (SBP) of plasma. Re-examination of rabbit SBP and comparison with the human, macaque and baboon proteins. *J. steroid Biochem.* **25**, 191–200.

————Titani, K., and Walsh, K. A. (1986b). Characterisation of the plasma sex steroid-binding protein. In *Binding Proteins of Steroid Hormones*, Colloque INSERM Vol. **149** (eds M. G. Forest and M. Pugeat). pp. 15–30. John Libbey Eurotext Ltd., London and Paris.

——Stanczyk, F. Z., Namkung, P. C., Fritz, M. A., and Novy, M. J. (1985). Direct effect of sex steroid-binding protein (SBP) of plasma on the metabolic clearance rate of testosterone in the rhesus macaque. *J. steroid Biochem.* **22**, 739–746.

————Senear, D. F., Namkung, P. C., Novy, M. J., Ross, J. B. A., Turner, E., and Brown, J. A. (1983). Current status of the molecular structure and function of the plasma sex steroid-binding protein (SBP). *J. steroid Biochem.* **19**, 699–706.

——Titani, K., Walsh, D. R., Hall, S. H., and French, F. S. (1986). Comparison of the amino acid sequence of the sex steroid-binding protein of human plasma (SBP) with that of the androgen-binding protein (ABP) of the rat testis. In *Binding Proteins of Steroid Hormones*, Colloque INSERM Vol. **149** (eds M. G. Forest and M. Pugeat) pp. 137–142, John Libbey Eurotext, Ltd., London and Paris.

Pike, M. C. and Ross, R. K. (1984). Breast Cancer. *Br. med. Bull.* **40**, 351–354.

Pirke, K. M., and Doerr, P. (1973). Age related changes and interrelationships between plasma testosterone, oestradiol and testosterone-binding globulin in normal adult males. *Acta endocr. Copenh.* **74**, 792–800.

Plymate, S. R., Fariss, B. L., Bassett, M. L., and Matej, L. (1981). Obesity and its role in polycystic ovary syndrome. *J. clin. Endocr. Metab.* **52**, 1246–1248.

—— Leonard, J. M., Paulsen, C. A., Farriss, B. L., and Karpas, A. E. (1983). Sex hormone binding globulin changes with androgen replacement. *J. clin. Endocr. Metab.* **57**, 645–648.

—— Moore, D. E., Cheng, C. Y., Bardin, C. W., Southworth, M. B., and Levinski, M. J. (1985). Sex hormone binding globulin changes during the menstrual cycle. *J. clin. Endocr. Metab.* **61**, 993–996.

—— Stutz, F. H., and Fariss, B. L. (1984). Relationship between sex hormone binding globulin and estrogen receptors in breast cancer. *J. clin. Oncol.* **2**, 652–654.

Pogmore, J. R., and Jequier, A. M. (1979). Sex hormone binding globulin capacity and postmenopausal hormone replacement therapy. *Br. J. Obstet. Gynaec.* **85**, 568–571.

Potts, G. O. (1977). Pharmacology of danazol. *J. Int. Med. Res.* **5**, Suppl. 3, 1–14.

Pugeat, M., Dunn, J. F., and Nisula, B. C. (1981). Transport of steroid hormones: interaction of 70 drugs with testosterone-binding globulin and corticosteroid-binding globulin in human plasma. *J. clin. Endocr. Metab.* **53**, 69–75.

—— Forest, M. G., Nisula, B. C., Corniau, J., de Peretti, E., and Tourniaire, J. (1982). Evidence of excessive androgen secretion by both the ovary and the adrenal in patients with idiopathic hirsutism. *Obstet. Gynec.* **59**, 46–51.

—— Nicolas, B., Tourniaire, J., and Forest, M. G. (1984). Interaction of gestrinone with human plasma steroid-binding proteins. In *Medical Management of Endometriosis* (eds J. P. Raynaud, T. Ojasoo, and L. Martin) pp. 183–192. Raven Press, New York.

Purifoy, F. E., Koopmans, L. H., and Mayes, D. M. (1981). Age differences in serum androgen levels in normal adult males. *Human Biology* **53**, 499–511.

Ratajczak, T., Twaddle, E., and Hahnel, R. (1980). Sex hormone-binding globulin and estrogen receptor in endometrial and cervical cancer. *Gynec. Oncol.* **10**, 162–266.

Reed, M. J., Baranek, P. A., Cheng, R. W., and James, V. H. T. (1986). Free fatty acids: A possible regulator of the available oestradiol fractions in plasma. *J. steroid Biochem.* **24**, 657–659.

—— Cheng, R. W., Beranek, P. A., Few, J. D., Franks, S., Gilchik, M. W., and James, V. H. T. (1986). The regulation of the biologically available fractions of oestradiol and testosterone in plasma. *J. steroid Biochem.* **24**, 317–320.

—— —— Noel, C. T., Dudley, H. A. F., and James, V. H. T. (1983). Plasma levels of estrone, estrone sulfate, and estradiol and the percentage of unbound estradiol in postmenopausal women with and without breast disease. *Cancer Res.* **43**, 2940–3943.

—— and Murray, M. A. F. (1979). The oestrogens. In *Hormones in Blood*, Vol. **3** (eds. C. H. Gray and V. H. T. James) pp. 263–340. Academic Press, London, New York and San Francisco.

Riad-Fahmy, D., Read, G. F., Walker, R. F., and Griffiths, K. (1982). Steroids in saliva for assessing endocrine function. *Endocrin. Rev.* **3**, 367–395.

Ridgway, E. C., Longcope, C., and Maloof, F. (1975). Metabolic clearance and blood production rates of estradiol in hyperthyroidism. *J. clin. Endocr. Metab.* **41**, 491–497.

—— Maloof, F., and Longcope, C. (1982). Androgen and oestrogen dynamics in hyperthyroidism. *J. Endocr.* **95**, 105–115.

Rogers, J. and Mitchell, G. W. (1952). The relation of obesity to menstrual disturbances. *New Engl. J. Med.* **247**, 53–55.

Rosner, W. (1972). A simplified method of the quantitative determination of testosterone-estradiol-binding globulin. *J. clin. Endocr. Metab.* **34**, 383–988.

—— (1986). Measurement of TeBG in biological fluids: evolution and problems. In *Binding Proteins of Steroid Hormones*, Colloque INSERM Vol. **149** (eds M. G. Forest and M. Pugeat) pp. 207–214, John Libbey Eurotext Ltd., London and Paris.

—— Khan, M. S., Breed, C. N., Fleisher, M., and Bradlow, H. L. (1985). Plasma steroid-binding proteins in the cysts of gross cystic disease of the breast. *J. clin. Endocr. Metab.* **61**, 200–203.

—— and Smith, R. N. (1975). Isolation and characterization of the testosterone-estradiol-binding globulin from human plasma. Use of a novel affinity column, *Biochemistry, N.Y.* **14**, 4813–4819.

Rudd, B. T., Duignan, N. M., and London, D. R. (1974). A rapid method for the measurement of sex hormone binding globulin capacity of sera. *Clin. Chim. Acta* **55**, 165–178.

Ruokonen, A., Alen, M., Bolton, N., and Vihko, R. (1985). Response of serum testosterone and its precursor steroids, SHBG and CBG to anabolic steroid and testosterone self-administration in man. *J. steroid Biochem.* **23**, 33–38.

—— and Kaar, K. (1985). Effects of desogestrel, levonorgestrel and lynestrenol on serum sex hormone binding globulin, cortisol binding globulin, ceruloplasmin and HDL-cholesterol. *Eur. J. Obstet. Gynec. reprod. Biol.* **20**, 13–18.

Saez, J. M. and Forest, M. G. (1979). Kinetics of human chorionic gonadotropin-induced steroidogenic response of the human testis. I. Plasma testosterone: implications for human chorionic gonadotropin stimulation test. *J. clin. Endocr. Metab.* **49**, 278–283.

Sakai, F., Cheix, F., Clavel, M., Colon, J., Mayer, M., Pommatau, E., and Saez, S. (1978). Increases in steroid binding globulins induced by tamoxifen in patients with carcinoma of the breast. *J. Endocr.* **76**, 219–226.

Santen, R. J. (1986). Determinants of tissue oestradiol levels in human breast cancer. *Cancer Surveys* **5**, 597–616.

Schneider, G., Kirschner, M. A., Berkowitz, R., and Ertel, N. H. (1979). Increased estrogen production in obese men. *J. clin. Endocr. Metab.* **48**, 633–638.

Schwarz, S., Tappeiner, G., and Hintner, H. (1981). Hormone binding globulin levels in patients with hereditary angiooedema during treatment with danazol. *Clin. Endocr.* **14**, 563–570.

Semmens, J., Rouse, I., Beilin, L. J., and Maserei, J. R. L. (1983). Relationship between plasma HDL-cholesterol to testosterone, estradiol and sex-hormone-binding globulin levels in men and women. *Metabolism* **32**, 428–432.

—— —— —— —— (1983a). Relationships between age, body weight, physical fitness and sex hormone binding globulin capacity. *Clin. Chim. Acta.* **133**, 295–300.

Siiteri, P. K. (1986). Plasma steroid-binding proteins—past, present and future. In *Binding Proteins of Steroid Hormones*, Colloque INSERM, Vol. **149** (eds M. G. Forest and M. Pugeat) pp. 593–610. John Libbey Eurotext Ltd., London and Paris.

—— Hammond, G. L., and Nisker, J. A. (1981). Increased availability of serum estrogens in breast cancer: a new hypothesis. In *Hormones and Breast Cancer* (eds M. C. Pike, P. K. Siiteri, and C. W. Welsch) pp. 87–101 Banbury Report 8, Cold Spring Harbor Laboratory, New York.

—— and MacDonald, R. C. (1973). Role of extraglandular estrogen in human endocrinology. In *Handbook of Physiology* Vol. **12** Pt **1** (ed. R. D. Greep) pp. 615–629. Am. Physiological Soc., Washington D.C. 12.

—— Murai, J. T., Hammond, G. L., Nisker, J. A., Raymoure, W. J., and Kuhn, R. W. (1982). The serum transport of steroid hormones. *Recent Prog. Horm. Res.* **38**, 457–510.

—— Simberg, N., and Murai, J. (1984). Estrogens and breast cancer. *Ann. N.Y. Acad. Sci.* **464**, 100–105.

Solomon, M., Iqbal, M. J., Dalton, M., Jeffcoate, S. L., and Ginsburg, J. (1979). Sex-hormone-binding globulin: an additional test for ovulatory function. *Lancet* **i**, 984.

Stanczyk, F., Hess, D. L., Senner, J. W., Petra, P. H., and Novy, M. J. (1986). Alterations in sex steroid-binding protein (SBP), corticosteroid-binding globulin (CBG), and steroid hormone concentrations during pregnancy in rhesus macaques. *Biol. Reprod.* **35**, 126–132.

—— Namkung, P. C., Fritz, M. A., Novy, M. J., and Petra, P. H. (1986). The influence of sex steroid-binding protein on the metabolic clearance rate of testosterone. In *Binding Proteins of Steroid Hormones*, Colloque INSERM, Vol. **149** (eds M. G. Forest and M. Pugeat) pp. 555–563, John Libbey Eurotext Ltd., London and Paris.

—— Petra, P. H., Senner, J. W., and Novy, M. J. (1985). Effect of dexamethasone treatment on sex steroid-binding protein, corticosteroid-binding globulin, and steroid hormones in cycling rhesus macaques. *Am. J. Obstet. Gynec.* **151**, 464–470.

Strel'chyonok, O. A., Avvakumov, G. V., and Survilo, L. I. (1984). A recognition system for sex-hormone-binding protein-estradiol complex in human decidual endometrium plasma membranes. *Biochem. Biophys. Acta* **802**, 459–466.

Sulkes, A., Fuks, Z., Gordon, A., and Gross, J. (1984). Sex hormone binding globulin (SHBG), in breast cancer: a correlation with obesity but not with estrogen receptor status. *Eur. J. Cancer clin. Oncol.* **20**, 19–23.

Suzuki, Y., Itagaki, E., Mori, H., and Hosoya, T. (1977). Isolation of testosterone-binding globulin from bovine serum by affinity chromatography and its molecular characterization, *J. Biochem.* **81**, 1721–1731.

Tait, J. F. and Burstein, S. (1964). *In vivo* studies of steroid dynamics in man. In *The Hormones*, Vol. **5** (eds G. Pincus, K. V. Thimann, and E. B. Astwood) pp. 441–557, Academic Press, New York.

Tardival-Lacombe, J., Egloff, M., Mazabrand, A., and Degrelle, H. (1984). Immunohistochemical detection of the sex steroid-binding plasma protein in human mammary carcinoma cells. *Biochem. Biophys. Res. Commun.* **118**, 488–494.

Tulchinsky, D., and Chopra, I. J. (1973). Competitive ligand-binding assay for measurement of sex hormone-binding globulin (SHBG). *J. clin. Endocr. Metab.* **37**, 873–881.

Umstot, E. S. and Andersen, R. N. (1986). Nonsteroidal substances that affect serum free testosterone. *J. steroid Biochem.* **25**, 225–229.

Uriel, J., Dupiers, M., Rimbaut, C., and Buffe, D. (1981). Maternal serum levels of sex steroid-binding protein during pregnancy. *Br. J. Obstet. Gynaec.* **88**, 1229–1232.

van Baelen, H., Heyns, W., Schonne, E., and De Moor, P. (1968). An estradiol binding globulin in human serum: partial purification. *Annls Endocr.* **29**, Suppl. 153–158.

van Kammen, E., Thijssen, J. H. H., Rademaker, B., and Schwarz, F. (1975). The influence of hormonal contraceptives on sex hormone binding globulin (SHBG) capacity. *Contraception* **11**, 53–59.

van Look, P. F. A. and Frolich, M. (1981). Effects of ethinyloestradiol on plasma levels of pituitary gonadotrophins, testicular steroids and sex hormone binding globulin in normal men. *Clin. Endocr.* **14**, 237–243.

Vermeulen, A. (1979). The androgens. In *Hormones in Blood*, Vol. 3 (eds C. H. Gray and V. H. T. James) pp. 356–405. Academic Press, London, New York and San Francisco.

——(1986). Human mammary cancer as a site of sex steroid metabolism. *Cancer Surveys* **5**, 585–596.

—— and Ando, S. (1979). Metabolic clearance rate and interconversion of androgens and the influence of the free androgen fraction. *J. clin. Endocr. Metab.* **48**, 320–325.

—— Rubens, R., and Verdonck, L. (1972). Testosterone secretion and metabolism in male senescence. *J. clin. Endocr. Metab.* **34**, 730.

—— Stoica, T., and Verdonck, L. (1971). The apparent free testosterone concentration, an index of androgenicity. *J. clin. Endocr. Metab.* **33**, 759–767.

—— Verdonck, L., van der Straeten, M., and Orie, N. (1969). Capacity of the testosterone-binding globulin in human plasma and influence of specific binding of testosterone on its metabolic clearance rate. *Metabolism*, **29**, 1470–1480.

Victor, A., and Johansson, E. D. B. (1977). Effects of d-norgestrel induced decreases in sex hormone binding globulin capacity on the d-norgestrel levels in plasma. *Contraception* **16**, 115–123.

—— Lundberg, P. O., and Johansson, E. D. B. (1977). Induction of sex hormone binding globulin by phenytoin. *Br. med. J.* **2**, 934–935.

—— Weiner, E., and Johansson, E. D. B. (1976). Sex hormone binding globulin: the carrier protein for d-Norgestrel. *J. clin. Endocr. Metab.* **43**, 244–247.

Wagner, R. K. (1978). Extracellular and intracellular steroid binding proteins. Properties, discrimination, assay and clinical applications. *Acta endocr.* Suppl. **218**, 5–73.

Walsh, K. A., Titani, K., Takio, K., Kumar, S., Hayes, R., and Petra, P. H. (1986). Amino acid sequence of the sex steroid-binding protein of human blood plasma. *Biochemistry, N.Y.* **25**, 7584–7590.

Wang, D. Y., Fantl, V. E., Habibollahi, F., Clark, G. M. G., Fentiman, I. S., Hayward, J. L., and Bulbrook, R. D. (1986). Salivary oestradiol and progesterone levels in premenopausal women with breast cancer. *Eur. J. Cancer clin. Oncol.* **22**, 427–433.

Weinberger, C., Hollenberg, S. M., Rosenfeld, M. G., and Evans, R. M. (1985). Domain structure of human glucocorticoid receptor and its relationship to the v-erb-A oncogene product. *Nature, Lond.* **318**, 670–672.

Wenn, R. V., Kamberi, I. A., Vossough, P., Kariminejad, M. H., Torabee, E., Ayoughi, F., Keyvanjah, M., and Sarberi, N. (1977). Human testosterone-oestradiol binding globulin in health and disease. *Acta endocr. Copenh.* **84**, 850–859.

Westphal, U. (1971). Steroid-protein interactions. *Monogr. Endocr.* **4**, Springer-Verlag, New York, Heidelberg, Berlin.

——(1986). Steroid-protein interactions II. *Monogr. Endocr.* **27**, Springer-Verlag, New York, Heidelberg, Berlin.

Wheeler, M. J., Crisp, A. H., Hsu, L. K. G., and Chen, C. N. (1983). Reproductive hormone changes during weight gain in male anorectics. *J. clin. Endocr. Metab.* **18**, 423–429.

Wilkinson, M. L., Iqbal, M. J., and Williams, R. (1983). A new sex-steroid binding protein in foetal liver. *IRCS Med. Sci.* **11**, 1123–1124.

Willemse, M. H. B., Sleijfer, D. T., Pratt, J. J., Sluiter, W. J., and Doorenbos, H. (1984). No change in plasma free testosterone ratio and plasma sex hormone-binding globulin concentration during hCG stimulation. *J. clin. Endocr. Metab.* **58**, 1193–1196.

Winters, S. J., and Troen, P. (1982). Episodic luteinizing hormone (LH) secretion and the response of LH and follicle-stimulating hormone to LH-releasing hormone in aged men: evidence for coexistent primary testicular insufficiency and an impairment in gonadotropin secretion. *J. clin. Endocr. Metab.* **55**, 560–565.

Wu, C. H. (1982). Plasma free and protein-bound testosterone in hirsutism. *Obstet, Gynec.* **60**, 188–194.

——Motohashi, T., Abdel-Rahman, H. A., Flickinger, G. L., and Mikhail, G. (1976). Free and protein bound plasma estradiol-17β during the menstrual cycle. *J. clin. Endocr. Metab.* **43**, 436–445.

Wynn, V. (1977). Metabolic effects of danazol therapy. *J. Int. Med. Res.* **5**, Suppl. 3, 25–35.

Yosha, S., Fay, M., Longcope, C., and Braverman, L. E. (1984). Effect of D-thyroxine on serum sex hormone binding globulin (SHBG), testosterone, and pituitary-thyroid function in euthyroid subjects. *J. Endocr. Invest.* **7**, 489–494.

Yu, M. C., Gerkins, V. R., Henderson, B. E., Brown, J. B., and Pike, M. C. (1981). Elevated levels of prolactin in nulliparous women. *Br. J. Cancer* **43**, 826–832.

6 Selective delivery of sex steroid hormones to tissues by albumin and by sex hormone-binding globulin

WILLIAM M. PARDRIDGE

I   Introduction

Sex steroids such as testosterone or oestradiol circulate in human serum in three physical states: free, albumin-bound, and sex hormone-binding globulin (SHBG)-bound (Westphal 1971). About 2 per cent of the total sex steroid is free and the remaining 98 per cent is distributed to either albumin or to SHBG in varying proportions, depending on the relative concentrations of the two plasma proteins. It has generally been regarded over the last thirty years that only the free fraction is biologically available for uptake by tissues *in vivo*, and this view has been called the free hormone hypothesis (Tait and Burstein 1964; Westphal 1971; Siiteri 1981). The free hormone hypothesis has both basic and clinical implications. With regard to basic endocrinology, the free hormone hypothesis predicts that the concentration of free cytoplasmic or nuclear hormone, that is, the hormone pool that drives nuclear receptor occupancy and hormone biological action, may be reliably determined by measurements of free hormone concentrations in human serum *in vitro* (Pardridge 1987). With regard to clinical endocrinology, the free hormone hypothesis predicts that clinical status in patients can also be assessed by measurements of *in vitro* concentrations of free hormone in human serum (Pardridge 1981). In addition, the free hormone hypothesis predicts that, in the absence of cellular metabolism, all tissues are exposed to an equal concentration of circulating free hormone and the function of hormone-binding plasma proteins, such as albumin or SHBG, is to sequester passively steroid hormone within the circulating plasma. As is true in other areas of science, theories that posit a passive role to a biologic process have little, if any, heuristic value and generally lead to little progress, often for long periods of time.

    This chapter will review and develop further the protein-bound hormone hypothesis (Pardridge 1981; Pardridge 1987), which assumes that plasma proteins such as albumin or SHBG selectively and dynamically deliver hormones to tissues via mechanisms that vary from tissue to tissue, from species to species, during development and among differing clinical conditions. Since it has been known for nearly thirty years that plasma proteins *per se* do not significantly leave the plasma compartment during a single circulatory passage through the organ (Dewey 1959), then it is clear that the finding that albumin-bound or SHBG-bound sex steroid is available to tissues *in vivo* indicates that the hormone is stripped away from the plasma protein as it courses through the organ microcirculation. In this way, the plasma protein-bound hormone is operationally available for transport into tissue without significant exodus of the plasma protein *per se*. Therefore, the important parameter is not the concentration of free hormone in serum measured *in vitro*, for example, by equilibrium dialysis but the concentration of exchangeable hormone measured *in vivo* in a particular organ (Pardridge 1981; Pardridge 1987). There are several implications of the protein-bound hormone hypothesis with respect to both basic and clinical endocrinology. The protein-bound hormone hypothesis predicts that the concentrations of free cytosolic and nuclear hormone, which have never been directly measured *in vivo*, may be log orders higher than the concentration of free hormone, but approximate the concentration of plasma exchangeable hormone measured with *in vivo* techniques. For example, if it is found that albumin-bound steroid hormone is readily available for transport into a particular organ, then the concentration of free cytosolic hormone in that organ, in the absence of significant cellular metabolism of hormone, will be equal to the free plus albumin-bound concentration as measured *in vitro* (Pardridge and Landaw 1985). The albumin-bound concentration is often ten-fold higher than the free concentration and is approximately equal to the dissociation constant (K_D) of testosterone or oestradiol nuclear receptors (Fig. 6.1). In order to generate 50 per cent nuclear receptor occupancy, the concentration of free cellular hormone must equal the K_D of the nuclear receptor. As shown in Fig. 6.1, the concentration of albumin-bound testosterone or oestradiol does, in fact, approximate the K_D of the respective receptor, whereas the concentration of free testosterone or oestradiol only leads to very low occupancy of the nuclear receptor and does not allow for saturation of hormone biological response (Siiteri, Murai, Hammond, Nisker, Raymoure, and Kuhn 1982). With regard to clinical endocrinology, the protein-bound hormone hypothesis predicts that either the concentration of albumin-bound hormone or the total concentration of hormone in plasma will reliably predict the hormone biological response in a particular organ for a given clinical condition. Whether the albumin-bound or the total concentration of hormone is used to predict clinical status depends on whether both albumin-bound and SHBG-bound hormone are available for uptake by a particular organ. The protein-bound hormone hypothesis has heuristic value since it leads to the discovery of organ selective transport mechanisms involving albumin or SHBG. In addition, this hypothesis should also lead to the discovery of the biochemical mechanisms mediating the catalysis of hormone dissociation from the binding proteins within the organ microcirculation (Pardridge 1987).

The need for a unified transport theory is illustrated by the diversity of opinion regarding the mechanisms of steroid hormone delivery to tissues *in*

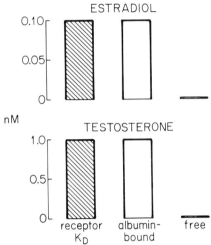

Fig. 6.1 Comparison of nuclear receptor dissociation constant (K_D), albumin-bound fraction, and free fraction for testosterone and oestradiol. These comparisons show that the concentration of free sex steroid is very small compared to the K_D, that is, the concentration of free nuclear steroid required to cause 50 per cent occupancy of the nuclear receptor. In contrast, the concentration of albumin-bound sex steroid approximates the K_D. Therefore, the availability of circulating plasma protein-bound sex steroid to tissues allows for significant occupancy of the nuclear sex steroid receptor and for saturation of hormone biological response. Data from Siiteri *et al.* (1982); Griffin and Wilson (1985); Pardridge (1986*b*).

vivo. Tait and Burstein (1964) recognized twenty years ago that albumin-bound steroid hormones are available for uptake by liver, since this organ extracts 10–50 per cent of circulating steroid hormone on a single passage. They proposed that globulin-bound hormone is not available for uptake by tissues, including liver. Implicit in their hypothesis is the idea that only free hormone is available for uptake by tissues that only minimally extract steroid hormone from the circulation. This view was explicitly stated by Riad-Fahmy and co-workers (1982) who proposed that free hormone concentration is important in tissues where metabolic clearance rate (MCR) is low, but concentrations of bound hormone are important in tissues where MCR is high. This view is actually a restatement of the free hormone hypothesis, since it predicts that the concentration of free cellular hormone cannot exceed the concentration of free hormone measured *in vitro*, and significant hormone nuclear receptor occupancy would not be attainable (Fig. 6.1). In contrast, the protein-bound hormone hypothesis assumes that the delivery of albumin-bound or SHBG-bound hormone to tissues is independent of the rate of hormone metabolism by the organ and is a function of conformational changes about the plasma protein hormone binding site caused by interactions between the protein and the surface of the microcirculation (see Biochemical Model below).

II Quantifiable physiological model

A quantifiable physiological model of sex steroid transport into tissues is shown in Fig. 6.2. The concentration of free cellular hormone (L_M) is believed to be the single-most important parameter determining either nuclear receptor occupancy and hormone biological response or hormone metabolism (Tait and Burstein 1964). For example, hormone biological response is a function of the degree of nuclear receptor occupancy, and this occupancy is proportional to the total concentration of nuclear receptor and to the concentration of free nuclear hormone, which is assumed to be in equilibrium with free cytoplasmic hormone (L_M). Similarly, the rate of steroid hormone metabolism, which determines the MCR, is proportional to the total concentration of cytosolic enzymes and to the concentration of free cytoplasmic hormone (Tait and Burstein 1964).

Despite the importance of the free cytoplasmic hormone concentration (L_M), this parameter has never been measured directly *in vivo* since this would require the development of a testosterone or oestradiol sensitive electrode that can be placed into the cellular cytoplasm. The concentration of free cytoplasmic hormone, however, can be measured indirectly by determination of the concentration of plasma exchangeable hormone (L_F, Fig. 6.2). Given the simplified case of a single plasma protein-binding system where hormone dissociation is rapid compared to capillary transit time, the concentrations of free cytoplasmic hormone (L_M) and capillary exchangeable hormone (L_F) are given by (Pardridge and Landaw 1985):

$$L_M = \frac{k_3}{k_4 + k_9}(V_P/V_T) \cdot L_F \tag{1}$$

$$L_F = \left[\frac{K_A^a}{A_F + K_A^a(1 + R)} \right] L_T^\circ \tag{2}$$

$$R = \frac{k_3 \cdot k_9}{(k_4 + k_9)k_{10}} \tag{3}$$

where k_3, k_4, k_9, and k_{10} are defined in Fig. 6.2 and Table 6.1, and V_P = the organ capillary plasma volume, V_T = the organ extravascular water volume, K_A^a = the apparent *in vivo* molar dissociation constant of ligand binding to the plasma protein within the microcirculation, and A_F = the concentration of unoccupied plasma protein binding sites (Table 6.1). The above equations are applicable to the rat since there is little SHBG circulating in this species (Corvol and Bardin 1973). However, in other species, such as humans, which possess SHBG, the above equations are simplifications of more complex relationships (Pardridge and Landaw 1985).

The above equations are simple illustrations of important principles. First, when organ metabolism is nil, for example, R is ≈ 0, plasma exchangeable hormone (L_F) = free cytosolic hormone (L_M). Although L_M is not presently

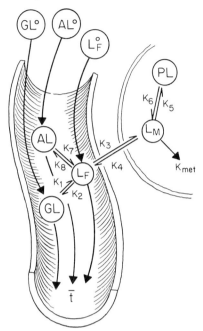

Fig. 6.2   Steady state model of testosterone transport through the brain capillary wall and into brain cells. Rate constants are defined in Table 6.1, where rate constant of hormone metabolism (k_{met}) is defined as k_9. Pools of globulin-bound, albumin-bound, and free ligand in the systemic circulation are denoted as GL°, AL°, and L_F°, and pools of globulin-bound, albumin-bound and exchangeable ligand in the brain capillary are GL, AL, and L_F, respectively. Pools of free and cytoplasmic-bound steroid in cells are given by L_M and PL, respectively; $\bar{t}$ is mean capillary transit time. (From Pardridge and Landaw 1985 with permission).

amenable to direct empiric measurement, L_F can be determined for a particular organ using the tissue sampling single injection technique (Pardridge 1981). Second, the above equations illustrate that the concentration of cytoplasmic free hormone is fully independent of cytoplasmic binding proteins and is simply a function of the concentration of capillary exchangeable hormone, the rates of membrane transport (k_3, k_4), and cytoplasmic metabolism (k_9). Third, as emphasized previously by Tait and Burstein (1964), hormone volume of distribution (V_D) and MCR are not linked if, as depicted in Fig. 6.2, cellular enzymes act only on free cytoplasmic hormone. Hormone volume of distribution is a function of hormone binding to cytoplasmic proteins. However, MCR is independent of hormone binding to cytoplasmic proteins, for example:

$$MCR = E^{SS} \cdot F \tag{4}$$

$$E^{SS} = \frac{f(R)}{1 + fR} \tag{5}$$

Table 6.1
Testosterone basal parameters

Parameter No.	Parameter symbol U	Parameter name	Parameter value
1	k_1 (s⁻¹)	Capillary globulin-ligand dissociation	0.03
2	k_2 (M⁻¹·s⁻¹)	Capillary globulin-ligand association	1.5×10^7
3	k_3 (s⁻¹)	Plasma to tissue influx	1.9
4	k_4 (s⁻¹)	Tissue to plasma efflux	0.0271
5	k_5 (M⁻¹·s⁻¹)	Cytosolic protein-ligand association	1.0×10^7
6	k_6 (s⁻¹)	Cytosolic protein-ligand dissociation	0.03
7	k_7 (s⁻¹)	Capillary albumin-ligand dissociation	2.5×10^3
8	k_8 (M⁻¹·s⁻¹)	Capillary albumin-ligand association	1.0×10^6
9	k_9 (s⁻¹)	Ligand tissue metabolism	0 or 0.1
10	k_{10} (s⁻¹)	Plasma transit	1.0
11	K_G (M)	Systemic globulin dissociation constant	2.0×10^{-9}
12	K_A (M)	Systemic albumin dissociation constant	5.3×10^{-5}
13	G_T^o (M)	Total globulin plasma concentration	2.8×10^{-8}
14	A_F (M)	Free (total) albumin plasma concentration	6.4×10^{-4}
15	P_T (M)	Total cytosolic protein concentration	2.5×10^{-9}
16	L_T^o	Total ligand plasma concentration	1.0×10^{-8}
17	V_P (1/kg)	Brain capillary plasma volume	0.01
18	V_T (1/kg)	Brain extravascular water volume	0.70

The assignments of k_7 and k_8 are somewhat arbitrary, because only the ratio of k_7/k_8 is known (Pardridge and Landaw, 1984). However, simulation studies showed that increasing k_7 and k_8 to 2.5×10^5 s⁻¹ and 1.0×10^8 M⁻¹·s⁻¹, respectively, had no effect on the results in Table 6.3. The range in varying k_8 was restricted to 10^6 to 10^8 (M⁻¹·s⁻¹) because this is the known range of hormone-plasma protein association rate constants (Westphal, 1978). However, decreasing k_7 and k_8 to 53 s⁻¹ and 21,200 (M⁻¹·s⁻¹), respectively, also had no significant effect on the results in Table 6.3. From (Pardridge and Landaw 1985 with permission).

where E^{ss} = the net steady state hormone extraction by the organ, F = organ blood flow, and $f = K_A^a/(A_F + K_A^a)$, that is, the fraction of capillary exchangeable hormone. Thus, MCR, like the free cytoplasmic hormone (and nuclear receptor occupancy), is strictly a function of the concentration of capillary exchangeable hormone, membrane transport, and organ metabolism, and is independent of tissue-binding proteins.

While it is possible to derive the relatively simple equations listed above that predict the concentrations of capillary exchangeable and cytoplasmic free hormone for a single, rapidly dissociating plasma protein system, the equations describing the two protein (albumin and SHBG) models, as shown in Fig. 6.2, are more complex and are best analyzed using a small computer. A program in Basic for performing such an analysis has recently been published (Pardridge and Landaw 1987), and this program was used in previous studies to estimate the concentration of free cytoplasmic testosterone in brain under a variety of conditions using the model parameters listed in Table 6.1 (Pardridge and Landaw 1985). The reader is encouraged to copy this program in Basic for use on any microcomputer since it provides a useful interactive approach to learning transport theory. The model is composed of eighteen parameters (Table 6.1) and it is important to understand each of these parameters since they all describe discrete physiological events.

1 STEROID DISSOCIATION RATES IN VIVO

The parameters k_1, k_2, k_7, and k_8 refer to the capillary-globulin or albumin-ligand dissociation or association rates *in vivo* (Table 6.1). Since previous studies have shown that SHBG-bound testosterone is not available for uptake by brain (Pardridge 1981), k_1 is set at 0.03 sec^{-1}, which is the measured rate of testosterone dissociation from SHBG *in vitro* at 37°C (Heyns and De Moor 1971). Since the SHBG dissociation constant (K_G) is 2 nM (Mickelson and Petra 1978) and since $k_2 = k_1/K_G$, then $k_2 = 1.5 \times 10^7$ $M^{-1}sec^{-1}$. Previous studies have shown that, owing to enhanced rates of dissociation of testosterone from albumin within the brain microcirculation, the capillary albumin dissociation constant for testosterone (K_A^a) = 2500 μM (Pardridge and Landaw 1984). Assigning $k_8 = 1 \times 10^6 M^{-1}sec^{-1}$, then $k_7 = K_A^a \times k_8 = 2500$ sec^{-1}. The assignments of k_7 and k_8 are somewhat arbitrary because only the ratio of k_7/k_8 is known. However, simulation studies have shown that increasing k_7 and k_8 by up to two log orders of magnitude has no effect on the results (Pardridge and Landaw 1985). The range in varying k_8 was restricted to $10^6 - 10^8$ $M^{-1}sec^{-1}$ because this is the known range of hormone-plasma protein association rate constants (Westphal 1978). The assignment of k_7 in Table 6.1 only pertains to brain since other studies have shown that the rate of testosterone dissociation from albumin varies from organ to organ (Table 6.2). As discussed below, the degree to which testosterone dissociation is enhanced in the organ microcirculation is

Table 6.2

In vivo dissociation constant (K_D^a) *of bovine albumin binding of testosterone or oestradiol in three organs*

Organ	K_D^a (μM)	
	Testosterone (T)	Oestradiol (E$_2$)
Brain	$2,500 \pm 700$	710 ± 100
Salivary Gland	602 ± 40	N.D.
Lymph Node	300 ± 90	$1,500 \pm 500$

The *in vitro* K_D is $53 \pm 1\,\mu$M and $23 \pm 1\,\mu$M, respectively, for T and E$_2$. N.D. = not determined. (From Pardridge and Landaw 1984; Cefalu, Pardridge, Chaudhuri, and Judd 1986; Cefalu and Pardridge 1987).

believed to be a function of the conformational change about the testo-sterone binding site that, in turn, is caused by interactions between the plasma protein and the microcirculatory surface (Pardridge and Landaw 1984). Since this interaction may differ among organs, the conformational change and the ultimate rates of testosterone dissociation may also vary from organ to organ (see **III Biochemical model**). It is assumed that interactions between the binding protein and the surface of the large systemic arteries is minimal and that the rates of hormone dissociation from binding proteins in the arterial circulation is approximately equal to the values found *in vitro*. Therefore, the systemic globulin or albumin dissociation constants (K_G or K_A) are fixed at the respective *in vitro* values of 2 nM (Mickelson and Petra 1978) and 53 μM (Pardridge 1981), respectively (Table 6.1).

2 ALBUMIN AND SHBG CONCENTRATIONS

The concentration of albumin (A_F) or SHBG (G_T°) is fixed at the normal male values of 640 μM or 28 nM, respectively (Pardridge and Landaw 1985). These values may vary among clinical conditions. For example, SHBG is ele-vated and albumin is decreased in cirrhosis (see **VII Clinical endocrinology**).

3 MEMBRANE TRANSPORT

The plasma to tissue influx and the tissue to plasma efflux of hormone is given by k_3 and k_4, respectively (Table 6.1). Previous studies have shown that the k_3 value for testosterone transport through the brain capillary endo-thelial wall, that is, the blood-brain barrier (BBB), is 1.9 sec^{-1} (Pardridge and Landaw 1984). It is assumed that cell membrane permeability to hormone is symmetric (that is, there is no active transport) such that the ratio of transport rate constants $k_3/k_4 = V_P/V_T$, where $V_P = 0.01$ L/kg and $V_T = 0.70$ L/kg, respectively (Pardridge and Landaw 1985). Therefore,

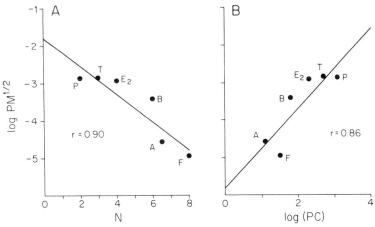

Fig. 6.3 (A) The log P (mol wt)1/2 for each hormone is plotted versus the number (N) of hydrogen bonds the steroid makes in aqueous solution, where P is the blood-brain barrier permeability constant. The N value was assigned according to the rules of Stein (1967): 2 for hydroxyl groups, 1 for carbonyl, aldehyde, or ketone groups, and 0 for ether moieties. (B) The log P (mol wt)1/2 per ionic is plotted versus the log of the 1-octanol/Ringer's partition coefficient. Steroids have been abbreviated as follows: P—progesterone, T—testosterone, E_2—oestradiol, B—corticosterone, A—aldosterone, F—cortisol. (From Pardridge and Mietus 1979b with permission).

$k_4 = k_3 (V_P/V_T) = 0.0271 \text{ sec}^{-1}$. Previous studies have shown that BBB transport of testosterone or oestradiol is nonsaturable up to 25 µM concentrations of steroid hormone (Pardridge 1981) and this transport process is presumed to be free diffusion. Although some studies have suggested that steroid hormones traverse cell membranes via carrier-mediated transport (Milgrom, Atger, and Baulieu 1973; Rao 1981), this has not been confirmed (Muller and Wotiz 1979; Kilvik, Furu, Haug, and Gautvik 1985), and would seem unlikely for steroid hormones that are highly lipid soluble. The major factor regulating cell membrane transport of steroid hormones appears to be the polarity of the molecule as determined by the hydrogen bond number (Pardridge 1981). As illustrated by the linear regression plot in Fig. 6.3, membrane transport of steroid hormones can be predicted by determining the hydrogen bond number, that is, the number of hydrogen bonds formed between the water solvent and the polar functional groups on the steroid hormone nucleus. The hydrogen bond number is computed on the basis of the steroid hormone structure (Fig. 6.4), given the rules of hydrogen bonding: hydroxyl group forms two hydrogen bonds, carbonyl, aldehyde, or ketone groups form one hydrogen bond, and ether moieties form no hydrogen bonds (Stein 1967).

Active transport of steroid hormones across the cell membrane would allow for an enrichment of the pool size of the cytoplasmic hormone relative to concentration of plasma exchangeable hormone, and would be repre-

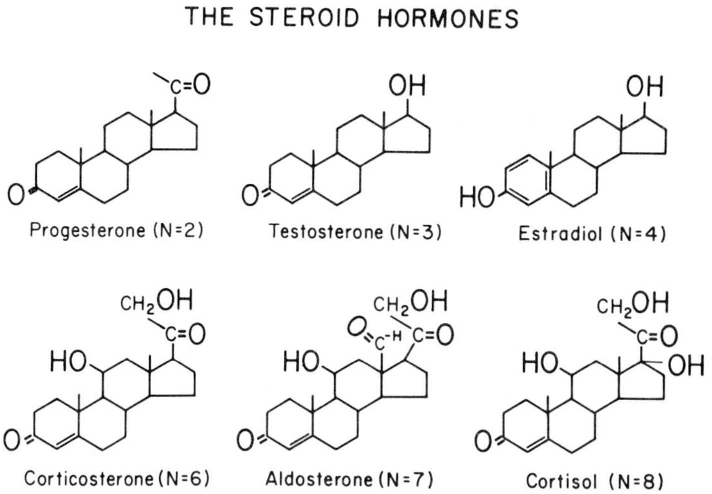

THE STEROID HORMONES

Fig. 6.4   Molecular structures for steroid hormones with emphasis placed on hydrogen bond forming functional groups. The addition of one hydroxyl to corticosterone results in the formation of cortisol and a log-order decrease in BBB permeability (see Fig. 6.3). N = number of hydrogen bonds. (From Pardridge 1981 with permission).

sented by a k_4 value that is disproportionately low given the known k_3, V_P, and V_T values. Thus far, there have been no studies published which support active transport of steroid hormones across cell membranes. Transport studies performed in isolated cells *in vitro* show saturable uptake of steroid hormones by the cells against the concentration gradient (Milgrom *et al.* 1973; Rao 1981). However, it is likely that these studies are actually measuring cytoplasmic binding and not membrane transport, since the latter process probably reaches equilibrium within a second in an isolated cell system where the extracellular water volume is log orders greater than the intacellular water volume. For example, initial studies suggesting active transport of thyroxine into isolated liver cells have been recently re-evaluated to show that thyroxine crosses the hepatocyte cell membrane by free diffusion (Rao and Rao 1983).

4   CAPILLARY TRANSIT TIME

The capillary transit time (t) = $1/k_{10}$, where k_{10} is the rate constant of plasma transit through the organ, equals 1.0 sec for brain (Pardridge and Landaw 1985). Organ blood flow = $V_P \times k_{10}$. Therefore, organ blood flow may increase owing to either increased plasma velocity (decreased transit time) or to increased plasma volume. There may be either a direct or inverse relationship between organ blood flow and capillary exchangeable hormone, depending on whether blood flow is increased owing to increased plasma volume or to decreased plasma transit time. An increased plasma volume would have no effect on the *fraction* of capillary exchangeable hormone, but would decrease the concentration of total hormone in plasma (L_T°) and, thus, would decrease the capillary exchangeable hormone in proportion. However, if organ blood flow is increased owing to decreased plasma transit time (that is, increased k_{10}), then the R fraction would decrease (equation #3) and the concentration of capillary exchangeable hormone would increase (see equation #2).

5   CYTOPLASMIC HORMONE BINDING

The values for the dissociation constant of testosterone binding to brain cytoplasmic binding systems (K_P) has not been measured directly *in vivo*. However, the P_T/K_P ratio, where P_T = the concentration of cytoplasmic binding systems, has been measured in rat brain *in vivo* and is 0.67 for testosterone (Pardridge, Moeller, Mietus, and Oldendorf 1980). Moreover, the rate constant of testosterone dissociation from the brain cytosolic protein has also been measured directly, that is, $k_6 = 0.03\ \mathrm{sec}^{-1}$ (Pardridge *et al.* 1980*b*). Because $K_P = k_6/k_5$ and a typical $k_5 = 10^7\ \mathrm{M}^{-1}\ \mathrm{sec}^{-1}$ (Westphal 1978), then $K_P = 3\ \mathrm{nM}$ and $P_T = 2\ \mathrm{nM}$ (Table 6.1). If k_5 were arbitrarily increased or decreased, then both K_P and P_T would change in proportion to each other. The individual k_5 and k_6 rate constants are not used in the mathematical model, as only the k_6/k_5 ratio is important (Pardridge and Landaw 1985).

The cytoplasmic binding systems should not be confused with nuclear hormone receptors, since the sequestration of hormone by the nuclear receptor is probably trivial compared to cytoplasmic binding. It is known that both testosterone and oestradiol are avidly bound by cytoplasmic fatty acid binding proteins such as ligandin or Z-protein (Ketterer, Tipping, and Hackney 1976; Ketterer, Carne, and Tipping 1978).

Owing to the cytoplasmic binding, the volume of distribution of steroid hormone may be greatly in excess of the cellular water volume, and this explains why the tissue concentration of steroid hormones is oftentimes many-fold greater than the total plasma concentration of hormone. The volume of distribution (V_D) is given by

$$V_D = \frac{(L_M + PL)V_T}{L_T^\circ} \tag{6}$$

where PL = the pool of cytoplasmic-bound steroid hormone (Fig. 6.2). Therefore, hormone volume of distribution is a function of both plasma and cytoplasmic binding systems, membrane transport (k_3, k_4), and organ metabolism (k_9). While a primary increase in MCR (for example, owing to an increase in k_9) may result in a decreased hormone volume of distribution, the reverse is not true. A primary increase in hormone volume of distribution (for example, owing to either increased P_T or decreased K_P) is predicted to have no effect on the concentration of cytoplasmic free hormone or MCR (Pardridge and Landaw 1985).

6   ORGAN METABOLISM

The rate constant of tissue metabolism of steroid hormone is given by k_9 (Table 6.1). If metabolic clearance of steroid hormone by a given organ is essentially nil, then $k_9 = 0$ and $R = 0$ (equation #3). When metabolic clearance by an organ is high, then k_9 is much greater than k_4 and R reduces to $k_3/k_{10} = k_3 t$. Under these conditions, steroid hormone transport into the organ from plasma is rate limiting for overall hormone metabolic clearance and the steady state MCR is equal to the unidirectional clearance determined by the tissue sampling single injection technique. Although the use of the single injection technique and measurements of hormone undirectional clearance are the best estimate of *in vivo* plasma protein binding effects, measurements of unidirectional clearance may also predict hormone MCR if hormone metablism is rate limited by hormone transport. That this is, in fact, the case, is suggested by the good correlation between androgen MCR in humans and androgen unidirectional clearance measurements in rats. For example, the ratio of testosterone to dihydrotestosterone MCR in humans is 1.2–1.9 (Vermeulen and Ando 1979), and the ratio of unidirectional testosterone clearance to unidirectional dihydrotestosterone clearance by rat brain is 0.8–2.0 (Pardridge, Mietus, Frumar, Davidson, and Judd 1980). Other studies have shown that oestradiol MCR in the rat is rate limited by the transport of oestrogen out of the plasma compartment (Larner and Hochberg 1985).

7   STEADY STATE MODEL PREDICTIONS

Table 6.3 shows the predicted concentrations of testosterone in the various tissue pools (Fig. 6.2) for both the free hormone model (simulations #2 and #11) and the protein-bound model (simulations #1 and #3–10). In simulations #1–9, the rate constant of testosterone metabolism (k_9) is set at 0 and k_9 is set at 0.1 for simulations #10 and #11. Simulation #1 is the basal state and is described by the parameters listed in Table 6.1. These results show that the concentration of cytoplasmic free hormone is equal to the concentration of capillary exchangeable hormone but is nine-fold greater than the free hormone concentration (Fig. 6.5). Simulation #2 is the situation wherein the albumin dissociation constant (k_7/k_8) in the capillary is identical to the albumin dissociation constant *in vitro*, and this results in the concentration of cytoplasmic free hormone equalling the concentration of free hormone *in vitro*. Simulation #3 represents the case in which blood flow is increased five-fold owing to a five-fold increase in the velocity of capillary

Table 6.3
Steady state model predictions

Simulation No.	Parameter* Change	Arterial†				Capillary			Cytosolic		
		G_L^0 nM	A_L^0 nM	L_F^0 nM	GL nM	AL nM	L_F nM	L_M nM	PL nM	V_D 1/kg	E^{ss}
1	basal	4.7	4.9	0.40	5.7	0.88	3.4	3.4	1.3	0.33	0
2‡	$k_7 = 53\ s^{-1}$	4.7	4.9	0.40	4.7	4.9	0.40	0.40	0.30	0.049	0
3	$k_{10} = 5\ s^{-1}$	4.7	4.9	0.40	5.0	1.0	4.0	4.0	1.4	0.38	0
4	$k_1 = 3\ s^{-1}$	4.7	4.9	0.40	1.8	1.7	6.5	6.5	1.7	0.58	0
5	$G_T^0 = 50\ nM$	6.3	3.5	0.29	7.4	0.54	2.1	2.1	1.0	0.22	0
6	$P_T = 25\ nM$	4.7	4.9	0.40	5.7	0.88	3.4	3.4	13.3	1.17	0
7	$G_T^0 = 50\ nM$	8.1	1.5	0.39	8.7	0.10	1.2	1.2	0.73	0.14	0
8	$A_F = 200\mu M$; $k_1 = 3\ s^{-1}$; $G_T^0 = 50\ nM$; $A_F = 200\ \mu M$	8.1	1.5	0.39	3.1	0.51	6.3	6.3	1.7	0.56	0
9	$G_T^0 = 0$	0	9.2	0.76	0	2.0	8.0	8.0	1.8	0.68	0
10	$k_9 = 0.1\ s^{-1}$	4.7	4.9	0.40	5.2	0.45	1.8	0.38	0.28	0.046	0.26
11‡	$k_7 = 53\ s^{-1}$; $k_9 = 0.1\ s^{-1}$	4.7	4.9	0.40	4.7	4.4	0.36	0.078	0.063	0.0098	0.054

*Basal parameters are listed in Table 6.1. Simulations #2–11 include basal parameters plus the respective change in individual parameters for each simulation.
†Because the total ligand concentration is 10 nM (Table 6.1), the percent albumin-bound, globulin-bound, or free hormone in the arterial plasma may be computed by multiplying GL^0, AL^0, and L_F^0, respectively, by 10. ‡These simulations assume hormone-binding protein dissociation constant in the microcirculation is identical to the *in vitro* constant. (From Pardridge and Landaw 1985 with permission).

FREE TESTOSTERONE
CONCENTRATIONS IN BRAIN

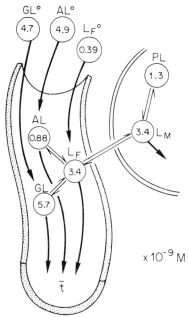

Fig. 6.5 Predicted steady state concentrations of testosterone in the various pools of the brain capillary and in brain cells that are depicted in Fig. 6.2. The pool sizes shown in this figure represent the basal state, which is simulation #1 in Table 6.3. The concentration of free cytosolic testosterone in brain cells is predicted to approximate the concentration of albumin-bound hormone in the circulation and to be more than ten-fold greater than the concentration of free hormone (L_F^o). (From Pardridge and Landaw 1985 with permission).

plasma flow (that is, five-fold decrease in capillary transit time). In the case of zero metabolism, intracellular hormone concentrations are increased 18 per cent over the basal state, which indicates both the capillary exchangeable and cytosolic free hormone are a function of the capillary transit time.

Simulation #4 is the case of enhanced dissociation of testosterone from the SHBG binding site within the brain microcirculation. In this situation, testosterone is operationally available for transport out of the brain capillary from the circulating SHBG-bound pool without significant exodus of the globulin *per se*. In this case, the concentration of cytoplasmic free testosterone is increased approximately two-fold above the concentration found in simulation #1. Simulation #5 represents the case in which the concentration of SHBG is nearly doubled and, in this case, the systemic free hormone (L_F^o), systemic albumin-bound hormone (A_L^o), capillary exchangeable hormone (L_F), and cytoplasmic free hormone (L_M) are all decreased proportionately relative to the basal state. Simulation #6 shows that a ten-fold increase in cellular hormone-binding protein results in a ten-fold increase in the concentration of cytoplasmic-bound hormone, but there is no change in the concentration of cytoplasmic free hormone. The concentration of bound hormone is directly related to the concentration of cytosolic binding sites and is inversely related to the dissociation constant (K_P).

Simulation #7 represents a case such as cirrhosis wherein the concentration of SHBG is increased and albumin concentrations are decreased (Sakiyama, Pardridge, and Judd 1982). In this case, the cytoplasmic free hormone is decreased to 35 per cent of control levels, although there is no change in the concentration of free hormone. On the other hand, if the SHBG concentration is increased, albumin concentration is decreased, and SHBG-bound testosterone is available for uptake by the tissue (for example, owing to an increase in k_1 over the basal state), then the concentration of cytoplasmic free hormone is actually increased. Simulation #9 represents the case wherein SHBG concentrations are zero. The concentration of free cytoplasmic testosterone is predicted to be ten-fold greater than the free hormone concentration and to approximate the concentration of albumin-bound testosterone.

Simulation #10 gives the hormone pool sizes when the metabolism of testosterone by the tissue is moderate, for example, $k_9 = 0.1$. Under these conditions, the steady state net extraction (E^{SS}) is 26 per cent. A marked concentration gradient exists across the cell membrane since the concentration of free cytoplasmic hormone is only 21 per cent of the concentration of plasma capillary exchangeable hormone. Moreover, increased hormone metabolism results in an 86 per cent decrease in hormone volume of distribution (Table 6.3). Simulation #11 represents the free hormone model (that is, no enhanced dissociation of testosterone from albumin-binding sites *in vivo*) and gives the pool sizes when $k_9 = 0.1$. In this situation, E^{SS} and V_D are approximately five-fold less than the values predicted for the protein-bound hormone model (simulation #10).

Simulations #2 and #11 expose two fundamental flaws of the free hormone hypothesis. First, if only free hormone is available for uptake by tissues, then the concentration of free cytosolic hormone does not exceed the concentration of free hormone measured *in vitro*, and this concentration is at least ten-fold lower than that needed to cause 50 per cent occupancy of nuclear steroid hormone receptors (Fig. 6.1). Second, in an organ of net metabolic clearance of testosterone (simulation #11), the net steady state extraction (E^{SS}) barely exceeds the dialyzable fraction and, thus, the high extraction values for testosterone in organs such as liver could not be generated (Tait and Burstein 1964).

The comparison of simulations #1 and #10 show that the free cytosolic concentration of testosterone in an organ such as brain, where hormone metabolism is minimal (simulation #1), is nearly tenfold freater than the free

Table 6.4

Species diversity of serum SHBG or TeBG

SHBG	TeBG	None
humans	rabbit	rat
baboon	dog	pig
chimpanzee	goat	horse
	bull	

SHBG = sex hormone-binding globulin (binds both testosterone and oestradiol with high affinity); TeBG = testosterone-binding globulin (binds only testosterone with high affinity). (From Corvol and Bardin 1973; Renoir, Mercier-Bodard, and Baulieu 1980).

cytosolic testosterone concentration in organs such as liver (simulation #10) that actively metabolizes testosterone. However, this discrepancy in free cytosolic hormone concentration could be rectified if, in the organ of high steroid metabolism, there is transport of globulin-bound steroid hormone into the tissue (for example, simulation #4). As will be discussed in Section **IV (Organ specificity)**, SHBG-bound oestradiol is available for transport into metabolic organs such as liver or salivary gland, and SHBG-bound testosterone is available for transport into tissues such as the testis or prostate. Thus, the selective delivery of steroid hormones to tissues by albumin and by SHBG allows for maintenance of high cytosolic concentrations of free hormone in the face of active organ metabolism of the hormone.

III   Biochemical model

1   SHBG

Human SHBG is an 88,000 Dalton heterodimer that is composed of a 47,000 Dalton heavy chain and a 41,000 Dalton light chain (Cheng, Musto, Gunsalus, and Bardin 1983; Joseph, Hall, and French 1987). The protein electrophoreses in the β-globulin fraction, as opposed to the α-globulin fraction wherein CBG or α-fetoprotein migrate (Swartz and Soloff 1974). SHBG in humans and primates binds both testosterone and oestradiol with high affinity (Table 6.4). In contrast, the globulin is primarily a testosterone-binding globulin (TeBG) in species such as the bull, goat, dog, or rabbit. The serum concentration of SHBG or TeBG is essentially negligible in species such as the rat, pig, horse, donkey, cat, or bird (Table 6.4). However, in the rat, and presumably other species, the testes secretes androgen-binding protein (ABP) and small amounts of this protein are secreted into the bloodsteam so that the circulating rat ABP concentration is about 1 nM (Gunsalus, Musto, and Bardin 1978). Recent studies have shown a 68 per cent homology between the amino acid sequences of human SHBG and rat ABP (Joseph *et al.* 1987).

Both the heavy and light chains of rat ABP are encoded by a single copy gene and a 1600 base pair mRNA transcript (Joseph *et al.* 1987). The heavy and light chains are formed by posttranslational modification of a 45,000 Dalton precursor. It is believed that a 4000 Dalton signal peptide is cleaved from the precursor to generate the 41,000 Dalton light chain and that the heavy chain is formed by the addition of carbohydrate moieties on asparagine 274 and asparagine 397 residues (Joseph *et al.* 1987). The estimates of the carbo-hydrate moiety range from 10 to 30 per cent of the total protein molecular weight (Westphal 1978). Heterogeneity in the carbohydrate added to the SHBG polypeptide accounts for several SHBG isoforms seen on isoelectric focusing gels (Petra, Stanczyk, Senear, Namkung, Novy, Ross, Turner, and Brown 1983). There appear to be species differences in the posttranslational modification of the protein, since the rabbit TeBG isoforms are more acidic than the human SHBG isoforms (Petra *et al.* 1983). These structural species differences may explain the functional species differences when comparing the activity of human SHBG or rabbit TeBG in delivering steroid hormones to tissues (see Section V. Species Differences).

2   ALBUMIN

Albumin is a 66,000 Dalton single polypeptide that is normally not glycosy-lated (Brown 1977). The 3-dimensional structure of albumin has not yet been determined since crystals appropriate for X-ray diffraction cannot be gener-ated with the albumin molecule. However, a model of the 3-dimensional structure of albumin has been deduced from the primary sequence by Brown (1977). Albumin is composed of three major domains that form six hemi-cylinders which are believed to constitute the six primary binding sites on albumin (Fig. 6.6). These binding sites have specific affinities for different classes of substrate. For example, it is known that four of the binding sites selectively bind free fatty acids, bilirubin, warfarin, or indole compounds/ diazepam (Muller.and Wollert 1979). The interior of the hemicylinder acts as the binding site. Albumin is a highly flexible molecule that literally "breathes" as it fluctuates back and forth between conformations (Kragh-Hansen 1981). The binding of ligands to albumin is mediated by an induced conformational fit (Muller and Wollert 1979). That is, the binding of the ligand to albumin causes a conformational change about the binding site that allows the ligand to stay within the binding hemicylinder. The concept that albumin exists in different conformations, depending on the ligand bound to it, has recently been confirmed when it was shown that the isoelectric point of bovine albumin carrying ^{125}I-D-T$_3$ is more acidic than the isoelectric point of bovine albumin carrying ^{125}I-L-T$_3$ (Terasaki and Pardridge 1987*a*). Moreover, stereospecificity of plasma protein-mediated transport of T$_3$ into liver was detected using the *in vivo* single injection technique, since D-T$_3$

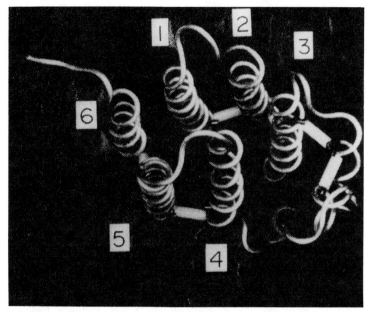

Fig. 6.6   Three-dimensional structure of albumin as deduced from the primary sequence by Brown (1977). Albumin is composed of three domains and six hemicylinders. The ligand binding sites are interiors of the six hemicylinders of the albumin molecule. This model illustrates the high flexibility of albumin and a marked increase in ligand dissociation is expected with a slight uncoiling caused by conformational changes about the binding site. Conformational changes are likely to be ligand- and tissue-specific. For example, steroid or thyroid hormone dissociation from albumin is enhanced in brain capillaries, whereas propranolol or lidocaine dissociation is not. Conversely, the dissociation of these two drugs from albumin is markedly enhanced in the liver microcirculation. (See Pardridge 1987).

bound to albumin was only partially available for uptake by liver. In contrast, albumin-bound L-T_3 was freely available for transport into liver (Terasaki and Pardridge 1987a).

Thermodynamic studies of ligand binding to albumin have shown that the free energy (ΔG), that is, the binding dissociation constant, changes little as the temperature is increased from 8°C to 37°C, and that enthalpy (ΔH) plays a minimal contribution in the overall free energy of the albumin-ligand interaction (Burton and Westphal 1972). The dominant term is a positive entropy (ΔS) that arises from a disordering of the water lattice surrounding amino acid residues comprising the interior of the binding site. This disordering of the water structure and the positive entropy is generated by the conformational change associated with ligand binding within the interior of the albumin binding site (Fig. 6.6). Moreover, relatively minor conformational changes about the ligand binding site can have rather drastic effects on the free energy and the dissociation constant of albumin binding of a particular

Table 6.5

Comparison of bovine albumin and human α_1-acid glycoprotein (AAG) dissociation constant in vivo in brain capillary (K_D^a) and in vitro

Plasma Protein	Ligand	K_D (μM)	K_D^a (μM)
Bovine albumin	testosterone	53 ± 1	2520 ± 710
	tryptophan	130 ± 30	1670 ± 110
	corticosterone	260 ± 10	1330 ± 90
	dihydrotestosterone	53 ± 6	830 ± 140
	oestradiol	23 ± 1	710 ± 100
	propranolol	290 ± 30	220 ± 40
	bupivacaine	141 ± 10	211 ± 107
	T_3	4.7 ± 0.1	46 ± 4
Human AAG	propranolol	3.3 ± 0.1	19 ± 4
	bupivacaine	6.5 ± 0.5	17 ± 4

T_3 = triiodothyronine. (From Pardridge and Landaw 1984; Pardridge 1986*b*; Terasaki, Pardridge, and Denson 1986).

steroid hormone or any other ligand. Moreover, this principle is true for other proteins such as SHBG or even enzymes. For example, recent X-ray diffraction studies showed that the loss of a single hydrogen bond between the ligand and an enzyme binding site results in 2–3 log order differences in binding dissociation constant (Bartlett and Marlowe 1987).

A conformational change about the albumin binding site (caused by interactions with the organ microcirculation, see below) would be expected to cause a partial uncoiling about the binding site that would result in a markedly increased rate of ligand dissociation (Pardridge 1987). The enhanced dissociation can be quantified *in vivo* for a particular organ by using tracer kinetic models to estimate the apparent dissociation constant of ligand-albumin interaction (Table 6.5). These experiments have shown that the conformational changes about the albumin binding site may be restricted to particular regions of the albumin molecule (Fig. 6.6), since some albumin-bound ligands do not undergo enhanced dissociation in some organs. For example, the *in vivo* dissociation constant (K_D^a) of lipophilic amine drug binding to bovine albumin is not significantly different from the K_D measured *in vitro* (Table 6.5). Secondly, conformational changes about a particular binding site may differ among organs (Table 6.2).

3   PLASMA PROTEIN-MEDIATED TRANSPORT MECHANISMS

There are at least three possible mechanisms by which plasma protein-bound hormone may be transported into tissues *in vivo*. First, the plasma protein-ligand complex, *per se*, may undergo net transport through the cellular barrier lining the microcirculation, for example, the capillary endothelium in most tissues or the hepatocyte plasma membrane in liver. Siiteri *et al.* (1982) have hypothesized this mechanism with little supportive experimental data. However, Rosner and co-workers (1985) have recently provided evidence for an SHBG receptor on human prostatic cell membranes, and Bordin and Petra (1980) have shown that SHBG is found intracellularly in several tissues of the monkey, for example, prostate, testis, adrenal, and liver, but not other organs. Therefore, if receptor-mediated uptake of SHBG does occur, this process may be restricted to a few organs. There are several observations that mitigate against receptor-mediated uptake of the binding protein *per se*, as the principle mechanism underlying the transport of plasma protein-bound hormones into tissues *in vivo*. First, it is known that the rate of egress of proteins from the plasma compartment is log orders slower than the rate of exodus of steroid hormones and other ligands from the circulating plasma protein-bound pool (Dewey 1959). Although a substantial amount of plasma protein is found extravascularly, this plasma protein resides in the lymphatic compartment and not in the tissue interstitial space (Reeve and Chen 1970). (The low concentration of plasma protein in the interstitium arises from the rapid uptake of plasma protein into the lymphatic compartment, compared to the relatively slow exodus of plama protein from the plasma compartment into the interstitium.) Second, previous studies have shown that the uptake of circulating binding globulins, such as CBG, by organs such as rabbit brain or rabbit uterus is immeasurably low compared to the transport of corticosterone from the circulating rabbit CBG-bound pool (Pardridge, Eisenberg, Fierer, and Kuhn 1986; Chaudhuri, Steingold, Pardridge, and Judd 1987). Similarly, the uptake of labelled albumin by tissues is trivial compared to the uptake of ligands from the circulating albumin-bound pool (Pardridge, Eisenberg, and Cefalu 1985). Therefore, while the receptor-mediated uptake of plasma proteins by tissues undoubtedly occurs, this process operates at rates that are log orders lower than the movement of circulating sex steroids into tissues from the plasma protein-bound pools.

There are two possible mechanisms by which plasma protein-bound ligands may be transported out of the plasma compartment without significant exodus of the plasma protein *per se*. First, the permeability of the ligand, when presented to the membrane in the bound form, may be much greater than the permeability of the membrane to the free ligand. Second, tissue-mediated conformational changes at the binding site may allow for markedly enhanced rates of ligand dissociation from the plasma protein within the organ microcirculation as compared to dissociation rates that exist in the unperturbed or *in vitro* state. Previously reported tracer kinetic modelling studies have shown that the enhanced membrane permeability mechanism cannot explain the experimental data, whereas the enhanced dissociation model does allow for a fit of the experimental data to the tracer kinetic model (Pardridge and Landaw 1984). Thus, the important parameter is the dissociation constant of the plasma protein-ligand binding interaction *in vivo* in a particular organ (Tables

6.2 and 6.5). The enhanced dissociation model also fits with the known thermo-dynamics and biophysics of albumin-ligand binding interactions and the importance of conformational changes about the ligand binding site in determining the rates of ligand dissociation from the plasma protein (see above).

The biochemical mechanism by which plasma proteins such as albumin or SHBG rapidly and transiently interact with the surface of the organ microcirculation to cause conformational changes about the binding site is at present unknown. Other workers have postulated the operation of receptors for plasma proteins such as albumin lining the organ microcirculation (Weisiger, Golland, and Ockner 1981; Forker and Luxon 1983). These putative receptors would transiently immobilize the circulating plasma protein, cause conformational changes, and increase rates of ligand dissociation from the plasma protein. However, the initial report of an albumin receptor has not been confirmed (Stollman, Gartner, Theilmann, Ohmi, and Wolkoff 1983; Stremmel, Potter, and Berk 1983). On the other hand, it is known that albumin normally transiently interacts with the surface of the organ microcirculation, possibly through electrostatic mechanisms involving positively charged residues (for example, arginine) on the plasma protein surface with negative charges on the surface of the organ microcirculation (Michel, Phillips, and Turner 1985). Indeed, the adsorption of plasma proteins such as albumin to the surface of the microcirculation as plasma traverses the organ is an important mechanism by which normal vascular permeability is maintained (Turner, Clough, and Michel 1983). Finally, recent biochemical studies using spin-labelling techniques have provided direct evidence for conformational changes in plasma proteins such as albumin that are mediated by interactions between the binding protein and the cells comprising the surface of the organ microcirculation (Mizuma, Horie, Hayashi, and Awazu 1986). The biochemical nature of the constituents lining the microcirculatory surface that interact with albumin and other plasma proteins is not known at present but may represent membrane carbohydrate or glycolipid moieties. Since it is known that the composition of carbohydrate moieties that line the endothelial glycocalyx differs from organ to organ (Ponder and Wilkinson 1983), it would be predicted that organ differences might exist in terms of plasma protein-mediated transport of steroid hormones and other ligands into tissues *in vivo*.

IV   Organ specificity of sex steroid transport

1   BRAIN

The *in vivo* dissociation constant (K_D^a) of bovine albumin binding of testosterone in the rat brain capillary was determined by fitting the unidirectional extraction data shown in Fig. 6.7 to the classical Kety-Renkin-Crone equation of capillary physiology (Pardridge and Landaw 1984),

$$E = 1 - e^{-f \cdot k_3 t} \tag{7}$$

$$f = \frac{K_D^a}{A_F + K_D^a} \tag{8}$$

where E = the first pass unidirectional extraction of ligand by the tissue, k_3 = the membrane permeability constant (Table 6.1), t = the capillary transit time, and A_F = the albumin concentration. The use of the Kety-Renkin-Crone equation to estimate the K_D^a parameter assumes maintenance of near equilibrium of the plasma protein-ligand binding reaction as the plasma protein traverses the organ microcirculation (Pardridge and Landaw 1984). This proviso only requires that the rates of ligand dissociation and/or association with the plasma protein are much faster than the rates of ligand movement through the endothelial membrane. Since testosterone dissociation from albumin (k_7, Table 6.1) ranges from 53–5300 sec^{-1} and since the k_8 (A_F) product (Table 6.1) ranges from 100–10 000 sec^{-1}, then the assumption of maintenance of near equilibrium *in vivo* is valid because the rate constant of testosterone transport through the brain capillary endothelium is only 1.9 sec^{-1} (Table 6.1).

The data in Figure 6.7 indicate that albumin-bound testosterone is readily available for transport through the rat brain capillary wall (Pardridge 1981). The transport of testosterone or oestradiol into brain from the circulating human SHBG-bound pool was assayed using serum obtained from nearly 80 patients representing seven different clinical conditions (Pardridge, Mietus, Frumar, Davidson, and Judd 1980). These seven conditions differed in the serum concentration of SHBG (Table 6.6). As the SHBG concentration decreased, the unidirectional clearance of either oestradiol or testosterone increased (Fig. 6.8). In addition, the oestradiol/testosterone unidirectional clearance ratio decreased as the concentration of SHBG decreased (Table 6.6). Thus, SHBG acts as an oestradiol amplifier relative to testosterone delivery to tissues, and the *in vivo* data in Figure 6.7 support the original concept put forward by Anderson (1974) regarding the role of SHBG as an oestradiol amplifier. The close relationship between oestradiol or testosterone clearance and the SHBG concentration is shown in Figure 6.9, which describes the linear relationship between the unidirectional clearance and the SHBG concentration for either of the two sex steroids. The relationship between steroid clearance and SHBG concentration is five-fold greater for testosterone as compared to oestradiol (see slopes of the plots, Fig. 6.8).

The data in Figures 6.7–6.9 indicate that only albumin-bound oestradiol or testosterone is available for transport into brain, whereas SHBG-bound sex steroid is not available. Since these studies were some of the first results published with the tissue sampling single injection technique, there is a tend-

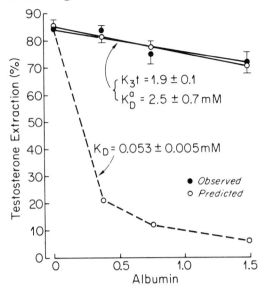

Fig. 6.7   The unidirectional extraction of ^{3}H-testosterone by rat brain is plotted versus the concentration of arterial bovine albumin. The experimentally observed values are given by the closed circles (mean ± S.E., n = 3–6 animals per point). The extraction values predicted on the basis of fitting the experimental data to equation #7 are shown by the open circles and the curve fitting gives the two parameters k_3t and K_D^a (see equation #7). The parameters k_3 and t are defined in Fig. 6.2 and Table 6.1, where $t = 1/k_{10}$. The dashed line represents the extraction values predicted by substituting into equation #7 the albumin concentration, the k_3t product, and the *in vitro* albumin-testosterone dissociation constant, $K_D = 53 ± 1 \mu M$. Therefore, the dashed curve gives the expected inhibition of testosterone transport caused by hormone binding to albumin *if* testosterone was not available for transport into brain from the circulating albumin-bound pool. However, since albumin-bound testosterone is available via an enhanced dissociation mechanism, the upper curve is observed and the K_D^a *in vivo* (2500 ± 700 μM) is much greater than the K_D *in vitro*. (From Pardridge 1986b with permission).

ency to extrapolate these data to other organs and to make the general conclusion that while albumin-bound sex steroids are available for uptake by tissues, SHBG-bound sex steroids are not available for transport into tissues. However, as other organs have been studied directly in subsequent experiments, it has become clear that such generalizations are false and there are marked organ specificities in terms of the transport of albumin or SHBG-bound sex steroid into tissues *in vivo*.

2   LIVER

Albumin-bound testosterone or oestradiol is freely available for transport

Table 6.6
Oestradiol amplifier function of SHBG[a]

Patient category	SHBG (nM)	E_2/T unidirectional clearance ration
Pregnancy (9)	323 ± 28	3.9 ± 0.5
Oral contraceptives (8)	126 ± 16	2.6 ± 0.3
Thin postmenopausal (9)	74 ± 9	2.1 ± 0.2
Normal female (9)	65 ± 9	2.1 ± 0.2
Obese postmenopausal (9)	43 ± 4	1.7 ± 0.2
Normal male (8)	28 ± 3	1.3 ± 0.1
Hirsute (5)	17 ± 2	1.3 ± 0.1

[a]From Pardridge (1981) with permission. Data are mean ± S.E.M. E_2, oestradiol, T, testosterone. Numbers in parentheses are number of patients.

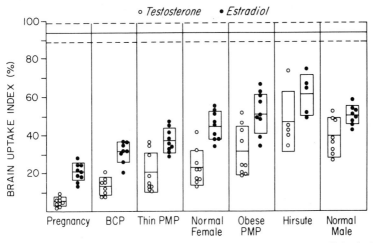

Fig. 6.8   Brain uptake index (BUI) for ^{3}H-testosterone and ^{3}H-oestradiol relative to ^{14}C-butanol is shown for 5–9 patients in seven clinical conditions. Vertical rectangles are mean ± S.D.; horizontal line is mean of testosterone or oestradiol BUI in absence of plasma proteins. BCP, birth control pill-treated women; PMP, postmenopausal women. (From Pardridge *et al.* 1980*a* with permission).

into liver (Baird, Longcope, and Tait 1969; Pardridge and Mietus 1979*a*). However, studies using human serum indicated that SHBG-bound oestradiol is also readily available for uptake by liver (Pardridge and Mietus 1979*a*). As shown in Figure 6.10, the concentration of bioavailable testosterone in human male serum in brain or liver is not significantly different from the frac-

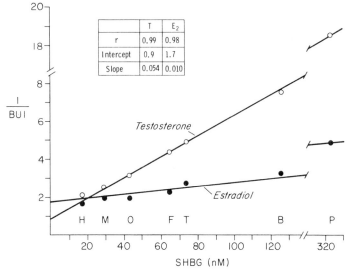

Fig. 6.9 Reciprocal of brain uptake index (BUI) for ³H-testosterone (T) and ³H-oestradiol (E₂) relative to ¹⁴C-butanol is plotted versus level of sex hormone-binding globulin (SHBG) in human serum. BUI data are shown in Fig. 6.8. P—pregnancy, B—birth control pills, T—thin postmenopausal female, F—normal follicular phase female, O—Obese postmenopausal female, M—normal male, H—hirsute female. Data obtained by linear regression are shown in inset for both plots. (From Pardridge *et al.* 1980*a* with permission).

tion of free plus albumin-bound hormone measured *in vitro*. Similarly, the concentration of brain bioavailable oestradiol or corticosterone is no different from the free plus albumin-bound steroid fraction *in vitro*. However, the hepatic bioavailable fraction of either oestradiol or corticosterone is markedly increased over the free plus albumin-bound fraction, indicating SHBG-bound oestradiol or CBG-bound corticosteroid is readily available for transport into rat liver *in vivo*. Thus, it is believed that transient interactions between SHBG and the hepatocyte surface cause conformational changes about the oestradiol binding site that result in markedly increased rates of oestradiol dissociation from SHBG within liver capillaries. Similarly, thyroid hormone-binding globulin (TBG)-bound T_3 is readily available for transport into rat liver, whereas human TBG-bound T_4 is not available (Pardridge 1981). The selective transport of SHBG-bound oestradiol or TBG-bound T_3 into liver would appear to be at odds with the conventional view that testosterone or oestradiol bind to a single competitive binding site on SHBG, or that T_3 or T_4 bind to a single competitive binding site on TBG. One would have to postulate an unusual conformational change that would allow for selective enhancement of oestradiol dissociation from the binding site without enhanced dissociation of testosterone. The heterogeneity of SHBG or TBG function in terms of hormone delivery to liver, however, also has an

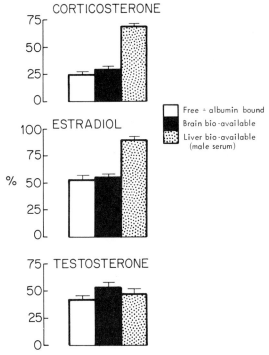

Fig. 6.10 The *in vitro* free plus albumin-bound, *in vivo* brain bioavailable, and liver bioavailable fractions for human male serum are shown for three steroid hormones. (From Pardridge 1986*b* with permission).

analogue in terms of structural heterogeneity of either the SHBG or TBG molecule. Recent studies have shown that the multiple isoforms of TBG in human serum do not bind T_3 and T_4 equally (Terasaki and Pardridge 1988). The most acidic isoforms of TBG selectively bind T_4 as compared to T_3. Whether a similar heterogeneity exists amongst the different SHBG isoforms in human serum in terms of selective binding of testosterone and oestradiol is not known at present but is being investigated in the author's laboratory.

The finding that SHBG-bound oestradiol is selectively available for transport into liver as compared to SHBG-bound testosterone provides a second order amplification of oestradiol delivery to tissues caused by the presence of SHBG in plasma. The first order of amplification is shown by the brain studies (Table 6.6) and arises from the combination of increased binding of oestradiol to albumin as compared to testosterone, and decreased binding of oestradiol to SHBG as compared to testosterone (Anderson 1974). Therefore, the albumin-bound fraction of testosterone decreases faster as SHBG increases. This allows for amplification of oestradiol delivery to tissues, even if SHBG-bound oestradiol is not available *per se*. However, in liver further amplification of oestrogen delivery occurs because the binding globulin selectively delivers this sex steroid to liver as compared to the androgen (Pardridge 1981).

3   SALIVARY GLAND

Since recent trends in endocrine practice involve the use of measurements of steroid hormone concentrations in human saliva as an index of the biologic-ally active steroid hormone in the circulation (Riad-Fahmy *et al.* 1982), studies were initiated to measure steroid hormone delivery mechanisms in rat salivary gland. This was felt important because other studies had shown that the concentration of testosterone in human saliva closely parallels the con-centration of free testosterone *in vitro* (Riad-Fahmy *et al.* 1982). Such a correlation is compatible with one of two explanations. First, only free testo-sterone may be available for transport across salivary gland capillaries (Riad-Fahmy *et al.* 1982). A second possible explanation, however, is that free plus albumin-bound testosterone is available for transport into salivary gland epithelium from the circulation, but the pool size of cytosolic free androgen in salivary gland is markedly restricted compared to the pool size of capillary exchangeable hormone owing to rapid androgen metabolism in this organ (for example, see simulation # 10, Table 6.3). As shown in Table 6.2, albumin-bound testosterone is partially available for transport into salivary gland as the rate of androgen dissociation from albumin is increased approximately twelve-fold in this tissue (Cefalu, Pardridge, Chaudhuri, and Judd 1986). Moreover, the salivary gland bioavailable testosterone using human serum approximates the free plus albumin-bound fraction measured *in vitro* (Figure 6.11). Thus, the marked discrepancy between the concentra-tion of free cytosolic testosterone in salivary gland (which is presumed to be in equilibrium with the salivary fluid testosterone), and the capillary exchangeable testosterone concentration suggested that rapid metabolism of testosterone in salivary tissue occurs. This prediction was confirmed by studies in which ^{3}H-testosterone was injected into the common carotid artery of rats followed by removal of the organ 60 seconds later and rapid process-ing of the tissue homogenate for thin-layer chromatography (Cefalu *et al.* 1986). As shown in Figure 6.12, testosterone is rapidly metabolized by saliv-ary gland tissue as compared to two other organs, lymph node or brain, in which no measurable metabolism of androgen within 60 seconds of pulse ad-ministration can be detected. Two-dimensional thin-layer chromatography studies showed that the major metabolite formed is androstenedione at 60 seconds and is 5α-androstane-3α,17β-diol (3α-diol) at 5 minutes after a single carotid injection of ^{3}H-testosterone (Cefalu *et al.* 1986). These studies sug-gest that enzymes such as 17β-hydroxysteroid dehydrogenase (which inter-converts testosterone and androstenedione), 5α-reductase (which converts testosterone to dihydrotestosterone or androstenedione to androstanedione), and 3α-hydroxysteroid dehydrogenase (which converts androstanedione to

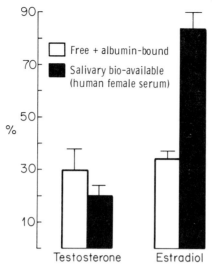

Fig. 6.11   The fractions of free plus albumin-bound testosterone and oestradiol *in vitro*, and testosterone and oestradiol bioavailable in rat salivary gland *in vivo* are compared and show that SHBG-bound oestradiol, but not testosterone, is readily available for transport into rat salivary gland. (From Pardridge 1986*b* with permission).

androsterone or dihydrotestosterone to 3α-diol) are present in salivary gland tissue.

The results in Figure 6.12 showing the organ diversity of androgen metabolic rates underscores the limited utility of MCR measurements on the whole organism. The whole body MCR is the average of interorgan differences in androgen metabolism and these differences can be profound. Tissues such as liver, prostate gland, or salivary gland rapidly metabolize androgen, and this situation may create marked concentration differences between the capillary exchangeable hormone and the cytosolic free hormone. On the other hand, organs such as brain, lymph node, and others metabolize testosterone at markedly reduced rates compared to testosterone transport into and out of the organ. Consequently, in these tissues, the concentration of free cytosolic hormone is nearly equal to the concentration of capillary exchangeable hormone. Moreover, one can envisage profound differences in individual organ MCRs that would not be detected in whole body MCR measurements. In addition to interorgan differences in sex steroid transport, similar interorgan differences most likely occur for the rates of hormone metabolism. Both of these processes, hormone transport from the plasma compartment and intracellular hormone metabolism, are the dominant forces controlling the cytosolic free steroid hormone concentration.

The finding that salivary gland, like liver or prostate, is an organ of active androgen metabolism contradicts the proposal of Riad-Fahmy *et al.* (1982),

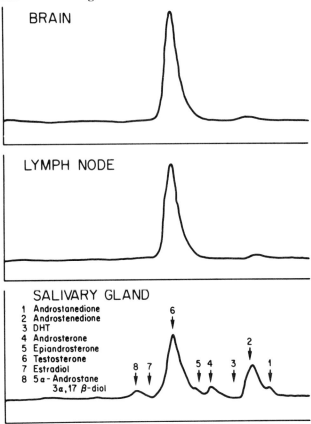

Fig. 6.12   One-dimensional thin-layer chromatographic separation of brain, cervical lymph node, and salivary gland homogenates of tissue obtained 60 seconds after a single carotid injection of ³H-testosterone (50 µCi/ml) in Ringer-Hepes buffer (0.1 g/dl bovine albumin) in the rat. Migration of testosterone or several other metabolites in the one-dimensional system is shown in the figure. The minor peak in the brain and lymph node studies that co-migrated with androstenedione (peak 2) represented an impurity in the isotope, as this was also found in the ³H-testosterone obtained from the manufacturer. (From Cefalu *et al.* 1986 with permission).

who proposed that only free testosterone is available for transport into salivary gland because androgen metabolism in this organ is low. These workers proposed that salivary gland is an organ of low androgen metabolism on the finding of low testosterone concentrations in saliva and the assumption that only free testosterone is available for transport into the tissue. As discussed previously, the low concentration of salivary fluid testosterone arises from active androgen metabolism in this organ.

Finally, the data in Figure 6.11 indicate that, like liver, SHBG-bound oestradiol is selectively available for transport into salivary gland, whereas

SHBG-bound testosterone is not (Cefalu *et al.* 1986). The finding of a high bioavailable oestradiol concentration in salivary gland indicates that if free salivary oestradiol concentrations are found to be low relative to the total concentration of serum hormone, it is likely that active organ metabolism of oestradiol occurs in this tissue.

4 LYMPH NODE

Albumin-bound testosterone is partially available for transport into lymph node (Cefalu and Pardridge 1987). However, the *in vivo* dissociation constants listed in Table 6.2 indicate the dissociation of testosterone from bovine albumin is only enhanced six-fold in lymph node capillaries as compared to a twelve-fold or fifty-fold enhancement of testosterone dissociation in salivary gland or cerebral capillaries. Conversely, oestradiol dissociation from albumin in lymph node capillaries is increased sixty-fold, but only about thirty-fold in brain capillaries. These findings indicate that albumin-bound oestradiol is freely available for transport into lymph node capillaries, but albumin-bound testosterone is only partially available for transport. The plasma protein *per se* does not undergo significant exodus from the capillary compartment as shown by the autoradiography data in Figure 6.13, whereas oestradiol rapidly escapes the organ microcirculation and distributes into the total organ water during a single circulatory passage (Cefalu and Pardridge 1987). Therefore, the transport of albumin-bound steroid hormone into the tissue represents a process of enhanced hormone dissociation from the plasma protein within the microcirculation.

With regard to the transport of testosterone or oestradiol into lymph node from the circulating SHBG-bound pool, recent studies have shown that the lymph node is similar to salivary gland or liver. SHBG-bound oestradiol, but not SHBG-bound testosterone, is readily available for transport into lymph node tissue (Cefalu and Pardridge 1987).

While this review is emphasizing organ diversity in regard to sex steroid transport and metabolism, it is also possible that there is a similar diversity in a particular organ over a variety of pathologic or physiologic conditions. For example, recent studies of testosterone transport and metabolism in lymphoid neoplasia have shown that both androgen transport and metabolism are enhanced in the neoplastic condition (Cefalu and Pardridge 1987). Although SHBG-bound testosterone is not available for transport into lymph nodes under normal conditions, SHBG-bound testosterone is partially available for transport into the neoplastic lymph node. Similarly, normal lymph node does not rapidly metabolize testosterone (Fig. 6.12). However, neoplastic lymph node rapidly metabolizes ³H-testosterone and the major metabolites formed 60 seconds after pulse administration of ³H-testosterone are epiandrosterone and androsterone, as well as dihydrotestosterone and 5α-androstane-3α,17β-diol (Cefalu and Pardridge 1987). These

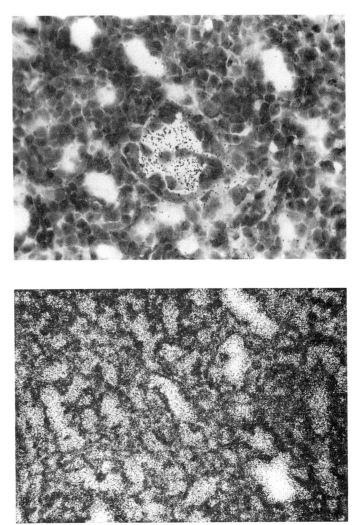

Fig. 6.13   Thaw-mount autoradiogram of control lymph node 15 seconds after a single arterial injection of either ^{3}H-albumin (upper) or ^{3}H-oestradiol (lower) in Ringer-Hepes buffer containing 0.1 g/dl bovine albumin. The albumin radioactivity is confined to the microvasculature compartment, showing that the plasma protein does not measurably cross the capillary wall on a single pass. Conversely, the ^{3}H-oestradiol radioactivity is found throughout the lymph node indicating complete extravascular distribution of the steroid hormone and absence of sequestation to the capillary endothelium. The tissue was counterstained after autoradiography with methyl green-pyronin. Magnification × 100. (From Cefalu and Pardridge 1987 with permission).

studies suggest that pivotable androgen metabolizing enzymes such as 17β-hydroxysteroid dehydrogenase, 5α-reductase, and 3α-hydroxysteroid dehydrogenase are activated in the lymphoid neoplastic state. If the products of testosterone metabolism, for example, dihydrotestosterone or 5α-androstane-3α-17β-diol, mediate in part the mitogenesis of the neoplastic state, then inhibitors of these enzymes, for example, 5α-reductase, may prove in the future to be of beneficial effect in controlling lymphoid neoplasia.

5   TESTIS/PROSTATE GLAND

Recent studies have investigated the bioavailability of testosterone and oestradiol in rat testis (Sakiyama, Pardridge, and Musto 1988). These studies afforded the opportunity to investigate the transport properties of the blood-testis barrier (BTB), which is comprised of the Sertoli cell epithelium surrounding the tubules in the testis (Setchell and Waites 1975). The results indicate both albumin-bound testosterone and oestradiol are readily available for transport into the testis or prostate gland (Sakiyama, Pardridge, and Musto 1988). Similarly, SHBG-bound oestradiol is freely available for uptake by the two organs. The surprising finding, however, was that SHBG-bound testosterone is also readily available for transport into the testis or prostate. The observation that both SHBG-bound testosterone and SHBG-bound oestradiol were transported through the BTB suggested that the SHBG-bound steroid complex may be transported from the testicular microcirculation into the Sertoli cell on a single pass through the organ. This was confirmed by showing the first pass extraction of ^{3}H-TeBG was approximately 75 per cent, as opposed to the extraction of ^{3}H-albumin or ^{113m}In-transferrin (that is, plasma space markers), which was approximately 35 per cent (Sakiyama, Pardridge, and Musto 1988). Thus, in some organs, the rapid uptake of circulating steroid hormones from the circulating SHBG-bound pool may occur via rapid movement of the plasma protein-hormone complex.

6   KIDNEY: SELECTIVE STEROID HORMONE CONJUGATE TRAFFICKING

Steroid hormones are inactivated in part by conjugation to either sulfate or glucuronate moieties (Pan, Woolever, and Bhavnani 1985). Recent studies have shown that oestrone sulfate or oestradiol glucuronate undergo selective trafficking between the plasma, liver, and kidney compartments (Chaudhuri, Verheugen, Pardridge, and Judd 1987). Albumin-bound oestrone sulfate or oestradiol glucuronate are both readily available for transport into liver. Similarly, oestradiol glucuronate is readily available for uptake by kidney from the circulation. However, albumin-bound oestrone sulfate is poorly, if at all, available for uptake by kidney (Fig. 6.14). Thus, the placement of a

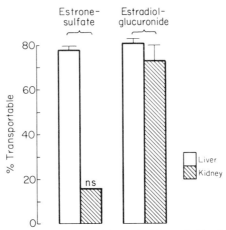

Fig. 6.14 Comparison of kidney and hepatic bioavailable fractions of oestrone sulfate or oestradiol glucuronide after aortic or portal vein injection of the labelled steroid mixed in 4 g/dl bovine albumin. The study shows that albumin-bound sulfate or glucuronate conjugate is readily available for transport into liver but only albumin-bound glucuronate is available for transport in the kidney. Therefore, sulfate conjugation causes selective trafficking of the steroid hormone to the liver. ns = not different from zero. (From Chaudhuri *et al.* 1987).

sulfate moiety on the oestrogen nucleus by peripheral tissues selectively trafficks this oestrogen to the liver, whereas the placement of a glucuronate moiety on the steroid nucleus allows for uptake by either kidney or liver and ultimately allows for either biliary or urinary excretion.

In tissues such as brain or uterus, the placement of a negatively charged sulfate or glucuronate moiety on the sex steroid molecule results in a markedly diminished microvascular permeability to the steroid hormone (Verheugen, Pardridge, Judd, and Chaudhuri 1984; Pardridge, Eisenberg, Fierer, and Musto 1987). Thus, oestrone sulfate, which is in high concentrations in postmenopausal serum (Noel, Ree, Jacobs, and James 1981), is poorly available for uptake by tissues other than liver (Verheugen *et al.* 1984). The high concentration of oestradiol or oestrone in breast cancer tissue (Santner, Feil, and Santen 1984) probably does not arise from the tissue uptake of oestrone sulfate. The low circulating concentrations of unconjugated oestrone or oestradiol may be taken up by breast cancer tissue and then sequestered within that tissue by the formation of oestrone or oestradiol sulfate. This would allow for the development of a large tissue pool of biologically inactive but sequestered oestrogen that could be slowly activated by conversion to free oestrogen due to the action of tissue sulfatases. Similar processes probably also occur in brain for dehydroepiandrosterone sulfate (DHEAS). Studies have shown that brain DHEAS is very high, particularly in olfactory lobe, compared to plasma concentrations (Corpechot, Robel,

Table 6.7

SHBG delivery of sex steroids to tissues: organ diversity in the rat

Organ	Oestradiol	Testosterone
Brain and uterus	−	−
Liver, salivary gland and lymph node	+	−
Testis	+	+

$(-)$ = little, if any, transport of hormone into tissue under normal conditions; $(+)$ = hormone is partially or freely transported into tissue under normal conditions. (From Pardridge and Mietus 1979*a*; Pardridge 1981; Verheugen, Pardridge, Judd, and Chaudhuri 1984; Cefalu, Pardridge, Chaudhuri, and Judd 1986; Cefalu and Pardridge 1987; Sakiyama and Pardridge 1988).

Axelson, Sjovall, and Baulieu 1981). However, DHEAS is poorly transported through the brain capillary endothelium (Pardridge *et al.* 1988). The unconjugated DHEA, however, is rapidly transported and may then be sequestered in brain tissue by formation of the DHEAS conjugate.

7　SUMMARY OF INTER-ORGAN DIVERSITY

Table 6.7 summarizes the inter-organ differences in terms of bioavailability of circulating SHBG-bound testosterone or oestradiol. Neither SHBG-bound testosterone nor oestradiol is available for transport into rat brain or uterus. However, SHBG-bound oestradiol is selectively available for uptake by tissues such as liver, salivary gland, or lymph node, and both SHBG-bound testosterone and SHBG-bound oestradiol are available for uptake by tissues such as testis or prostate. Thus, the presence of SHBG in human serum allows for selective amplification of either oestradiol or testosterone delivery across the entire spectrum of organs in the body. This allows for enhanced delivery of sex steroids to organs that are more sex steroid dependent than others and also allows for maintenance of cytosolic pool sizes of free hormone in the face of active organ metabolism of testosterone or oestradiol.

V　Species differences

The rabbit, unlike the rat, has high serum concentrations of circulating TeBG which selectively bind testosterone and not oestradiol, as opposed to human SHBG, which avidly binds both sex steroids (Table 6.4). Although the rat is a species in which high concentrations of SHBG or TeBG do not exist in the circulation, this species has proved to be a good model for assessing the biological functions of human sex steroid-binding proteins. This is probably true because, as reviewed in **Section III (Biochemical Model)**, spe-

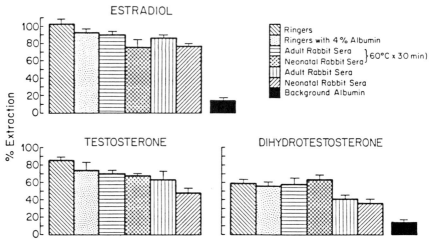

Fig. 6.15   The mean uterine extractions ± S.E.M. of ^{3}H-oestradiol (E$_2$), testosterone (T), dihydrotestosterone (DHT), and ^{3}H-bovine albumin by rabbit uterus are shown for different injection vehicles. Adult rabbit serum, or neonatal rabbit serum in some experiments, was heat treated at 60 C for 30 minutes prior to aortic injection of the serum. These studies show that the presence of rat serum causes no inhibition of the unidirectional extraction of E$_2$, T, or DHT by rabbit uterus as compared to the control injections using Ringer's solution. Therefore, the binding of androgens to TeBG causes no inhibition of the clearance of the hormone, and all three hormones are cleared by the rabbit uterus at approximately the same rate. (From Chaudhuri *et al.* 1988 with permission).

cific albumin or SHBG receptors do not appear to mediate the transport of hormones into tissues from the circulating plasma protein-bound pools in most tissues. Moreover, the rat does have high concentrations of CBG and the same general patterns have been observed for gonadal steroid and corticosteroid transport processes. Nevertheless, recent studies of steroid hormone transport in rabbits have been completed in order to investigate the role of rabbit TeBG or CBG.

The unidirectional clearance of oestradiol, testosterone, or dihydrotestosterone by rabbit uterus is high and is not significantly inhibited by either bovine albumin or by rabbit serum (Fig. 6.15). These studies predict that the MCR of the three sex steroid hormones in rabbits might be approximately the same. Indeed, this is what is observed, as two different studies have shown the MCR of oestradiol, testosterone, or dihydrotestosterone are not significantly different in the rabbit (Bourget, Flood, and Longcope 1984; Mahoudeau, Corvol, and Bricaire 1973). The studies shown in Fig. 6.15, and other results (Chaudhuri, Steingold, Pardridge, and Judd 1988) have shown that rabbit TeBG-bound testosterone or dihydrotestosterone is readily available for transport into rabbit tissues, since the bioavailable fraction is much greater than the free plus albumin-bound fraction. Similarly, the bioavailable corticosterone in rabbit brain is much greater than the free plus albumin-bound fraction of corticosteroid in rabbit serum (Pardridge, Eisenberg, Fierer, and Kuhn 1986). These results indicate that CBG-bound corticosteroid is readily available for uptake by rabbit brain. The findings on unidirectional clearance are in agreement with other studies which show that the concentration of CBG exerts no inhibitory influence on the MCR of corticosterone in rabbits (Daniel, Leboulenger, Vaudry, Floch, and Assenmacher 1982). Moreover, the unidirectional clearance of corticosterone by rabbit uterus is six-fold greater than the unidirectional clearance of cortisol by rabbit uterus in the presence of rabbit serum (Chaudhuri *et al.* 1988). These measurements of unidirectional clearance are also in agreement with MCR estimates, as other workers have shown that the MCR of corticosterone in rabbits is six-fold greater than the MCR of cortisol (Daniel *et al.* 1982). The much lower unidirectional clearance of cortisol in rabbit uterus is due to the combined effects of reduced membrane permeability (owing to increased polar functional groups, see Fig. 6.3 and 6.4) and to a greater inhibitory effect of rabbit CBG on cortisol transport as compared to corticosterone (Chaudhuri *et al.* 1988).

The possible role of CBG receptors on brain capillaries in the transport of CBG-bound corticosterone into rabbit brain was investigated recently (Pardridge, Eisenberg, Fierer, and Kuhn 1986). A ^{3}H-CBG preparation that retains normal ^{14}C-cortisol binding properties was prepared by reductive tritiation of the protein. Saturable CBG receptors in isolated brain capillaries were not detectable. These types of experiments further support the proposal that specific globulin receptors do not mediate the enhanced dissociation of steroid hormones from the circulating globulin-bound pools.

The propensity for rabbit TeBG to deliver testosterone to tissues, whereas human SHBG retards testosterone uptake by most, but not all, tissues may be related to the differential kinetics of androgen binding to the two plasma proteins. For example, the K_D of androgen binding to the two proteins is approximately the same at 4°C: $K_D = 0.4$ nM for human SHBG and 0.5 nM for rabbit TeBG (Tabei, Mickelson, Neuhaus, and Petra 1978). However, the rate of androgen dissociation from rabbit TeBG at 4°C is 27-fold faster than the rate of androgen dissociation from human SHBG. Similarly, the rate of androgen association with rabbit TeBG is 21-fold faster than the rate of androgen association with human SHBG (Tabei *et al.* 1978). Therefore, while the two plasma proteins have approximately the same equilibrium dissociation constant, the kinetics are dramatically different between the two binding proteins. There is a 68 per cent homology between the primary amino acid sequence of rat ABP and human SHBG (Joseph *et al.* 1987), and the binding kinetics of rat ABP and rabbit TeBG are similar (Kotite and Musto 1982). These considerations suggest that minor amino acid substitutions in the binding proteins about the androgen binding site results in markedly different conformational states and steroid binding kinetics.

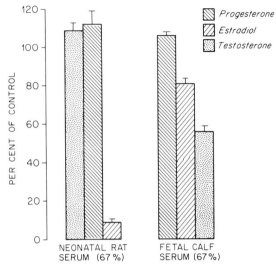

Fig. 6.16   The bioavailable fraction (shown on the y-axis as percent of control) of ³H-progesterone, ³H-oestradiol, or ³H-testosterone is shown for adult rat brain. The bioavailable fraction was determined from the ratio of the first pass steroid extraction after arterial injection of hormone mixed in either 67% neonatal rat serum or 67% fetal calf serum divided by the first pass extraction of hormone after arterial injection in Ringer's solution. Data are mean ± S.E.M. (n = 4–6 rats). Confidence limits for the neontal rat serum: progesterone and testosterone (NS), oestradiol (p < 0.0005) and for fetal calf serum: progesterone (NS), oestradiol (p < 0.05), testosterone (p < 0.0005). (From Pardridge and Mietus 1979b with permission).

VI   Developmental modulations

1   PLASMA PROTEIN BINDING

The development of sexually dimorphic behaviour in rodents and, possibly, in humans is believed to be the result of uptake and action in brain of circulating testosterone (Jacobson, Scernus, Shryne, and Gorski 1981; Holloway 1982). Testosterone action in developing rat brain may be mediated via aromatase conversion of testosterone to oestradiol (MacLusky and Naftolin 1981). The masculinization of the female brain by circulating oestradiol is believed not to occur owing to the avid oestradiol binding properties of α-fetoprotein (MacLusky and Naftolin 1981). The α-fetoprotein of developing rats, but not of developing humans, avidly binds oestradiol and oestrone but not testosterone (Raynaud, Mercier-Bodard, and Baulieu 1971; Crandall 1978). Moreover, the α-fetoprotein in the serum of developing rats markedly impairs the transport of oestradiol but not testosterone into adult rat brain (Fig. 6.16). However, human cord serum results in a slight depression of testosterone transport into adult rat brain as compared to oestradiol transport owing to the relatively low concentration of SHBG in human serum. Human α-fetoprotein, which is in high concentrations in human cord serum, does not retard oestradiol transport into brain (Pardridge 1982).

2 CYTOPLASMIC BINDING

As reviewed in **Section II (Quantifiable physiological model)**, an important determinant of hormone volume of distribution is binding to nonreceptor cytoplasmic proteins. The kinetics of steroid hormone binding to these cytoplasmic proteins can be conveniently quantified *in vivo* using the tissue sampling single injection technique (Pardridge *et al.* 1980*b*). This method was originally used to quantify the avid cytosolic binding of testosterone and oestradiol in adult rat brain, and preliminary studies indicated there was no measurable cytosolic sequestration of sex steroids by newborn rabbit brain (Pardridge *et al.* 1980*b*). These studies have recently been extended to developing rabbits of varying postnatal ages. The brain sequestration index (BSI) is a semi-quantitative measure of the activity of the cytoplasmic binding of testosterone by rabbit brain. As shown in Fig. 6.17, there is no measurable sequestration by newborn rabbit brain but the binding of testosterone to cytoplasmic proteins is induced between 10–15 days postnatally and reaches adult levels by three weeks of development (Pardridge, Eisenberg, Fierer, and Musto 1988). The increased sequestration of testosterone by cytoplasmic binding systems in rabbit brain parallels the time course of increased serum binding of testosterone in rabbits (Fig. 6.17), owing to a developmental increase in circulating TeBG (Hansson, Ritzen, Weddington, McLean, Tindall, Nayfeh, and French 1974). The postnatal induction of cytoplasmic and plasma binding systems of testosterone also parallel the developmental induction of the rat brain nuclear androgen receptor. This receptor is low in newborn rats but is greatly increased between 10–15 days postnatally (Lieberburg, MacLusky, and McEwen 1980). Thus, the androgen binding systems in plasma, brain cytoplasm, and brain nucleus appear to be orchestrated in parallel at a critical stage of development in rats and rabbits.

3 CAPILLARY TRANSPORT OF ANDROGEN-BINDING PROTEINS

The possibility that androgen-binding proteins such as albumin or TeBG are actually transported through the brain capillary wall in developing rabbits was investigated with isolated brain capillaries and a preparation of ^{3}H-TeBG that retains normal testosterone binding properties. Capillaries were isolated from 28-day old rabbit brain (Fig. 6.18) and uptake of peptides and plasma proteins by the isolated brain capillary preparation have been shown in other studies to predict reliably peptide or plasma protein transport through the BBB *in vivo* (Pardridge 1986*a*). Since iodination of hormone-binding plasma proteins invariably leads to denaturation of the protein, puri-

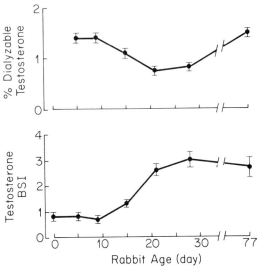

Fig. 6.17 *Bottom*: The brain sequestration index (BSI) of ³H-testosterone is shown in rabbits of varying postnatal ages. A BSI of 1 suggests there is no sequestration (that is, binding) of testosterone by brain, whereas a rising BSI indicates the development of a testosterone sequestration system in brain. *Top*: The percent dialyzable ³H-testosterone in rabbit serum of various ages is shown. The dialyzable percentage was determined by equilibrium dialysis at 37°C. (From Pardridge *et al.* 1988 with permission).

fied rabbit TeBG was radiolabelled by reductive methylation using ³H-sodium borohydride (Pardridge *et al.* 1988). As shown in Figure 6.19, this preparation of ³H-TeBG retained testosterone binding properties. This non-denatured preparation of TeBG was then used in *in vitro* uptake studies using the isolated brain capillaries. The results indicated that ³H-TeBG and, to a lesser extent, ³H-albumin were rapidly taken up by 28-day old rabbit brain capillaries via a process that was time- and temperature-dependent and was saturable by serum. Moreover, the uptake of the plasma proteins was much faster than the uptake of fluid phase markers such as ¹⁴C-sucrose (Pardridge *et al.* 1988). The uptake of the plasma proteins appeared to be a function of development, since the uptake of ³H-TeBG was several-fold greater by capillaries isolated from 28-day old rabbit brain as compared to capillaries isolated from adult rabbit brain. Why the developing rabbit brain takes up TeBG from the circulation is at present unknown. Owing to the rapid rates of androgen uptake by brain from the circulation via the enhanced dissociation mechanism, it would appear unlikely that the function of brain uptake of circulating TeBG is androgen delivery to brain cells. It may be that the apo-protein, *per se*, has a specific neuromodulator role in brain development. For example, transferrin, which mediates the brain uptake of circulating iron, is believed to have a neuromodulator role in brain that is separate from its iron

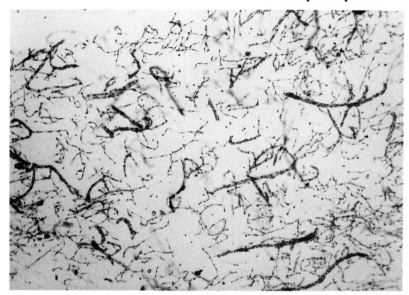

Fig. 6.18 Photomicrograph of isolated microvessels obtained from 28-day old rabbit brain. These capillaries were used in radioreceptor assays with ^{3}H-TeBG (see Fig. 6.19) to detect developmental modulations in uptake systems for TeBG in rabbit brain capillaries. Magnification × 25. (From Pardridge *et al.* 1988 with permission).

delivery function. Transferrin receptors in brain show little topographic correlation with iron distribution in brain cells (Hill, Ruff, Weber, and Pert 1985).

VII Clinical endocrinology

1 CIRRHOSIS

The gynaecomastia and feminization seen in cirrhotic men is a general observation, but the aetiology of this process is poorly understood (van Thiel 1979). Generally, either the total or free oestradiol level is normal in cirrhotic men, although the total testosterone level is decreased owing to hypogonadism (van Thiel 1979). A part of the feminization process includes a marked increase in the serum concentration of SHBG (van Thiel, Gavaler, Lester, Loriaux, and Braunstein 1975). The increased SHBG results in a decrease in the free plus albumin-bound fraction of testosterone in cirrhotic serum, and this correlates with a 35 per cent decrease in the unidirectional clearance of testosterone by rat brain using cirrhotic serum (Pardridge *et al.* 1980*a*), and a 35 per cent decrease in the MCR of testosterone in cirrhotic men (Gordon, Olivo, Rafii, and Southren 1975). The combinatiòn of increased SHBG and decreased albumin also results in an increase in the free plus albumin-bound fraction of oestradiol (Pardridge *et al.* 1980*a*). However,

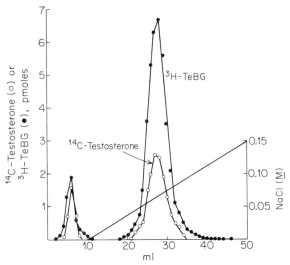

Fig. 6.19 Elution of ^{14}C-testosterone bound to ^{3}H-rabbit testosterone-binding globulin (TeBG) from a DEAE agarose column with a linear NaCl gradient. Testosterone co-eluted with labelled TeBG at 4°C indicating that the ^{3}H-TeBG molecule retained testosterone binding properties after the tritiation procedure. Less than 10 per cent of either the ^{3}H-TeBG or the ^{14}C-testosterone eluted in the low salt void volume of the column. (From Pardridge *et al.* 1988 with permission).

there is no decrease in the unidirectional clearance of oestradiol by rat brain using cirrhotic serum (Fig. 6.20). These measurements of unidirectional clearance confirm previous studies showing the MCR of oestradiol is not decreased in cirrhosis despite the elevation in SHBG (Olivo, Gordon, Rafii, and Southren 1975). The results in Figure 6.20 suggest the SHBG molecule in cirrhosis is structurally modified by the disease process and that this modification results in a change in the SHBG function with regard to oestradiol delivery to tissues. The hypothesis of structural differences in the SHBG molecule in cirrhosis has apparently not been tested to date. However, differences in glycosylation of TBG are known to occur in cirrhosis and this results in a change in the distribution of TBG isoforms in cirrhotic serum (Gartner, Henze, Horne, Pickardt, and Scriba 1981). Whether a similar modulation of the SHBG isoforms occurs in cirrhosis is not known but is presently being investigated in the author's laboratory. However, the data in Figure 6.20 and the study showing no decrease in the MCR of oestradiol in cirrhosis (Olivo *et al.* 1975) suggest that the increased SHBG in cirrhosis abnormally delivers oestradiol to tissues, and this may account for the feminization in this disease.

2 HYPERTHYROIDISM

Hyperthyroid men also show evidence of feminization and gynaecomastia

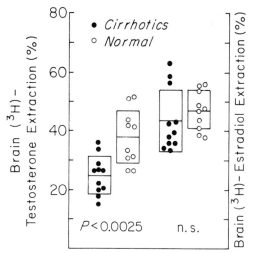

Fig. 6.20 The percentage of brain extraction of ^{3}H-testosterone or ^{3}H-oestradiol is shown after arterial injection of the labelled hormone mixed in either male cirrhotic or control human male serum. Rectangles represent the mean (horizontal line) ± S.D. The data show the unidirectional testosterone clearance by rat brain is decreased 33 per cent using cirrhotic serum, and this parallels a 2.6-fold increase in SHBG and a 40 per cent decrease in serum albumin in cirrhosis. However, the unidirectional clearance of oestradiol by rat brain was not decreased using cirrhotic serum, despite the marked increase in SHBG, decrease in albumin, and 41 per cent decrease in the non-SHBG-bound fraction of oestradiol. The brain bioavailable oestradiol using cirrhotic serum, 54 ± 4 per cent, is substantially greater than the non-SHBG-bound oestradiol, 37 ± 6 per cent, in cirrhotic serum, indicating SHBG-bound oestradiol is available for transport into brain from cirrhotic serum but not from control male serum. (From Sakiyama *et al.* 1982 with permission).

Table 6.8
Clinical alterations in serum SHBG

Increase	Decrease
Pregnancy	dexamethasone androgens
Estrogens	obesity
Hyperthyroidism	hirsutism
Cirrhosis	hyperprolactinemia
Hypogonadism	acromegaly

From DeMoor and Joossens (1970); Anderson (1974); Vermeulen (1977); Lobo and Kletzky (1983).

and an increase in the concentration of SHBG (Tulchinsky and Chopra 1973). Other conditions that result in an increase in serum SHBG in humans include hypogonadism, pregnancy, and oestrogen administration (Table 6.8). Serum SHBG concentrations are decreased in hyperprolactinemia, acromegaly, hirsutism, obesity, dexamethasone administration, and danazol treatment (Table 6.8). There are conflicting reports as to whether the MCR of oestradiol is decreased in hyperthyroidism (Ruder, Corvol, Mahoudeau, Ross, and Lipsett 1971; Olivo *et al.* 1975). Therefore, hyperthyroidism may resemble cirrhosis, and the feminization of these conditions may arise from enhanced oestradiol delivery to tissues owing to posttranslational modifications (for example, in the carbohydrate moiety) of the SHBG molecule.

3   CRITICAL ILLNESS

Acute illness is known to be associated with a rapid decrease in the serum testosterone concentration, similar to the low T_4 syndrome seen in nonthyroidal illness (Goussis, Pardridge, and Judd 1983). However, unlike nonthyroidal illness, which is associated with an increase in the dialyzable fraction of circulating thyroid hormones (Pardridge 1983), the low testosterone syndrome of illness is not associated with an inhibition of plasma protein binding. In fact, serum SHBG concentrations are slightly elevated (Goussis *et al.* 1983). Since serum luteinizing hormone concentrations are increased in the late phase of anesthesia and surgery (Nakashima, Koshiyama, Uozumi, Monden, Hamanaka, Kurachi, Aono, Mizutani, and Matsumoto 1975), the low testosterone observed with acute illness may reflect a primary testicular effect of stress (Moberg 1987).

4   SPIRONOLACTONE ADMINISTRATION

A side effect of spironolactone administration for the treatment of hypertension is gynaecomastia. This feminization appears to arise from a drug inhibition of the androgen receptor. Spironolactone and its active metabolite, canrenone, also inhibit androgen binding to serum SHBG (Manni, Pardridge, Cefalu, Nisula, Bardin, Santner, and Santen 1985). The brain bioavailable fraction of testosterone is increased proportionately with increasing concentrations of either spironolactone or canrenone (Fig. 6.21). Thus, these drugs inhibit androgen binding both to SHBG and to the androgen receptor. The effect on the androgen receptor is more profound since the inhibition of androgen binding to SHBG would be expected to increase cytoplasmic free testosterone and, thus, increase androgen receptor occupancy, whereas the drug actually inhibits the receptor.

5   DANAZOL ADMINISTRATION

Danazol is used in the treatment of endometriosis since the drug causes a "medical oophorectomy", and results in a marked diminution in serum oestradiol levels (Meldrum, Pardridge, Karow, Rivier, Vale, and Judd 1983). The androgen-like effects of this drug also result in a marked decrease in

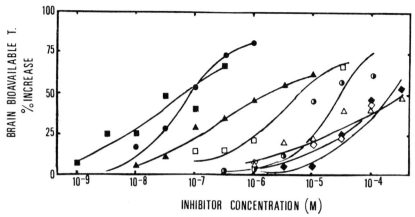

Fig. 6.21 The percent increase in brain bioavailable testosterone is plotted versus the concentration of inhibitors added to male serum ●, T(testosterone) ■, dihydro-testosterone; ▲, oestradiol; □, canrenone; △, SC24813; ◇, SC26519; ○, SC26962; ◆, spironolactone. Increasing concentration of these compounds result in gradual displacement of ^{3}H-testosterone from SHBG to human albumin and result in a progressive increase in the brain bioavailable steroid fraction. (From Manni *et al.* 1985 with permission).

SHBG production by the liver (Fig. 6.22). The diminished SHBG concentrations would be expected to result in a marked increase in testosterone MCR and the maintenance of normal plasma testosterone levels in danazol therapy (Fig. 6.22) indicate androgen production is increased in proportion to the increase in MCR. Recent studies indicate the "medical oophorectomy" may also be induced by chronic administration of gonadotropin-releasing hormone agonists, and that this therapy markedly lowers serum oestradiol concentrations without affecting the serum concentration of SHBG (Fig. 6.22).

6 DIETHYLSTILBESTROL ADMINISTRATION

Diethylstilbestrol (DES) was widely used in the past as a synthetic oestrogen (Steingold, Cefalu, Pardridge, Judd, and Chaudhuri, 1986). This oestrogenic drug does not bind to SHBG or α-fetoprotein, but does bind to albumin (Sheehan and Young 1979). In fact, the albumin binding of DES is about four-fold greater than albumin binding of oestradiol (Sheehan and Young 1979). However, the importance of measuring ligand binding to albumin with *in vivo* techniques is illustrated in Fig. 6.23, which shows the extraction of DES, oestradiol, oestrone, oestrone sulfate, and ethinyl oestradiol by rat uterus *in vivo* after aortic injection of labelled steroid hormone in either Ringer's solution, 4 g/dl bovine albumin, human pregnancy serum, or post-menopausal serum (Steingold *et al.* 1986). The results show that bovine albumin inhibits oestradiol extraction by more than 50 per cent but has no

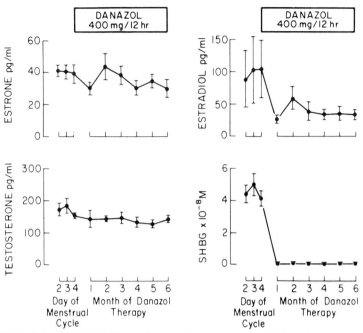

Fig. 6.22   Mean ± S.E. 8:00 a.m. levels of oestrone, oestradiol, testosterone, and sex hormone-binding globulin before and after treatment at the end of each month of danazol therapy. (From Meldrum *et al.* 1983 with permission).

measurable effect on uterine uptake of DES. Conversely, the brain extraction of DES was inhibited about 60 per cent by 4 g/dl bovine albumin, whereas the brain extraction of oestradiol was inhibited only about 25 per cent by this concentration of albumin. Therefore, albumin tends to retard DES delivery to an organ such as brain but mediates the avid uptake of DES by an organ such as the uterus.

VIII   Other sex steroid-binding proteins

1   PROGESTERONE-BINDING GLOBULIN

Progesterone-binding globulin (PBG) is a protein that is found in the serum of the pregnant guinea pig, but is absent in the non-pregnant guinea pig. This protein, like SHBG, is a hormone-binding globulin secreted by the liver that has up to a 70 per cent carbohydrate content (Westphal 1980). Because PBG has a very high affinity for steroids such as progesterone or androgens, and is found in nearly micromolar concentrations in pregnant guinea pig serum, the dialyzable fraction of these sex steroids in pregnant guinea pig serum is extremely low (Westphal 1980). Moreover, the unidirectional clearance of progesterone and androgens by rat brain is greatly retarded by pregnant guinea

UTERUS EXTRACTION

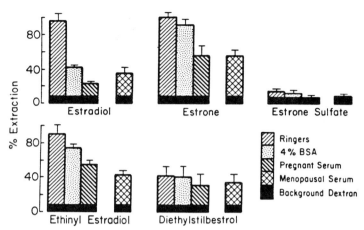

Fig. 6.23.   Mean unidirectional extraction of various oestrogens by the rat uterus during a single pass through the organ microvasculature using the different aortic injection vehicles shown. The data show that albumin-bound diethylstilbesterol is readily available for transport into the uterus (i.e. the extraction with Ringers or 4 per cent BSA is comparable), whereas albumin-bound oestradiol is only partially available *in vivo* (i.e. the extraction with 4 per cent BSA is about 40 per cent of the Ringer's extraction). *In vitro*, however, albumin binds diethylstilbesterol several-fold greater than the plasma protein binds oestradiol. (From Steingold *et al.* (1986) with permission).

pig serum (Fig. 6.24). Although PBG greatly restricts the brain uptake of progesterone, dihydrotestosterone, testosterone, and 17-hydroxyprogesterone, but not oestradiol (Fig. 6.24), a substantial fraction of plasma progesterone (10 per cent) or testosterone (25 per cent) is still available for transport into the brain (Pardridge and Mietus 1980). These results indicate that PBG-bound testosterone or progesterone is partially available for transport from the circulating globulin-bound pool. Since the rate of steroid hormone dissociation from PBG *in vitro* is particularly fast, $t_{\frac{1}{2}} \approx 1$ second (Westphal 1980), it is not surprising that the rates of steroid hormone dissociation from PBG *in vivo* are also sufficiently fast so that a portion of the PBG-bound hormone is operationally available for uptake by tissues (Pardridge and Mietus 1980).

2 α_1-ACID GLYCOPROTEIN (OROSOMUCOID)

α_1-acid glycoprotein (AAG), like PBG, binds both progesterone and testosterone (Burton and Westphal 1972), and exists normally in human serum in relatively high concentrations, for example, about 10 μM in normal human serum (Pardridge, Sakiyama, and Fierer 1983). However, the affinity of SHBG for testosterone is much higher than the affinity of AAG and the

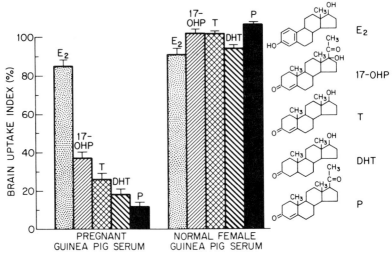

Fig. 6.24   The brain uptake index (BUI), mean ± S.E.M., n = 3–4 rats, of five ³H-labelled steroid hormones is shown after carotid injection of the hormone mixed in either 67 per cent pregnant guinea pig serum or 67 per cent normal female guinea pig serum. Abbreviations used: oestradiol, E2; 17-hydroxyprogesterone, 17OHP; testosterone, T; dihydrotestosterone, DHT; progesterone, P. (From Pardridge and Mietus 1980 with permission).

capacity of albumin binding of testosterone is much higher than that of AAG. Consequently, very little testosterone normally circulates in the AAG-bound pool (Westphal 1971). However, the recent introduction of the progesterone antagonist, RU-486 (Nieman, Choate, Chrousos, Healy, Morin, Renquist, Merrian, Spitz, Bardin, Baulieu, and Loriaux 1987), as a contraceptive agent signals the need for physiologic studies regarding the role of AAG in tissue uptake of this progesterone antagonist. RU-486 is a lipophilic amine and, like lidocaine, propranolol, or bupivacaine (Table 6.5), probably actively binds AAG (Moguilewski and Thilibert 1985). While AAG binding of drugs such as propranolol or lidocaine results in little inhibition of drug transport to brain *in vivo* (Pardridge *et al.* 1983), AAG binding of other lipophilic amine drugs such as bupivacaine does result in a marked decrease in brain uptake of the drug (Terasaki, Pardridge, and Denson 1985). That is, enhanced dissociation from AAG binding sites does not occur for all lipophilic amines in all organs. Conceivably, AAG binding of RU-486 may facilitate hepatic uptake of this drug (and account for reduced bioavailability of orally administered drug) but markedly restrict uptake of RU-486 by other tissues such as brain, uterus, or ovary.

IX   Conclusions

(1) The concentration of free cytosolic steroid hormone is believed to be the

driving force of both steroid hormone nuclear receptor occupancy and steroid hormone MCR. The free cytosolic testosterone or oestradiol has never been directly measured'in vivo since this would require the development of a testosterone or oestradiol sensitive electrode that could be placed into the living cytoplasm. Consequently, there is a great need in endocrine research for a quantifiable physiological model of steroid hormone transport so that the factors regulating the concentration of free cellular hormone may be understood. The free hormone hypothesis is inadequate as a transport theory since the pool of free hormone is too small to explain saturation of nuclear receptor occupancy and is too small to allow net extractions of 10–50 per cent of steroid hormones by organs such as liver. The free cellular hormone is a predictable function of membrane transport, hormone metabolism, and the concentration of plasma exchangeable hormone. The plasma exchangeable hormone is a predictable function of *in vivo* hormone dissociation rates from albumin or SHBG, capillary transit time, and membrane transport kinetics (Pardridge and Landaw 1985; Pardridge and Landaw 1987).

(2) The plasma capillary exchangeable hormone is many-fold greater than the free hormone measured *in vitro*, owing to enhanced rates of steroid hormone dissociation from plasma proteins *in vivo* within the organ microcirculation. The enhanced rates of dissociation are believed to arise from conformational changes about the ligand binding site, which are initiated by transient interactions between the binding proteins and the surface of the organ microcirculation. Enhanced dissociation of steroid hormones from albumin binding sites occurs in all organs studied thus far, but to a quantitatively different extent among the organs. The molecular basis of the interaction between the binding protein and the microcirculatory surface is unknown at present but may be electrostatic in nature and, possibly, related to carbohydrate moieties on either the surface of the microcirculation or the binding globulins.

(3) There is a marked organ specificity underlying the transportability of testosterone or oestradiol into tissues from the circulating SHBG-bound pool. SHBG-bound testosterone and SHBG-bound oestradiol are not available for transport into tissues such as brain or uterus in the rat. However, SHBG-bound oestradiol but not SHBG-bound testosterone is selectively available for transport into liver, salivary gland, and lymph node in the rat. In tissues such as testes, both SHBG-bound oestradiol and SHBG-bound testosterone are readily available for transport. Thus, the presence of SHBG in serum allows for varying degrees of amplification of hormone delivery to tissues. In this way, sex steroid delivery is amplified to sex steroid dependent tissues and cytosolic pools of free hormone may be maintained in the face of active steroid hormone metabolism by particular organs such as liver, prostate gland, or salivary gland. Oestrogen sulfate or glucuronate conjugates are readily available for uptake by liver, but only oestrogen-glucuronate conjugates, and not sulfate conjugates, are freely transported into kidney. There-

fore, the conjugation of steroid hormones with a sulfate moiety selectively traffics this compound for biliary excretion as opposed to urinary excretion. This selective trafficking of conjugates occurs because oestrogen sulfates undergo enhanced dissociation from the sulfate binding site on albumin in the hepatic microcirculation but not in the renal microcirculation.

(4) In addition to organ specificities, there are also species and developmental differences in the transport of plasma protein-bound hormones into tissues. Rabbit TeBG and CBG deliver testosterone and corticosterone to rabbit tissues *in vivo*, and this explains why oestradiol, testosterone, and dihydrotestosterone MCR in the rabbit are all the same, despite marked differences in *in vitro* binding of these steroid hormones to TeBG. The functional differences between human SHBG and rabbit TeBG have kinetic and structural counterparts. For example, the multiple isoforms of rabbit TeBG are more acidic than the human SHBG isoforms. Although the equilibrium dissociation constant of androgen binding to human SHBG and rabbit TeBG are approximately the same, the rates of androgen association and dissociation with rabbit TeBG are more than twenty-fold greater than the reaction rates with human SHBG.

(5) Albumin or SHBG may be modified in pathologic states in humans. For example, cirrhotic men are feminized and the MCR of oestradiol is not diminished in cirrhosis despite the marked elevation in serum SHBG. Transport studies indicate the SHBG molecule in cirrhosis is modified in such a way that the binding globulin delivers oestradiol to organs such as brain, where normally SHBG retards oestradiol delivery to this organ in the rat. Therefore, pathologic delivery of oestradiol to tissues may lead to feminization.

Acknowledgements

This work is supported by NIH grant R01-DK-25744-07. Dawn Brown skillfully prepared the manuscript. The author is indebted to Drs. Gautam Chaudhuri, Howard Judd, and Elliot Landaw for many valuable discussions.

References

Anderson, D. C. (1974). Sex-hormone-binding globulin. *Clin. Endocr.* **3**, 69–96.

Baird, D. T., Longscope, H. C., and Tait, J. F. (1969). Steroid dynamics under steady state conditions. *Recent Prog. Horm. Res.* **25**, 611–664.

Bartlett, P. A. and Marlowe, C. K. (1987). Evaluation of intrinsic binding energy from a hydrogen bonding group in an enzyme inhibitor. *Science, N.Y.* **235**, 569–571.

Bordin, S. and Petra, P. H. (1980). Immunocytochemical localization of the sex steroid-binding protein of plasma in tissues of the adult monkey *Macaca nemestrina*. *Proc. natn. Acad. Sci. U.S.A.* **77**, 5678–5682.

Bourget, C., Flood, C., and Longcope, C. (1984). Steroid dynamics in the rabbit. *Steroids* **43**, 225–233.

Brown, M. R. (1977). Serum albumin: Amino acid sequence. In *Albumin Structure,*

Function, and Uses (eds V. M. Rosenoer, M. Oratz, and M. A. Rothschild) pp. 27–51. Pergamon Press, New York.

Burton, R. M. and Westphal, U. (1972). Steroid hormone-binding proteins in blood plasma. *Metabolism* **21**, 253–276.

Cefalu, W. T. and Pardridge, W. M. (1987). Augmented transport and metabolism of sex steroids in lymphoid neoplasia in the rat. *Endocrinology* **120**, 1000–1009.

——— Chaudhuri, G., and Judd, H. L. (1986). Serum bioavailability and tissue metabolism of testosterone and estradiol in rat salivary gland. *J. clin. Endocr. Metab.* **63**, 20–28.

Chaudhuri, G., Steingold, K. A., Pardridge, W. M., and Judd, H. L. (1988). TeBG-bound and CBG-bound steroid hormones in rabbits are available for influx into uterus in vivo. *Am. J. Physiol.* **254**, E79–E83.

—— Verheugen, C., Pardridge, W. M., and Judd, H. L. (1987). Selective availability of estrogen and estrogen conjugates to the rat kidney. *J. Endocr. Invest.* **10**, 283–290.

Cheng, C. Y., Musto, N. A., Gunsalus, G. L., and Bardin, C. W. (1983). Demonstration of heavy and light protomers of human testosterone-estradiol-binding globulin. *J. steroid Biochem.* **19**, 1379–1389.

Corpechot, C., Robel, P., Axelson, M., Sjovall, J., and Baulieu, E.-E. (1981). Characterization and measurement of dehydroepiandrosterone sulfate in rat brain. *Proc. natn. Acad. Sci. U.S.A.* **78**, 4704–4707.

Corvol, P. and Bardin, C. W. (1973). Species distribution of testosterone-binding globulin. *Biol Reprod.* **8**, 277–282.

Crandall, B. F. (1978). The identification of developmental defects: The AFP model. In *Prevention of Neural Tube Defects: The Role of Alpha Fetoprotein.* (eds B. F. Crandall and M. A. B. Brazier) pp. 1–5. Academic Press.

Daniel, J.-Y., Leboulenger, F., Vaudry, H., Floch, H. H., and Assenmacher, I. (1982). Interrelations between binding affinity and metabolic clearance rate for the main corticosteroids in the rabbits. *J. steroid Biochem.* **16**, 379–387.

DeMoor, P. and Joossens, J. V. (1970). An inverse relation between body weight and the activity of the steroid binding β-globulin in human plasma. *Steroidologia* **1**, 129–136.

Dewey, W. C. (1959). Vascular-extravascular exchange of I^{131} plasma proteins in the rat. *Am. J. Physiol.* **197**, 423–431.

Forker, E. L. and Luxon, B. A. (1983). Albumin-mediated transport of rose bengal by perfused rat liver. Kinetics of the reaction at the cell surface. *J. clin. Invest.* **72**, 1764–1771.

Gartner, R., Henze, R., Horne, K., Pickardt, C. R., and Scriba, P. C. (1981). Thyroxine-binding globulin: Investigation of microheterogeneity. *J. clin. Endocr. Metab.* **52**, 657–664.

Gordon, G. G., Olivo, J., Rafii, F., and Southren, A. L. (1975). Conversion of androgens to estrogens in cirrhosis of the liver. *J. clin. Endocr. Metab.* **40**, 1018–1026.

Goussis, O., Pardridge, W. M., and Judd, H. L. (1983). Critical illness and low testosterone: Effects of human serum on testosterone transport in rat brain and liver. *J. clin. Endocr. Metab.* **56**, 710–714.

Griffin, J. E., and Wilson, J. D. (1985). Disorders of the testes and male reproductive tract. In *William's Textbook of Endocrinology* (eds J. D. Wilson and D. W. Foster) pp. 259–311. Saunders, Pennsylvania.

Gunsalus, G. L., Musto, N. A., and Bardin, C. W. (1978). Immunoassay of androgen binding protein: A new approach to the study of the seminiferous tubule. *Science, N.Y.* **200**, 65–67.

Hansson, V., Ritzen, E. M., Weddington, S. C., McLean, W. S., Tindall, D. J.,

Nayfeh, S. N., and French, F. S. (1974). Preliminary characterization of a binding protein for androgen in rabbit serum. Comparison with the testosterone-binding globulin (TeBG) in human serum. *Endocrinology* **95**, 690.

Heyns, W. and De Moor, P. (1971). Kinetics of dissociation of 17β-hydroxysteroids from the steroid binding β-globulin of human plasma. *J. clin. Endocr.* **32**, 147–154.

Hill, J. M., Ruff, M. R., Weber, R. J., and Pert, C. B. (1985). Transferrin receptors in rat brain: Neuropeptide-like pattern and relationship to iron distribution. *Proc. natn. Acad. Sci. U.S.A.* **82**, 4553–4557.

Holloway, R. L. (1982). Sexual dimorphism in the human corpus callosum. *Science, N. Y.* **216**, 1431.

Hryb, D. J., Khan, M. S., and Rosner, W. (1985). Testosterone-estradiol-binding globulin binds to human prostatic cell membranes. *Biochem. Biophys. Res. Commun.* **128**, 432–440.

Jacobson, C. D., Csernus, V. J., Shryne, J. E., and Gorski, R. A. (1981). The influence of gonadectomy, androgen exposure, or gonadal graft in the neonatal rat on the volume of the sexually dimorphic nucleus of the preoptic area. *J. Neurosci.* **1**, 1142–1157.

Joseph, D. R., Hall, S. H., and French, F. S. (1987). Rat androgen-binding protein: Evidence for identical subunits and amino acid sequence homology with human sex hormone-binding globulin. *Proc. natn. Acad. Sci. U.S.A.* **84**, 339–343.

Ketterer, B., Carne, T., and Tipping, E. (1978). Ligandin and protein A: Intracellular binding proteins. In *Transport by Proteins* (eds Gl Blauer and H. Sund) pp. 79–92. Walter de Gruyter & Co., New York.

—— Tipping, E., and Hackney, J. F. (1976). A low molecular weight protein from rat liver that resembles ligandin in its binding properties. *Biochem. J.* **155**, 511–521.

Kilvik, K., Furu, K., Haug, E., and Gautvik, K. M. (1985). The mechanism of 17β-estradiol uptake into prolactin-producing rat pituitary cells (GH_3 cells) in culture. *Endocrinology* **117**, 967–975.

Kotite, N. J. and Musto, N. A. (1982). Subunit structure of rabbit testosterone estradiol-binding globulin. *J. biol. Chem.* **257**, 5118–5124.

Kragh-Hansen, U. (1981). Molecular aspects of ligand binding to serum albumin. *Pharmac. Rev.* **33**, 17–53.

Larner, J. M. and Hochberg, R. B. (1985). The clearance and metabolism of estradiol and estradiol-17-esters in the rat. *Endocrinology* **117**, 1209–1214.

Lieberburg, I., MacLusky, N., and McEwen, B. S. (1980). Androgen receptors in the perinatal rat brain. *Brain Res.* **196**, 125–134.

Lobo, R. A. and Kletzky, O. A. (1983). Normalization of androgen and sex hormone-binding globulin levels after treatment of hyperprolactinemia. *J. clin. Endocr. Metab.* **56**, 562–566.

MacLusky, N. J. and Naftolin, F. (1981). Sexual differentiation of the central nervous system. *Science, N.Y.* **211**, 1294–1299.

Mahoudeau, J. A., Corvol, P., and Bricaire, H. (1973). Rabbit testosterone-binding globulin. II. Effect on androgen metabolism *in vivo*. *Endocrinology* **93**, 1120–1125.

Manni, A., Pardridge, W. M., Cefalu, W. T., Nisula, B. C., Bardin, W., Santner, S. J., and Santen, R. J. (1985). Bioavailability of albumin-bound testosterone. *J. clin. Endocr. Metab.* **61**, 705–710.

Meldrum, D. R., Pardridge, W. M., Karow, W. G., Rivier, J., Vale, W., and Judd, H. L. (1983). Hormonal effects of danazol and medical oophorectomy in endometriosis. *Obstet. Gynec.* **62**, 480–485.

Michel, C. C., Phillips, M. E., and Turner, M. R. (1985). The effects of chemically modified albumin on the filtration coefficient of single frog mesenteric capillaries. *J. Physiol., Lond.* **360**, 333–346.

Mickelson, K. E. and Petra, P. H. (1978). Purification and characterization of the sex steroid-binding protein of rabbit serum. *J. biol. Chem.* **253**, 5293–5298.

Milgrom, E., Atger, M., and Baulieu, E.-E. (1973). Studies on estrogen entry into uterine cells and on estradiol-receptor complex attachment to the nucleus—is the entry of estrogen into uterine cells a protein-mediated process? *Biochim. Biophys. Acta* **320**, 267–283.

Mizuma, T., Horie, T., Hayashi, M., and Awazu, S. (1986). Albumin-mediated uptake mechanism of drug by isolated hepatocytes. *J. Pharmacobio-Dyn.* **9**, 244–248.

Moberg, G. P. (1987). Influence of the adrenal axis upon the gonads. In *Oxford Reviews of Reproductive Biology* **Vol. 9** (ed. J. R. Clarke) pp. 456–496. Clarendon Press, Oxford.

Moguilewski, M. and Thilibert, D. (1985). Biochemical profile of RU-486. In *The Antiprogestin Steroid RU-486 and Human Fertility Control* (eds E. E. Baulieu and S. J. Segal) pp. 87–97, Plenum Press, New York.

Muller, R. E. and Wotiz, H. H. (1979). Kinetics of estradiol entry into uterine cells. *Endocrinology* **105**, 1107–1114.

Muller, W. E. and Wollert, U. (1979). Human serum albumin as a "silent receptor" for drugs and endogenous substances. *Pharmacology* **19**, 59–67.

Nakashima, A., Koshiyama, K., Uozumi, T., Monden, Y., Hamanaka, U., Kurachi, K., Aono, T., Mizutani, S., and Matsumoto, K. (1975). Effects of general anaesthesia and severity of surgical stress on serum LH and testosterone in males. *Acta endoc. Copenh.* **78**, 258–269.

Nieman, L. K., Choate, T. M., Chrousos, G. P., Healy, D. L., Morin, M., Renquist, D., Merriam, G. R., Spitz, I. M., Bardin, C. W., Baulieu, W.-E., and Loriaux, D. L. (1987). The progesterone antagonist. A potential new contraceptive agent. *New Engl. J. Med.* **316**, 187–191.

Noel, C. T., Ree, M. J., Jacobs, H. S., and James, V. H. T. (1981). The plasma concentration of estrone sulphate in postmenopausal women: Lack of diurnal variation, effect of ovariectomy, age and weight. *J. steroid Biochem.* **14**, 1101–1107.

Olivo, J., Gordon, G. G., Rafii, F., and Southren, A. L. (1975). Estrogen metabolism in hyperthyroidism and in cirrhosis of the liver. *Steroids* **26**, 47–56.

Pan, C. C., Woolever, C. A., and Bhavnani, B. R. (1985). Transport of equine estrogens: Binding of conjugated and unconjugated equine estrogens with human serum proteins. *J. clin. Endocr. Metab.* **61**, 499–507.

Pardridge, W. M. (1981). Transport of protein-bound hormones into tissues *in vivo*. *Endocr. Rev.* **2**, 103–123.

—— (1982). Brain uptake of steroid hormones. In *Androgens and Sexual Behaviour* (W. M. Pardridge, moderator). *Ann. Intern. Med.* **96**, 488–501.

—— (1983). Bioavailability of plasma protein-bound thyroid hormones *in vivo*. In *Thyroid Function in Nonthyroidal Illnesses* (I. J. Chopra, moderator). *Ann. Int. Med.* **98**, 946–957.

—— (1986*a*). Receptor-mediated transport through the blood-brain barrier. *Endocr. Rev.* **7**, 314–330.

—— (1986*b*). Serum bioavailability of sex steroid hormones. *Clin. Endocr. Metab.* **15**, 259–278.

—— (1987). Plasma protein-mediated transport of steroid and thyroid hormones. *Am. J. Physiol.* **252**, E157–E164.

—— Eisenberg, J., and Cefalu, W. T. (1985). Absence of albumin receptor on brain capillaries *in vivo* or *in vitro*. *Am. J. Physiol.* **249**, E264–E267.

—— —— Fierer, G., and Kuhn, R. W. (1986). CBG does not restrict blood-brain barrier corticosterone transport in the rabbit. *Am. J. Physiol.* **251**, E204–E208.

———— ——— and Musto, N. A. (1988). Developmental changes in brain and serum binding of testosterone and in brain capillary uptake of testosterone binding serum proteins in the rabbit. *Dev. Brain Res.* **38**, 245–254.

—— and Landaw, E. (1984). Tracer kinetic model of blood-brain barrier transport of plasma protein-bound ligands. Empiric testing of the free hormone hypothesis. *J. clin. Invest.* **74**, 745–752.

—— and Landaw, E. M. (1985). Testosterone transport in brain: Primary role of plasma protein-bound hormone. *Am. J. Physiol.* **249**, E534–E542.

———— (1987). Steady state model of T_3 transport in liver predicts high cellular exchangeable hormone concentration relative to *in vitro* free hormone concentration. *Endocrinology* **120**, 1059–1068.

—— and Mietus, L. J. (1979*a*). Transport of protein-bound steroid hormones into liver *in vivo*. *Am. J. Physiol.* **237**, E367–E372.

———— (1979*b*). Transport of steroid hormones through the rat blood-brain barrier. Primary role of albumin-bound hormone. *J. clin. Invest.* **64**, 145–154.

———— (1980). Effects of progesterone binding globulin vs. a progesterone antiserum on steroid hormone transport through the blood-brain barrier. *Endocrinology* **106**, 1137–1141.

———— Frumar, A. M., Davidson, B., and Judd, H. L. (1980*a*). Effects of human serum on the transport of testosterone and estradiol into rat brain. *Am. J. Physiol.* **239**, E103–E108.

—— Moeller, T. L., Mietus, L. J., and Oldendorf, W. H. (1980*b*). Blood-brain barrier transport and brain sequestration of the steroid hormones. *Am. J. Physiol.* **239**, E96–E102.

—— Sakiyama, R., and Fierer, G. (1983). Transport of propranolol and lidocaine through the rat blood-brain barrier. Primary role of globulin-bound drug. *J. clin. Invest.* **71**, 900–908.

Petra, P. H., Stanczyk, F. Z., Senear, D. F., Namkung, P. C., Novy, M. J., Ross, J. B. A., Turner, E., and Brown, J. A. (1983). Current status of the molecular structure and function of the plasma sex steroid-binding protein (SBP). *J. steroid Biochem.* **19**, 699–706.

Ponder, B. A. J. and Wilkinson, M. M. (1983). Organ-related differences in binding of dolichos biflorus agglutinin to vascular endothelium. *Devl Biol.* **96**, 535–544.

Rao, G. S. (1981). Mode of entry of steroid and thyroid hormones into cells. *Mol. Cell. Endocr.* **21**, 97–108.

—— and Rao, M. L. (1983). L-thyroxine enters the rat liver cell by simple diffusion. *J. Endocr.* **97**, 277–282.

Raynaud, J.-P., Mercier-Bodard, C., and Baulieu, E. E. (1971). Rat estradiol binding plasma protein (EBP). *Steroids* **18**, 767–788.

Reeve, E. B. and Chen, A. Y. (1970). Regulation of interstitial albumin. In *Plasma Protein Metabolism* (eds M. A. Rothschild and T. Waldmann) pp. 89–109. Academic Press, New York.

Renoir, J.-M., Mercier-Bodard, C., and Baulieu, E.-E. (1980). Hormonal and immunological aspects of the phylogeny of sex steroid binding plasma protein. *Proc. natn. Acad. Sci. U.S.A.* **77**, 4578–4582.

Riad-Fahmy, D., Read, G. F., and Griffiths, K. (1982). Steroids in saliva for assessing endocrine function. *Endocr. Rev.* **3**, 367–395.

Ruder, H., Corvol, P., Mahoudeau, J. A., Ross, G. T., and Lipsett, M. B. (1971). Effects of induced hyperthyroidism on steroid metabolism in man. *J. clin. Endocr.* **33**, 382–387.

Sakiyama, R., Pardridge, W. M., and Musto, N. A. (1988). Influx of TeBG and

TeBG-bound sex steroid hormones into rat testis and prostate. *J. clin. Endocr. Metab.*, in press.

—— —— and Judd, H. L. (1982). Effects of human cirrhotic serum on estradiol and testosterone transport into rat brain. *J. clin. Endocr. Metab.* **54**, 1140–1144.

Santner, S. J., Feil, P. D., and Santen, R. J. (1984). *In situ* estrogen production via the estrone sulfatase pathway in breast tumors: Relative importance versus the aromatase pathway. *J. clin. Endocr. Metab.* **59**, 29–33.

Setchell, B. P. and Waites, G. M. H. (1975). The blood-testis barrier. In *Handbook of Physiology.* Section 7: *Endocrinology* Vol. **5** *Male Reproductive System* (eds R. O. Greep and E. B. Astwood), Vol. **V**, pp. 143–172. American Physiological Society, Washington, D. C.

Sheehan, D. M. and Young, M. (1979). Diethylstilbestrol and estradiol binding to serum albumin and pregnancy plasma of rat and human. *Endocrinology.* **104**, 1442–1446.

Siiteri, P. K. (1981). Extraglandular oestrogen formation and serum binding of oestradiol: Relationship to cancer. *J. Endocr.* **89**, 119P–129P.

—— Murai, J. T., Hammond, G. L., Nisker, J. A., Raymoure, J. W., and Kuhn, R. W. (1982). The serum transport of steroid hormones. *Recent Prog. Horm. Res.* **38**, 457–503.

Stein, W. D. (1967). *The Molecular Basis of Diffusion Across Cell Membranes*, pp. 65–124. Academic Press, Inc., New York.

Steingold, K. A., Cefalu, W., Pardridge, W. M., Judd, H. L., and Chaudhuri, G. (1986). Enhanced hepatic extraction of estrogens used for replacement therapy. *J. clin. Endocr. Metab.* **62**, 761–766.

Stollman, T. R., Gartner, R. U., Theilmann, L., Ohmi, N., and Wolkoff, A. W. (1983). Hepatic bilirubin uptake in the isolated perfused rat liver is not facilitated by albumin binding. *J. clin. Invest.* **72**, 718–723.

Stremmel, B. J., Potter, B. J., and Berk, P. D. (1983). Studies of albumin binding to rat liver plasma membranes. Implications for the albumin receptor hypothesis. *Biochim. Biophys. Acta* **756**, 20–27.

Swartz, S. K. and Soloff, M. S. (1974). The lack of estrogen binding by human α-fetoprotein. *J. clin. Endocr. Metab.* **39**, 589–591.

Tabei, T., Mickelson, K. E., Neuhaus, S., and Petra, P. H. (1978). Sex steroid binding protein (SBP) in dog plasma. *J. steroid Biochem.* **9**, 983–988.

Tait, J. F. and Burstein, S. (1964). *In vivo* studies of steroid dynamics in man. In *The Hormones* (eds G. Pincus, K. V. Thiman, and E. B. Astwood) pp. 441–557. Academic Press, New York.

Terasaki, T. and Pardridge, W. M. (1987a). Stereospecificity of triiodothyronine transport into brain, liver, and salivary gland: Role of carrier- and plasma protein-mediated transport. *Endocrinology* **121**, 1185–1191.

—— —— (1988). Differential binding of T_4 and T_3 to acidic isoforms of thyroid hormone binding globulin in human sera. *Biochemistry*, in press.

—— —— and Denson, D. D. (1986). Differential effects of plasma protein binding of bupivacaine on its *in vivo* transfer into the brain and salivary gland of rats. *J. Pharmacol. exp. Ther.* **239**, 724–729.

Tulchinsky, D. and Chopra, I. J. (1973). Competitive ligand-binding assay for measurement of sex hormone-binding globulin (SHBG). *J. clin. Endocr. Metab.* **37**, 873–881.

Turner, M. R., Clough, G., and Michel, C. C. (1983). The effects of cationised ferritin and native ferritin upon the filtration coefficient of single frog capillaries. Evidence that proteins in the endothelial cell coat influence permeability. *Microvasc. Res.* **25**, 205–222.

Van Thiel, D. H. (1979). Feminization of chronic alcoholic men: A formulation. *Yale J. Biol. Med.* **52**, 219–225.

—— Gavaler, J. S., Lester, R., Loriaux, D. L., and Braunstein, G. D. (1975). Plasma estrone, prolactin, neurophysin, and sex steroid-binding globulin in chronic alcoholic men. *Metabolism.* **24**, 1015–1019.

Verheugen, C., Pardridge, W. M., Judd, H. L., and Chaudhuri, G. (1984). Differential permeability of uterine and liver vascular beds to estrogens and estrogen conjugates. *J. clin. Endocr. Metab.* **59**, 1128–1132.

Vermeulen, A. (1977). Transport and distribution of androgens at different ages. In *Androgens and Antiandrogens* (eds L. Martini and M. Motta) pp. 53–65. Raven Press, New York.

—— and Ando, S. (1979). Metabolic clearance rate and interconversion of androgens and the influence of the free androgen fraction. *J. clin. Endocr. Metab.* **48**, 320–326.

Weisiger, R., Golland, J., and Ockner, R. (1981). Receptor for albumin on the liver cell surface may mediate uptake of fatty acids and other albumin-bound substances. *Science Wash. D.C.* **211**, 1048–1051.

Westphal, U. (1971). In *Steroid-Protein Interactions*. Springer-Verlag, New York.

—— (1978). Receptor and Hormone Action. In *Steroid-Binding Serum Globulins: Recent Results* (eds B. W. O'Malley and L. Birnbaumer) pp. 443–472. Academic Press, New York.

—— (1980). Mechanism of steroid binding to transport proteins. In *Pharmacological Modulation of Steroid Action* (eds E. Genazzani, F. DiCarlo, and W. I. P. Mainwaring) pp. 33–47. Raven Press, New York.

7   Molecular action of progesterone

J. F. SAVOURET, M. MISRAHI, AND E. MILGROM

I   Introduction

The investigation of the female genital tract dates back to the early works of Fallopius in the 16th century. However truly rigorous use of experimental methods only began at the beginning of this century, if one excepts the pioneering discovery by Von Baer about corpus luteum formation (von Baer

1827). From his works stemmed a long controversy about the relationship of corpus luteum with pregnancy and the menstrual cycle that was finally resolved by Meyer. Another German scientist, Gustav Born, proposed that there is absolute requirement for an active corpus luteum during pregnancy, which was demonstrated after his death by his collaborator Ludwig Fraenkel (Fraenkel 1903). These fundamental studies made possible the discovery of progesterone and the elucidation of its biological role by such scientists as Corner, Allen, Doisy, Butenandt, and Clauberg. The credit for the first crystallization of biologically active progestin goes to Fevold and Hisaw (1932). This pioneering era was transformed in the 1950s by Pincus who initiated the study of the pharmacology of progestins in relation to human reproduction and contraception (Pincus 1965; Greep 1977). Jensen and Jacobsen (1962) have opened the field of molecular biology through their discovery of steroid receptors.

All subsequent work showed that progesterone controls many aspects of reproduction in female mammals from the uterus to the hypothalamus. Progesterone also controls egg white protein synthesis in the oviduct of oviparous animals and oocyte maturation in amphibia. In all except the latter case an intracellular receptor protein mediates hormone action and most available data show that all progesterone receptors are closely related in terms of structure, function, and regulation. As an exception to that general system, the interaction between a membrane bound adenylate-cyclase and progesterone in *Xenopus* oocytes takes place without classical receptor mediation, though a membrane receptor might be involved. Although receptor studies using labelled steroids date back 25 years, direct structural studies could only be initiated recently with the development of monoclonal antibodies against receptor proteins. The molecular structure of receptors has already been firmly assessed by cloning studies. However there is still very little understanding of the mechanism of action at the chromatin level and much remains to be done in the field of tissue and subcellular localization.

II   Progesterone metabolism in mammals

Progesterone is synthesized by the corpus luteum during the second half of the menstrual cycle (for review see Short 1964) and by the placenta and/or the corpus luteum during gestation. Minor amounts are also synthesized by testis and the adrenal cortex. The control of progesterone secretion in the corpus luteum by luteotrophins such as luteinizing hormone (LH) is mediated through intracellular modification of cyclic AMP (Marsch, Butcher, Savard, and Sutherland 1966; Jordan, Caffrey, and Niswender 1978). Progesterone secretion by the corpus luteum is submitted to several hormonal controls in addition to LH, including FSH, prolactin, prostaglandins and β adrenergic agents (Hsueh, Adashi, Jones, and Welsh 1984).

Upon fertilization of the ovum, implantation takes place, and in certain species, the trophoblast sustains the activity of the corpus luteum through the secretion of luteotrophic hormones. Starting around day 60 of gestation (in humans) the placental unit replaces the corpus luteum as the major source of progesterone secretion up to the time of delivery (Simpson and McDonald 1981).

Blood-borne progesterone is carried by transcortin (CBG) in many species including man (Diamond, Rust, and Westphal 1969). Some variations in binding proteins can be seen between species. Progesterone binding protein (PBP) exists only in hystricomorphs (such as the guinea pig) and appears predominantly during pregnancy. It is synthesized by the placental tissue (Milgrom, Allouch, Atger, and Baulieu 1973a; Perrot-Applanat and David-Ferreira 1982).

In the uterine fluid of the rabbit, progesterone is carried by a dimeric protein, uteroglobin, present only between day 3 to day 12 of pregnancy. The meaning of that interaction and its physiological role are unclear (see Savouret, Guiochon-Mantel, and Milgrom 1984 for review).

Progesterone is metabolized essentially to pregnanediol by the liver; 20α and 20β dihydro derivatives seem to be also synthesized in the corpus luteum itself (Wiest 1959). Metabolism of progesterone has also been found in the hypothalamus (Krause and Karavolas 1980), endometrium (Garzon, Aznar, Olivera, and Gallegos 1977) and lung (Milewich, Smith, and McDonald 1979).

By analogy to other steroids, progesterone is believed to go through the cytoplasmic membrane by passive diffusion. A "facilitated transport" has also been proposed, but conclusive evidence is still to be shown (Milgrom, Atger, and Baulieu 1973b).

III   The progesterone receptor

1   TISSUE LOCALIZATION

There is now general agreement in the fact that progesterone exerts most of its activity through binding to a specific intracellular protein, termed the "receptor" (Baulieu, Binart, Buchou, Catelli, Garcia, Gasc, Groyer, Joah, Moncharmont, Radanyi, Renoir, Tuohimaa, and Mester 1983; Weigel, Minghetti, Stevens, Schrader, and O'Malley 1983; Wrange, Okret, Radojcic, Carlstedt-Duke, and Gustafsson 1983; Sherman, Tuazon, Stevens, and Niu 1984). Initial discoveries were made in the two main experimental models: the chick oviduct (Sherman, Corvol, and O'Malley 1970; Toft and O'Malley 1972; Sherman, Atienza, Shansky, and Hoffman 1973) and the mammalian endometrium (mostly human and rodent) (Milgrom, Atger, and Baulieu 1970; Feil, Glasser, Toft, and O'Malley 1972; Rao, Wiest, and Allen 1973).

Following their discovery in guinea-pig uterus, progesterone receptors were detected in the uterus of other mammals where their concentration is under the control by oestradiol. Uterine progesterone receptors have been described in dogs, rats, cows and primates including man. Progesterone receptors are also present in the myometrium, Fallopian tubes, pituitary, hypothalamus, the cerebral cortex (reviewed in Baulieu 1983a), vagina (Batra and Losif 1985), testis (Terner 1977), and the thymus (Pearce, Khalid, and Funder 1983). Data on specific binding and subsequent action of progesterone in the pineal gland are controversial at the present time. However, most of the latter studies were done with progesterone, and not with stable synthetic progestins (for review see Preslock 1984). Progesterone receptors are also found in the mammary gland of virgin and pregnant mice but not during lactation (Haslam and Shyamala 1979).

Studies on human cancers have detected progesterone receptors in breast, kidney and prostate tumours, as well as in endometrial carcinoma (Terenius 1973; Pichon and Milgrom 1977; Concolino, Marocchi, Renaglia, DiSilverio, and Sparano 1978; McGuire and Horwitz 1978; Pollack, Irvin, Block, Lipton, Storer, and Claflin 1982; Clarke and Satyaswaroop 1985).

Progesterone receptors have been analysed in primary culture of endometrial cells (Gurpide, Fleming, Fridman, Haustrecht, and Holinka 1981) and in induced prolactin-producing pituitary tumours in the rat (Naess, Haug, Attramadal, and Gautvik 1982). However some established cell lines that retain progesterone receptors in culture appear to be of greater value in the study of hormonal control. Among breast cancer cell lines, MCF-7 cells (Michigan Cancer Foundation) and T47-D, the main experimental models, contain progesterone receptors in variable amounts depending on the particular sublines of the initial cell culture (Horwitz, Costloro, and McGuire 1975; Keydar, Chen, Karby, Weiss, Delarca, Radu, Chaitrik, and Brenner 1979; Horwitz, Mockus, and Lessey 1982). Other breast cancer cell lines containing progesterone receptors are Ia-270 (Loh, Chaman, McIndoe, White, Urdaneta, Hukku, and Peterson 1985), ZR-75-1 (Allegra, Korat, Do, and Lippman 1981), CAMA-1 (Yu, Leung, and Gao 1981), T-61 (Brunner, Spang-Thomsen, Skovgard-Poulsen, Engelholm, Nielsen, and Vindel 1985) and PMC-42 (Whitehead, Quirk, Vitali, Funder, Sutherland, and Murphy 1984). Progesterone receptors were also detected in HHUA cells from human endometrial adenocarcinoma (Ishiwata, Ishiwata, Soma, Arai, and Ishikawa 1984), mouse fibroblast L cells (Gal and Venetianer 1983), and rat pituitary tumour cell lines such as C-29 RAP (Lee, Davies, Soto, and Sonnenschein 1981) and GH3 (Roos, Strittmotter, Fabbro, and Eppenberger 1980).

In birds, progesterone receptors were detected in the oviduct (Toft and O'Malley 1972; Mester and Baulieu 1977) and in the bursa of Fabricius (Ylikomi, Gasc, Isola, Baulieu, and Tuohima 1985).

Finally, a saturable binding of progesterone in gram-negative bacteria has been reported (Miller and Norse 1977).

2 SUBCELLULAR LOCALIZATION

Receptor biochemistry started with the pioneering works of Jensen and Jacobsen (1962) with oestradiol.

Soon identical studies were performed for all steroids including progesterone, and led to the definition of "steroid receptors" as saturable cytosolic binding proteins with high and specific affinity for their ligand, and a salt-dependent behaviour in sucrose or glycerol gradients. The hormone bound its receptor upon entry into the cytoplasm, "activated" it and induced its transfer to the nucleus where subsequent regulation of specific genes would take place. That classic model was recently made obsolete when Gorski (1984) and King and Greene (1984) reported that unoccupied oestradiol receptors are loosely bound to the nucleus and only released in the cytosol by extraction. The new model also fits the progesterone receptor (PR). Initial studies on uterine stromal cells confirmed the exclusive nuclear localization of PR even in the absence of endogenous hormone. Non target tissues showed no staining. A marked heterogeneity in staining was consistently observed among identical cells in several target tissues. This suggests the existence of differences, or asynchronicity, in the hormonal sensitivity of those cells (Perrot-Applanat, Logeat, Groyer-Picard, and Milgrom 1985). The development of an immunogold technique allowed further refinements in subcellular localization by electron microscopy. PR was found almost in totality in the nucleus of uterine stromal cells, randomly associated with condensed chromatin in the absence of ligand. A very small amount of staining localized on the rough endoplasmic reticulum, which can be interpreted as newly synthesized receptor or may suggest post-transcriptional functions for PR. When the hormone was administered, the chromatin was extensively modified and PR migrated towards the borders of condensed chromatin and also within dispersed chromatin. These areas are known to be the most active sites of gene transcription (Fakan and Puvion 1980; Perrot-Applanat, Groyer-Picard, Logeat, and Milgrom 1986). Several monoclonal antibodies against rabbit PR cross-reacted with the human PR. Immunohistochemical studies on breast cancer cell lines (T47D), human myometrium and endometrium revealed PR levels in accordance with data from hormone binding assays. Similar studies were also performed on breast cancer biopsies (Perrot-Applanat, Groyer-Picard, Lorenzo, Jolivet, Vu Hai, Pallud, Spyratos, and Milgrom 1987). The nuclear localization of progesterone receptor has also been reported for the chick oviduct (Gasc *et al.* 1984) and the chick bursa of Fabricius (Ylikomi *et al.* 1985). In the case of glucocorticoids, the receptor is still consistently found in the cytosol in the absence of hormone, as shown by immunocytological mapping of neural cells using anti-receptor monoclonal antibodies (Fuxe, Winkstrom, Okret, Agnati, Harfstrand, Yu, Granholm, Zoli, Vale, and Gustafsson 1985), as well as enucleation experiments (Lukola, Akerman, and Patra 1985). Several authors have reported partial cystosolic localization for oestradiol and progesterone receptors but this may be due to artefacts caused by the use of polyclonal antibodies, as discussed by Perrot-Applanat *et al.* (1985). The nuclear localization of all steroid hormone receptors has been extensively reviewed by Walters (1985).

Binding studies indicate that the progesterone receptor has a narrow specificity for progesterone in all cases. No variations in the properties of progesterone receptor can be found between different tissues of the same animal or between normal and malignant tissues. On the other hand progesterone receptors exhibit strong differences in their binding affinities between species.

In rat and dog, the progesterone receptor has a lower affinity for progesterone (K_d 10 nM) than in other mammals (K_d 1–2 nM). Therefore in the rat uterus interference from tissue CBG has to be avoided for precise determinations of progesterone receptor (Milgrom and Baulieu 1970). This problem was solved by the use of 20 per cent glycerol which stabilizes the interaction between progesterone and its receptor, and by the availability of slow dissociating, high affinity synthetic progestins such as R5020 and ORG 2058 (Philibert and Raynaud 1973). These compounds have nanomolar K_d's for progesterone receptor in all situations and do not recognize CBG (Milgrom, Luu Thi, and Baulieu 1973*e*; Philibert and Raynaud 1977; Philibert, Ojasoo, and Raynaud 1977). On the other hand they bind to the glucocorticoid receptor. The differential binding of steroidal components to the progesterone receptor has been studied by Kontula, Janne, Vinkho, de Jager, de Visser, and Zeelen (1975).

As a general feature of all steroid hormone receptors, hormone binding markedly transforms the progesterone receptor. Initial studies focused on physicochemical properties induced by the binding such as sedimentation constant and receptor affinity for the nucleus (Jensen and DeSombre 1973; Gorski and Gannon 1976). Changes in these attributes, termed "activation", have been extensively studied *in vivo* and *in vitro* (Baxter, Rousseau, Benson, Garcea, Ito, and Tomkins 1973; Higgins, Rousseau, Baxter, and Tomkins 1973; Kalimi, Beato, and Feigelson 1973; Milgrom, Atger, and Baulieu 1973*c*; see Milgrom 1981 for a review). Hormone binding and activation are not simultaneous but rather successive events. Exposure of the receptor to the hormone in cells kept at low temperature results in a delayed activation that takes place only when the temperature is raised. This can be reproduced *in vitro* (Milgrom, *et al.* 1973*c*). It is therefore possible to observe complexed receptors in either non-activated or activated form, and study the transition (Bailly, Le Fèvre, Savouret, and Milgrom 1980). In the process of such studies several inhibitors of activation have been discovered. Transition metal oxyanions such as molybdate are now widely used to stabilize steroid-receptor complexes in the non-activated form (for review see Dahmer, Housley, and Pratt 1984). A small heat stable endogenous inhibitor (Bailly, Sallas, and Milgrom 1977) and several macromolecular heat labile factors that inhibit nuclear translocation have also been found in cytosols from a variety of tissues (Milgrom and Atger 1975; Dougherty and Toft 1980). The small factor has been reported in the glucocorticoid receptor model, but further studies suggest it exists in all steroid receptor models (Sato, Noma, Nishizawa, Nakoo, Matsumoto, and Yamamura 1980).

RU38486 (also called RU486) a potent progesterone antagonist has been recently described (Herrmann, Wyss, Riondel, Philibert, Teutsch, Sakiz, and Baulieu 1982; Philibert, Deraedt, Teutsch, Tournemine, and Sakiz 1982). The RU486 receptor-complex appears slightly different from the native progesterone-receptor complex at the level of physicochemical properties in T47-D cells in culture: its sedimentation constant in high salt sucrose gradients is 6S and not 4S. This heavier form reverts to the 4S form following treatment by urea or thioglycerol suggesting that disulfide bonds and hydrophobic interactions contribute to the integrity of the 6S form. The 6S RU486-receptor complex may result from interactions between receptors, or with a different protein such as the 90 K heatshock protein for example (Mullick and Katzenellenbogen 1986). Although there is a general agreement that RU486 does not interfere with the process of activation of the PR nor with its high affinity for DNA (Horwitz, 1985; Rauch, Loosfelt, Philibert, and Milgrom 1985), some contradictory data have been recently published (Mougdil and Hund 1987).

Initially, most data obtained in either the chick oviduct or the mammalian endometrium were in good agreement. The next logical step was the purification of the receptor by analysis of the purified molecular species and antibody preparation. Early purification protocols took advantage of the affinity of activated receptor for nuclei and DNA. The purified progesterone receptor from rabbit uterine cytosol was considered 20–30 per cent pure, with a specific activity of 2 nmol of bound hormone per mg of protein, and a 6 per cent yield. That receptor was used to produce a polyclonal antibody in the goat (Logeat, Vu Hai, and Milgrom 1981). However the purification method was extremely tedious, and polyclonal antibodies can generate problems of specificity in immunocytochemical and immunoblot experiments. The advent of monoclonal antibody technology solved most of these problems. A monoclonal antibody against progesterone receptor was obtained by Logeat, Vu Hai, Fournier, Legrain, Buttin, and Milgrom (1983) after immunization of a mouse by the semi-purified progesterone receptor. Previous works on progesterone receptor structure proposed a dimeric form composed of two different subunits A and B in the chick oviduct. Subunit A has a molecular weight of 79,000 and binds to DNA whereas subunit B has a molecular weight of 109,000 and binds to chromatin (Schrader, Birnbaumer, Hughes, Wiegel, Grody, and O'Malley 1981). That model was also extended to the mammalian progesterone receptor (Lessey, Alexander, and Horwitz 1983; Lamb and Bullock 1984). Two subunits were also detected by photoaffinity labelling of the chick oviduct progesterone receptor with R5020. Those studies emphasized the high degree of immunological similarity between both subunits

(Gronemeyer, Govindan, and Chambon 1985). Steroid receptor photo-affinity labelling has been recently reviewed by Gronemeyer and Govindan (1986).

In the case of oestrogen, androgen and glucocorticoid receptors, a single subunit was generally reported (Molinari, Medici, Moncharmont, and Puca 1977; Wrange, Carlstedt-Duke, and Gustafsson 1979; Agrawal 1983; DeBoer 1986). The question then was to ascertain if progesterone receptors were really different from the other steroid receptors.

Using the monoclonal antibody against progesterone receptor described by our laboratory (Logeat et al. 1983), Loosfelt, Logeat, Vu Hai, and Milgrom (1984) studied the rabbit uterine receptor by immunoblot methods. The progesterone receptor consists mainly of a 110,000 Daltons (110 kD) species together with a 79,000 Daltons (79 kD) and other smaller forms. The latter appeared in variable proportions between different preparations, suggesting they arose from proteolytic degradation. The in vitro translation product of the progesterone receptor mRNA was also detected as a single 110 kD species, allowing enrichment of mRNA preparations prior to molecular cloning (Loosfelt et al. 1984). As a further indication, monoclonal antibodies were sensitive enough to detect the receptor in unfractionated cytosols, in conditions where proteolysis and artefactual losses were minimized by using fresh tissue and a cocktail of protease inhibitors. In these conditions, only the 110 kD species is found. About 70 monoclonal antibodies have been prepared in our laboratory (Lorenzo, Jolivet, and Milgrom, unpublished results) and all recognize the 110 kD and 79 kD species or the 110 kD alone, but none recognize the 79 kD alone. Finally, proteolytic studies using 3 different enzymes showed the expected homology if the 79 kD species is derived from the intact 110 kD progesterone receptor (Loosfelt et al. 1984).

The monoclonal antibody was also used to prepare an immunosorbent column that made possible a one step purification of the active progesterone receptor (Logeat, Pamphile, Loosfelt, Jolivet, Fournier, and Milgrom 1985a). Purification yield was 49 per cent, and the receptor had a specific activity of 6.71 ± 0.79 nmol of bound steroid per mg of protein. The purified receptor consisted of a mixture of 110 kD or 79 kD species. The latter appeared during purification since immunoblots showed that initial cytostolic material consisted chiefly of the 110 kD form (85–90 per cent). The relatively mild conditions of elution from the immunosorbent (pH 10.5) allowed the recovery of active hormone-devoid receptor that retains its original properties: affinity for the hormone, DNA-binding and density gradient sedimentation behaviour. Further enzymatic proteolysis studies confirmed the uniqueness of the 110 kD species as the bonafide progesterone receptor in the rabbit uterus. Monoclonal antibodies against the chick oviduct progesterone receptor have been obtained by Sullivan, Beito, Proper, Krco, and Toft (1986). These authors report 4 monoclonals against the two forms A (80 kD) and B (110 kD) of the receptor and one specific for the B form. These results can be interpreted as supporting the hypothesis that there is a unique 110 kD species of receptor. Moreover, recent physico-chemical studies of the rabbit progesterone receptor using protease digestion, photoaffinity labelling and reversible denaturation strongly support that point of view (Lamb, Kima, and Bullock 1986).

Initial experiments of *in vitro* translation of the progesterone receptor mRNA, showed the receptor to be synthesized in a form apparently identical to the physiological molecule, suggesting the absence of any major maturation process. These experiments also showed that oestrogen regulation of progesterone receptors is mediated by an increase in progesterone receptor mRNA transcription (Loosfelt *et al.* 1984). A rabbit progesterone receptor cDNA has been cloned in λgt11 vector following random primer driven synthesis. The fusion protein was recognized by monoclonal and polyclonal antibodies to the progesterone receptor, and competed out by purified progesterone receptor. Northern blot experiments showed a mRNA of 5900 nucelotides mRNA, present under oestradiol control in the uterus and vagina and absent in the liver, spleen, kidney and muscle (Loosfelt, Atger, Logeat, and Milgrom 1986*a*).

The sequences of the rabbit (Fig. 7.1) and human (Fig. 7.2) progesterone receptors cDNAs have been recently established in our laboratory (Loosfelt, Atger, Misrahi, Guiochon-Mantel, Mériel, Logeat, Benarous, and Milgrom 1986*b*; Misrahi, Atger, d'Auriol, Loosfelt, Mériel, Fridlansky, Guiochon-Mantel, Galibert, and Milgrom 1987). The comparison of the amino acid sequence of the progesterone receptor with other steroid hormones receptors (such as glucocorticoids and oestrogens) and v-*erb*-A oncogene reveals a common cysteine-rich central region, which is involved in the DNA binding function of receptors (Weinberger, Hollenberg, Rosenfeld, and Evans 1985; Krust, Green, Argos, Kumar, Walter, Bornert, and Chambon 1986). This was confirmed by deletion experiments in the case of the oestradiol receptor (Kumar, Green, Staub, and Chambon 1986). Other experiments involving oestradiol receptor deletions mutants showed that hormone binding and DNA binding activities can be dissociated and even recombined in an heterologous manner between oestradiol and glucocorticoid receptors (Green and Chambon 1987). This independence between the two activities makes unlikely the existence of any transconformational mechanism. On the contrary, deletion of the DNA binding activity is always associated with the loss of regulation on specific transcription. Although similar data have not been published yet in the case of PR, it is highly probable that PR behaves like the other steroid receptors.

Deletion mutants of the glucocorticoid receptor, lacking the hormone binding domain, display a permanent, unregulated induction of transcription (Godowski, Rusconi, Miesfeld, and Yamamoto 1987). This can be explained either by deletion of the attachment sites to the stabilizing factor or by unmasking of pre-existent DNA binding sites. The peculiar structure of the

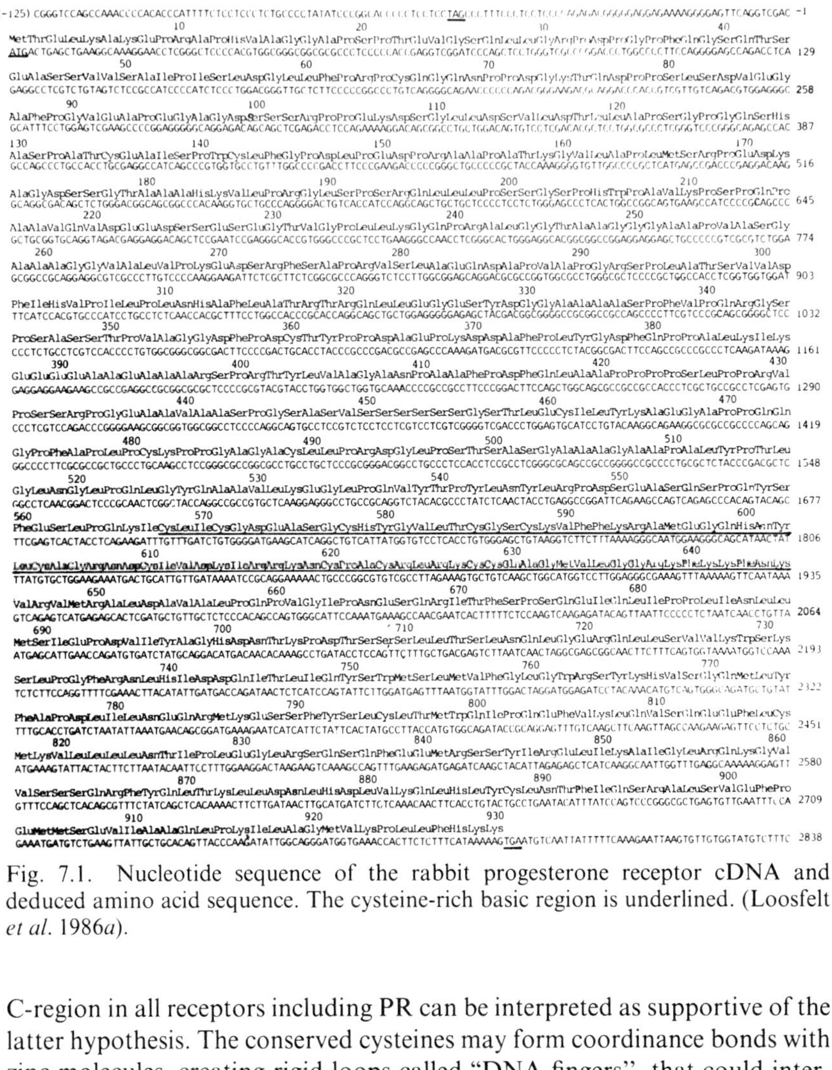

Fig. 7.1. Nucleotide sequence of the rabbit progesterone receptor cDNA and deduced amino acid sequence. The cysteine-rich basic region is underlined. (Loosfelt *et al.* 1986*a*).

C-region in all receptors including PR can be interpreted as supportive of the latter hypothesis. The conserved cysteines may form coordinance bonds with zinc molecules, creating rigid loops called "DNA-fingers", that could interact with the grooves of the double helix, in a similar manner to the *Xenopus laevis* transcription factor TF IIIA. Indeed, at least two DNA-fingers can be drawn from the C-region of the PRs (Krust *et al.* 1986; Kumar *et al.* 1986). The identification of that particular region delineates two other parts: the C-terminal and N-terminal regions. The C-terminal regions are very similar in all three receptors and v-*erb*-A, while the N-terminal appears fundamentally different (Fig. 7.3). That latter segment is responsible for the size difference between oestrogen (595 amino acids) and progesterone (930 amino acids) receptors. It has a high proline content, and this structure of a central DNA-binding region surrounded by proline-rich segments can also be found

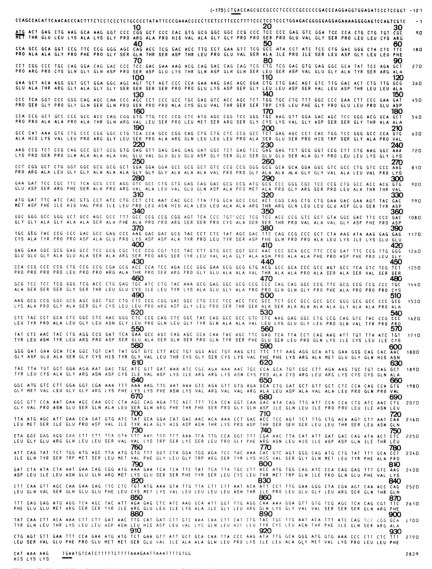

Fig. 7.2. Nucleotide sequence of the human progesterone receptor cDNA and deduced amino-acid sequence. (Misrahi *et al.* 1987).

in the "fushi tarazu locus" gene in *Drosophila*, thus suggesting a possible link between steroid hormone receptors and homeogene-coded proteins (Weinberger *et al.* 1985).

The gene for the human PR has been mapped to chromosome 11q22–11q23 (Rousseau-Merck, Misrahi, Loosfelt, Milgrom, and Berger 1987), and not 11q13 as previously reported (Law, Kao, Wei, Hartz, Greene, Zarucki-

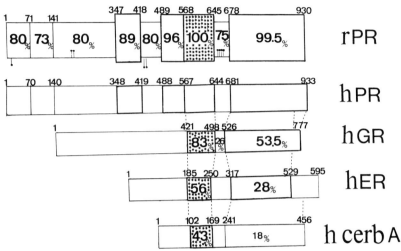

Fig. 7.3.   Schematic comparison between human progesterone receptor (hPR) and rabbit progesterone receptor (rPR), human glucocorticoid receptor (hGR), human oestrogen receptor (hER) and human c-*erb*-A. Receptors and c-*erb*-A sequences were aligned by their cysteine-rich basic region (dark boxes). Regions of major homology are boxed with thick lines. Added (♦) and deleted (♦) amino acids in the rabbit receptor when compared with the human receptor are indicated.

Schulz, Conneely, Jones, Puck, O'Malley, and Horwitz 1987). Law and coworkers speculated about the localization in 11q13 of c-*int*-2, a proto-oncogene that has been implicated in mouse mammary cancers, but the localization at 11q22–q23 has no relation with any breast cancer related oncogene. The c-*ets* oncogene is the nearest neighbour at 11q23–24 (de Taisne, Gegonne, Stehelin, Bernheim, and Berger, 1984).

Partial clones for the chicken oviduct progesterone receptor have also been obtained by Jeltsch, Krosowski, Quirin-Stricker, Gronemeyer, Simpson, Garnier, Krust, Jacob, and Chambon (1986) and Conneely, Sullivan, Toft, Birnbaumer, Cook, Maxwell, Zarucki-Schulz, Greene, Schrader, and O'Malley (1986) from λGT11 cDNA expression libraries using immuno-screening.

When purified in the presence of molybdate, hormone receptor complexes sediment at 8–10 S. These stable aggregates contain a 90 kD protein associated with the receptor (Gasc, Renoir, Radanyi, Jaob, Tudrimaa, and Baulieu 1984; Joab, Radanyi, Renoir, Buchou, Catelli, Binart, Mester, and Baulieu 1984), later found to be a heatshock protein (Catelli, Binart, Jung-Testas, Renoir, Baulieu, Feramisco, and Welch 1985). That protein is essentially cytoplasmic and present at concentrations a 100-fold higher than that of the receptor. The physiological relevance of these data is still debated.

During immunoblot studies of the various forms of progesterone receptor, it was noted that the "cystosol-extractable" species (the receptor that does not contain endogenous progesterone and is therefore loosely bound to the nucleus) and the nuclear "hormone loaded" receptor migrate as distinct species in polyacrylamide gels. The nuclear form shows an upshift in the size of the 110 kD as well as the 79 kD species. This was previously described for phosphorylated proteins (Wegener and Jones 1984). Incubation of uterine slices with ^{32}P orthophosphate with or without R5020, showed the cytostolic progesterone receptor is slightly phosphorylated and undergoes further phosphorylation upon hormone administration, which causes the mobility upshift. Scanning of the autoradiographs showed a 4-fold increase in phosphorylation due to the hormone. Thus the change is quantitative as well as qualitative. Similar results were obtained *in vivo* by injecting ^{32}P phosphate intraperitoneally into rabbits (Logeat, Le Cunff, Pamphile, and Milgrom 1985).

Studies in our laboratory have shown that a kinase activity co-purifies with the progesterone receptor on immunosorbent columns. Glucocorticoid and oestrogenic receptors as well as progesterone receptors from non-mammalian species are also phosphorylated *in vivo* and seem to be also associated with kinases (Dougherty, Puri, and Toft 1985). At least in the case of glucocorticoid receptors, the receptor itself is not responsible for the kinase activity (Sanchez and Pratt 1986). Several groups have reported *in vitro* phosphorylation of hen oviduct progesterone receptors by purified cAMP-dependent protein kinase (Weigel, Tash, Means, Schrader, and O'Malley 1981; Singh, Eliezer, and Mougdil 1986). However, *in vivo* phosphorylation experiments performed with analogs of cAMP on breast cancer cells in culture failed to confirm these data (Rao, Peralta, Greene, and Fox 1987). It is interesting that hormone binding transforms the progesterone receptor into its chromatin-bound, putatively active form through phosphorylation, since the intranuclear receptor appears similar in that respect to various membrane polypeptide receptors which are phosphorylated upon ligand binding. Moreover, studies on the beta adenergic receptor show that hormone-induced phosphorylation is responsible for its "desensitization" and altered electrophoretic mobility. Desensitization in that instance occurs through uncoupling from adenylate cyclase (Lefkowitz, Stadel, and Caron 1983).

IV Regulation of the progesterone receptor

Oestrogens increase progesterone receptor synthesis in mammals such as the guinea-pig (Milgrom *et al.* 1970; Milgrom and Baulieu 1970) and rat (Vu Hai, Logeat, Warembourg, and Milgrom 1977) as well as birds (Toft and O'Malley 1972). Following oestradiol injection in ovariectomized guinea-pigs, uterine progesterone receptor concentrations rose roughly 10-fold in the first 24 hours, The half life of the following decrease was 5 days (Milgrom *et al.* 1973d; Milgrom *et al.* 1973e). Maximal concentration was 40,000 sites per cell. That phenomenon has been observed in all target organs for progesterone action *in vivo* (Brenner and West 1975; Katzenellenbogen 1980) and *in vitro* using primary cultures of rat uterine cells or stable mammary cancer cell lines.

If the mechanism of action of oestradiol on progesterone receptor levels is now elucidated (through regulation studies of progesterone receptor mRNA as previously discussed in the text), quantitative kinetics of oestradiol receptor are still subject to controversy.

Upon oestradiol administration, oestradiol receptors shift to a "tight nuclear association". Following hormone dissociation, nuclear receptors apparently disappear and there is a long lag before "cytosolic" receptors return to their initial level, thus causing the total receptor concentration to decrease. Some authors have reported that "processing" is necessary in oestradiol stimulation of progesterone receptor in the rat uterus (Sica, Weiz, Petrillo, Armetto, and Puca 1981) while others have refuted it (Kassis, Walent, and Gorski 1986). Similarly, it was described (Horwitz and McGuire 1978) and refuted (Kasid, Huff, Greene, and Lippman 1984) for MCF-7 cells *in vitro*.

Progesterone receptors are controlled by their own ligand: progesterone decreases progesterone receptor concentration in guinea-pigs previously stimulated by oestradiol (Milgrom *et al.* 1973*d*, 1973*e*). It was suspected that an accelerated inactivation rate might be responsible for the effect. Later studies on progesterone-dependent progesterone receptor phosphorylation further sustain this idea, since it is known for several proteins that phosphorylation increases the turnover rate (Logeat *et al.* 1985). Strikingly, progesterone appears to autolimit its own action by decreasing its binding sites.

As reported by Vu Hai *et al.* (1977), during gestation in the rat, the concentration of "nuclear" receptor varies inversely to the "cytosolic" concentration. Nuclear progesterone receptor concentrations are low at the beginning, rising sharply to a plateau (2600 sites/cell) until day 15 and then decreasing sharply, showing wide individual variations. It was then assumed that low nuclear progesterone receptor concentration was in accordance with the abolition of progesterone block preliminary to parturition. Data on the final stage of pregnancy have been reassessed using R-5020, notably in the rabbit. Progesterone decreases both the progesterone receptor and oestradiol receptor concentrations in endometrium and myometrium while the fall in plasma progesterone at day 29 of gestation induced a doubling of progesterone receptor concentration that remained elevated until post-partum (day 32) (Quirk and Currie 1984).

Apart from general regulation, local effects should also be taken into consideration: "nuclear" progesterone receptors are present in two-fold higher concentration in implantation sites than in other endometrial regions of 6-days pregnant rats. Decidualization in the absence of embryo did not reproduce that effect. The increase persisted after ovariectomy of the mother, suggesting a control by steroids from the embryo (Logeat, Sartor, Vu Hai, and Milgrom 1980).

In mammalian cancer cells in culture, MCF-7 cells have been used as the main model, and results agree with the general model of double regulation (oestrogen and progesterone) of the progesterone receptor (Horwitz and McGuire 1978). In these cells, as in other oestrogen dependant breast cancer cell lines (T47-D and ZR-75-1) growth factors have been proposed as intermediates of oestrogen action. Indeed, upon inactivation of epidermal growth factor (EGF) receptors by phorbol esters, progesterone receptors decreased while oestrogen receptors stayed unchanged (Roos, Fabbro, Kurg, Costa, and Eppenberger 1986).

T47-D cells have very low oestrogen receptor and very high, constitutively expressed progesterone receptor levels (300,000 sites/cell). The progesterone receptor is not oestrogen-dependent although it retains some features of inducible proteins (inhibition by BUdR and sodium butyrate) (Horwitz et al. 1982). Variant T47-D$_{co}$ cells are totally depleted of oestrogen receptors and allow for the study of isolated effects of progesterone on breast cancer growth (Horwitz and Freidenberg 1985). It may be noted that progesterone receptors in rat pituitary tumour Mt TF4 are induced by oestradiol, when growth of that tumour is inhibited by oestradiol (Albaladejo and André 1986).

V   Physiological effects of progesterone

Progesterone exerts a general metabolic effect, with an increase of basal and glucose-induced insulin production. Neither the metabolism of lipids nor the level of blood glucose is modified by progesterone (Beck 1977). The variation in the type of beta-adenergic receptors (beta 1 versus beta 2) in the endometrium during the menstrual cycle clearly shows an influence of progesterone. During the luteal phase, beta 1 receptors totally disappear and are replaced by beta 2 receptors (percentage ratio 20:80 in follicular phase and 0:100 in luteal phase) (Lefkowitz et al. 1983). However, the most prominent function of progesterone in mammals and oviparous species is on the physiology of the genital tract: decidualization, and the establishment and maintenance of pregnancy. Only eutherian mammals require progesterone throughout gestation unlike reptiles and marsupials (Yaron 1977; Young and Renfree 1979).

Accordingly, extensive literature covers the subject, both at the level of physiology and histology and at the finer level of molecular studies dealing with specific marker proteins of progesterone action (see Rothchild 1983 for review).

1   ENDOMETRIUM IN THE LUTEAL PHASE

Progesterone is essential for the development of decidual tissue. Cytodifferentiation and proliferation of luminal and glandular epithelia have been studied *in vivo* (Conner, Murai, and Gerschenson 1978; Shapiro, Dyer, and Colas 1980; Conti, Conner, Gimenez-Conti, Silverberg, and Gerschenson 1981*a*; Conti, Gimenez-Conti, Zerbe, and Gerschenson 1981*b*; Boomsma, Jaffe, and Verhage 1982; Gerschenson, Conti, Depaoli, Liezberman, Lynch, Orlicky, and Rivas-Berrios 1984; Cheng, McDonald, Clark, and Pollard 1985) and *in vitro* on explants and primary cultures of endometrial cells, with focus on induction of glycogenesis (Demers, Feil, and Bardin 1977; Shapiro *et al.* 1980), cyclic nucleotide metabolism (Beatty, Bocek, Herrington, Young, and Brenner 1979), regulation of the cell cycle (Gerschenson, Conner, and Murai 1977) and protein synthesis and secretion.

A great number of endometrial proteins are under the control of progesterone. This has been demonstrated *in vivo* and *in vitro*. Most of the recorded effects deal with secretory proteins from endometrial glands, since their secretion is largely under progesterone control (McLaughlin, Sylvan, and Richardson 1981; Salamonsen, Wai, Doughton, and Findlay 1985). In rabbit endometrium progesterone stimulates several proteolytic activities involved in the digestion of blastocyst coverings (Kirchner 1980), arylamidase I (Denker 1982), neuraminidase (Srivastava and Farooqui 1980), relaxin (Zarrow and Brennam 1959; Too, Bryant-Greenwood, and Greenwood 1984) and cytoplasmic poly A polymerase (Orawa, Isomaa, and Janne 1979). In rats, cytoplasmic 17β-hydroxysteroid dehydrogenase (Amr, Faye, Bayard, and Kreitmann 1980), nicotinamide adenine dinucleotide kinase (Cummings and Yochim 1983), lactic dehydrogenase (Gershbein and Raikoff 1978), ornithine decarboxylase (Levy, Glikman, Vegh, Mester, Baulieu, and Suto 1984), gammaglutamyl transpeptidase (Tarachand and Eapen 1982), and glucosamine 6-phosphate synthetase (Rukmini and Reddy 1981) are stimulated. Other reported stimulations concern oestrogen sulfotransferase (Meyers, Lozon, Corombos, Saunders, Hunter, Christiensen, and Brooks 1983) and inhibitor of the plasminogen activator from pig uterus (Mullins, Bazer, and Roberts 1980).

In the human endometrium, two basic groups of cellular products have been implicated as under progesterone control (presumably receptor-mediated): intracellular and secretory. Among intracellular components are receptors for progesterone and oestrogens themselves, enzymes of oestradiol metabolism (Tseng and Gurpide 1975; Tseng and Gurpide 1979; Clarke, Adams, and Wren 1982), alphafucosidase (Vesce, Biondi, Gulinati, Zeni, Simonetti, and Salvatorelli 1985), type II cyclic AMP dependent protein kinase (Miyazaki, Miyamoto, Maryama, and Uchido 1980). Secreted proteins under progesterone control comprise various enzymes for carbohydrates, proteins and prostaglandins metabolism, hydrolases, phosphatases (see McLaughlin, Sylvan, and Richardson 1981 for review), the progestagen-associated endometrial protein (Joshi 1983), a 24 kD protein originally discovered in a breast cancer cell line (Ciocca, Asch, Adams, and McGuire

1983) and prolactin (Taii, Ihara, and Mori 1984). The progestagen-associated endometrial protein is a 47,000 Dalton glycoprotein. It is found also in peripheral blood, thus being of interest as a clinical marker for progesterone status. It has plasmin-inhibiting properties. Prolactin is produced in the decidualized endometrium from the late secretory phase in non-fertile cycles (Riddick and Kusmick 1977; De Ziegler and Gurpide 1982; Bischof, Sizonenko, and Herrmann 1986) and pregnancies (Brawerman, Bagni, de Ziegler, Den, and Gurpide 1984). RU486, a potent antiprogesterone inhibits prolactin production by the decidua (Bischof *et al.* 1986). The decidual prolactin is identical to the pituitary hormone (Takahashi, Nakeshima, Oyata, and Takerchi 1984).

Prolactin release in the endometrium is inhibited by a decidual peptide suggesting local control of prolactin action in the endometrium (Markoff, Howell, and Handwerger 1983). Surprisingly, myometrium also secretes prolactin, and that secretion is inhibited by progesterone (Walters, Daly, Chapitis, Kusus, Prior, Kusmick, and Riddick 1983). A new progesterone-induced mRNA has been recently detected in the rabbit endometrium at the luteal phase. It is strictly progesterone dependent and barely sensitive to oestrogen administration in contrast with uteroglobulin. Northern blot analysis using the cloned cDNA as probe revealed the presence of this mRNA in other tissues, such as ovaries, Fallopian tubes, and in the human endometrium at the luteal phase. It was also detected in the rabbit liver where it is constitutively expressed, and in rabbit endometrium during early pregnancy. A molecular weight of 41 kD can be ascribed to the related protein by calculation from the coding sequence (Misrahi, Atger, and Milgrom 1987).

2 THE IMPLANTATION AND EARLY PREGNANCY PERIOD

In mammals, when the blastocyst reaches the uterus, implantation is immediate or delayed according to the species' particular physiology. The delay may be facultative as in the rat where it occurs due to lactation, or obligatory. Depending on the species, progesterone treatment of an ovariectomized pregnant animal will protect the blastocyst and prevent implantation as in rats (Martel and Psychoyos 1980) or induce it as in rabbits, and guinea-pigs (Martin 1980). Mouse embryos can implant and grow outside the uterus without progesterone treatment (Kirby 1962). Progesterone must therefore be necessary only when normal pregnancy occurs in the uterus. Since the aspects of implantation differ between species, it is difficult to pin-point the essential roles of progesterone on that sequence of events (reviewed in Rothchild 1983). Activation of uterine proteases by progesterone to lyse the zona pellucida is an event common to all species, but not primordial since lysis of the zona does not induce implantation by itself (Renfree 1980). Progesterone seems to inhibit implantation in the rat while maintaining the egg in the lumen at the preimplantation stage. The blastocyst then secretes oestrogen and/or unknown factors which in turn release the progesterone block, permitting implantation (Mallonee and Yochin 1979; Rothchild 1983). Oestrogen secretion by the blastocyst has been reported in several species including pig, rat and mouse (Dickmann 1979; Gadsby, Heap, and Burton 1980).

Endometrial proteins during early pregnancy have received much attention (Beato 1979; Beier and Mootz 1979; Beier 1982). The protein composition of uterine fluids varies in a time-specific manner and is influenced by the balance between oestrogens and progesterone. Several experimental models are currently under study. The inhibitor of plasminogen activator from pig uterus is a small protein (15,000 Daltons) produced under progesterone control in luteal phase and early pregnancy. Since pig trophoblast is not invasive, in contrast to some other mammals, it has been suggested that the plasminogen activator inhibitor may play a role in preventing invasive implantation (Mullins et al. 1980; Fazleabas, Bazer, and Roberts 1982). Conversely, rat and rabbit trophoblasts are highly invasive and provoke multiple interactions with the endometrium that result in observable biochemical modifications at the level of steroid receptors (Logeat et al. 1980), and specific proteins at the surface of endometrial cells (Ricketts, Scott, and Bullock 1984). Surprisingly, progesterone also inhibits prostaglandin production and elicits precursor accumulation (Abel and Baird 1980; Alwachi, Bland, and Poyser 1981). This is paradoxical since prostaglandins seem to be required for implantation (Kennedy 1980), but may only reflect the stepwise organization of hormone action during the successive phases of implantation, as previously described.

A few proteins are inhibited by progesterone in the uterus: oestrogen receptor, progesterone receptor, type I cAMP dependent protein kinase (Miyazaki et al. 1980), a small inhibitor of embryonic development (Richardson, Hahmer, and Oliphant 1980), and in the rat myometrium, the angiotensin II receptors (Schirar, Capponi, and Catt 1980a). The latter effect follows a three-fold increase of these receptors in early pregnancy (day 2 to day 12) with a peak at day 9, under the influence of oestradiol. The modifications in angiotensin II receptors concentration are exclusively located in the implantation area (Schirar, Capponi, and Catt 1980b) and there is little or no modification in the adjacent myometrium. The low levels of angiotensin II receptors in the second half of gestation have prompted these authors to deny a role for angiotensin II in the parturition process. Its inhibition by progesterone can be considered a part of the anti-contractility properties of progesterone.

In the immediate period after implantation in the rabbit (day 2 to day 12) progesterone considerably stimulates the endometrial production of uteroglobin. This protein is one of the most extensively studied models of progesterone action (see Savouret and Milgrom 1983; Savouret et al. 1984 for review), due to its extreme convenience since uteroglobin (also called blastokinin) represents 40–60 per cent of the total proteins in the uterine fluid at its peak on the fifth day after mating. This protein is secreted only between day 1 and day 12 of gestation and then totally disappears throughout the rest of the gestation, despite continuously increasing hormonal concentrations.

Uteroglobin was first believed to be a uterine-specific protein, regulated only by progesterone. However, further studies have shown specific effects of oestradiol on the synthesis of uteroglobin and its mRNA. The presence of an identical protein has been reported in several rabbit tissues including oviduct, blastocyst and lung, where it is constitutively expressed. Some authors have proposed that uteroglobin exerts a stimulatory effect on the blastocyst (hence the other name of blastokinin), together with retarding egg implantation (Beier 1968). Proposed mechanisms of uteroglobin action range from protease inhibition to cross-linking with histocompatibility surface antigens of the blastocyst to prevent rejection by maternal lymphocytes (see **Section V, 3**). Uteroglobin has also been found in the rabbit seminal fluid, demonstrating that it is not a sex-specific component. However, the most striking feature of uteroglobin is its ability to bind progesterone. Unlike the progesterone receptor, uteroglobin binds synthetic progestins poorly but binds 5α pregnan-3,20-dione, a progesterone metabolite, three-fold better than progesterone itself (Fridlansky and Milgrom 1976). Other steroid hormones, such as cortisol, testosterone and oestrogen, show practically no affinity for uteroglobin. Uteroglobin binds diethylstilbestrol slightly better than synthetic progestins.

This ability to bind its own regulator has prompted some investigators to consider that the mode of action of uteroglobin might be to modulate progesterone activity and transport within the blastocyst. High concentrations of progesterone in the uterus are necessary to maintain pregnancy; however, these amounts of hormone may be detrimental to the blastocyst, and hence uteroglobin might sequester a fraction of endo-uterine progesterone. Uteroglobin has not been found in other mammals, except other members of the order Lagomorpha. Similar research in the human species has shown a protein bearing electrophoretic and immunological similarities, but conclusive evidence remains to be produced.

The uteroglobin model is valuable in that the same gene undergoes completely different regulation in two different organs of the same animal: lung uteroglobin can be studied as a marker for glucocorticoid-induced developmental regulation and differentiation, while endometrial uteroglobin can be a source of information on steroid regulation of genomic expression, implantation mechanisms, blastocyst development, and blastocyst protection from the maternal hormonal and immune systems (Beier 1982). Uteroglobin is also the only steroid-binding protein to have been analysed in detail by X-ray diffraction, and thus it provides a unique model for the study of steroid protein interactions at the atomic level (Mornon, Fridlansky, Bailly, and Milgrom 1980).

The stimulation of uteroglobin synthesis by progesterone occurs through enhancement of specific transcription (Kumar, Chandra, Woo, and Bullock 1982). This has been shown to be a direct effect of progesterone, and does not require oestrogen-priming. A peculiar amplification of this progesterone effect at the post transcriptional level has also been described in the endometrium (Loosfelt, Fridlansky, Atger, and Milgrom 1981). Oestradiol also stimulates uteroglobin production, but does not present the post-transcriptional effect. RU486, an antagonist of progesterone action totally inhibits the stimulation of uteroglobin mRNA transcription in the endometrium (Rauch *et al.* 1985). As uteroglobin disappears at day 12 of gestation, it is relevant to consider it in relation to implantation and separately from *bona fide* pregnancy proteins present until parturition.

3   PREGNANCY AND PREGNANCY-ASSOCIATED PROTEINS

Progesterone has several different roles in the maintenance of gestation: inhibition of myometrial contractility, immunological protection of the embryo, inhibition of prostaglandin synthesis and maintenance of uterine growth and plasticity (Rothchild 1983). Inhibition of myometrial contractions has been related to prostaglandin inhibition (Rothchild 1983) or calcium metabolism in the sarcolemma (Currie and Jeremy 1979). Inhibition of myometrial contractions by progesterone is not common to all mammals, and some species present an active uterus during pregnancy (Rothchild 1983). The immunological protection of the embryo is not well understood. The involvement of progesterone has been suggested in several ways (see **Section V**, 4, iv, and Carter 1984). A simple idea is that progesterone has anti-inflammatory capacity and the large amount of progesterone present during gestation might be sufficient to block any inflammatory response from the endometrium. Again prostaglandins have been suggested to interact in that mechanism (Siiteri, Febres, Clemens, Chang, Gondos, and Stites 1977; Kincl and Ciaccio 1980). The involvement of uteroglobin, under progesterone control has also been invoked at that level (Beier 1982; Mukherjee, Ulane, and Agrawal 1982).

The suppression of endometrial prostaglandin production and the concomitant increase in stored precursors such as arachidonic acid enables the uterus to be quiescent but still able to respond immediately to the different signals at the onset of parturition (Abel and Baird 1980; Kennedy 1980; Rothchild 1983).

Finally, the maintenance of uterine growth and plasticity may be the most important feature of pregnancy protection by progesterone, although ovariectomy experiments show that this protection requires oestradiol. It appears, therefore, that some oestradiol-progesterone cooperation occurs (Rothchild 1983).

About 20 proteins have been detected during gestation which are induced by progesterone though only a few appear interesting as markers of pregnancy evolution (Grudzinskas, Teisner, and Seppala 1982). Pregnancy associated plasma protein A (PAPP-A) was discovered in the plasma and trophoblast of pregnant women but it is also secreted by the endometrium of non-pregnant women during the late phase of the cycle (Bischof, Duberg, Schindler, Obradovic, Weil, Faigaux, Herrmann, and Sizonenko 1982), where it is under progesterone control (Sjoberg, Wahlstrom, and Seppala 1984). *In vitro* experiments on trophoblast and decidual explants showed that progesterone induction of PAPP-A is inhibited by RU486 (Bischof *et al.* 1986). PAPP-A prevents fibrinolysis by inhibiting plasmin and is involved in coagulation processes (Klopper 1982) as is placental protein 5, another progesterone induced protein, which has been identified as placental antithrombin III (Salem, Seppala, and Chard 1981). Placental protein 12 (PP12) is a glycoprotein found in the trophoblast, decidua, amniotic fluid and maternal serum, under progesterone control (Wahlstrom and Seppala 1984). Recent data by Koistinen, Kalkkinen, Huhtala, Seppala, Bohn, and Rutanen (1986) show that PP12 binds somatomedin and bears partial homology with a somatomedin-binding protein from human amniotic fluid.

Schwangerschaft Protein I (SPI) can be detected within 14 days of pregnancy (Grudzinskas, Gordon, Jeffery, and Chard 1977). It is in fact two related but different proteins (Teisner, Westegaard, Folkersen, Husby, and Svehag 1978).

Besides oestrogen and progesterone, relaxin is a third humoral component of pregnancy (Weiss 1984). It is a small peptide (6000 Daltons) consisting of two chains linked by disulfide bridges, that is produced by a cleavage of an internal peptide, as is the case for insulin. Insulin and relaxin have almost identical three-dimensional structures (James, Niall, Kwok, and Bryant-Greenwood 1977).

Relaxin was initially discovered in pig corpus luteum and later in rat, mouse, rabbit, human and other mammals (Zarrow and Brennam 1959). Relaxin is also synthesized by granulosa cells from preovulatory follicles. *In vitro* stimulation of relaxin production by cultured rat granulosa cells has been activated by a combination of prolactin, placental lactogen and progesterone. This *in vitro* effect has not been obtained with human cells. Circulating relaxin is considered a local hormone of the second half of pregnancy in rats, whereas in women plasma levels are maximal at the third month, then decrease about 20 per cent and stays at a plateau until parturition. Basically, the role of relaxin is to modify the plasticity and structure of the reproductive tract (at the level of interpubic ligament, uterus, myometrium and cervix) to accommodate pregnancy and facilitate delivery, by remodelling the connective tissue at the level of collagen and elastin (Too *et al.* 1984). Thus the relation between progesterone and relaxin appears much more complex than a simple induction of expression, which may in fact be mediated by progesterone-induced placental hormones such as placental lactogen or chorionic gonadotrophin (Quagliarello, Goldsmith, Steinetz, Lustig, and

Weiss 1980). The main relationship between relaxin and progesterone might be their synergistic action to protect pregnancy, particularly at the level of inhibition of uterine contractions. In males, relaxin has been found in seminal fluid, originating from the prostatic gland. Relaxin increases activity and life-span of spermatozoa (Essig, Schoenfeld, Amelar, Dubin, and Weiss 1982a; Essig, Schoenfeld, D'Eletto, Amelar, Dubin, Steinetz, O'Byrne, and Weiss 1982b). Thus, as an exocrine factor in males and an endocrine factor in females, relaxin is central to several aspects of the reproductive process.

Relaxin and progesterone are also both involved in the growth of the mammary gland by stimulating the growth of mammary glands ducts and alveola (Harness and Anderson 1977).

4   OTHER ORGANS

i   Chick oviduct

Schimke and Palmiter initiated the study of gene regulation in the chick oviduct by steroids (Palmiter, Christensen, and Schimke 1970; Schimke, McKnight, Shapiro, Sullivan, and Palacios 1975; Schimke, Pennequin, Robins, and McKnight 1977; O'Malley and Schrader 1978). Since then, a considerable amount of data has been published on the subject, as reviewed by Chambon, Dierich, Gaub, Jakowlev, Jongstra, Krust, Lepennec, Oudet, and Reudelhuber (1984) and O'Malley (1984).

Administration of progesterone to "withdrawn" immature chickens results in the stimulation of expression of a set of gene products termed "egg white proteins": ovalbumin, conalbumin (ovotransferrin), ovomucoid and lysozyme, in the tubular gland cells of the magnum part of the oviduct. "Withdrawn" chickens are immature animals that have been treated by injections of oestrogens for a short time and then left untreated until all biochemical parameters of oestrogen action were back within control values; when treated again with steroids these animals will respond in an amplified manner. Conalbumin is also called ovotransferrin due to its identity with the protein moiety of liver transferrin which is not hormonally regulated. This makes transferrin a "comparative" model, akin to rabbit uteroglobin (Lee, McKnight, and Palmiter 1977). The progesterone stimulation of these genes takes place at the level of transcription (Palmiter, Oka, and Schimke 1971; Palmiter 1972; McKnight, Pennequin, and Schimke 1975; Palmiter 1975; Tsai, Tsai, Schwartz, Kalimi, Clark, and O'Malley 1975; Palmiter, Moore, and Mulvihill 1976; Palmiter, Mulvihill, McKnight, and Senear 1977; Sutherland, Mester, and Baulieu 1977; McKnight 1978; Pennequin, Robins, and Schimke 1978; Seaver, Van Eys-Fuchs, Hoffman, and Coulson 1980; Sutherland, Greynet, Binart, Catelli, Schmelk, Mester, Lebeau, and Baulieu 1980; Palmiter, Mulvihill, Shephard, and McKnight 1981; Chambon et al. 1984).

Mediation of steroid hormone effects by a somatomedin factor has been invoked in the case of oestradiol stimulation of ovalbumin transcription. In contrast, no evidence for such mechanism has ever been shown in oestradiol induction of conalbumin or in the effect of progesterone on those two proteins (Evans, Hager, and McKnight 1981).

The genes coding for these proteins have all been cloned, both as cDNA and genomic sequences. They have been sequenced and subjected to extensive experimentation by deletion, truncation and site directed mutagenesis to ascertain the precise location of steroid-receptor complex action on the gene (see Chambon *et al.* 1984 for review). The molecular basis for receptor-gene interaction is discussed in **Section VII**.

ii   Mammary glands

Mammary glands are a major target for progesterone together with other mammotropic hormones (oestradiol, prolactin, growth hormone, insulin, glucocorticoids and thyroxin) (Houdebine, Teyssot, Devinoy, Olivier-Bousquet, Djiane, Kelly, Delouis, Kann, and Ferre 1983). Generally speaking progesterone acts as an anti-prolactin in the mammary gland *in vivo* (Houdebine 1976). *In vitro*, progesterone does not inhibit mammary explant growth stimulation, but amplifies the stimulatory effect of oestradiol.

The drop in progesterone levels at the end of pregnancy in the rat induces polymerization of tubulin in the mammary gland thus facilitating prolactin induced lactogenesis (Loizzi 1985). It is surprising that progesterone inhibits that effect of prolactin but not its mammotrophic effect. The molecular basis of these interactions is still a matter of debate due to contradictory reports, but evidence suggests a progesterone block of prolactin release, as discussed by Houdebine *et al.* (1983).

Regulation studies of the synthesis of rat casein and lactalbumin show that progesterone inhibits transcription and mRNA accumulation of milk proteins (Nakashi and Qasba 1979; Ganguly, Majumder, Ganguly, and Banerjee 1982). The initiation of casein synthesis in the rabbit at mid-pregnancy is concomitant with the drop in circulating progesterone. It is not the only stimulus: the decline in progesterone levels must be associated with a rise in prolactin to ensure the enhanced transcription of milk proteins (Harrington and Rothermel 1977). But progesterone is not able totally to counteract high doses of prolactin, at least at the level of casein synthesis. Thus progesterone has itself an antiprolactin effect on lactogenesis, independent of prolactin levels (Devinoy, Houdebine, and Olivier-Bousquet 1979).

Besides its antiprolactin effects, progesterone also inhibits the increase in prolactin receptors induced by prolactin itself (Djiane and Durand 1977).

iii   Central nervous system

The influence of steroid hormones on the brain, resulting in behavioural modifications has been known for many years. However the molecular analysis of that influence is only 10 years old (reviewed in Ehrardt and Meyer-Balhburg 1981; McEwen, Bigon, Davies, Kreg, Luine, McGinnis, Paden, Parsons, and Rainbow 1981). Progesterone receptors have been mapped in the brain and pituitary (McEwen, Davies, Gerlach, McLusky, McGinnis, Parsons, and Rainbow 1983). In contrast to other target organs, brain progesterone receptors are controlled by oestradiol only in certain parts of the brain, while other zones show non-inducible progesterone receptor concentrations. The hypothalamus and preoptic area also show testosterone inducible progesterone receptors, due to the presence of testosterone conversion into oestradiol by aromatase (McEwen *et al.* 1981). These regions also show noradrenergic positive control of progesterone receptor levels in guinea pigs. The mechanism of this action has not been elucidated. In the same area, progesterone has been shown to reduce the inhibitory influence of endogenous opioid peptides on LH secretion in oestradiol-primed ovariectomized rats and proestrous rats. The advancing and enhancement of the LH surge by progesterone may be due to that phenomenon (Gabriel, Simpkins, and Kalva 1983). In the classic hypothesis of sex-dimorphism, androgens were considered the major determinant. However progesterone is a *bona fide* anti-androgen and therefore may protect the female fetus brain (in certain species) from the effect of available circulating androgens. Later in adult life, progesterone contributes in concert with oestrogen to the differentiation of male and female sexual behaviour (McEwen *et al.* 1981). However, data from the literature appear sketchy and even contradictory in some cases, due to the complexity of experimental models and the probable multiplicity of interactive control mechanisms in brain functions. Takahashi and Lisk (1985) reported the action of progesterone at the level of different diencephalic areas in the golden hamster to inhibit aggression in oestrogen-primed females and induction of sexual receptivity. Richmond and Clemens (1986) reported that the cholinergic mediation of sexual receptivity in oestrogen-treated, ovariectomized rats does not involve progesterone receptors, since animals pretreated with RU486 can still be made receptive by intracerebral injections of the cholinergic agonist oxotremorine, 1-[4-(1-pyrrolidinyl)-2-butynyl]-2-pyrrolidinone). Conversely Brown and Braustein (1985) report that RU486 terminates the period of sexual behaviour in female guinea-pigs through a decrease in nuclear concentrations of hypothalamic progesterone receptors.

At the level of 5-hydroxytryptamine (5HT), progesterone influence on sexual behaviour is still a matter of debate. Wilson and Hunter (1985) reported that progesterone stimulates sexual behaviour in female rats by increasing 5HT concentration in the brain, and its availability for 5-HT_2 receptors.

However, it is now largely accepted that some neural circuits are permanently modified by sex hormones such as progesterone. This view has arisen from accumulating evidence of brain sexual dimorphism in most animals. In contrast with those persistent organizational effects, activation effects of progesterone appear to be mediated by rapid reversible inductions of gene expression in neurons.

Progesterone action on mating behaviour requires protein synthesis, as can be seen by the inhibitory effect of anisomycin. This inhibition takes place in less than one hour, suggesting an extremely fast turnover of the proteins involved (McEwen *et al.* 1981). Moreover, the effects of progesterone on gonadotropin secretion have been inhibited by actinomycin D, thus indicating *de novo* transcription (Jackson 1985). Hormones may also modulate behaviour by rapidly modifying the levels of synaptic components, thereby altering the response of neural circuits to triggering stimuli. The well-known study of oestrogen-induced lordosis in rats at the morphological and molecular level, has not yet been repeated for progesterone (McEwen *et al.* 1981).

iv   The immune system

Progesterone receptors have been detected in the mammalian thymus (Pearce *et al.* 1983) and in the chick bursa of Fabricius (Sullivan and Wira 1979; Ylikomi *et al.* 1985). Kincl and Ciaccio (1980) reported that progesterone increases skin graft survival in rodents. It was also reported that progesterone inhibits immune spleen cell function *in vivo* and *in vitro*. The mechanism has not been elucidated totally but seems to involve an increase in suppressor cell functions. Progesterone is also known to inhibit the phytohemagglutinin-induced lymphocyte transformation. As the progesterone doses utilized in these studies match physiological levels, progesterone could reasonably be considered as a native immunosuppressive agent acting to prevent fetal rejection during pregnancy, as discussed in **Section V**, 3 (see also Grossman 1984).

5   CANCER CELLS IN CULTURE

Rat pituitary cells (GH3) contain progesterone receptors (Naess *et al.* 1982). Administration of progesterone to these cells caused a 60 per cent decrease in prolactin production. Progesterone also inhibited the stimulatory effect of oestradiol production (Haug and Gautvik 1976).

Many human breast cancer cell lines contain progesterone receptors. They have been the subject of extensive work, with the purpose of detecting specific progesterone and oestradiol markers as tools for breast cancer prognosis, and surveillance (see **Section III**, 1, above). Most work has been done on MCF-7 (Brooks, Locke, and Soule 1973) and T47-D cells (Engel and Young 1978). MCF-7 progesterone receptors are oestrogen inducible and also respond to the agonist effect of low doses of tamoxifen ($< 0.1 \mu$M) (Horwitz, Koselei, and McGuire 1978). The T47-D line contains many subclones with different responses to hormonal stimuli *in vitro*. Generally speaking T47-D have high levels of progesterone receptors, very low levels of oestrogen receptors and are considered oestrogen- and antioestrogen-resistant (Horwitz *et al*. 1982). The T47-D$_{co}$ variant line contains constitutive high levels of progesterone receptor, while oestrogen receptors are absent. Progesterone exerts different general effects on these cell lines, such as a decrease in cell proliferation [MCF-7: Horwitz, Mockus, Pike, Fenresey, and Steridon (1983); and T47-D: McEwen *et al*. (1981)] and an increase in triglyceride biosynthesis (T47-D) (Judge and Chatterton 1983). The latter effect occurs at micromolar concentrations of progesterone and its significance is not understood.

Vignon, Bardon, Chalbos, and Rochefort (1983) report that R5020 has no effect of its own on cell growth but inhibits oestrogen induced growth stimulation. This was found even in the tamoxifen resistant T47-D subline, showing the two antioestrogenic effects to be different. RU486 also inhibits cell growth in MCF-7 but this effect is paradoxically counteracted by low doses of R5020. Cell lines devoid of progesterone receptors are insensitive to RU486 and the observed effect of RU486 on sensitive cells correlates with unoccupied progesterone receptor level. RU486 has no progestin-agonist activity on the induction of the 48kD and 250kD proteins (Bardon, Vignon, Chalbos, and Rochefort 1985). Parallel studies have shown that R5020 inhibits the growth of T47-D$_{co}$ cells while increasing insulin binding sites in the cell membrane (Horwitz 1985). In contrast, RU486 also inhibits cell growth but fails to stimulate insulin receptors, and even partially impedes their induction by R5020. Thus RU486 displays both agonist and antagonist properties, depending on the mode of biological evaluation (Horwitz and Freidenberg 1985).

Cell culture data should be considered with caution since it has been recently shown that phenol red, the commonly used pH indicator in cell culture, has oestrogenic activity at the concentration used. Knowing the multiple physiological interactions between oestradiol and progestins on cell metabolism and proliferation it is reasonable to assume that the definition of "steroid-free control experiments" should be thoroughly reassessed in experiments involving oestrogen-responsive cell lines. As an example, phenol red increases the progesterone receptor concentration by 300 per cent in MCF-7 cells (Berthois, Katzenellenbogen and Katzenellenbogen 1980).

Several proteins induced by progestins have been detected in either MCF-7 or T47-D cells: a 48kD protein in T47-D cells (Chalbos and Rochefort 1984*a*) and a 250kD protein in MCF-7 and T47-D (Chalbos and Rochefort 1984*b*). The cDNA clone for the 250kD protein has recently been isolated. As shown by dose-response studies, R5020 elicits a quick accumulation of the 250kD protein mRNA prior to protein increase and secretion (Chalbos, Westley, May, Alibert, and Rochefort 1986). In this latter case, *in vitro* effects of progestins on cultured cells appear to involve a genomic interaction of the receptor and subsequent increase of transcription, similar to that observed *in vivo*.

Lactate dehydrogenase is induced by progestins in T47-D cells but not in MCF-7 cells. The difference lies probably in the higher content of progesterone receptor in T47-D cells (Hagley and Moore 1985). Progestins also increase more than twofold membrane binding sites for the various peptide hormones involved in human breast development such as prolactin (Murphy, Stead, Sutherland, and Lazarus 1986a), growth hormone (Murphy, Sutherland, and Lazarus 1985) and EGF (Murphy, Sutherland, Stead, Murphy, and Lazarus 1986b).

Again, variant cell lines show different results since T47-D$_{co}$ respond to progestins by an increase in insulin receptors (Horwitz and Freidenberg 1985) while wild type T47-D do not (Murphy et al. 1986a). Those increases in membrane receptors occur without modification of affinity for the ligands. The case for EGF is particularly interesting since this hormone stimulates DNA synthesis and cell proliferation while inhibiting milk protein synthesis, as with progesterone. However, the presence of progesterone receptors in breast cancer is considered a marker of differentiation of the tumour, and an indication for better prognosis of tumour evolution (notably in terms of metastases, and response to hormonal therapy) (Pichon, Pallud, Brunet, and Milgrom 1984). The relationship between hormones and their eventual co-operation or antagonism in breast cancer development remains an enigma. For example, it is surprising that progesterone decreases prolactin binding in the mammary gland *in vivo* (Djiane and Durand 1977) yet stimulates that binding in cancer cell lines *in vitro* (Murphy et al. 1986a).

6 MODULATION OF OTHER STEROID HORMONES

Progesterone exerts antagonistic activities against other steroid hormones (for review see Bardin, Milgrom, and Mauvais-Jarvis 1983). The action on oestrogens is explained by a reduction of oestrogen receptor replenishment, thus decreasing the total number of binding sites for oestradiol (Hsueh, Peck, and Clark 1976; Koligian and Stormshak 1977; West and Brenner 1985). Another explanation is given by the induction of 17β hydroxysteroid dehydrogenase that converts oestradiol into oestrone. This increased enzymatic activity is associated with a lower uptake of oestradiol. Some authors favour this latter explanation (Gurpide 1978). As a consequence progesterone inhibits the oestrogen induction of protein synthesis in the uterus (Bhakoo and Katzenellenbogen 1977) including the glucose-6-phosphate dehydrogenase (Swanson and Barker 1983), and of prolactin in GH$_3$ cells in culture (Haug 1979).

At the level of androgens, progesterone inhibits the 5α reduction of testosterone by competition for the reductase, and specific binding of dihydrotestosterone by competition for the receptor in human skin homogenates (Wright, Kirchhoffer, and Giacomini 1980). Indeed, progesterone appears as a natural antiandrogen, and has been suspected of protecting the female fetus from the influence of available androgens, notably at the brain level (Ehrhardt and Meyer-Bahlburg 1981).

Similarly, the discovery of the competition by progesterone with aldo-sterone binding to the tubular mineralocorticoid receptor has led to the synthesis of spironolactones (Wambach and Higgins 1978). The result of that competition is an increased natriuresis under progesterone stimulation.

Depending on the experimental model, progesterone exerts antagonistic (Obinata, Takata, Kawada, Hirano, and Endo 1984) or partial agonistic activities (Hackney, Holbrook, and Grasso 1981) on glucocorticoids, mostly by interacting with the glucocorticoid receptor. Elevated concentrations of progesterone provoke a glucocorticoid-like inhibition of bone collagen and protein synthesis *in vitro* (Canalis and Raisz 1978).

VI   Membrane-bound activities of progesterone

The reinitiation of meiosis in *Xenopus* oocytes is induced by progesterone secretion from the surrounding follicle cells. It is to be noted that very high concentrations of progesterone are required for this effect, and that it is also brought about by androgens, glucocorticoids, progesterone precursors or metabolites and antiandrogens such as cyproterone. Oestrogen activity is weak or non-existent (see Baulieu 1983*b* for review). *In vitro* experiments on that model are performed on oocytes freed from follicles and in a medium to which micromolar amounts of progesterone have been added. These conditions allow observation of all meiotic events. During progesterone induction of meiosis, cell division and protein synthesis are stimulated. This activity occurs in enucleated oocytes or upon treatment by transcription inhibitors. No nuclear receptor could be detected, but a membrane "receptor" for progesterone (M.W. 30 kD, K_d 4 μM) has been recently reported (Blondeau and Baulieu 1984). Similar results have been presented for other amphibians (Cloud and Schuetz 1977; Kostellow, Weinstein, and Morrill 1982; Morrill, Ziegler, Kumar, Weinstein, and Kostellow 1984). Progesterone inhibits vitellogenin uptake by the oocyte, probably as a prerequisite for maturation induction (Schuetz 1977). Among the progesterone induced proteins, the "maturation promoting factor" (MPF) appears fundamental in the maturation process, by causing germinal vesicle breakdown, chromosome and spindle anchorage, and finally polar body expulsion. Progesterone also inhibits the Na/K dependent ATPase (Weinstein, Kostellow, Ziegler, and Morrill 1982) and membrane bound adenylate cyclase (Sadler and Maller 1981; Finidori, Hanoune, and Baulieu 1982; Jordana, Olate, Allende, and Allende 1984). The latter effect causes a decrease in cyclic AMP concentration, which is generally observed in the occurrence of cell division (Pastan, Johnson, and Anderson 1975).

This effect is dose-dependent and does not require protein synthesis but its mechanism remains unknown. Recently Blondeau and Baulieu (1985) reported that progesterone inhibits the phosphorylation of a membrane protein of 48kD in *Xenopus* oocytes. Insulin and insulin-like growth factors strongly potentiate progesterone activity at nanomolar levels. The mechanism of the thousand-fold potentiation of adenylate cyclase inhibition has not been elucidated (Bauleiu and Schorderet-Slatkine 1983).

VII Action of progesterone at the genomic level

In the course of transcription-stimulation of specific genes by progesterone, the hormone-receptor complex interacts with active chromatin, essentially with DNA at particular places in or around the stimulated gene (reviewed in Chambon *et al.* 1984; O'Malley 1984; Yamamoto 1985). The actual mechanism in unknown but is likely to be complex as one can observe extensive modifications of chromatin throughout the nucleus upon hormone administration (Perrot-Applanat, Groyer-Picard, Logeat, and Milgrom 1986). The steroid-receptor complex has been reported to associate with almost every component of chromatin outside the DNA (review in Anderson 1984), especially with subclasses of non-histone proteins (Thrall and Spelsberg 1980). Most of the data obtained so far are too sketchy to provide anything but hypothesis, although tripartite interactions "Non-Histone Proteins/DNA/Receptor" are a definite possibility. However, some experimental results may be put together to create a plausible working hypothesis: RNA is transcribed at the nuclear matrix in a closely associated manner (Jackson, McReady, and Cook 1981), and steroid-receptor complexes have been reported to interact with the nuclear matrix (Barrack and Coffey 1982). These data bring out the hypothesis that steroid hormones may tightly link the inducible promoter to tissue specific matrix determinants resulting in the stabilized initiation of transcription. This hypothesis, if correct, would also imply that receptor-DNA interaction should carry gene specificity of transcription while receptor-matrix interaction should be responsible for tissue specificity.

There was new interest in DNA-receptor interaction studies when the purification of receptors allowed the study of their interaction *in vitro* with purified DNA from specific hormone-regulated genes (Payvar, Wrang, Carlstedt-Duke, Okret, Gustafsson, and Yamamoto 1981). High affinity binding sites were detected upstream to the initiation of transcription and within the gene sequence (Payvar, De Franco, Firestone, Edgar, Wrange, Okret, Gustafsson, and Yamamoto 1983). These data fit Yamamoto's model that hormone-regulated genes should be modified in totality by hormone-receptor complexes prior to activation of transcription (Yamamoto and Alberts 1976). Apart from acellular *in vitro* binding experiments of purified receptor with specific DNA, the basis of our knowledge lies in mapping regions of genes that bind receptors and confer hormone regulation when linked to a previously hormone-unrelated gene. However these studies have been done *in vivo*, using microinjected or calcium-phosphate-transferred DNA, and to date no *in vitro* system has ever reproduced the stimulating effect of progestin-receptor complexes on transcription.

1   THE UTEROGLOBIN GENE IN RABBIT ENDOMETRIUM

With the purification of progesterone receptor by Logeat *et al.* (1981) the filter binding studies used by Payvar *et al.* (1981) have been adapted to the progesterone receptor.

The preparation of monoclonal antibodies and subsequent immunopurification of the progesterone receptor in our group enabled us to prepare highly purified progesterone receptor, under forms that had never been obtained before, that is, ligand free or antagonist bound. The nuclear bound, *in vivo* phosphorylated progesterone-receptor complexed with the hormone was also purified by this technique. These different preparations of receptor have been used to map binding sites in the uteroglobin gene. These studies did not show any binding at the promoter region but detected it very far upstream (-2700 to -2500) and within the first intron ($+300$ and $+1000$) (Bailly, Le Page, Rauch, and Milgrom 1986). The hormone-dependency of PR-DNA interaction is observed only *in vivo* and in crude cellular extracts but not *in vitro* with purified receptor. In all cases, using ligand bound, ligand free, antagonist bound and phosphorylated receptor, identical footprints were obtained. All sites contain a common TGTTCACT sequence. These data suggest that *in vivo* the unliganded receptor is stabilized in an "unactivated form" by some nuclear component(s) while *in vitro* it can freely activate and bind DNA even in the absence of hormone. These factors are unknown at present: a possible candidate is the 90 kD heatshock protein (Catelli *et al.* 1985). One can also deduce that antihormone binding (and possibly receptor phosphorylation) elicits a transcriptional effect through a mechanism different from the original interaction between the receptor and the gene, and probably posterior to it.

One should however keep in mind that naked DNA is not the physiological substrate of hormone action, and whatever is missing *in vitro* must be present in the chromatin (Anderson 1984; Thrall and Spelsberg 1980). Further studies will certainly have to cope with supplementary parameters such as tissue-specific factors, chronology of events and topological state of the substrate.

None of these binding sites contain sequences analogous to the "consensus binding sequence" described by Mulvihill *et al.* in the chick oviduct ovalbumin (Mulvihill, Lepennec, and Chambon 1982).

Using the chicken lysozyme gene, Von der Ahe, Janich, Scheidereit, Renkawitz, Schultz, and Beato (1985) showed that the progesterone receptor and the glucocorticoid receptor bind at similar places, closely upstream to the promoter (-400 to -200). These data were obtained with partially (30 per cent) purified receptor-hormone complexes, prepared by affinity chromatography and phosphocellulose chromatography. The detected sequences were homologous to the glucocorticoid responsive element consensus (GRE).

Similar studies on the uteroglobin gene were performed by Cato, Geisse,

Wenz, Westphal, and Beato (1984), using the glucocorticoid receptor. Specific DNA binding was detected in an upstream 650 bp fragment (-3250 to -2600). Three binding sites were further defined by DNAseI footprinting and methylation protection experiments. All three contained the hexanucleotide 5'-TGTTCT-3' and were found in close homology with the consensus derived from other glucorcorticoid induced genes (Payvar, De Franco, Firestone, Edgar, Wrange, Okret, Gustafsson, and Yamamoto 1983). These sites correspond to the distal half of the binding region described by Bailly, Le Page, Rauch, and Milgrom (1986) for the progesterone receptor.

Unfortunately, uteroglobin is poorly expressed when transfected into hormone responsive cell lines, and does not show any regulation (Cato *et al.* 1984). Therefore, mutational studies have not been pursued to the point which has been attained in the case of egg white protein genes.

2 EGG WHITE PROTEIN GENES

Using a filter binding assay, Mulvihill *et al.* (1982) have detected a preferential association of the progesterone receptor complex to sequences immediately upstream from transcription initiation (around -200), but also within the gene. A consensus sequence for progesterone receptor binding was deduced from the comparison of these sites.

O'Malley and his collaborators have used DNAse protection (Schrader *et al.* 1981) and filter binding assays (Compton, Schrader, and O'Malley 1983) to assess the binding regions of the progesterone receptor in the ovalbumin promoter. They report a main binding site around -150 to -190 bases upstream from the initiation start, but also several other sites throughout the gene. These results should be considered with caution since the same group later reported that their semi-purified preparation of progesterone receptor is extensively contaminated by distinct 108 kD, 90 kD and 78 kD heatshock proteins (Kulomaa, Weigel, Kleinsek, Beattie, Conneely, March, Zarucki-Schulz, Schrader, and O'Malley 1986; Sargan, Tsai, and O'Malley 1986). Extensive discussion of these data can be found throughout the relevant literature. However, similar results were obtained *in vivo* by transient expression experiments with the lysozyme gene, where binding takes place between -200 and -80 upstream of transcription initiation (Von der Ahe *et al.* 1985). When primary cultures of chick oviduct cells are transfected with the ovalbumin gene, the gene is regulated by progesterone during the transient expression period. The responsive region of the gene has been mapped to the -220 to -95 region and it also corresponds to the oestradiol responsive region (Dean, Knoll, Riser, and O'Malley 1983; Dean, Gope, Knoll, Riser, and O'Malley 1984). When the chicken lysozyme gene was microinjected into oviduct cells, the responsive region was similarly located between -140 and -220, as reported by Renkawitz, Beug, Graf, Matthias, Grez, and Schutz (1982).

Chambon and his colleagues have addressed the problem of relationship between hormonal regulation and tissue specificity, by microinjecting the ovalbumin gene (intact or mutated) into various cell lines and primary cultures (Chambon *et al.* 1984). They report that egg white protein genes (ovalbumin, ovotransferrin) do not function in non-chicken cells, including the hormone responsive MCF-7 cells. They function in primary cultures of chicken hepatocytes and oviduct cells, but not in fibroblasts from the same organs. Ovotransferrin is constitutively expressed in hepatocytes and partially regulated by steroids in oviduct cells. The ovalbumin gene contains a specific element in the -420 to -295 region of its promoter that blocks basal expression in oviduct cells, while hepatocytes allow constitutive expression. Upon progesterone or oestradiol administration to oviduct cells, the block is relieved and the gene is transcribed. When deleting upstream sequences down to -295, the blockage disappears, and thus progesterone "derepression" is not observed any more.

This is even more puzzling when one considers that the native ovalbumin gene is not expressed in the liver where vitellogenin is oestrogen-regulated, while ovalbumin is oestrogen dependent in the oviduct where vitellogenin is not expressed. Chambon's results clearly suggest the existence of a tissue specific factor which inhibits constitutive expression and is "repressed" in some way by the steroid-receptor complex. The crossed inactivation of ovalbumin and vitellogenin in the two receptor-containing organs cannot be explained by this mechanistic hypothesis. It must be supposed that superior orders of developmental regulation exist at the level of entire chromatin domains (Chambon *et al.* 1984).

Another field of interest would be the eventual effect of those putative repressors on gene repression following hormone withdrawal and their location during withdrawal.

3   MOUSE MAMMARY TUMOUR VIRUS, MMTV

The promoter in the LTR (Long Terminal Repeat) of MMTV (Mouse Mammary Tumour Virus) is under hormonal control by glucocorticoids. The high affinity binding of glucocorticoid receptor complexes to the upstream regions of this promoter has been the first of such interactions to be documented (Payvar *et al.* 1981). It has since been reported that the binding region elicits hormone-responsive transcription of the MMTV genome in transfection experiments (Hynes, Kennedy, Rahmsdorf, and Groner 1981) and that it can be dissociated from the viral genome and retain its hormone dependent-enhancer capabilities even in heterologous constructions (Ponta, Kennedy, Skroch, Hynes, and Groner 1985). The progesterone receptor from rabbit uterus and the rat liver glucocorticoid receptor bind very similarly to the lysozyme and the MMTV promoters, as shown by the footprinting experiments of Von der Ahe *et al.* (1985). This line of evidence suggested that

MMTV may be an interesting model in the study of progesterone action through transfection of receptor containing mammalian cells by reconstructed genes (Cato *et al.* 1984). Indeed MCF-7 and T47-D cells have been successfully transfected with MMTV-TK-CAT constructions (MMTV-Thymidine Kinase promoter-Chloramphenicol Acetyl Transferase coding region). The transient expression of chloramphenicol acetyl transferase was enhanced by dexamethasone alone in MCF-7 cells and by progesterone and dexamethasone in T47-D cells. The hormonal effect was observed in both orientations of the hormone binding domain while removal of the proximal half of the binding region resulted in loss of hormonal control (Cato, Miksicek, Schutz, Arnemann, and Beato 1986). Other authors have reported that deletion of the proximal half of the binding site decreased hormonal effect to 30 per cent of controls and could be restored in totality by duplication of the remaining distal part (Kuhnel, Buetti, and Diggelman 1986).

4   THE CURRENT MODEL

The overlapping of binding sites for glucocorticoid and progesterone receptors in the MMTV genome (but also in the uteroglobin gene) has prompted Yamamoto to coin the term "Hormone Responsive Elements" or HRE for these pluri-sensitive sites (Yamamoto 1985).

The great distance between these HRE and the corresponding promoters is reminiscent of enhancers, and authors such as Parker (1983) have also called them "Hormone-dependent Enhancers". HRE have been shown to be frequently associated with basal enhancers, as is the case in the glucocorticoid HRE of metallothionein (Scholer, Haslinger, Heguy, Holtgreve, and Karin 1985), the oestradiol HRE of vitellogenin associated with the elegans box (Klein-Hitpass, Schorpp, Wagner, and Ryffel 1986), and the HRE of MMTV (Kuhnel *et al.* 1966). Exact "core sequences" of the SV40 enhancer (Yaniv 1984) have also been found within the upstream HRE of the uteroglobin gene (Savouret, unpublished observations). Electron microscopy experiments have shown that progesterone receptors (PR) oligomers bind to the HRE of MMTV and uteroglobin. The interaction of purified PR with uteroglobin HRE, organized in doublets in the native DNA molecule caused the intervening DNA to form loops. The size of proteins bridging the DNA loops was approximately twice that of oligomers of PR involved in single site interactions, indicating that loop formation was caused by protein-protein interactions rather than DNA interactions. The MMTV LTR promoter displayed a similar situation, but in this case the two sites were too close together and it was difficult to decide whether the receptors were strongly interacting or merely side by side apposed. The first hypothesis appeared more likely because no receptor was ever observed to be in a "trans" position, and the presence of high concentrations of DNA caused the PR oligomers to bridge independent DNA fragments (Théveny, Bailly, Rauch, Delain, and Milgrom

1987). The existence of cooperative processes in these interactions is currently under study in our laboratory. A similar situation has been described by Ptashne (1984) in the case of prokaryotic regulations such as the transcription switch of bacteriophage lambda. The regulatory proteins bind to adjacent sites and interact together to provoke protein-protein interactions which are responsible for the subsequent binding of transcription factors and/or RNA polymerase to DNA. During these interactions, the DNA between the sites is constrained and loops out. The loop model gives a plausible explanation to the presence of several HREs scattered upstream and within all steroid-regulated genes (Klein-Hitpass *et al.* 1986). It is supported by transfection experiments which show that deletions of one HRE out of a naturally occurring pair decreases the efficiency of hormonal regulation on transcription (Kuhnel *et al.* 1986).

The function of such a receptor-receptor association following binding to the HREs might be to induce a subsequent protein-protein interaction with transcription factors and/or RNA polymerase. This may happen at any distance from the promoter since the intervening DNA loops out. Cordingley, Riegel, and Hager (1987) report that glucocorticoid receptor binding to the HRE of MMTV occurs simultaneously with the binding of Nuclear Factor 1 (NF-1) on the promoter.

Mutagenesis experiments on the progesterone receptor are now in progress to elucidate these interactions and correlate them with biological activities.

VIII   Conclusion

Although substantial progress has been achieved in the last years, much remains to be done, especially at the level of the molecular basis of gene stimulation by progesterone-receptor complexes. To date, no *in vitro* system has been developed that could reproduce hormone action through stepwise reconstitution of the nuclear environment. Transient expression experiments using MMTV binding sites coupled to promiscuous eukaryotic promoter to transfect receptor-containing cell lines provide an interesting alternative, currently under study in several laboratories including ours. The discrepancies bertween DNA binding studies performed *in vivo* and *in vitro*, where in the latter case, both hormone binding and phosphorylation are not required for an effective association of the receptor to its specific sites, raise numerous questions. What is the exact role of the hormone? What nuclear components stabilize the unbound receptor away from the DNA? What is the meaning of the phosphorylation? This latter line of investigation leads naturally to the identification and isolation of receptor associated kinases, which may provide a further level of regulation within hormone action. At present, a common hypothesis favours the concept of desensitization and processing of the phosphorylated receptor, by analogy to membrane bound receptors.

Moreover, since hormone antagonists do not modify the interaction between receptors and DNA, at least in terms of localization, their mode of action will have to be searched for throughout the path of hormone-receptor complexes in the nucleus.

Finally, sequence analysis of the extensive homologies between the different hormone receptors and certain oncogenes may also contribute to a greater understanding of hormone action through evolutionary studies of the primary structure of hormone binding proteins.

Abbreviations and trivial names

R5020:   Promegestone, 17α 21-dimethyl-19-nor 4,9-pregnadiene-3,20-dione.

RU38486:   17β-hydroxy-11β-(4-dimethyl aminophenyl) 17α-(1-propyl) estra-4,9-diene-3-one.

ORG5058:   16α-ethyl-21-hydroxy-19-nor pregn-4-ene-3,20-dione.

MCF-7:   Michigan Cancer Foundation breast cancer cell line. A pleural metastatic cell line from a human breast cancer.

T47-D: Breast cancer cell line. A peritoneal metastatic cell line from a human breast cancer.

Acknowledgements

This work was supported by the Institut National de la Santé et de la Recherche Médicale (INSERM), the Centre National de la Recherche Scientifique (CNRS), the Unité d'Enseignement et de Recherche Kremlin-Bicêtre, the Association pour la Recherche sur le Cancer (ARC) and the Foundation pour la Recherche Médicale Française (FRMF). N. Malpoint has typed the manuscript.

References

Abel, M. H. and Baird, D. T. (1980). The effect of 17 beta-oestradiol and progesterone on prostaglandin production by human endometrium maintained in organ culture. *Endocrinology* **106**, 1599–1606.

Agrawal, A. (1983). *Principles of Receptorology*. Walter de Gruyter Press, Berlin.

Albaladejo, V. and André, J. (1986). Induction of progestin receptors in the rat MtTF4 tumor of pituitary origin whose growth is inhibited by oestradiol. *Molec. Cell. Endocr.* **44**, 109–115.

Allegra, J. C., Korat, O., Do, H. M., and Lippman, M. (1981). The regulation of progesterone receptor by 17 beta-oestradiol and tamoxifen in the Zr-75-1 human breast cancer cell line in defined medium. *J. Recept. Res.* **2**, 17–27.

Alwachi, S. N., Bland, K. P., and Poyser, N. L. (1981). Prostaglandin production by the progesterone-dominated sheep uterus. *Prostaglandins Med.* **6**, 571–576.

328   J. F. Savouret, M. Misrahi, and E. Milgrom

Amr, S., Faye, J. C., Bayard, F., and Kreitmann, O. (1980). Induction of rat endometrial oestradiol-17 beta dehydrogenase activity by oestradiol and progesterone. *Biol. Reprod.* **22**, 159–163.

Anderson, N. S. (1984). The effect of steroid hormones on gene transcription. In *Biological Regulation and Development* Vol. 3B (eds. R. F. Goldberger and K. R. Yamamoto) pp. 169–212. Plenum Press, New York.

Anderson, W. A., Desombre, E. R., and Kang, Y. H. (1977). Oestrogen-progesterone antagonism with respect to specific marker protein synthesis and growth by the uterine endometrium. *Biol. Reprod.* **16**, 409–419.

Bailly, A., Sallas, N., and Milgrom, E. (1977). A low molecular weight inhibitor of steroid receptor activation. *J. biol. Chem.* **252**, 858–863.

—— Lefevre, B., Savouret, J. F., and Milgrom, E. (1980). Activation and changes in sedimentation properties of steroid receptors. *J. biol. Chem.* **255**, 2729–2734.

—— Atger, M., Atger, P., Cerbon, M. A., Alizon, M., Vu Hai, M. T., Logeat, F., and Milgrom, E. (1983). The rabbit uteroglobin gene. Structure and interaction with the progesterone receptor. *J. biol. Chem.* **258**, 10384–10389.

—— Le Page, C., Rauch, M., and Milgrom, E. (1986). Sequence specific DNA binding of the progesterone receptor to uteroglobin gene. Effects of hormone, antihormone and receptor phosphorylation. *E.M.B.O. J.* **5**, 3235–3241.

Bardin, C. W., Milgrom, E., and Mauvais-Jarvis, P. (eds.) (1983). *Progesterone and Progestins.* Raven Press, New York.

Bardon, S., Vignon, F., Chalbos, D., and Rochefort, H. (1985). RU486, A progestin and glucocorticoid antagonist, inhibits the growth of breast cancer cells via the progesterone receptor. *J. clin. Endocr. Metab.* **60**, 692–697.

Bareither, M. L. and Verhage, H. G. (1980). Effect of oestrogen and progesterone on secretory granule formation and release in the endometrium of the ovariectomized cat. *Biol. Reprod.* **22**, 635–643.

Barrack, E. R. and Coffey, D. S. (1982). Biological properties of the nuclear matrix: steroid hormone binding. *Recent Prog. Horm. Res.* **38**, 133–189.

Batra, S. and Losif, S. (1985). Progesterone receptors in human vaginal tissue. *Am. J. Obstet. Gynec.* **153**, 524–528.

Baulieu, E. E. (1983a). The progesterone receptor. In *Progestogens in Therapy.* (ed. G. Benagiano) pp. 27–38. Raven Press, New York.

—— (1983b). Steroid membrane adenylate-cyclase interactions during *Xenopus laevis* oocyte meiosis reinitiation. In *Progesterone and Progestins* (eds. C. Wayne-Bardin, E. Milgrom and P. Mauvais-Jarvis) pp. 91–108. Raven Press, New York.

—— Atger, M., Best-Belpomme, M., Corvol, P., Courvalin, J. C., Mester, J., Milgrom, E., Robel, P., Rochefort, H., and de Catalogne, D. (1975). Steroid hormones receptors. *Vitams. Horm.* **33**, 649–736.

—— Binart, N., Buchou, T., Catelli, M. G., Garcia, T., Gasc, J. M., Groyer, A., Joab, I., Moncharmont, B., Radanyi, C., Renoir, M., Tuohimaa, P., and Mester, J. (1983). Biochemical and immunological studies of the chick oviduct cytosol progesterone receptor. In *Steroid Hormone Receptors: Structure and Function.* Nobel Symposium n° 57 (eds. H. Eriksson and J. A. Gustafsson) pp. 45–72. Elsevier, Amsterdam.

—— and Schorderet-Slatkine, S. (1983). Steroid and peptide control mechanisms in membranes of *Xenopus laevis* oocytes resuming meiotic division. In *Molecular Biology of Egg Maturation. Ciba Fdn. Symp.* Vol. 98, 137–158. Pitman Press, London.

Baxter, J. D., Rousseau, G. G., Benson, M. L., Garcea, R. L., Ito, J., and Tomkins, G. M. (1973). Role of DNA and specific cytoplasmic receptors in glucocorticoid action. *Proc. natn. Acad. Sci. U.S.A.* **69**, 1892–1896.

Beato, M. (1979). *Steroid Induced Uterine Proteins*. Elsevier Publishing Co., Amsterdam.

Beatty, C. H., Bocek, R. M., Herrington, P. T., Young, M. K., and Brenner, R. M. (1979). Effects of oestradiol-17 beta and progesterone on cyclic nucleotide metabolism in myometrium of macaques. *Biol. Reprod.* **21**, 309–318.

Beck, P. (1977). Effects of progestins on glucose and lipid metabolism. In *Biochemical Actions of Progesterone and Progestins. Ann. N.Y. Acad. Sci.* **286**, 434–445.

Beier, H. M. (1968). Uteroglobin: a hormone sensitive endometrial protein involved in blastocyst development. *Biochem. Biophys. Acta.* **160**, 289–291.

—— (1982). Uteroglobin and other endometrial proteins: biochemistry and biological significance in beginning pregnancy. In *Proteins and steroids in early pregnancy* (eds. H. M. Beier and P. Karlson) pp. 39–71. Springer-Verlag, Berlin.

—— and Mootz, U. (1979). Significance of maternal uterine proteins in the establishment of pregnancy. In *Maternal Recognition of Pregnancy. Ciba Fdn. Symp.* Vol. 64, 111–140, Excerpta Medica, Amsterdam.

Berthois, Y., Katzenellenbogen, J. A., and Katzenellenbogen, B. S. (1986). Phenol red in tissue culture media is a weak oestrogen: implications concerning the study of oestrogen-responsive cells in culture. *Proc. natn. Acad. Sci. U.S.A.* **83**, 2496–2500.

Bhakoo, H. S. and Katzenellenbogen, B. S. (1977). Progesterone antagonism of oestradiol stimulated uterine induced protein synthesis. *Molec. Cell. Endocr.* **8**, 105–112.

Bischof, P., Duberg, S., Schindler, A. M., Obradovic, D., Weil, A., Faigaux, R., Herrmann, W. L., and Sizonenko, P. C. (1982). Endometrial and plasma concentrations of pregnancy associated plasma protein A (PAPP-A). *Br. J. Obstet. Gynaecol.* **89**, 701–703.

—— Sizonenko, M. T., and Herrmann, W. L. (1986). Trophoblastic and decidual response to RU486: effects on human chorionic gonadotrophin, human placental lactogen, prolactin and pregnancy associated plasma protein A production *in vitro. Human Reprod.* **1**, 3–6.

Blondeau, J. P. and Baulieu, E. E. (1984). Progesterone receptor characterized by photoaffinity labelling in the plasma membrane of *Xenopus laevis* oocytes. *Biochem. J.* **219**, 785–792.

—— —— (1985). Progesterone-inhibited phosphorylation of an unique Mr 48,000 protein in the plasma membrane of *Xenopus laevis* oocytes. *J. biol. Chem.* **260**, 3617–3625.

Boomsma, R. A., Jaffe, R. C., and Verhage, H. G. (1982). The uterine progestational response in cats: changes in morphology and progesterone receptors during chronic administration of progesterone to oestradiol-primed and non-primed animals. *Biol. Reprod.* **26**, 511–521.

Brawerman, B. A., Bagni, A., de Ziegler, D., Den, B., and Gurpide, E. (1984). Isolation of prolactin-producing cells from first and second trimester decidua. *J. clin. Endocr. Metab.* **58**, 521–525.

Brenner, R. M. and West, N. B. (1975). Hormonal regulation of the reproductive tract in female mammals. *A. Rev. Physiol.* **37**, 273–302.

Brooks, S. C., Locke, E. R., and Soule, H. D. (1973). Estrogen receptor in a human cell line (MCF-7) from breast carcinoma. *J. biol. Chem.* **248**, 6251–6253.

Brown, T. J. and Braustein, J. D. (1986). Abbreviation of the period of sexual behaviour in female guinea-pigs by the progesterone antagonist RU486. *Brain Res.* **373**, 103–113.

Brunner, N., Spang-Thornsen, M., Skovgard-Poulsen, H., Engelholm, S. A., Nielsen, A., and Vindel, L. (1985). Endocrine sensibility of the receptor-positive T-61 human breast carcinoma serially grown in nude mice. *Int. J. Cancer* **35**, 59–64.

Canalis, E. M. and Raisz, L. G. (1978). Effect of sex steroids on bone collagen synthesis *in vitro. Calcif. Tissue Res.* **25**, 105–110.

Carter, J. (1984). The maternal immunological response during pregnancy. In *Oxford Reviews of Reproductive Biology* Vol. **6** (ed. J. R. Clarke) pp. 47–128. Clarendon Press, Oxford.

Catelli, M. G., Binart, N., Jung-Testas, I., Renoir, J. M., Baulieu, E. E., Feramisco, J. R., and Welch, W. J. (1985). The common 90 Kd protein component of non transformed 8 S steroid-receptors is a heat-shock protein. *E.M.B.O. J.* **4**, 3131–3135.

Cato, A. C. B., Geisse, S., Wenz, M., Westphal, H. M., and Beato, M. (1984). The nucleotide sequences recognized by the glucocorticoid receptor in the rabbit utero-globin gene region are located far upstream from the initiation of transcription. *E.M.B.O. J.* **3**, 2771–2778.

—— Miksicek, R., Schutz, G., Arnemann, J., and Beato, M. (1986). The hormone regulatory element of mouse mammary tumor virus mediates progesterone induction. *E.M.B.O. J.* **5**, 2237–2240.

Chalbos, D. and Rochefort, H. (1984*a*). Dual effects of the progestin R5020 on proteins released by the T47-D human breast cancer cells. *J. biol. Chem.* **259**, 1231–1238.

—— —— (1984*b*). A 250 kilodalton cellular protein is induced by progestins in two human breast cancer cell lines MCF-7 and T47-D. *Biochem. Biophys. Res. Commun.* **121**, 421-427.

—— Westley, B., May, F., Alibert, C., and Rochefort, H. (1986). Cloning of cDNA sequences of a progestin regulated mRNA from MCF-7 human breast cancer cells. *Nucleic Acids Res.* **14**, 965–982.

Chambon, P., Dierich, A., Gaub, M. P., Jakowlev, S., Jongstra, J., Krust, A., Lepennec, J. P., Oudet, P., and Reudelhuber, T. (1984). Promoter elements of genes coding for proteins and modulation of transcription by estrogens and progesterone. *Recent Prog. Horm. Res.* **40**, 1–42.

—— —— —— Astinotti, D., Lepennec, J. P., and Touitou, Y. (1985). Steroid hormones relieve repression of the ovalbumin gene promoter in chick oviduct tubular gland cells. In *Proc. 7th Int. Congr. Endocr.*, Quebec, July 1984, (eds. F. Labrie and L. Proulx) pp. 3–10. Elsevier Sciences Publications, Amsterdam.

Cheng, S. V., McDonald, B. S., Clark, B. F., and Pollard, J. W. (1985). Cell growth and cell proliferation may be dissociated in the mouse uterine luminal epithelium treated with female sex steroids. *Expl. Cell Res.* **160**, 459–470.

Ciocca, D. R., Asch, R. H., Adams, D. J., and McGuire, W. D. (1983). Evidence for modulation of a 24 K protein in human endometrium during the menstrual cycle. *J. clin. Endocr. Metab.* **57**, 496–506.

Clarke, C. L., Adams, J. B., and Wren, B. G. (1982). Induction of oestrogen sulfotransferase in the human endometrium by progesterone in organ culture. *J. clin. Endocr. Metab.* **55**, 70–75.

—— and Satyaswaroop, P. G. (1985). Photoaffinity labelling of the progesterone receptor from human endometrial carcinoma. *Cancer Res.* **45**, 5417–5420.

Cloud, J. G. and Scheutz, A. W. (1977). Interaction of progesterone with all or isolated portions of the amphibian (*Rana pipiens*) oocyte surface. Physical and biological characteristics. *Devl. Biol.* **60**, 359–370.

Compton, J. G., Schrader, W. T., and O'Malley, B. W. (1983). DNA sequence preference of the progesterone receptor. *Proc. natn. Acad. Sci. U.S.A.* **80**, 16–20.

Concolino, G., Marocchi, A., Renaglia, R., Di Silverio, F., and Sparano, F. (1978). Specific progesterone receptor in human renal cancer. *J. steroid Biochem.* **9**, 399–402.

Conneely, O. M., Sullivan, W. P., Toft, D. O., Birnbaumer, M., Cook, R. G., Maxwell, B. L., Zarucki-Schulz, T., Greene, G. L., Schrader, W. T., and O'Malley, B. W. (1986). Molecular cloning of the chicken progesterone receptor. *Science, N.Y.* **233**, 767–770.

Conner, E. A., Murai, J. T., and Gerschenson, L. E. (1978). Histodifferentiation and cytodifferentiation in rabbit endometrium. A comparison of the effects of progesterone and various progesterone metabolites. *Life Sci.* **22**, 1015–1020.

Conti, C. J., Conner, E. A., Gimenez-Conti, I. B., Silverberg, S. G., and Gerschenson, L. E. (1981a). Regulation of ciliogenesis and proliferation of uterine epithelium by 20 alpha-hydroxy-pregn-4-en-3-one administration and withdrawal in ovariectomized rabbits. *Biol. Reprod.* **24**, 903–911.

—— Gimenez-Conti, I. B., Zerbe, G. O., and Gerschenson, L. A. (1981b). Differential effects of oestradiol 17 beta and progesterone on the proliferation of glandular and luminal cells of rabbit uterine epithelium. *Biol. Reprod.* **24**, 643–648.

Cordingley, M. G., Riegel, A. T., and Hager, G. L. (1987). Steroid-dependent interaction of transcription factors with the inducible promoter of mouse mammary tumor virus *in vivo*. *Cell* **48**, 261–270.

Cummings, A. M. and Yochim, J. M. (1983). Nicotinamide adenine dinucleotide kinase in the rat uterus: regulation by progesterone and decidual induction. *Endocrinology* **112**, 1412–1419.

Currie, W. B. and Jeremy, S. Y. (1979). *In vitro* action of progesterone on myometrium I. Reversible modulation of the resistance of rabbit uterus to excitation-contraction uncoupling. *Biol. Reprod.* **21**, 945–952.

Dahmer, M. K., Housley, P. R., and Pratt, W. B. (1984). Effects of molybdate and endogenous inhibitors on steroid-receptor inactivation, transformation and translocation. *A. Rev. Physiol.* **46**, 67–81.

Dean, D. C., Knoll, B. J., Riser, M. E., and O'Malley, B. W. (1983). A 5'-flanking sequence essential for progesterone regulation of an ovalbumin fusion gene. *Nature, Lond.* **305**, 551–554.

—— Gope, R., Knoll, B. J., Riser, M. E., and O'Malley, B. W. (1984). A similar 5'-flanking region is required for oestrogen and progesterone induction of ovalbumin gene expression. *J. biol. Chem.* **259**, 9967–9970.

De Boer, W., Lindh, M., Boet, L., Brinkmann, A., and Mulder, E. (1986). Characterization of the calf uterine androgen receptor and its activation to the deoxyribonucleic acid-binding site. *Endocrinology* **118**, 851–861.

Demers, L. M., Feil, P. D., and Bardin, C. W. (1977). Factors that influence steroid induction of endometrial glycogenesis in organ culture. *Ann. N.Y. Acad. Sci.* **286**, 249–259.

Denker, H. W. (1982). Proteases of the blastocyst and of uterus. In *Proteins and steroids in early pregnancy* (eds. H. M. Beier and P. Karlson) pp. 183–208. Springer-Verlag, Berlin.

Devinoy, E., Houdebine, L. M., and Olivier-Bousquet, M. (1979). Role of glucocorticoids and progesterone in the development of rough endoplasmic reticulum involved in casein biosynthesis. *Biochimie* **61**, 453–461.

De Ziegler, D. and Gurpide, E. (1982). Production of prolactin by cultures of cells from human decidua. *J. clin. Endocr. Metab.* **55**, 511–516.

Diamond, M., Rust, N., and Westphal, V. (1969). High affinity binding of progesterone, testosterone and cortisol in normal and androgen treated guinea-pig during various reproductive states: relationship to masculinization. *Endocrinology* **84**, 1143–1151.

Dickmann, Z. (1979). Blastocyst oestrogen: an essential factor for the control of implantation. *J. steroid. Biochem.* **11**, 771–773.

Djiane, J. and Durand, P. (1977). Prolactin-progesterone antagonism in self regulation of prolactin receptors in the mammary gland. *Nature, Lond.* **266**, 641–643.

Dougherty, J. J. and Toft, D. O. (1980). A macromolecular inhibitor of progesterone-receptor activation in the chick oviduct. In *Abstr. 62nd Ann. Meeting Endocr. Soc.* p. 132.

——— Puri, R. K., and Toft, D. O. (1985). Phosphorylation of steroid receptors. *Trends Pharmacol. Sci.* **6**, 83–85.

Dunbar, B. S. and Daniel, J. C. Jr. (1979). High molecular weight components of rabbit uterine fluids. *Biol. Reprod.* **21**, 723–733.

Ehrardt, A. K. and Meyer-Balhburg, H. F. L. (1981). Effects of prenatal sex hormones on gender-related behavior. *Science, N.Y.* **211**, 1312–1324.

Engel, L. W. and Young, N. A. (1978). Human breast carcinoma cells in continuous culture: A review. *Cancer Res.* **38**, 4327–4339.

Essig, M., Schoenfeld, C., Amelar, R. D., Dubin, L., and Weiss, G. (1982*a*). Stimulation of human sperm motility by relaxin. *Fertil. Steril.* **38**, 339–343.

Essig, M., Schoenfeld, C., D'Eletto, R., Amelar, R., Dubin, L., Steinetz, B. G., O'Byrne, E. M., and Weiss, G. (1982*b*). Relaxin in human seminal plasma. *Ann. N. Y. Acad. Sci.* **380**, 224–230.

Evans, M. I., Hager, L. J., and McKnight, G. S. (1985). A somatomedin-like peptide hormone is required during the oestrogen-mediated induction of ovalbumin gene transcription. *Cell* **25**, 187–193.

Fakan, S. and Puvion, E. (1980). The ultrastructural visualization of nucleolar and extranucleolar RNA synthesis and distribution. *Int. Rev. Cytol.* **65**, 255–299.

Fazleabas, A. T., Bazer, F. W., and Roberts, R. M. (1982). Purification and properties of a progesterone-induced plasmin/trypsin inhibitor from uterine secretions of pigs and its immunocytochemical localization in the pregnant uterus. *J. biol. Chem.* **25**, 6886–6897.

Feil, P. D., Glasser, S. R., Toft, D. O., and O'Malley, B. W. (1972). Progesterone binding in the mouse and rat uterus. *Endocrinology* **91**, 738–746.

Fevold, H. L. and Hisaw, F. L. (1932). Purification of corporin. *Proc. Soc. exp. Biol. Med.* **29**, 620–621.

Finidori, J., Hanoune, J., and Baulieu, E. E. (1982). Adenylate cyclase in *Xenopus laevis* oocytes: characterization of the progesterone-sensitive, membrane-bound form. *Mol. Cell. Endocr.* **28**, 211–227.

Fraenkel, L. (1903). Die Function des Corpus-Luteum. *Arch. Gynakol.* **68**, 438–455.

Fridlansky, F. and Milgrom, E. (1976). Interaction of uteroglobin with progesterone, 5-alpha pregnane 3,20 dione and oestrogens. *Endocrinology* **99**, 1244–1251.

Fuxe, K., Winkstrom, A. C., Okret, S., Agnati, L. F., Harfstrand, A., Yu, Z. Y., Granholm, L., Zoli, M., Vale, W., and Gustafsson, J. A. (1985). Mapping of glucocorticoid receptor immunoreactive neurons in the rat tel- and diencephalon using a monoclonal antibody against rat liver glucocorticoid receptor. *Endocrinology* **117**, 1803–1812.

Gabriel, S. M., Simpkins, J. W., and Kalva, S. P. (1983). Modulation of endogenous opoid influence on LH secretion by progesterone and oestrogen. *Endocrinology* **113**, 1806–1811.

Gadsby, J. E., Heap, R. B., and Burton, R. D. (1980). Oestrogen production by blastocyst and early embryonic tissue of various species. *J. Reprod. Fert.* **60**, 409–417.

Gal, A. and Venetianer, A. (1983). Progesterone receptor in cultured mouse fibroblast L-cells. *Biochem. Pharmacol.* **32**, 919–921.

Ganguly, R., Majumder, P. K., Ganguly, N., and Banerjee, M. R. (1982). The mechanism of progesterone-glucocorticoid interaction in regulation of casein gene expression. *J. biol. Chem.* **257**, 2182–2187.

Garcia, M., Domergue, J., and Rochefort, H. (1984). Marqueurs circulants régulés par les hormones stéroides sexuelles en pathologie humaine. *Horm. Reprod. Metab.* **1**, 19–25.

Garzon, P., Aznar, R., Olivera, E., and Gallegos, A. J. (1977). Progesterone metabolism by human proliferative and secretory endometria. *Contraception* **16**, 79–87.

Gasc, J. M., Renoir, J. M., Radanyi, C., Joab, I., Tuohimaa, P., and Baulieu, E. E. (1984). Progesterone receptor in the chick oviduct. An immunohistochemical study with antibodies to distinct receptor components. *J. Cell. Biol.* **99**, 1193–1201.

Gerschenson, L. E., Conner, E., and Murai, J. T. (1977). Regulation of the cell cycle by diethylstilbestrol and progesterone in cultured endometrial cells. *Endocrinology* **100**, 1468–1471.

—— Conti, C. J., Depaoli, J. R., Lieberman, R., Lynch, M., Orlicky, D., and Rivas-Berrios, A. (1984). Oestrogen and progesterone regulation of proliferation and differentiation of rabbit uterine epithelium. *Prog. Clin. Biol. Res.* **142**, 119–132.

Gershbein, L. L. and Raikoff, K. G. (1978). Uterine total lactic dehydrogenase and isozymes of rats administered steroids. *Enzyme* **23**, 64–69.

Green, S. and Chambon, P. (1987). Oestradiol induction of a glucocorticoid-responsive gene by a chimaeric receptor. *Nature, Lond.* **325**, 75–78.

Gorski, J. and Gannon, F. (1976). Current models of steroid hormone action: acritique. *A. Rev. Physiol.* **38**, 425–450.

Greep, R. O. (1977). The genesis of research in the progestins. In *Biochemical Actions of Progesterone and Progestins. Ann. N.Y. Acad. Sci.* **286**, 1–9.

Gronemeyer, H., Govindan, M. V., and Chambon, P. (1985). Immunological similarity between the chick oviduct progesterone receptor forms A and B. *J. biol. Chem.* **260**, 6919–6925.

—— —— (1986). Affinity labelling of steroid hormone receptors. *Molec. Cell. Endocr.* **46**, 1–19.

Grossman, C. J. (1984). Regulation of the immune system by sex steroids. *Endocr. Rev.* **5**, 435–455.

Grudzinskas, J. G., Gordon, Y. B., Jeffery, D., and Chard, T. (1977). Specific and sensitive determination of pregnancy-specific β_1-glycoprotein by radioimmunoassay. *Lancet* **i**, 333–335.

—— Teisner, B., and Seppala, M. (1982). *Pregnancy Proteins. Biology, Chemistry and Clinical Applications.* Academic Press, Sydney.

Gurpide, E. (1978). Metabolic influences of the action of oestrogens: therapeutic implications. *Pediatrics* **62**, 1114–1120.

—— Fleming, H., Fridman, O., Hausknecht, V., and Holinka, C. (1981). Receptors, enzymes and hormonal responses of endometrial cells. *Prog. Clin. Biol. Res.* **74**, 427–446.

Hackney, J. F., Holbrook, N. J., and Grasso, R. J. (1981). Progesterone as a partial glucocorticoid agonist in L929 mouse fibroblasts: effects on cell growth, glutamine synthetase inductions and glucocorticoid receptors. *J. steroid Biochem.* **14**, 971–977.

Hagley, R. M. and Moore, M. R. (1985). A progestin effect on lactate dehydrogenase in the human breast cancer cell line T47-D. *Biochem. Biophys. Res. Commun.* **128**, 520–524.

Hansen, P. J., Bazer, F. W., and Roberts, R. M. (1985). Appearance of beta-hexosaminidase and other lysosomal-like enzymes in the uterine lumen of gilts, ewes and mares in response to progesterone and oestrogens. *J. Reprod. Fert.* **73**, 411–424.

Harness, J. and Anderson, R. (1977). Effect of relaxin and somatropin in combination with ovarian steroids on mammary glands in rats. *Biol. Reprod.* **17**, 599–603.

Harrington, F. E. and Rothermel, J. D. (1977). Daily changes in peripheral plasma progesterone concentrations in pregnant and pseudo-pregnant rabbits. *Life Sci.* **20**, 1333–1340.

Haslam, S. Z. and Shyamala, G. (1979). Effect of oestradiol on progesterone receptors in normal mammary glands and its relationship with lactation. *Biochem. J.* **182**, 127–131.

Haug, E. (1979). Progesterone suppression of oestrogen-stimulated prolactin secretion and oestrogen receptor levels in rat pituitary cells. *Endocrinology* **104**, 429–437.

—— and Gautvik, K. M. (1986). Effects of sex-steroids on prolactin secreting rat pituitary cells in culture. *Endocrinology* **118**, 1482–1489.

Herrmann, W., Wyss, R., Riondel, A., Philibert, D., Teutsch, G., and Sakiz, E. (1982). RU38486 a new lead for steroid antihormones. *64th Ann. Meeting Endocr. Soc.* San Francisco, CA. Abstract n° 668.

Higgins, S. J., Rousseau, G. G., Baxter, J. D., and Tomkins, G. M. (1973). Nuclear binding of steroid receptors: comparison in intact cells and cell free systems. *Proc. natn. Acad. Sci. USA* **70**, 3415–3418.

Horwitz, K. B. (1985). The antiprogestin RU38486: receptor-mediated progestin versus antiprogestin action screened in oestrogen insensitive T47-D$_{co}$ human breast cancer cells. *Endocrinology* **116**, 2236–2245.

—— Costloro, M. E., and McGuire, W. L. (1975). MCF-7: a human breast cancer cell line with oestrogen, androgen, progesterone and glucocorticoid receptors. *Steroids* **26**, 785–795.

—— and McGuire, W. L. (1978). Oestrogen control of progesterone receptor in human breast cancer. *J. biol. Chem.* **253**, 2223–2228.

—— Koselei, O., and McGuire, W. L. (1978). Oestrogen control of progesterone receptor in human breast cancer: role of oestradiol and antioestrogen. *Endocrinology* **103**, 1742–1751.

—— Mockus, M. B., and Lessey, B. A. (1982). Variant T47-D human breast cancer cells with high progesterone receptor levels despite oestrogen and antioestrogen resistance. *Cell* **28**, 633–642.

—— —— Pike, A. W., Fenressey, P. V., and Steridon, P. L. (1983). Progesterone receptor replenishment in T47-D human breast cancer cells. *J. biol. Chem.* **258**, 7603–7610.

—— and Freidenberg, G. R. (1985). Growth inhibition and increase of insulin receptors in antioestrogen resistant T47-D$_{co}$ human breast cancer cells by progestins: implications for endocrine therapies. *Cancer Res.* **45**, 167–173.

Houdebine, L. M. (1976). Effect of prolactin and progesterone on expression of casein genes. *Eur. J. Biochem.* **68**, 219–225.

—— Teyssot, B., Devinoy, E., Olivier-Bousquet, M., Djiane, J., Kelly, P. A., Delouis, C., Kann, G., and Ferre, J. (1983). Role of progesterone in the development and the activity of the mammary gland. In *Progesterone and Progestins* (eds. C. W. Bardin, E. Milgrom, and P. Mauvais-Jarvis) pp. 297–319. Raven Press, New York.

Hsueh, A. J. W., Peck, A. J. Jr., and Clark, J. H. (1976). Control of uterine oestrogen receptor levels by progesterone. *Endocrinology* **98**, 438–444.

—— Adashi, E. Y., Jones, P. B. C., and Welsh, T. H. Jr. (1984). Hormonal regulation of the differentiation of cultured ovarian granulosa cells. *Endocr. Rev.* **5**, 76–127.

Hynes, N. E., Kennedy, N., Rahmsdorf, U., and Groner, B. (1981). Hormone-responsive expression of an endogenous provival gene of mouse mammary tumor virus after molecular cloning and gene transfer into cultured cells. *Proc. natn. Acad. Sci. U.S.A.* **78**, 2038–2042.

Ishiwata, I., Ishiwata, C., Soma, M., Arai, J., and Ishikawa, H. (1984). Establishment of human endometrial adenocarcinoma cell line containing oestradiol 17 beta and progesterone receptors. *Gynecol. Oncol.* **17**, 281–290.

Jackson, D. A., McCready, S. J., and Cook, P. R. (1981). RNA is synthesized at the nuclear cage. *Nature, Lond.* **292**, 552–555.

Jackson, G. L. (1975). Blockage of progesterone-induced release of LH by intrabrain implants of actinomycin D. *Neuroendocrinology* **17**, 236–244.

James, R., Niall, H., Kwok, S., and Bryant-Greenwood, G. (1977). Primary structure of porcine relaxin: homology with insulin and related growth factors. *Nature, Lond.* **267**, 544–546.

Jeltsch, J. M., Krosowski, Z., Quirin-Stricker, C., Gronemeyer, H., Simpson, R. J., Garnier, J. M., Krust, A., Jacob, F., and Chambon, P. (1986). Cloning of the chicken progesterone receptor. *Proc. natn. Acad. Sci. U.S.A.* **83**, 5424–5428.

Jensen, E. V. and Jacobsen, H. I. (1962). Basic guides to the mechanism of oestrogen action. *Recent Prog. Horm. Res.* **18**, 387–414.

——and De Sombre, E. R. (1973). Estrogen-receptor interaction. Estrogenic hormones effect transformation of specific receptor proteins to a biologically functional form. *Science, N.Y.* **182**, 126–134.

Joab, I., Radanyi, C., Renoir, J. M., Buchou, M., Catelli, M. G., Binart, N., Mester, J., and Baulieu, E. E. (1984). Immunological evidence for a common non-hormone binding component in non-transformed chick oviduct receptors of four steroid hormones. *Nature, Lond.* **308**, 850–853.

Jordan, A. W., Caffrey, J. L., and Niswender, G. D. (1978). Catecholamine-induced stimulation of progesterone and adenosine 3′,5′-monophosphate production by dispersed ovine luteal cells. *Endocrinology* **103**, 385–392.

Jordana, X., Olate, J., Allende, C. C., and Allende, J. E. (1984). Studies on the mechanism of inhibition of amphibian oocyte adenylate cyclase by progesterone. *Archs. Biochem, Biophys.* **228**, 379–387.

Joshi, S. G. (1983). A progestagen associated protein of the human endometrium. Basic studies and potential clinical applications. *J. steroid. Biochem.* **19**, 751–757.

Judge, S. M. and Chatterton, R. T. Jr. (1983). Progesterone-specific stimulation of triglyceride biosynthesis in a breast cancer cell line (T47-D). *Cancer Res.* **43**, 4407–4412.

Kalimi, M., Beato, M., and Feigelson, P. (1973). Interaction of glucocorticoids with rat liver muclei. I. Role of the cytosol proteins. *Biochemistry* **12**, 3365–3371.

Kasid, A., Huff, K., Greene, G. L., and Lippman, M. E. (1984). A novel nuclear form of oestradiol receptor in MCF-7 human breast cancer cells. *Science, N.Y.* **225**, 1162–1165.

Kassis, J. A., Walent, J. H., and Gorski, J. (1986). Oestrogen receptors in cultured rat uterine cells: induction of progesterone receptors in the absence of oestrogen receptor processing. *Endocrinology* **118**, 603–608.

Katzenellenbogen, B. S. (1980). Dynamics of steroid hormone receptor action. *A. Rev. Physiol.* **42**, 17–35.

Kennedy, T. G. (1980). Prostaglandins and the endometrial vascular permeability changes preceding blastocyst implantation and decidualization. *Prog. Reprod. Biol.* **7**, 234–243.

Keydar, I., Chen, L., Karby, S., Weiss, F. R., Delarca, J., Radu, M., Chaitrik, S., and Brenner, H. J. (1979). Establishment and characterization of a cell line of human breast carcinoma origin. *Eur. J. Cancer* **15**, 659–670.

Kincl, F. A. and Ciaccio, L. A. (1980). Suppression of immune response by progesterone. *Endocr. Exp.* **14**, 27–33.

King, W. J. and Greene, G. L. (1984). Monoclonal antibodies localize oestrogen receptor in the nuclei of target cells. *Nature, Lond.* **307**, 745–747.

Kirby, D. R. S. (1962). The effect of uterine environment on the development of mouse eggs. *J. Embryol. exp. Morphol.* **10**, 496–506.

Kirchner, C. (1980). Non-uteroglobin proteins in the rabbit. In *Steroid induced uterine proteins*. (ed. M. Beato) pp. 69–86. Elsevier/North-Holland Biochemical Press, Amsterdam.

Klein-Hitpass, L., Schorpp, M., Wagner, U., and Ryffel, G. U. (1986). An estrogen-responsive element derived from the 5′ flanking region of the *Xenopus* vitellogenin A2 gene functions in transfected human cells. *Cell* **46**, 1053–1061.

Klopper, A. (1982). Newly discovered pregnancy-associated plasma proteins. *Br. J. Obstet. Gynaecol.* **89**, 687–693.

Koistinen, R., Kalkkinen, N., Huhtala, M. J., Seppala, M., Bohn, H., and Rutanen, E. M. (1986). Placental protein 12 is a decidual protein that binds somatomedin and has an identical N-Terminal amino-acid sequence with somatomedin-binding protein from human amniotic fluid. *Endocrinology* **118**, 1375–1378.

Koligian, K. B. and Stormshak, F. (1977). Progesterone inhibition of estrogen receptor replenishment in ovine endometrium. *Biol. Reprod.* **17**, 412–416.

Kontula, K., Janne, O., Vihko, R., de Jager, E., de Visser, J., and Zeelen, P. (1975). Progesterone binding proteins: in vitro binding and biological activity of different steroidal ligands. *Acta endocr. Copenh.* **78**, 574–592.

Kostellow, A. B., Weinstein, S. P., and Morrill, G. A. (1982). Specific binding of progesterone to the cell surface and its role in the meiotic divisions in *Rana* oocytes. *Biochim. Biophys. Acta* **720**, 356–363.

Krause, J. E. and Karavolas, H. J. (1980). Subcellular location of hypothalamic progesterone metabolizing enzymes and evidence for distinct NADH- and NADPH-linked 3-alpha-hydroxysteroid oxidoreductase activities. *J. steroid Biochem.* **13**, 271–280.

Krust, A., Green, S., Argos, P., Kumar, V., Walter, P., Bornert, J. M., and Chambon, P. (1986). The chicken oestrogen receptor sequence: homology with v-*erb*-A and the human oestrogen and glucocorticoid receptors. *EMBO J.* **5**, 891–897.

Kuhnel, B., Buetti, E., and Diggelmann, H. (1986). Functional analysis of the glucocorticoid regulatory elements present in the mouse mammary tumor virus long terminal repeat. A synthetic distal binding site can replace the proximal binding domain. *J. mol. Biol.* **190**, 367–378.

Kulomaa, M. S., Weigel, N. L., Kleinsek, D. A., Beattie, W. G., Conneely, O. M., March, C., Zarucki-Schulz, T., Schrader, W. T., and O'Malley, B. W. (1986). Amino acid sequence of a chicken heat shock protein derived from the complementary DNA nucleotide sequence. *Biochemistry* **25**, 6244–6251.

Kumar, N. M., Chandra, T., Woo, S. L. C., and Bullock, D. W. (1982). Transcriptional activity of the uteroglobin gene in rabbit endometrial nuclei during early pregnancy. *Endocrinology* **3**, 1115–1120.

Kumar, V., Green, S., Staub, A., and Chambon, P. (1986). Localisation of the oestradiol-binding and putative DNA-binding domains of the human oestrogen receptor. *E.M.B.O. J.* **5**, 2231–2236.

Lamb, D. J. and Bullock, D. W. (1984). Heterogenous deoxyribonucleic acid-binding forms of rabbit uterine progesterone receptor. *Endocrinology* **114**, 1833–1840.

—— Kima, P. E., and Bullock, D. W. (1986). Evidence for a single steroid-binding protein in the rabbit progesterone receptor. *Biochemistry* **25**, 6319–6324.

Law, M. L., Kao, F. T., Wei, Q., Hartz, J. A., Greene, G. L., Zarucki-Schulz, T.,

Conneely, O. M., Jones, C., Puck, T. T., O'Malley, B. W., and Horwitz, K. B. (1987). The progesterone receptor gene maps to human chromosome band 11q13, the site of the mammary oncogene *int-2*. *Proc. natn. Acad. Sci. U.S.A.* **84**, 2877–2881.

Lee, D. C., McKnight, G. S., and Palmiter, R. D. (1977). The action of oestrogen and progesterone on the expression of the transferrin gene. *J. biol. Chem.* **253**, 3494–3503.

Lee, H., Davies, I. J., Soto, A. M., and Sonnenschein, C. (1981). Oestrogen induction of progesterone receptor and its relationship to cell multiplication rate in the rat pituitary tumor cell line C29 RAP. *Endocrinology* **108**, 990–995.

Lefkowitz, R. J., Stadel, J. M., and Caron, M. G. (1983). Adenylate cyclase-coupled beta adrenergic receptors: structure and mechanisms of activation and desensitization. *A. Rev. Biochem.* **52**, 159–186.

Lejeune, B., Lamy, F., Lecocq, R., Deschacht, J., and Leroy, F. (1985). Patterns of protein synthesis in endometrial tissues from ovariectomized rats treated with oestradiol and progesterone. *J. Reprod. Fert.* **73**, 223–228.

Lessey, B. A., Alexander, P. S., and Horwitz, K. B. (1983). The subunit structure of human breast cancer progesterone receptors: characterization by chromatography and photoaffinity labeling. *Endocrinology* **112**, 1267–1274.

Levy, C., Glikman, P., Vegh, I., Mester, J., Baulieu, E. E., and Suto, R. (1984). Oestradiol, progesterone and tamoxifen regulation of ornithine decarboxylase (ODC) in rat uterus and chick oviducts. *Prog. Clin. Biol. Res.* **142**, 133–144.

Logeat, F., Sartor, P., Vu Hai, M. T., and Milgrom, E. (1980). Local effect of the blastocyst on oestrogen and progesterone receptors in the rat endometrium. *Science, N.Y.* **207**, 1083–1085.

—— Vu Hai, M. T., and Milgrom, E. (1981). Antibodies to rabbit progesterone receptor: crossreaction with human receptor. *Proc. natn. Acad. Sci. U.S.A.* **78**, 1426–1430.

—— —— Fournier, A., Legrain, P., Buttin, G., and Milgrom, E. (1983). Monoclonal antibodies to rabbit progesterone receptors. *Proc. natn. Acad. Sci. U.S.A.* **80**, 6456–6459.

—— Pamphile, R., Loosfelt, H., Jolivet, A., Fournier, A., and Milgrom, E. (1985*a*). One-step immunoaffinity of purification of active progesterone receptor. Further evidence in favor of the existence of a single steroid binding subunit. *Biochemistry* **24**, 1029–1035.

—— Le Cunff, M., Pamphile, R., and Milgrom, E. (1985*b*). The nuclear-bound form of the progesterone receptor is generated through hormone dependent phosphorylation. *Biochem. Biophys. Res. Commun.* **131**, 421–427.

Loh, P. M., Chamon, G., McIndoe, J., White, M., Urdaneta, L., Hukku, B., and Peterson, W. D. (1985). Development of a new human breast cancer cell line Ia-270. *Breast Cancer Res. Treat.* **5**, 23–29.

Loizzi, R. F. (1985). Progesterone withdrawal stimulates mammary gland tubulin polymerization in pregnant rats. *Endocrinology* **116**, 2543–2547.

Loosfelt, H., Fridlansky, F., Atger, M., and Milgrom, E. (1981). A possible non-transcriptional effect of progesterone. *J. steroid. Biochem.* **15**, 107–110.

—— Logeat, F., Vu Hai, M. T., and Milgrom, E. (1984). The rabbit progesterone receptor. Evidence for a single steroid binding subunit and characterization of receptor mRNA. *J. biol. Chem.* **259**, 14196–14202.

—— Atger, M., Logeat, F., and Milgrom, E. (1986*a*). Cloning of rabbit progesterone receptor complementary DNA. In *Program Abstr. 68th Ann. Meeting Endocr. Soc. Anaheim, CA.* p. 138, abstract 430.

—— —— Misrahi, M., Guiochon-Mantel, A., Meriel, C., Logeat, F., Benarous, R., and Milgrom, E. (1986b). Cloning and sequence analysis of rabbit progesterone-receptor complementary DNA. *Proc. natn. Acad. Sci. U.S.A.* **83**, 9045–9049.

Lukola, A., Akerman, K., and Patra, T. (1985). Human lymphocyte glucocorticoid receptors reside mainly in the cytoplasm. *Biochem. Biophys. Res. Commun.* **131**, 877–882.

Maathuis, J. B. and Aitken, R. J. (1978). Protein patterns of human uterine flushings collected at various stages of the menstrual cycle. *J. Reprod. Fert.* **53**, 343–348.

McEwen, B. S., Bigon, A., Davies, P. G., Kreg, L. C., Luine, V. L., McGinnis, M. Y., Paden, C. M., Parsons, B., and Rainbow, T. C. (1981). Steroid hormones: hormonal signals which alter brain cell properties and functions. *Recent Prog. Horm. Res.* **38**, 41–85.

——Davies, P. G., Gerlach, J. L., McLusky, N. J., McGinnis, M. Y., Parsons, B., and Rainbow, T. C. (1983). Progestin receptors in the brain and pituitary gland. In *Progesterone and Progestins* (eds. C. W. Bardin, E. Milgrom, and P. Mauvais-Jarvis) pp. 59–76. Raven Press, New York.

McGuire, W. L. and Horwitz, K. B. (1978). Progesterone receptors in breast cancer. *Prog. Cancer. Res. Ther.* **10**, 31–42.

McKnight, G. S. (1978). The induction of ovalbumin and conalbumin mRNA by oestrogen and progesterone in chick oviduct explant cultures. *Cell* **14**, 403–413.

——Pennequin, P., and Schimke, R. T. (1975). Induction of ovalbumin mRNA sequences by oestrogen and progesterone in chick oviduct as measured by hybridization to complementary DNA. *J. biol. Chem.* **250**, 8105–8110.

McLaughlin, D. T., Sylvan, P. E., and Richardson, G. S. (1982). The search for progesterone-dependent proteins secreted by human endometrium. *Adv. exp. Med. Biol.* **138**, 113–131.

Mallonee, R. C. and Yochim, J. M. (1979). The uterus during progestation: hormonal modulation of pyridine nucleotide activity in relation to decidual sensitivity. *J. steroid. Biochem.* **11**, 745–755.

Markoff, E., Howell, S., and Handwerger, S. (1983). Inhibition of decidual prolactin release by a decidual peptide. *J. clin. Endocr. Metab.* **57**, 1282–1286.

Marsch, J. M., Butcher, R. W., Savard, K., and Sutherland, E. W. (1966). The stimulatory effect of luteinizing hormone on adenosine 3′5′ monophosphate accumulation in corpus luteum slices. *J. biol. Chem.* **24**, 5436–5440.

Martel, D. and Psychoyos, A. (1980). Behavior of uterine steroid receptors at implantation. *Prog. Reprod. Biol.* **7**, 216–223.

Martin, L. (1980). What roles are fulfilled by uterine epithelial components in implantation? *Prog. Reprod. Biol.* **7**, 54–69.

Mester, J. and Baulieu, E. E. (1977). Progesterone receptors in the chick oviduct: determination of the total concentration of binding sites in the cytosol and nuclear fraction and effect of progesterone on their distribution. *Eur. J. Biochem.* **72**, 405–411.

Meyers, S. A., Lozon, M. M., Corombos, J. D., Saunders, D. E., Hunter, K., Christensen, C., and Brooks, S. C. (1983). Induction of porcine uterine oestrogen sulfotransferase activity by progesterone. *Biol. Reprod.* **28**, 1119–1128.

Milewich, L., Smith, S. L., and McDonald, J. C. (1979). Metabolism of pregnenolone and progesterone by the human lung *in vitro*. In *Program Abstr. 61st Ann. Meeting Endocr. Soc.*, *Anaheim, CA*. p. 214. Abstract 565.

Milgrom, E. (1981). Activation of steroid receptor complexes. In *Biochemical Action of Hormones* Vol. **VIII** (ed. G. Litwack) pp. 465–492. Academic Press, New York.

——Atger, M., and Baulieu, E. E. (1970). Progesterone in uterus and plasma. IV-progesterone receptor(s) in guinea-pig uterus cytosol. *Steroids* **16**, 741–754.

——and Baulieu, E. E. (1970). Progesterone in the uterus and the plasma. II. The role of hormone availability and metabolism on selective binding to uterus proteins. *Biochem. Biophys. Res. Commun.* **40**, 723–730.

—— Allouch, P., Atger, M., and Baulieu, E. E. (1973*a*). Progesterone-binding plasma protein of pregnant guinea-pig. Purification and characterization. *J. biol. Chem.* **248**, 1106–1114.

—— Atger, M., and Baulieu, E. E. (1973*b*). Studies on oestrogen entry into uterine cells and on oestradiol-receptor complex attachment to the nucleus. Is the entry of oestrogen into uterine cells a protein-mediated process? *Biochim, Biophys. Acta* **320**, 267–283.

—————— (1973*c*). Acidophilic activation of steroid hormone receptors. *Biochemistry* **12**, 5198–5205.

—— Luu Thi, M., Atger, M., and Baulieu, E. E. (1973*d*). Mechanisms regulating the concentration and the conformation of progesterone receptors in the uterus. *J. biol. Chem.* **248**, 6366–6374.

—— —— and Baulieu, E. E. (1973*e*). Control mechanism of steroid hormone receptors in the reproductive tract. *Acta. endocr. Suppl.* **180**, 380–403.

—— and Atger, M. (1975). Receptor translocation inhibitor and apparent saturability of the nuclear acceptor. *J. steroid. Biochem.* **6**, 487–492.

Miller, R. D. and Norse, S. A. (1977). Binding of progesterone to *Neisseria gonorrheae* and other gram-negative bacteria. *Infect. Immun.* **16**, 115–123.

Misrahi, M., Atger, M., d'Auriol, L., Loosfelt, H., Mériel, C., Fridlansky, F., Guiochon-Mantel, A., Galibert, F., and Milgrom, E. (1987). Complete amino acid sequence of the human progesterone receptor deduced from cloned cDNA. *Biochem. Biophys. Res. Commun.* **143**, 740–748.

Misrahi, M., Atger, M., and Milgrom, E. (1987). A novel progesterone-induced messenger RNA in rabbit and human endometria. Cloning and sequence analysis of the complementary DNA. *Biochemistry* **26**, 3975–3982.

Miyazaki, K., Miyamoto, E., Maeyama, M., and Uchida, M. (1980). Specific regulation by steroid hormones of protein kinases in the endometrium. 1. Alteration by oestrogen and progesterone in levels of protein kinases in rabbit endometrium. *Eur. J. Biochem.* **104**, 535–452.

Molinari, A. M., Medici, N., Moncharmont, B., and Puca, G. A. (1977). Oestradiol receptor of calf uterus: interaction with heparin-agarose and purification. *Proc. natn. Acad. Sci. U.S.A.* **74**, 4886–4890.

Mornon, J. P., Fridlansky, F., Bally, R., and Milgrom, E. (1980). X-ray crystallographic analysis of a progesterone binding protein. The C222 crystal form of oxidized uteroglobin at 2.2Å resolution. *J. mol. Biol.* **137**, 415–429.

Morrill, G. A., Ziegler, D. H., Kunar, J., Weinstein, S. P., and Kostellow, A. B. (1984). Biochemical correlates of progesterone-induced plasma membrane depolarization during the first meiotic division in *Rana* oocytes. *J. membr. Biol.* **69**, 41–48.

Moudgil, V. K. and Hurd, C. (1987). Transformation of calf uterine progesterone receptor: analysis of the process when receptor is bound to progesterone and RU38486. *Biochemistry* **26**, 4993–5001.

Mukherjee, A. B., Ulane, R. E., and Agrawal, A. K. (1982). Role of uteroglobin and transglutaminase in masking the antigenicity of implanting rabbit embryos. *Am. J. Reprod. Imm.* **2**, 135–141.

Mullick, A. and Katzenellenbogen, B. S. (1986). Antiprogestin-receptor complexes: differences in the interaction of the antiprogestin RU38486 and the progestin R5020 with the progesterone receptor of human breast cancer cells. *Biochem. Biophys. Res. Commun.* **135**, 90–97.

Mullins, D. E., Bazer, F. W., and Roberts, R. M. (1980). Secretion of a progesterone induced-inhibitor of plasminogen activator by the porcine uterus. *Cell* **20**, 865–872.

Mulvihill, E. R., Lepennec, J. P., and Chambon, P. (1982). Chicken oviduct progesterone receptor. Localization, of specific regions of high affinity binding in cloned DNA fragments of hormone responsive genes. *Cell* **24**, 621–632.

Murphy, L. J., Sutherland, R. L., and Lazarus, L. (1985). Regulation of growth hormone and epidermal growth factor receptors by progestins in breast cancer cells. *Biochem. Biophys. Res. Commun.* **131**, 763–773.

—— Murphy, L. C., Stead, B., Sutherland, R. L., and Lazarus, L. (1986a). Modulation of lactogenic receptors by progestins in cultured human breast cancer cells. *J. clin. Endocr. Metab.* **62**, 280–287.

—— Sutherland, R. L., Stead, B., Murphy, L. C., and Lazarus, L. (1986b). Progestin regulation of epidermal growth factor receptor in human mammary carcinoma cells. *Cancer Res.* **46**, 728–734.

Naess, O., Haug, E., Attramadal, A., and Gautvik, K. M. (1982). Progestin receptors in prolactin and growth hormone producing tumours in rats. *Acta endocr. Copenh.* **100**, 25–30.

Nakashi, H. L. and Qasba, P. K. (1979). Quantitation of milk proteins and their mRNA in rat mammary gland at various stages of gestation and lactation. *J. biol. Chem.* **254**, 6106–6025.

Obinata, A., Takata, K., Kuwada, M., Hirano, H., and Endo, H. (1984). Glucocorticoid receptor in chick embryonic epidermis. Inhibition by progesterone of both the binding of glucocorticoid to the receptor and glucocorticoid-induced keratinization. *J. invest. Dermatol.* **83**, 363–369.

O'Malley, B. W. (1984). Steroid hormone action in eukaryotic cells. *J. clin. Invest.* **74**, 307–312.

—— and Schrader, W. T. (1978). Molecular structure and analysis of progesterone receptors. In *Receptors and hormone action* Vol. 2 (eds. B. W. O'Malley and L. Birnbaumer) pp. 189–224. Academic Press, New York.

Orawa, M. M., Isomaa, V. V., and Janne, O. A. (1979). Nuclear poly(A) polymerase-activities in the rabbit uterus. *Eur. J. Biochem.* **101**, 195–203.

Palmiter, R. D. (1972). Regulation of protein synthesis in the chick oviduct I. Independent regulation of ovalbumin, conalbumin, ovomucoid and lysozyme induction. *J. biol. Chem.* **247**, 6450–6461.

—— (1975). Quantitation of parameters that determine the rate of ovalbumin synthesis. *Cell* **4**, 189–197.

—— Christensen, A. K., and Schimke, R. T. (1970). Organisation of polysomes from pre-existing ribosomes in chick oviduct by a secondary administration of either oestradiol or progesterone. *J. biol. Chem.* **245**, 833–845.

—— Oka, T., and Schimke, R. T. (1971). Modulation of ovalbumin synthesis by oestradiol 17β and actinomycin D as studied in explants of chick oviduct in culture. *J. biol. Chem.* **246**, 724–732.

—— Moore, P. B., and Mulvihill, E. R. (1976). A significant lag in the induction of ovalbumin messenger RNA by steroid hormones: a receptor translocation hypothesis. *Cell* **8**, 557–572.

—— Mulvihill, E. R., McKnight, G. S., and Senear, A. W. (1977). Regulation of gene expression in chick oviduct by steroid hormones. *Cold Spring Harbor Symp. quant. Biol.* **42**, 639–647.

—— —— Shepherd, J. H., and McKnight, G. S. (1981). Steroid hormone regulation of ovalbumin and conalbumin gene transcription. A model based upon multiple regulatory sites and intermediary proteins. *J. biol. Chem.* **256**, 7910–7916.

Parker, M. (1983). Enhancer elements activated by steroid hormones? *Nature, Lond.* **304**, 687–688.

Pastan, I. H., Johnson, G. S., and Anderson, W. B. (1975). Role of cyclic nucleotides in growth control. *A. Rev. Biochem.* **44**, 491–522.

Payvar, F., Wrange, O., Carlstedt-Duke, J., Okret, S., Gustafsson, J. A., and Yamamoto, K. R. (1981). Purified glucocorticoid receptors bind selectively *in vitro* to a cloned DNA fragment whose transcription is regulated by glucocorticoids *in vitro*. *Proc. natn. Acad. Sci. U.S.A.* **78**, 6628–6632.

—— De Franco, D., Firestone, G. L., Edgar, B., Wrange, O., Okret, S., Gustafsson, J. A., and Yamamoto, K. R. (1983). Sequence-specific binding of glucocorticoid receptor to MTV DNA at sites within and upstream of the transcribed region. *Cell* **35**, 381–392.

Pearce, P., Khalid, B. A. K., and Funder, J. W. (1983). Progesterone receptors in rat thymus. *Endocrinology* **113**, 1287–1291.

Pennequin, P., Robins, D. M., and Schimke, R. T. (1978). Regulation of translation of ovalbumin messenger RNA by oestrogens and progesterone in oviduct of withdrawn chicks. *Eur. J. Biochem.* **90**, 51–59.

Perrot-Applanat, M. and David-Ferreira, J. F. (1982). Immunocytochemical localization of progesterone-binding protein (PBP) in guinea-pig placental tissue. *Cell Tissue Res.* **223**, 627–639.

—— Logeat, F., Groyer-Picard, M. T., and Milgrom, E. (1985). Immunocytochemical study of mammalian progesterone receptor using monoclonal antibodies. *Endocrinology* **116**, 1473–1484.

—— Groyer-Picard, M. T., Logeat, F., and Milgrom, E. (1986). Ultrastructural localization of the progesterone receptor by an immunogold method: effect of hormone administration. *J. Cell Biol.* **102**, 1191–1199.

—— Groyer-Picard, M. T., Lorenzo, F., Jolivet, A., Vu Hai, M. T., Pallud, C., Spyratos, F., and Milgrom, E. (1987). Immunocytochemical study with monoclonal antibodies to progesterone receptor in human breast tumors. *Cancer Res.* **47**, 2652–2661.

Philibert, D. and Raynaud, J. P. (1973). Progesterone binding in the immature mouse and rat uterus. *Steroids* **22**, 89–98.

—— —— (1977). Cytoplasmic progestin receptors in mouse uterus. In *Progesterone receptors in normal and neoplastic tissues.* (ed. W. L. McGuire) pp. 227–243. Raven Press, New York.

—— Ojasoo, T., and Raynaud, J. P. (1977). Properties of the cytoplasmic progestin-binding protein in the rabbit uterus. *Endocrinology* **101**, 1850–1860.

Philibert, D., Dereadt, R., Teutsch, G., Tournemine, C., and Sakiz, E. (1982). RU38486 a new lead for steroid antihormones. *Program 64th Ann. Meeting Endocr. Soc.*, San Francisco, CA. Abstract 668.

Pichon, M. F. and Milgrom, E. (1977). Characterization and assay of progesterone receptor in human mammary carcinoma. *Cancer Res.* **37**, 464–471.

—— Pallud, C., Brunet, M., and Milgrom, E. (1984). Prognostic value of progesterone receptors in primary breast cancer. *Recent Results Cancer Res.* **91**, 186–191.

Pincus, G. (1965). *The control of fertility.* Academic Press, New York.

Pollack, A., Irvin, G. L., Block, N. L., Lipton, R. M., Storer, B. J., and Claflin, A. J. (1982). Hormone sensitivity of the R3327-G rat prostate adenocarcinoma: growth rate, DNA content and hormone receptors. *Cancer Res.* **42**, 2184–2190.

Ponta, H., Kennedy, N., Skroch, P., Hynes, N. E., and Groner, B. (1985). Hormonal response region in the mouse mammary tumor virus long terminal repeat can be dissociated from the provial promoter and has enhancer properties. *Proc. natn. Acad. Sci. U.S.A.* **82**, 1020–1024.

Preslock, J. P. (1984). The pineal gland: basic implications and clinical correlations. *Endocrine Rev.* **5**, 282–308.

Quagliarello, J., Goldsmith, L., Steinetz, B., Lustig, D. S., and Weiss, G. (1980). Induction of relaxin secretion in non-pregnant women by human chorionic gonadotropin. *J. clin. Endocr. Metab.* **51**, 74–77.

Quirk, S. M. and Currie, W. B. (1984). Uterine steroid receptor changes associated with progesterone withdrawal during pregnancy and pseudo-pregnancy in rabbits. *Endocrinology* **114**, 182–191.

Rao, B. R., Wiest, N. S., and Allen, W. M. (1973). Progesterone receptor in rabbit uterus. I. Characterization and oestradiol 17β augmentation. *Endocrinology* **92**, 1229–1240.

Rao, K. V. S., Peralta, W. D., Greene, G. L., and Fox, C. F. (1987). Cellular progesterone receptor phosphorylation in response to ligands activating protein kinases. *Biochem. Biophys. Res. Commun.* **146**, 1357–1365.

Rauch, M., Loosfelt, H., Philibert, D., and Milgrom, E. (1985). Mechanism of action of an antiprogesterone RU486 in the rabbit endometrium. Effects of RU486 on the progesterone receptor and on the expression of the uteroglobin gene. *Eur. J. Biochem.* **148**, 213–218.

Renfree, M. B. (1980). Signals exchanged between blastocyst and endometrium in the period leading to implantation. *Prog. Reprod. Biol.* **7**, 1–13.

Renkawitz, R., Bcug, H., Graf, T., Matthias, P., Grez, M., and Schutz, G. (1982). Expression of a chicken lysozyme recombinant gene is regulated by progesterone and dexamethasone after microinjection in oviduct cells. *Cell* **31**, 167–176.

Richardson, L. L., Hahmer, C., and Oliphant, G. (1980). Some characteristics of an inhibitor of embryonic development from rabbit oviductal fluid. *Biol. Reprod.* **22**, 553–559.

Richmond, G. and Clemens, L. G. (1986). Cholinergic mediation of feminine sexual receptivity: demonstration of progesterone independence using a progestin receptor antagonist. *Brain Res.* **373**, 159–163.

Ricketts, A. P., Scott, D. W., and Bullock, D. W. (1984). Radioiodinated surface proteins of separated cell types from rabbit endometrium in relation to the time of implantation. *Cell Tissue Res.* **236**, 421–429.

Riddick, D. H. and Kusmick, W. F. (1977). Decidua: a possible source of amniotic fluid prolactin. *Am. J. Obstet. Gynecol.* **127**, 187–192.

Roos, W., Strittmatter, B., Fabbro, D., and Eppenberger, U. (1980). Progesterone and oestrogen receptors in GH3 cells. *Horm. Res.* **12**, 324–332.

——Fabbro, D., Kung, W., Costa, S. D., and Eppenberger, U. (1986). Correlation between hormone dependency and the regulation of epidermal growth factor receptor by tumor promoters in human mammary carcinoma cells. *Proc. natn. Acad. Sci. U.S.A.* **83**, 991–995.

Rothchild, I. (1983). Role of progesterone in initiating and maintaining pregnancy. In *Progesterone and Progestins* (eds. C. W. Bardin, E. Milgrom, and P. Mauvais-Jarvis) pp. 219–229. Raven Press, New York.

Rousseau-Merck, M. F., Misrahi, M., Loosfelt, H., Milgrom, E., and Berger, R. (1987). Localization of the human progesterone receptor gene to chromosome 11q22–q23. *Human Gen.* in press.

Rukmini, V. and Reddy, P. R. (1981). Effect of oestradiol and progesterone on glucosamine 6-phosphate synthetase in the uterus of rat. *Steroids* **37**, 573–579.

Sadler, S. E. and Maller, J. L. (1981). Progesterone inhibits adenylate cyclase in *Xenopus* oocytes. Action on the guanine nucleotide regulatory protein. *J. biol. Chem.* **256**, 6368–6373.

Salamonsen, L. A., Wai, S. O., Doughton, B., and Findlay, J. K. (1985). The effects of oestrogen and progesterone *in vivo* on protein synthesis and secretion by cultured epithelial cells from sheep endometrium. *Endocrinology* 117, 2148–2159.

Salem, H., Seppala, M., and Chard, T. (1981). The effect of thrombin on placental protein 5 (PP5): is PP5 the naturally occurring antithrombin III of the human placenta? *Placenta* 2, 205–209.

Sanchez, E. R. and Pratt, W. B. (1986). Phosphorylation of L-cell glucocorticoid receptors in immune complexes: evidence that the receptor is not a protein kinase. *Biochemistry* 25, 1378–1382.

Sargan, D. R., Tsai, M. J., and O'Malley, B. W. (1986). hsp108, a novel heat shock inducible protein of chicken. *Biochemistry* 25, 6252–6258.

Sato, B., Noma, K., Nishizawa, Y., Nakoo, K., Matsumoto, K., and Yamamura, Y. (1980). Mechanism of activation of steroid receptors: involvement of low molecular weight inhibitor in activation of androgen glucocorticoid and oestrogen receptor systems. *Endocrinology* 106, 1142–1148.

Savouret, J. F. and Milgrom, E. (1983). Uteroglobin: a model for the study of progesterone action in mammals. *DNA* 2, 99–104.

—— Guiochon-Mantel, A., and Milgrom, E. (1984). The uteroglobin gene, structure and interaction with the progesterone-receptor. In *Oxford Survey of Eukaryotic Genes* Vol. 1 (ed. N. McLean) pp. 192–214. Oxford University Press.

Schatz, F., Markiewicz, L., Barg, P., and Gurpide, E. (1985). *In vitro* effects of ovarian steroids on prostaglandin F2 alpha output by human endometrium and endometrial epithelial cells. *J. clin. Endocr. Metab.* 61, 361–367.

Schimke, R. T., McKnight, G. S., Shapiro, D. J., Sullivan, D., and Palacios, R. (1975). Hormonal regulation of ovalbumin synthesis in the chick oviduct. *Recent Prog. Horm. Res.* 31, 175–209.

—— Pennequin, P., Robins, D., and McKnight, G. S. (1977). Effects of oestrogen and progesterone on ovalbumin synthesis in the withdrawn oviduct. In *Biochemical Action of Progesterone and Progestins*. (ed. E. Gurpide). *Ann. N.Y. Acad. Sci.* 286, 116–124.

Schirar, A., Capponi, A., and Catt, K. J. (1980a). Regulation of uterine angiotensin II receptors by oestrogen and progesterone. *Endocrinology* 106, 5–12.

—— —— —— (1980b). Elevation of uterine angiotensin II receptors during early pregnancy in the rat. *Endocrinology* 106, 1521–1527.

Scholer, H., Haslinger, A., Heguy, A., Holtgreve, H., and Karin, M. (1987). *In vivo* competition between a metallothioneim regulatory element and the SV40 enhancer. *Science, N.Y.* 232, 76–80.

Schrader, W. T., Birnbaumer, M. E., Hughes, M. R., Wiegel, N. L., Grody, W. W., and O'Malley, B. W. (1981). Studies on the structure and function of chicken progesterone receptor. *Recent Prog. Horm. Res.* 37, 583–633.

Schuetz, A. W. (1977). Induction of oocytic maturation and differentiation: mode of progesterone action. In *Biochemical Actions of Progesterone and Progestins*. (ed. E. Gurpide) *Ann. N.Y. Acad. Sci.* 286, 408–420.

Seaver, S. S., Van Eys-Fuchs, D. C., Hoffmann, J. F., and Coulson, P. B. (1980). Ovalbumin messenger ribonucleic acid accumulation in the chick oviduct during secondary stimulation: influence of combinations of steroid hormones and circannual rhythm. *Biochemistry* 19, 1410–1416.

Shapiro, S. S. and Forbes, S. H. (1978). Alterations in human endometrial protein synthesis during the menstrual cycle and in progesterone-stimulated organ culture. *Fertil. Steril.* 30, 175–180.

—— Dyer, S. D., and Colas, A. E. (1980). Progesterone-induced glycogen accumulation in human endometrium during organ culture. *Am. J. Obstet. Gynecol.* 15, 419–425.

Sherman, M. R., Corvol, P. H., and O'Malley, B. W. (1970). Progesterone-binding components of the chick oviduct. Preliminary characterization of cytoplasmic components. *J. biol. Chem.* **245**, 6085–6096.

—— —— Atienza, S. B. P., Shansky, J. R., and Hoffman, L. M. (1973). Progesterone receptors of chick oviduct. Steroid-binding subunit formed with divalent cations. *J. biol. Chem.* **249**, 5351–5363.

—— Tuazon, T. B., Stevens, Y. W., and Niu, E. M. (1984). Oligomeric steroid receptor forms and the products of their dissociation and proteolysis. In *Steroid Hormone Receptors: Structure and Function*. Nobel Symposium n° 57 (eds. H. Eriksson, and J. A. Gustafsson) pp. 3–24. Elsevier, Amsterdam.

Short, R. V. (1964). Ovarian steroid synthesis and secretion *in vivo*. *Recent Prog. Horm. Res.* **20**, 303–340.

Sica, V., Weisz, A., Petrillo, A., Armetto, J., and Puca, G. A. (1981). Assay of total oestradiol receptor in tissue homogenate and tissue fractions by exchange with sodium thiocyanate at low temperature. *Biochemistry* **20**, 686–691.

Siiteri, P. K., Febres, F., Clemens, C. E., Chang, R. J., Gondos, B., and Stites, B. (1977). Progesterone and the maintenance of pregnancy: is progesterone's nature immunosuppressant? In *Biochemical Actions of Progesterone and Progestins* (ed. E. Gurpide) *Ann. N.Y. Acad. Sci.* **286**, 384–396.

Singh, V. B., Eliezer, N., and Moudgil, V. K. (1986). Transformation and phosphorylation of purified molybdate-stabilized chicken oviduct progesterone receptor. *Biochim. Biophys. Acta* **888**, 237–248.

Simpson, E. R. and McDonald, P. C. (1981). Endocrine physiology of the placenta. *A. Rev. Physiol.* **43**, 163–188.

Sjoberg, J., Wahlstrom, T., and Seppala, M. (1984). Pregnancy associated plasma protein A in the human endometrium is dependent on the effect of progesterone. *J. clin. Endocr. Metab.* **58**, 359–362.

Smith, S. K., Abel, M. H., and Baird, D. T. (1984). Effects of 17 beta-oestradiol and progesterone on the levels of prostaglandins F2 alpha and E in human endometrium. *Prostaglandins* **27**, 591–597.

Smith, T. C. (1984). The action of relaxin on mammary gland growth in the rat. *Endocrinology* **54**, 59–67.

Srivastava, P. N. and Farooqui, A. A. (1980). Studies on neuraminidase activity of the rabbit endometrium. *Biol. Reprod.* **22**, 858–863.

Strinden, S. T. and Shapiro, S. S. (1983). Progesterone-altered secretory proteins from cultured human endometrium. *Endocrinology* **112**, 862–870.

Sullivan, D. A. and Wira, C. R. (1979). Sex hormone and glucocorticoid receptors in the bursa of Fabricius in immature chicks. *J. Immunol.* **122**, 2617–2625.

Sullivan, W. P., Beito, T. G., Proper, J., Krco, C. J., and Toft, D. O. (1986). Preparation of monoclonal antibodies to the avian progesterone receptor. *Endocrinology* **119**, 1549–1557.

Sutherland, R. L., Mester, J., and Baulieu, E. E. (1977). Hormonal regulation of sex steroid hormone receptor concentration and subcellular distribution in chick oviduct. In *Hormones and Cell Regulation* Vol. 1 (eds. J. Dumont and J. Nunez) pp. 31–48. North Holland Publishing Company, Amsterdam.

—— Geynet, C., Binart, N., Catelli, M. G., Schmelk, P. H., Mester, J., Lebeau, M. C., and Baulieu, E. E. (1980). Steroid receptors and effects of oestradiol and progesterone on chick oviduct proteins. *Eur. J. Biochem.* **107**, 155–164.

Swanson, L. V. and Barker, K. L. (1983). Antagonistic effects of progesterone on oestradiol-induced synthesis and degradation of uterine glucose-6-phosphate dehydrogenase. *Endocrinology* **112**, 459–465.

Takahashi, H., Nakeshima, Y., Oyata, K., and Takenchi, S. (1984). Molecular cloning and nucleotide sequences of DNA complementary to human decidual pro-lactin mRNA. *J. Biochem.* **95**, 1491–1799.

—— and Lisk, R. D. (1985). Diencephalic sites of progesterone action for inhibiting aggression and facilitating sexual receptivity in oestrogen primed golden hamsters. *Endocrinology* **116**, 2393–2399.

Taii, S., Ihara, Y., and Mori, T. (1984). Identification of the mRNA coding for pro-lactin in the human decidua. *Biochem. Biophys. Res. Commun.* **124**, 530–537.

Tarachand, U. and Eapen, J. (1982). Influence of estrogen, progesterone and estrous cycle on gamma-glutamyl-transpeptidase of rat endometrium. *FEBS Lett.* **17**, 210–212.

Teisner, B., Westeigaard, J. G., Folkersen, J., Husby, S., and Svehag, S. E. (1978). Two pregnancy associated serum proteins with pregnancy specific glycoprotein determinants. *Am. J. Obstet. Gynecol.* **131**, 262–266.

Terenius, L. (1973). Oestrogen and progestagen binders in human and rat mammary carcinoma. *Eur. J. Cancer* **9**, 291–294.

Terner, C. (1977). Progesterone and progestins in the male reproductive system. In *Biochemical Actions of Progesterone and Progestins* (ed. E. Gurpide). *Ann. N.Y. Acad. Sci.* **286**, 313–320.

Theveny, B., Bailly, A., Rauch, C., Rauch, M., Delain, E., and Milgrom, E. (1987). Association of DNA-bound progesterone receptor. *Nature, Lond.* **329**, 79–81.

Thrall, C. L. and Spelsberg, T. C. (1980). Factors affecting the binding of chick ovi-duct progesterone receptor to deoxyribonucleic acid: evidence the deoxyribo-nucleic acid alone is not the nuclear acceptor site. *Biochemistry* **19**, 4130–4138.

Toft, D. O. and O'Malley, B. W. (1972). Target tissue receptors for progesterone: the influence of estrogen treatment. *Endocrinology* **90**, 1041–1045.

Too, C. K. L., Bryant-Greenwood, G. D., and Greenwood, F. C. (1984). Relaxin in-creases the release of plasminogen activator collagenase and proteoglycanase from rat granulosa cells in vitro. *Endocrinology* **115**, 1043–1050.

Tsai, S. Y., Tsai, M. J., Schwartz, R., Kalimi, M., Clark, J. H., and O'Malley, B. W. (1975). Effect of oestrogen on gene expression in chick oviduct: nuclear receptor levels and initiation of transcription. *Proc. natn. Acad. Sci. U.S.A.* **72**, 4228–4232.

Tseng, L. (1984). Effect of oestradiol and progesterone on human endometrial aroma-tase activity in primary cell culture. *Endocrinology* **115**, 833–835.

—— and Gurpide, E. (1975). Induction of human endometrial oestradiol dehydro-genase by progestins. *Endocrinology* **97**, 825–833.

—— and Gurpide, E. (1979). Stimulation of various 17 beta- and 20 alpha-hydroxysteroid dehydrogenase activities by progestins in human endometrium. *Endocrinology* **104**, 1745–1748.

Vesce, F., Biondi, C., Gulinati, A. M., Zeni, C., Simonetti, V., and Salvatorelli, G. (1985). *In vitro* effect of progesterone on alpha-L-fucosidase activity of human endometrium. *Eur. J. Gynaecol. Oncol.* **6**, 154–156.

Vignon, F., Bardon, S., Chalbos, D., and Rochefort, H. (1983). Antioestrogen effect of R5020, a synthetic progestin in human breast cancer cells in culture. *J. clin. Endocr. Metab.* **56**, 1124–1130.

Von Baer, K. E. (1827). *De ovi mammalium et hominis genesi. Epistolam ad Academiam imperialem Scientiarum Petropolitanam*. Lipsiae, L. Vossii.

Von der Ahe, D., Janich, S., Scheidereit, C., Renkawitz, R., Schutz, G., and Beato, M. (1985). Glucocorticoid and progesterone receptors binds to the same sites in two hormonally regulated promoters. *Nature, Lond.* **313**, 706–709.

Vu Hai, M. T., Logeat, F., Warembourg, M., and Milgrom, E. (1977). Hormonal control of progesterone receptors. In *Biochemical actions of progesterone and pro-gestins* (ed. E. Gurpide). *Ann. N.Y. Acad. Sci.* **286**, 199–208.

Wahlstrom, T. and Seppala, M. (1984). Placental protein 12 (PP12) is induced in the endometrium by progesterone. *Fertil. Steril.* **41**, 781–786.

Walters, M. R. (1985). Steroid hormone receptors and the nucleus. *Endocrine Rev.* **6**, 512–543.

Walters, M. A., Daly, D. C., Chapitis, J., Kuslis, S. T., Prior, J. C., Kusmick, W. F., and Riddick, D. M. (1983). Human myometrium: a new potential source of prolactin. *Am. J. Obstet. Gynecol.* **147**, 639–645.

Wambach, G. and Higgins, J. (1978). Antimineralocorticoid action of progesterone in the rat. Correlation of the effect on electrolyte excretion and interaction with renal mineralocorticoid receptor. *Endocrinology* **102**, 1686–1696.

Wegener, A. D. and Jones, L. R. (1984). Phosphorylation-induced mobility shift in phospholamban in sodium dodecyl sulfate-polyacrylamide gels. *J. biol. Chem.* **259**, 1834–1841.

Weigel, N. L., Tash, J. S., Means, A. R., Schrader, W. T., and O'Malley, B. W. (1981). Phosphorylation of hen progesterone receptor by cAMP dependent protein kinase. *Biochem. Biophys. Res. Commun.* **102**, 513–519.

—— Minghetti, P. P., Stevens, B., Schrader, W. T., and O'Malley, B. W. (1983). The structure of the chicken progesterone receptor. In *Steroid Hormone Receptors: Structure and Function.* Nobel Symposium n° 57. (eds. H. Eriksson, and J. A. Gustafsson) pp. 25–44. Elsevier, Amsterdam.

Weinberger, C., Hollenberg, S. M., Rosenfeld, M. G., and Evans, R. E. (1985). Domain structure of human glucocorticoid receptor and its relationship to the v-*erb*-A oncogene product. *Nature, Lond.* **318**, 670–672.

Weinstein, S. P., Kostellow, A. B., Ziegler, D. H., and Morrill, G. A. (1982). Progesterone-induced down-regulation of electrogenic Na/K ATPase during the first meiotic division in amphibian oocytes. *J. Membr. Biol.* **69**, 41–48.

Weiss, G. (1984). Relaxin. *A. Rev. Physiol.* **46**, 43–52.

Welshons, W. V., Lieberman, M. E., and Gorski, J. (1984). Nuclear localization of unoccupied estrogen receptors. *Nature, Lond.* **307**, 747–749.

West, N. B. and Brenner, R. M. (1985). Progesterone-mediated suppression of oestradiol receptors in *Cynomolgus* macaque cervix, endometrium and oviduct during sequential oestradiol-progesterone treatment. *J. steroid Biochem.* **22**, 29–37.

Whitehead, R. H., Quirk, S. J., Vitali, A. A., Funder, J. W., Sutherland, R. L., and Murphy, L. C. (1984). A new human breast carcinoma cell line (PMC42) with stem cells characteristics. III. Hormone receptor status and responsiveness. *J. Natl Cancer Inst.* **73**, 643–648.

Wiest, W. G. (1959). Conversion of progesterone to Δ4 pregnen-20α-ol 3-one by rat ovarian tissue in vitro. *J. biol. Chem.* **234**, 3115–3121.

Wilson, C. A. and Hunter, A. J. (1985). Progesterone stimulates sexual behaviour in female rats by increasing 5-HT activity and 5-HT$_2$ receptors. *Brain Res.* **333**, 223–229.

Wira, C. R. and Sullivan, D. A. (1981). Effect of oestradiol and progesterone on the secretory immune system in the female genital tract. *Adv. exp. Med. Biol.* **138**, 99–111.

Wrange, O., Carlstedt-Duke, J., and Gustafsson, J. A. (1979). Purification of the glucocorticoid receptor from rat liver cytosol. *J. biol. Chem.* **254**, 9284–9290.

—— Okret, S., Radojcic, M., Carlstedt-Duke, J., and Gustafsson, J. A. (1983). Characterisation of the purified glucocorticoid receptor from rat liver cytosol. In *Steroid Hormone Receptors: Structure and Function.* Nobel Symposium n° 57. (eds. H. Eriksson, and J. A. Gustafsson) pp. 73–92. Elsevier, Amsterdam.

Wright, F., Kirchhoffer, M. O., and Giacomini, M. (1980). Antiandrogenic activity of progesterone in human skin. In *Percutaneous Absorption of Steroids* (eds. P.

Mauvais-Jarvis, C. F. H. Vickers and Wepierre, J.) pp. 123–137. Academic Press, London.

Yamamoto, K. R. (1985). Steroid receptor regulated transcription of specific genes and gene networks. *A. Rev. Genet.* **19**, 209–252.

—— and Alberts, B. M. (1976). Steroid receptors: elements for modulation of eucaryotic transcription. *A. Rev. Biochem.* **45**, 421–746.

Yaniv, M. (1984). Regulation of eukaryotic gene expression by transactivating proteins and cis acting DNA elements. *Biol. Cell* **50**, 203–216.

Yaron, Z. (1977). Embryo-maternal interrelations in the lizard *Xanthusia vigilis*. In *Reproduction and Evolution* (eds. J. H. Calaby and C. H. Tyndale-Biscoe) pp. 271–277. Australian Academy of Science.

Ylikomi, T., Gasc, J. M., Isola, J., Baulieu, E. E., and Tuohimaa, P. (1985). Progesterone receptor in the chick bursa of Fabricius: characterization and immunohistochemical localization. *Endocrinology* **117**, 155–160.

Young, I. R. and Renfree, M. B. (1979). The effect of corpus luteum removal during gestation on parturition in the Tammar Wallaby (*Macropus eugenii*). *J. Reprod. Fert.* **56**, 249–254.

Yu, W. C., Leung, B. S., and Gao, Y. L. (1981). Effects of 17-beta-oestradiol on progesterone receptors and the uptake of thymidine in human breast cancer cell line CAMA-1. *Cancer Res.* **2**, 17–27.

Zarrow, M. X. and Brennam, D. M. (1959). The action of relaxin on the uterus of the rat, mouse and rabbits. *Ann. N.Y. Acad. Sci.* **75**, 981–990.

8 Countercurrent transfer in the ovarian pedicle and its physiological implications

N. EINER-JENSEN

I Introduction

Countercurrent transfer is a well established technology. It is used to transfer heat and chemical substances from one fluid to another or from a fluid to a gas. The transfer may take place without any mixing or direct contact between the two phases. A simple model (Fig. 8.1) (Einer-Jensen 1976a), demonstrates the principle. Transfer is a passive process and takes place without energy expenditure. The efficiency of the system is dependent on the area of contact, the difference in temperature or concentrations between the two tubes, the flow rate in the tubes and the resistance against the transfer. The resistance is dependent upon the physicochemical properties of the substance and of the separating tubes or membranes. The efficiency of a countercurrent transfer system can be close to 100 per cent in engineering systems, for example in heat exchange.

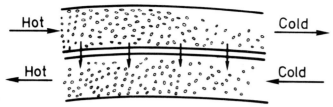

Fig. 8.1.   Counter current transfer of heat (small circles) between fluids in two tubes with opposite directions of flow.

Countercurrent transfer is recognized to occur in many animal species and in many different organs. The transfer normally involves blood or lymph vessels.

A long, but incomplete list of examples in which countercurrent transfer plays a part outside the area of reproductive biology includes the following: protection against cooling and overheating of the brain; heat and water conservation in the nose of the reindeer; low leg temperatures but normal body temperature in birds standing on ice; high oxygen concentration in the swim bladder of some deep-sea fishes; maintenance of an osmolar gradient in the kidney by the vasa recta and absorption of water and electrolytes in the intestinal mucosa. Further information and references can be found in Schmidt-Nielsen (1972); Einer-Jensen (1976*b*); Lundgren (1976); Steen (1976); Sund (1976); Baker (1979); Bendz (1982*a*). The exchange may occur to some extent between any closely apposed pair of arteries and veins (Sejrsen 1970, 1985).

II   Anatomy

1   MACROSCOPIC ANATOMY

i   History

About 300 years ago Blancardi (1687, Fig. 8.2) presented a drawing of the blood vessels of the genital organs in a woman. The ovarian veins can be seen to form a plexus together with the tortuous artery on each side. The tortuosity leads to a marked elongation of the gonadal arteries, as was observed in various domestic animals 50–100 years ago (see Vollmerhaus, 1964). Robinson (1903) showed very instructive pictures of *arteria uterina ovarica* that forms an arcade between *a. ovarica* and *a. uterina*. The arcade adds flexibility to the blood supply of the uterus and ovary. Many minor arterial anastomoses were clearly illustrated by Calza (1807, see Reynolds 1973) between the two *a. uterinae*, and these further add to the flexibility of the blood supply.

Loeb (1908) was the first to describe an effect of the ovary on the uterus in the guinea pig. He reported in 1923 that the reverse situation also existed; the

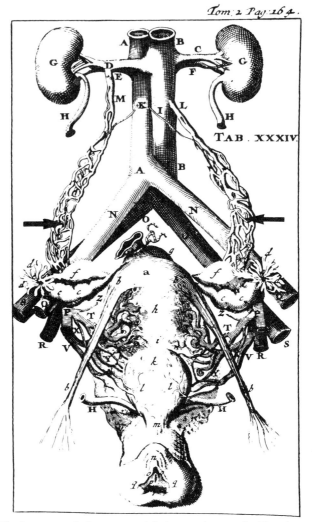

Fig. 8.2   The human genital organs and their vascular supply. The plexus formed by the ovarian artery and vein can be seen (arrows). (From Blancardi 1687).

uterus had an effect upon the ovary since hysterectomized guinea-pigs did not show cyclic regression of the corpora lutea. The first suggestion of a counter-current exchange between the female gonadal vessels was published for the sheep in the late 1960s, a time when several workers (see Anderson, Bland, Melampy, 1969; Ginther 1974) described the anatomical basis for such a mechanism between blood vessels supplying and draining the female genital organs: uterus, oviduct and ovaries. In this review I shall concentrate on the transfer that takes place between the cranially directed utero-ovarian vein/lymph system and the ovarian artery.

ii   Sheep, cow and sow

Domestic animals (sheep, cow and sow) form a group in which the ovary and the greater part of the uterus are drained by a common utero-ovarian vein. The vein and its branches are in close apposition with the ovarian artery. The ovarian vein consists of several interconnected branches forming an intimate network. The appearance is strikingly similar to the male pampiniform plexus (Harrison and Weiner, 1949; Dierschke, Walsh, Mapletoft, Robinson, and Ginther, 1975). The ovarian artery is tortuous and ascends to the ovary over the surface of the utero-ovarian vein. It supplies the ovary, the oviduct and the proximal parts of the uterine horn (via an anastomosis to the uterine artery and through the uterine branch of the ovarian artery: see below). The appearance of the vessel can be influenced by ovarian steroids in the cow (Oxenreider, McClure, and Day, 1965) and sheep (Dobrowolski and Hafez, 1970). Its length is increased to many times more than the actual distance between the aorta and the ovary, and the vessel runs in and out between the branches of the vein thereby increasing the surface area of contact. A diagram of the blood vessels in the ewe is shown in Figure 8.3, and photos of latex injected vessels can be found in Del Campo and Ginther (1973). Several authors have described the vascular pattern (Goding, McCracken, and Baird, 1969; Ginther and Bisgaard, 1972; Lee and O'Shea, 1976) even in the camel (Ghazi, 1981). An overview of the anatomy of the blood vessels to the reproductive organs with very instructive colour pictures has been published by Ginther (1976).

The ovarian artery of the horse has very little contact with the utero-ovarian vein (see Fig. 2 in Ginther 1974). As might be expected from the anatomical description the horse does not belong to the group of animals in which a local pathway between the uterus and the ovary has been demonstrated (Del Campo and Ginther 1973; Ginther 1974).

iii   Laboratory animals

The arteries and veins to the genital organs are generally closely apposed in the small laboratory animals where indications of a local pathway have been found (guinea-pig, hamster, rat, mouse: Einer-Jensen 1974a). Judged from the anatomical findings much of the uterine venous blood passes cranially into a common trunk, which drains both the uterine horn and the ovary in guinea-pig, rat and hamster, whereas in the rabbit most of the uterine blood seems to drain caudally (see Fig. 8 in Carter, Göthlin, and Olin 1971, or Fig. 1 in Del Campo and Ginther 1972).

In many mammals, there are anastomoses between branches of the ovarian and uterine arteries (Del Campo and Ginther, 1972; Mossman and Duke 1973; Reynolds 1973). The uncertainties about direction and magnitude of

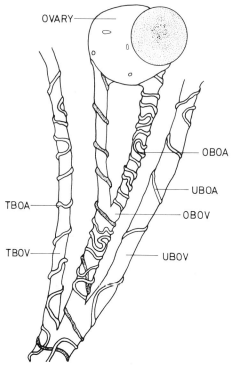

Fig. 8.3   Schematic drawing of the genital blood vessels in sheep. The ovary is shown with a large corpus luteum (stippled). TBOA, Tubal Branch of the Ovarian Artery; TBOV, Tubal Branch of the Ovarian Vein; OBOA, Ovarian Branch of the Ovarian Artery (normally 2–4 branches); OBOV, Ovarian Branch of the Ovarian Vein; UBOA, Uterine Branch of the Ovarian Artery; UBOV, Uterine Branch of the Ovarian Vein. (From Schramm *et al.* 1986*a*, with permission.)

flow in the utero-ovarian arterial anastomoses increase the difficulties in evaluating the importance of the countercurrent system. Ford and Chenault (1981) found evidence in the cow that a portion of the blood flowing to the ovary containing a fully functional corpus luteum is contributed by the ipsi-lateral uterine artery. The measurements were performed with 3 blood flow transducers on unanaesthetized animals throughout an oestrous cycle. In horse as in sheep the ovarian artery appears to contribute to the uterine blood supply through the uterine branch of the ovarian artery (Del Campo and Ginther 1973). The authors' statements were based on anatomical find-ings because the direction of flow was not measured. Chaichareon, Rankin, and Ginther (1976) found by the microsphere technique that the ovarian artery contributed 90 per cent of the arterial blood supply to the cranial and 42 per cent to the middle third of the uterus in the guinea-pig. Hossian and O'Shea (1983) also found that blood flowed from the ovarian artery towards the tip of the uterus in the guinea pig while Peepler (1976) found the opposite direction of flow in the rat. A dual arterial supply was also present in the pregnant guinea-pig (Egund and Carter 1974).

The arterial blood supply to the ovary is complex. In summary, blood may be supplied by the ovarian and the uterine arteries, and the fraction supplied by each vessel may change during the ovarian cycle and during pregnancy. The blood vessels within the ovarian adnex have functional arteriovenous shunts. Mettner, Brown, and Hales (1981) found evidence in the sheep for functional arterio-venous anastomoses within the ovary itself. One must therefore distinguish between total blood flow to the ovary and functional capillary blood flow. The capillary flow is not the same in the corpus luteum as in non-luteal tissue, and is correlated with the functional state of the tissues. A figure for total ovarian blood flow may thus have limited value for understanding ovarian function. Changes in body position may induce changes in flow. Finally, there are significant problems in making reliable measurements on an organ positioned deep within the abdomen.

iv   Primates including the human female

The basic description of the course followed by arteries and veins can be found in anatomical textbooks (Woman: Paturet 1958; Benninghoffs and Goerttler 1975; see also Hartman and Straus 1933; monkeys: Eddy and Pauerstein 1980). Arterio-venous anastomoses were described by Clara (1956). The anatomical basis for a possible countercurrent transfer mechanism in the human adnex was investigated by Bendz (1977; 1982a,b) (Fig. 8.4). He found the vascular anatomy to be similar to that of the sheep; spiral arteries run in close contact to or even within the walls of the veins in the mesovarian region. The finding was puzzling at the time of publication, as no functional aspects could be connected to the structure. For example, in contrast to many non-primates, hysterectomy has been postulated to have no influence on the life span of the human corpus luteum (Doyle, Barclay, Duncan, and Kirton 1971).

A similar anatomical structure had been found a few years previously in the rhesus monkey (Ginther, Dieschke, Walsh, and Del Campo 1974). The initial one-fourth of the ovarian artery was relatively straight, but the remainder was very tortuous or tightly spiralled and located between the utero-ovarian or ovarian veins. The same authors also suggested that the ovarian artery is functionally unimportant because the uterine artery contributes the major part of the ovarian blood supply.

Tortuosity of the uterine arteries has been observed in several species, including the rabbit, rhesus monkey and man (Carter 1975). The vessels are more tightly coiled (Hansel and Asdell 1951) and have a greater diameter (Yamauchi and Sadaki 1968) in non-pregnant cows than in heifers. The arteries of the ovine uterus may straighten out as gestation advances and the uterus grows (Barcroft and Barron 1946). Similar changes in the ovarian artery caused by pregnancy have to my knowledge not been described.

2   SUBMACROSCOPIC ANATOMY

Little information seems to exist with regard to the submacroscopic anatomy. In cattle, the wall of the ovarian artery has been reported to be thinnest in areas in close contact with the vein (Vollmerhaus 1964). Del Campo and Ginther (1974) found that the walls of the vessels were closely connected and significantly thinner in the area of apposition. The connective tissue bundles formed a single stratum so that the demarcation between the artery and vein was poorly defined. Bendz (1977, Fig. 4) shows a microscopical section of the human adnex with close contact between the ovarian artery and a vein. The

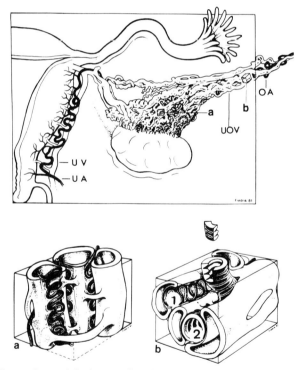

Fig. 8.4   Illustrations of the human female reproductive organs and extrinsic utero-ovarian blood vessels based upon a corrosion cast. OA, ovarian artery; UA, uterine artery; UOV, utero-ovarian vein; UV, uterine vein. a, detail from the mesovarium showing the winding arteries (dark) jammed in between larger veins (light); b, detail from the lateral part of the ovarian pedicle demonstrating both the tortuous (1) and the "inner helical folded" (2) artery close to veins. (From Bendz 1982b, with permission).

muscular medial layer of the vein wall facing the arteries appears to be absent thus decreasing the distance between the lumina.

A double capillary system, a true portal system, was described in the sow by a group in Olsztyn, Poland (Krzymowski, Kotwica, Stefanczyk, Debek, and Czarnocki 1981; Krzymowski, Kotwica, Stefanczyk, Czarnocki, and Debek 1982). The first part of the portal system was observed on the surface of the uterine veins. The veins from this "*vasa vasorum*" later formed a venous mesh on the surface of the ovarian artery.

The Olsztyn group also described changes in the microscopic appearance of the ovarian artery and vein in the paraovarian plexus (Janowicz, Doboszynska, Stefanowski, and Zamojska 1981) and of their branches in the ovary during the oestrous cycle (Doboszynska, Janowics, Stefanowski, and Zamojska 1980; Doboszynska and Ziecik 1986). Branches of the ovarian artery locally had characteristic thickenings in their walls ("muscular pegs") between Day 15 and Day 21 of the cycle. Doboszynska (personal communication) suggests that these changes are part of the explanation for the decreased blood flow to the regressing corpora lutea. The muscular pegs have not been observed in early pregnant or oestradiol-treated pigs (Doboszynska and Ziecik 1986; Ziecik, Doboszynski, and Dusze 1986), but do exist during late pregnancy.

III   Physiology

1   EXPERIMENTAL EVIDENCE FOR LOCAL TRANSFER

The development of the concept of countercurrent exchange in the female reproductive tract arose from studies in which specific organs were removed and transplanted, and the oestrous cycle and fertility were found to be disrupted in certain species (see Donovan 1967; Ginther 1967; Anderson *et al.* 1969, for early reviews on utero-ovarian relationships).

Loeb (1923, 1927) performed total hysterectomy in guinea-pigs and observed an increase in cycle length from the normal 15–18 days to 80–110. His choice of species was fortunate, because Bradbury (1937) did not observe such changes when he repeated the experiment on rats. A prolonged cycle was also observed in the sheep (Wiltbank and Casida 1956; Kiracofe and Spies 1963; Rowson and Moor 1965), pig (du Mesnil du Buisson and Dauzier 1959; Spies, Zimmerman, Self, and Casida 1960; Anderson, Butcher, and Melampy 1961, 1963) and cow (Wiltbank and Casida 1956; Anderson, Neal, and Melampy 1962). The life span of the corpus luteum often increased 10 times.

The minimum amount of residual uterine tissue required to maintain the normal oestrous cycle varies from species to species. The guinea pig requires the whole of one horn (Butcher, Chu, and Melampy 1962), the pig about one quarter of one horn (Anderson *et al.* 1961), and the cow even less (Wiltbank and Casida 1956; Anderson, Neal, and Melampy 1962). Experiments involving the removal of fetuses from one uterine horn were also made (Dhindsa and Dziuk 1968). Little change occurs in the length of the cycle in the human female (Andreoli 1965) or monkey (Hartman 1932; Burford and Diddle 1936; Jones and Telinde 1941; Wagenen and Catchpole 1941) after hysterectomy. As discussed later, in the primate, and maybe also in other species, it seems to be of some importance whether the tubes are removed as well, but this information is not always available from the publications.

More and Nalbandov (1953) changed the cycle length by distension of the uterine horns and abolished the change by surgical denervation, but they may have overlooked the damage done to the vascular and lymphatic systems. The possible existence of neural utero-ovarian communication has, to my knowledge, not been proved convincingly.

In the 1920s, Grafenberg inserted rings of silk or silver into the uteri of women in an attempt to decrease fertility (see Southam 1964). The technique gained a bad reputation some years later because other clinicians found serious side effects (infections and penetrations). There was a renaissance of the method after 1959 (Ishihama 1959; Oppenheimer 1959) and intrauterine devices are now widely used in many countries.

The IUD has been used as an investigative tool in laboratory and farm animals. Placing a plastic bead into the uterus of the sheep on day 3 of the oestrous cycle reduced that and subsequent cycles to 9–13 days (Moore and Nalbandov 1953; Oloufa, Inskeep, Pope, and Casida 1961; Inskeep, Oloufa, Howland, Pope, and Casida 1962). Ginther, Pope, and Casida (1965, 1966) showed that insertion of a plastic coil in one horn on day 4 of the oestrous cycle caused the corpora lutea in the ipsilateral ovary to regress more rapidly than those in the opposite ovary. The regression could, however, be counteracted by hCG injections (Stormshak and Hawk 1966; Stormshak, Lehmann, and Hawk 1967).

Similar results were obtained with IUDs in cows and heifers (Hansel and Wagner 1960; Anderson, Bowerman, and Melampy 1965; Hawk, Conley, Brinsfield, and Righter 1965; Ginter, Woody, Janakiraman, and Casida 1966). IUDs, however, failed to influence the cycle in the pig (Anderson 1962; Anderson and Melampy 1962; Gerrits and Hawk 1966).

Rathmacher, Anderson, Kawata, and Melampy (1967) found the effect to be local in the pig, an IUD in one horn prevented pregnancy in that horn but did not change the cycle length. There is also some indication of the importance of a local pathway in connection with the *post partum* period in the cow (Casida and Venzke 1936). The interval from parturition until a follicle was about 12 mm in diameter was comparatively shorter if the follicle appeared in the ovary adjacent to the previously non-pregnant horn.

The local pathway between the ovary and uterus could be associated with the utero-ovarian vein and the ovarian artery. Without knowing the substances exchanged, this could only be investigated by surgical manipulation of the organs or the blood vessels. Often one uterine horn was removed and/ or the source of the ovarian blood supply changed. The basis for the experiments was the assumption that the putative messenger substance would be present in the uterine venous blood from which it would reach the ovarian arterial blood by some local mechanism.

Selective ligation of blood vessels showed that a vascular pathway was essential for normal luteal regression in the sheep (Kiracofe, Menzies, Gier, and Spies 1966; McCracken and Baird 1969; Abdel Rahim, Bland, and Poyser 1984), guinea pig (Howe 1965; Fischer 1967; Bland and Donovan 1969), rat (Clemens, Minaguchi, and Meites 1968; Butcher, Barley, and Inskeep 1969) and hamster (Orsini 1968).

Experiments with ovarian and utero-ovarian transplants in sheep showed that utero-ovarian transplants had a normal ovulatory cycle while the corpus luteum persisted in ovarian transplants (Goding, McCracken, and Baird 1967; McCracken 1971). Cross circulation experiments between donor sheep bearing utero-ovarian transplants and recipient sheep with only the ovarian transplant showed that uterine vein blood from animals in the late luteal phase could induce luteolysis in the recipient ovary (McCracken, Carlson, Glew, Goding, Baird, Green, and Samuelsson 1972). Surgical transposition of the blood vessels, with the organs left *in situ*, yielded essentially the same results (Ginther 1974; Mapletoft, Del Campo, and Ginther 1975; Mapletoft and Ginther 1975; Mapletoft, Lapin, and Ginther 1976).

2   PROSTAGLANDIN $F_{2\alpha}$

The final step was to identify the luteolytic substance. Prostaglandin $F_{2\alpha}$ was the candidate suggested by Pharriss and Wyngarden (1969) (see Pharriss 1971), because high doses of $PGF_{2\alpha}$ administered subcutaneously to rats (1 mg/kg/day) caused a significant shortening in the duration of pseudo-pregnancy. $PGF_{2\alpha}$ was found in the endometrium (Egliton, Raphael, Smith, Hall, and Pickles 1963; Hunter and Poyser 1972; Wilson, Cenedella, Butcher, and Inskeep 1972), and was able to decrease progesterone production (to be luteolytic) when infused into the ovarian artery supplying an active corpus luteum (McCracken, Glew, and Scaramuzzi 1970; Barrett, Blickey, Brown, Cumming, Goding, Mole, and Obst 1971; McCracken 1971). It was found in uterine vein blood during the days of the cycle when luteolysis takes place (see Fig. 7 in McCracken *et al.* 1972). The concentration in peripheral arterial blood, and therefore in blood reaching the ovary, was, however, too low to induce luteolysis (McCracken 1971; Pexton, Weams, and Inskeep 1975). The low peripheral concentration was only partly due to dilution, because $PGF_{2\alpha}$ was found to be almost completely metabolised in lung tissue during one passage in the cat, dog, rat and guinea pig (see Piper, Wane, and Wyllie 1970; Piper and Vane 1971). Therefore, if $PGF_{2\alpha}$ was the luteolytic substance, a local pathway had to be proved.

The final step was taken at the Worcester Foundation, U.S.A., by infusion of ^{3}H-PGF$_{2\alpha}$ into the uterine vein and collection of parallel samples from the ovarian artery and contralateral carotid artery in ewes where previous auto-transplantation to the neck had been performed on the ovary and ipsilateral uterine horn. Later the experiment was performed in intact animals with infusion of the labelled hormone into the uterine vein and collection of parallel blood samples from the iliac artery and ovarian artery (McCracken 1971; McCracken, Schramm, Barcikowski, and Wilson 1981; McCracken 1984). When the radioactivity in the parallel samples was compared (Fig. 8.5, McCracken *et al.* 1972) there were much higher concentrations in ovarian blood than in the parallel iliac samples. This result can at present only be explained by the theory of local transfer from the utero-ovarian vein into the ovarian artery. Similar results have been obtained in the ewe (Land, Baird, and Scaramuzzi 1976) as well as in the sow (Lindloff, Holtz, Elsaisser, Kreikenbaum, and Schmidt 1976; Krzymowski, Kotwica, Okrasa, Doboszynska, and Ziecik 1978; Kotwica 1980) and guinea pig (see Poyser 1976). Not all studies have demonstrated local transfer: Coudert, Phillips, Faiman, Cherncki, and Palmer (1974b) did not find it in sheep, and conflicting results have been reported in the cow (Hixon and Hansel 1974; Milvae and Hansel 1980). There is as yet no adequate explanation for these contradictory reports.

Intrauterine application of a microgram dose of PGF$_{2\alpha}$ led to luteolysis in the rhesus monkey (Einer-Jensen 1973). This observation indicated that the hormone was transferred locally also in the primate, because milligram doses were needed to induce luteolysis when injected intramuscularly (Kirton, Duncan, Oesterling, and Forbes 1971).

A strange and still unexplained part of the original results from the group at the Worcester Foundation is the slow transfer of PGF$_{2\alpha}$. As Figure 8.5 indicates, it took 20 minutes before the difference between the arterial concentrations was significant, and the transfer went on for two hours after the hormone infusion was stopped. The slow transfer may possibly be explained by the vasoconstrictive properties of PGF$_{2\alpha}$, the hormone may reduce the blood flow locally and thus inhibit its own transfer. On the other hand, PGF$_{2\alpha}$ did not change the speed by which ^{85}krypton was transferred from the utero-ovarian vein to the ovarian artery (see below).

Luteolysis is supposed to be induced by pulsatile secretion of PGF$_{2\alpha}$ and the corpus luteum is certainly very sensitive to pulsatile infusion of small PGF$_{2\alpha}$ doses *in situ* (Schramm, Bovaird, Glew, Schramm, and McCracken 1983). Five doses (increasing from 0.2 µg/h to 1 µg/h), each dose infused for 1 hour at intervals of 6 hours, induced CL regression. The slow transfer of PGF$_{2\alpha}$ seems to be inconsistent with the requirement for a pulsatile pattern. To my knowledge, evidence for a pulsatile pattern of PGF$_{2\alpha}$ in the ovarian arterial blood is still lacking.

Pharriss and Wyngarden (1969) and Pharriss (1970) originally suggested

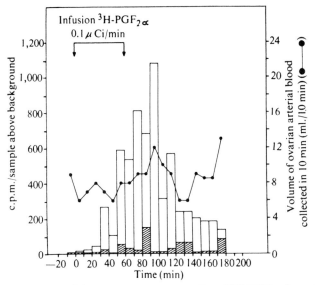

Fig. 8.5   Time course of the countercurrent transfer of ^{3}H-PGF$_{2\alpha}$ from the utero-ovarian vein to the adherent ovarian artery is illustrated: onset of the transfer was slow and the peak value occurred after stopping the infusion, indicating that equilibrium was not achieved. Unshaded bars, radioactivity in ovarian arterial plasma; shaded bars: radioactivity in iliac arterial plasma. (From McCracken *et al*. 1972, with permission).

that luteolysis was caused by constriction (Ducharme, Weeks, and Montgomery 1968) of the utero-ovarian vein, causing a decreased ovarian blood flow, a functional hypoxaemia. The theory was found improbable as organ blood flow regulation is expected to be found on the arterial side of the vascular system. The capillary blood flow through the corpus luteum and the progesterone production from the corpus luteum was therefore measured simultaneously in intact ewes as well as in animals with an ovary transplanted to the neck (Einer-Jensen and McCracken 1976 and 1981*b*). The progesterone production started to decrease before the change in blood flow was observed. The conclusion was that the effect of PGF$_{2\alpha}$ must be explained as a biochemical event in the corpus luteum cells. Reduced blood flow as an initiator of luteal regression was also excluded in the rabbit (Bruce and Hillier 1974) and guinea pig (Wehrenberg, Dierschke, and Wolf 1979).

3   HEAT AND INERT GASES

The temperature in the testes is 2–3°C lower than the body temperature (Harrison and Weiner 1949; Setchell and Waites 1967). This is due to heat loss through the scrotal wall and heat exchange from the testicular artery to the venous pampiniform plexus. Heat loss from the ovaries cannot be expected because of their intra-abdominal position and very little if any work has been published about heat exchange in the ovarian adnex. A Danish group found evidence of a decreased temperature within the ovarian follicles. The temperature in the follicles was 2.8°C lower than the body temperature and 1.4°C lower than in the ovarian stroma (Grindsted, Blendstrup, Andreasen, and Byskov 1980). The same group found similar results in women during surgery (personal communication.) The significance of the colder follicular temperature in relation to the resumption of meiosis or survival of the oocytes has yet to be evaluated.

[133]Xenon and [85]krypton are inert gases. They have been used as indicators of passive exchange; if they are exchanged without any transport mechanism being involved, other substances may be too. Their lipophilicity may result in a high rate of exchange but does not change its passive nature. [133]Xenon was found in the ovaries after instillation into the uterine lumen in the mouse, rat, hamster and guinea pig. No transfer was found in the rabbit (Einer-Jensen 1974*b*). These findings indicated that a substance could diffuse into the uterine blood or lymph capillaries from the uterine lumen, pass from the venous blood or lymph into arterial blood and diffuse into the ovarian tissue, thus for the first time establishing a chain of local transfer from the uterus lumen to the ovary.

[85]Krypton could be measured with a miniature Geiger-Müller tube within the corpus luteum after infusion into the uterine vein in ewes (Einer-Jensen and McCracken 1977, 1981*b*), confirming that a substance present in the uterine vein blood could reach the corpus luteum by a local path of transport. The appearance time of [85]krypton was very short—20 seconds compared to 20 minutes for $PGF_{2\alpha}$. The blood supply to the Fallopian tube must participate in the countercurrent transfer, because [85]krypton was found in the tubal lumen after being infused into the uterine vein (Fig. 8.6). Coudert, Phillips, Fairman, Chernnecki, and Palmer (1974*a*) wrote that they "found no indication of a counter-current transfer of [85]krypton", although their krypton results (see their Fig. 4) do seem to indicate transfer.

The first results indicating a transfer *in vivo* in the female human adnex were found by inserting the miniature Gieger-Müller probe in the ovary and measuring the radioactivity after infusion of [85]krypton into the uterine vein in patients undergoing hysterectomy (Fig. 8.7). Transfer of the gas was observed within 20 seconds (Bendz, Einer-Jensen, Lundgren, and Janson 1979). The Gothenburg group also found transfer under *in vitro* conditions of radiolabelled $PGF_{2\alpha}$, antipyrine and progesterone (see Bendz 1982*a*).

4   STEROID HORMONES

Tritium labelled progesterone was infused into an ovarian vein branch in ewes and parallel blood samples obtained from the aorta and a side branch of the ovarian artery (McCracken and Einer-Jensen 1976). The radioactivity

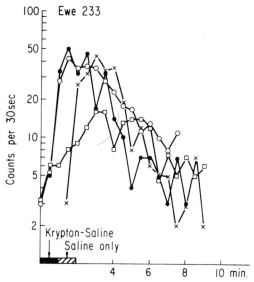

Fig. 8.6   The appearance and disappearance of radioactivity in the corpus luteum (○, ×) and Fallopian tube (○, □) after the injection of [85]krypton into a branch of the adjacent uterine vein. Radioactivity was monitored with a miniature Geiger-Müller probe inserted into the tissue of the corpus luteum and the lumen of the Fallopian tube. Radioactivity reached a peak within 1 to 3 minutes and had returned to background (2–5 counts per 30 sec) by 10 minutes post infusion. No increase in radioactivity was observed in the contralateral ovary during and after the infusion in control experiments. (From Einer-Jensen and McCracken 1977, with permission).

was higher in the ovarian samples than in the aortic samples from 3 to 9 minutes after the start of the infusion (Fig. 8.8) indicating a transfer of the hormone (Einer-Jensen and McCracken 1981a). The efficiency of the exchange was calculated to be in the order of 0.5 to 2 per cent. Walsh, Yutrzenkan, and Davis (1979) infused labelled progesterone into the uterine vein in the sheep and subsequently found a smaller transfer (0.05 per cent) of hormone into the ovarian arterial blood.

Countercurrent transfer has been found with several steroids. Three pairs of steroids, one ^{3}H the other ^{14}C labelled, (oestrone and oestradiol, C-18; androstenedione and testosterone, C-19; and progesterone and 20α-dihydroprogesterone, C-21), each pair differing by one hydroxyl group, were infused separately into a side branch of an ovarian vein near the hilus of the ovary in anaesthetized sheep. All the steroids were transferred; the less polar (ketonic) member of each steroid pair more efficiently than its hydroxy counterpart (McCracken, Schramm, and Einer-Jensen 1984).

Testosterone, progesterone and oestradiol were found to be transferred to the ovarian artery after infusion into an ovarian or uterine vein in the cow (Kotwica, Williams, and Machello 1982) and the sow (Krzymowski, Kotwica, and Stefanczyk 1979, 1981; Kotwica, Krzymowski, Stefanczyk,

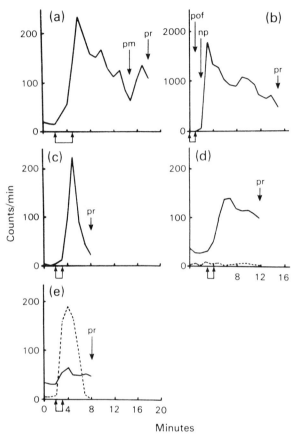

Fig. 8.7   Radioactivity determined in ovaries of women during and after 1–3 min infusion into a uterine vein (between double arrows) of ⁸⁵krypton (200–400 μCi in 2–5 ml saline). Determinations made in the ipsilateral ovary (a, b, c) or in both ovaries (d, e; —— ipsilateral, ----- contralateral). Geiger-Müller probe was inserted into a corpus luteum in (b) and (c). High radioactivity in the contralateral ovary in case e was probably caused by redistribution of venous blood flow and transfer in the contralateral plexus due to a haematoma. np, new probe inserted; pm, probe moved; pof, probe out of function; pr, probe removed. (From Bendz *et al.* 1979, with permission).

Nowicka, Debek, Czarnocki, and Kuznia 1980; Krzymowski, Kotwica, Stefanzcyk, Debek, and Czarnocki 1981, 1982; Krzymowski, Stefanczyk, Kotwica, Czarnocki, Glazer, Janowski, and Chmiel 1981). As an example, ³H-testosterone was infused for 30 minutes into the ovarian vein below the ovarian hilus of cycling sows. The venous blood was not returned to the animal, as the whole of the vascular pedicle was isolated by ligatures from the systemic circulation except for the ovarian artery. Radioactivity was found in the ovarian arterial blood, but not in peripheral blood, indicating transfer of testosterone.

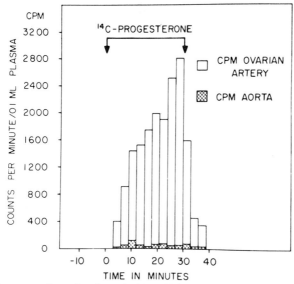

Fig. 8.8 Concentration of radioactively labelled progesterone in plasma collected simultaneously from an ovarian artery and the aorta after infusion of ^{14}C-progesterone into an adjacent ovarian vein. Plasma extracted with organic solvents, and residue further purified by thin layer silica gel chromatography (two different solvent systems). (From Einer-Jensen and McCracken 1981a, with permission).

Transfer of ^{14}C-progesterone was found under *in vitro* conditions in the human adnex after surgical removal of the vessels associated with utero-ovarian extirpations (Bendz, Lundgreen, and Hamberger 1982). The transfer was also found in *in vivo* studies in patients where ^{13}C-progesterone was infused into the uterine vein and parallel samples obtained from the ovary (stab incisions) and a peripheral vessel. The ^{13}C isotope was used instead of ^{14}C because it is not radioactive, and it was measured by a gas chromatographic-mass spectrometric technique (Halket, Leidenberger, Einer-Jensen, and Bendz 1985).

An effort was made to demonstrate the transfer of endogenous hormones in the female, which had already been shown to occur in the male. The concentration in the ovarian vein was high because of steroidogenesis. The concentration of progesterone during the luteal phase ought thus to be higher in the ovarian arterial blood than in any other artery. Many blood samples were taken, radioimmunoassays were run with 4 replicate samples to diminish the variation caused by the assay and still the difference could not be established. FSH and LH secretory peaks (Walters, Schams, and Schallenberger 1984) as well as handling of the ovary (Bendz and Einer-Jensen, personal communication) may induce secretory peaks of the hormones and thus induce a huge variation during an experiment. It was concluded that the difference was so small that it would be indistinguishable from assay variation, and that the pulsatile nature of the secretion and time delay would minimize or conceal the apparent transfer. However there is some reason to believe that even a small increase in the total content may be physiologically important, as discussed below (protein binding).

A turning point came when Hunter, Cook, and Poyser (1983) investigated preovulatory sows in oestrus. They found transfer from the ovarian vein to the ovarian artery of endogenous progesterone, androstendione and oestradiol. The mean increase was "frequently 5- to 10-fold". In one animal the progesterone concentration increased 25 times.

The degree of exchange of endogenous hormones may vary throughout the cycle and/or pregnancy and the transfer of endogenous hormones has clearly to be evaluated under different but specified conditions. For example, the uterine branch of the ovarian artery of the female tammar wallaby ramifies extensively very close to the ovary, giving a plexiform arrangement with the ovarian veins. Progesterone concentrations in plasma from the mesometrial side of the uterine branch of the ovarian vein are markedly higher than in tail vein plasma, especially during the 'Day 5 peak' early in pregnancy, and also at full term (Towers, Shaw, and Renfree 1986).

5   PEPTIDES

Transfer was for some years thought to be limited to low molecular weight, lipid soluble, molecules like krypton, $PGF_{2\alpha}$ and certain steroids. Then the exchange of ^{14}C-methylantipyrine was found in the human adnex under *in vitro* conditions (Bendz *et al.* 1982). Evidence for *in vivo* transfer of amino acids and peptides in sheep is now beginning to appear. Infusion of iodinated tyrosine into an ovarian vein resulted in an increased concentration in ovarian artery branches. The amount transferred was about 1 per cent of the amount infused (Schramm, Einer-Jensen, and Schramm 1986). Infusion of iodinated oxytocin (M_v 1000) showed a transfer of 1–3 per cent (Schramm, Einer-Jensen, Schramm, and McCracken 1986). A much larger substance, iodinated relaxin (M_v 6000), was also transferred in the same experimental model, but the rate of transfer decreased to 0.1–0.75 per cent (Schramm *et al.* 1986). Using chromatographic purification, the results in all three cases were proved to be due to the transfer of intact iodinated hormone.

Both oxytocin (see Hansel and Dowd 1986) and relaxin (see Bryant-Greenwood 1985) are produced in the ovaries in some species and influence the function of the ovaries, oviduct and/or uterus. The transfer found may therefore be of functional importance.

Every compound mentioned above was shown to be transferred, but relevant control substances were of course employed. The Polish group (Krzymowski *et al.* 1981) infused 51-chromium labelled red blood cells and found no transfer. Nor was albumin transferred *in vivo* in the ewe (Einer-Jensen and

McCracken 1981*a*), or in the human adnex under *in vitro* conditions (Bendz *et al.* 1982).

6   MECHANISM OF EXCHANGE

The mechanism behind the transfer is unknown. There seems, however, to be general agreement that the transfer is based on diffusion, although no proof exists. The inert gases, xenon and krypton, as well as certain steroids and prostaglandins, are lipophilic and are supposed to diffuse through and/or along the cell membranes. The hydrophilic peptides may not diffuse through the cells but through the intercellular space or be transferred by receptor-mediated endocytosis.

Free, Jaffe, Jain, and Gomes (1973) found support for the theory of diffusion. During *in vitro* perfusion of the testicular artery and veins in the pampiniform plexus, they found bidirectional transfer of labelled testosterone: from the vein to the artery as well as from the artery to the vein. An active mechanism would as a general rule be expected to be unidirectional. There is also some evidence for a dependency between the rate of transfer and the vein blood concentration. Land *et al.* (1976) found a decreasing efficiency of the transfer mechanism when the venous concentration was increased. On the other hand, when Einer-Jensen and Waites (1977) measured the transfer of endogenous testosterone in male rhesus monkeys, transfer was only found in animals with a high endogenous concentration of the hormone.

The speed by which the transfer takes place is remarkable; [85]krypton passes from the venous blood, through the walls of the vein and artery, into the arterial supply and reaches the corpus luteum tissue within 20 seconds. Steroid hormones and the small peptides pass within minutes, while $PGF_{2\alpha}$ seems to be a special case with a delay of 20 minutes.

7   TRANSFER FROM LYMPH

Prominent lymphatic vessels drain tissue from each uterine horn in the pig (Andersen 1927; Anderson *et al.* 1969; Yoffey and Courtice 1970) and are closely associated with the ipsilateral uterine vasculature (Hoggan and Hoggan 1882). Staples, Fleet, and Heap (1982) investigated the vessels in the sheep. They found a complex lymphatic network in the region of the utero-ovarian pedicle draining both the uterus and the ipsilateral ovary. The network was intimately connected with the ovarian artery, supporting the idea of an additional pathway for counter current diffusion of $PGF_{2\alpha}$ as suggested by Kotwica (1980).

Kotwica, Krzymowski, Stefanczyk, Koziorowski, Czarnocki, and Ruszczyk (1983) surgically isolated and perfused both ovaries, their pedicles and part of the adjacent uterine horn as two separate preparations. The only connection between the two preparations was that one of the ovaries (B) was

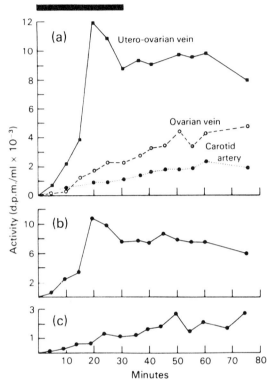

Fig. 8.9   Rate of occurrence of ³H-labelled compound in the adjacent utero-ovarian vein and ovarian vein after infusion of ³H-PGF$_{2\alpha}$ (1.1 µCi per min.) into a uterine afferent lymphatic (black bar) in sheep, 15 days after oestrus. (a) Concentration of radioactivity in the adjacent utero-ovarian and ovarian vein and carotid arterial plasma; (b) veno-arterial difference across the adjacent uterine horn and ovary (utero-ovarian vein—carotid artery in (a)); (c) veno-arterial difference across the adjacent ovary (ovarian vein—carotid artery). (From Heap *et al.* 1985, with permission).

covered with the mesosalpinx from preparation A. When tritiated PGF$_{2\alpha}$ was injected into the myometrium of A, it was found in the ovarian venous blood, interstitial fluid, and ovarian and pedicle tissue of preparation B. The data indicated an extravascular penetration into the ovary through the mesosalpinx (Krzymowski, Stefanczyk, Koziorowski, Czarocki, Ruszczyk, and Nowicka 1982). The transfer of PGF$_{2\alpha}$ was later quantified in sheep (Fig. 8.9) by Heap, Fleet, and Hamon (1985). The hormone was infused into a uterine lymphatic vessel in a sheep 15 days after oestrus. A 3-fold increase was observed within 10 minutes, the transfer was thus as fast as when PGF$_{2\alpha}$ was infused into the uterine vein. The lymphatic pathway may be of physiological importance since release of ovarian oxytocin was found after infusion of prostaglandin F$_{2\alpha}$ into a uterine lymphatic or uterine vein in the sheep (Heap, Fleet, Flint, Sheldrick, and Goode 1986).

The content of steroid hormones (oestrone, oestradiol-17β and progesterone) was measured in 12 hour samples of uterine lymph collected throughout the porcine cycle (Magness and Ford 1982, 1983). Concentrations of oestrone, oestradiol and progesterone in uterine lymph were correlated with concentrations of the same steroid in systemic blood. The oestrogens were, however, 2–5 times higher and progesterone 2–5 times lower in the lymph than in blood. Afferent lymph was collected in sheep and goats after chronic cannulation of utero-ovarian ducts ipsilateral to an ovary bearing a corpus luteum. It contained a mean progesterone concentration which was 10- to 1000-fold higher than that in jugular vein plasma between 15 and 45 days of gestation (Staples et al. 1982). Uterine lymph collected after cannulation of ducts followed by ipsilateral ovariectomy had a progesterone value equivalent to that in plasma. No data from women exist, to my knowledge, in this area.

Until now the whole discussion has been about countercurrent transfer to the ovarian arterial blood. If transfer occurs mainly from the utero-ovarian vein then all material transferred will be found in the ovarian artery. If, however, the lymph is an important source, some of the material transferred may also appear in the uterine arterial blood. This possibility has recently been investigated experimentally by the Polish group. Krzymowski, Kotwica, Stefanczyk, Czarnocki, and Koziorowski (1986) found back transfer of $PGF_{2\alpha}$ from the uterus into the uterus via the mesometrium during early pregnancy and oestrogen-induced pseudopregnancy in the sow. As oestrogens induce changes in the uterine blood flow, an interaction at the uterine level between oestrogens, oxytocin and $PGF_{2\alpha}$ may well exist (Krzymowski, Czarnocki, Koziorowski, Kotwica, and Stefanzyk-Krzymowska 1986, Krzymowski et al. 1988, Koziorowski et al. 1988, Kotwica and Krzymowski 1988).

8   BINDING TO PLASMA PROTEINS

Certain steroid hormones bind in plasma to sex hormones binding globulin (Mercier, Alfsen, and Baulien 1966) and albumin (Hoffman, Forbes, and Westphal 1969). Several other proteins have been described and some may have more than one name (see Westphal 1978). Protein binding is substantial, only 1–2 per cent of testosterone, progesterone and oestradiol are unbound in adult men and women (Hammond, Nisker, Jones, and Siiteri 1980). The small fraction of unbound hormones is, however, considered to be the physiologically active part, while the protein bound fraction may be a protected depot (see Anderson 1976).

It is not possible to measure rapid changes in plasma after a sudden increase in the unbound fraction with conventional methods (for example equilibrium dialysis, ultrafiltration, or steady state gel filtration) and the rates of binding have not evoked much interest from a functional point of view with respect to the general circulation (Dunn, Nisula, and Rodbard 1981; Dunn 1984) as they were considered non-limiting.

One set of experimental data does, however, indicate that the time lapse before equilibrium is reached between plasma proteins and some steroids may be in the order of seconds in bovine and human plasma even at body temperature (Einer-Jensen 1984*a,b*, 1986).

A non-steady state situation of potential physiological importance may therefore exist in the ovarian arterial blood as the steroid hormone transferred probably arrives in the arterial lumen in an unbound state.

IV   Physiological implications

The physiological implications of countercurrent transfer in the female genital system may be summarized as follows (the summary is elaborated below):

● The blood vessels of the ovaries, Fallopian tubes and uterus and the lymph vessels draining them participate in a countercurrent transfer. The compounds transferred are present in the blood of the utero-ovarian vein, or in the lymph, and are transferred to the ovarian artery including branches to the Fallopian tube and proximal portion of the uterine horn, probably down a concentration gradient. Lymph vessels going caudally may also participate in a local transfer system with the uterine artery.

● Countercurrent transfer of $PGF_{2\alpha}$ in animals is associated with luteolysis at the end of a three week non-fertile cycle. Transfer of oxytocin from the ovarian vein to the uterine branch of the ovarian artery has also been found. The two hormones may be proven to form a double, but still local, positive feed back loop as oxytocin of ovarian origin influences the uterine secretion of $PGF_{2\alpha}$ and *vice versa*.

● Countercurrent transfer of steroid hormones has been found. Local high concentrations of (unbound) steroid hormones may prove to have local stimulatory or regulatory effects on ovarian, tubal and uterine tissues.

In many,.but not all, species the local transfer system includes the arterial blood supply to the ovary, the oviduct and sometimes even the proximal part of the uterus. The venous blood (and lymph) from these organs as well as that from the uterus are also included in this local transfer network. Compounds (steroids, peptides, prostaglandins) produced and secreted by any of these organs can in principle be transferred locally and returned to an organ by the local pathway.

Experimental evidence supports the theory. Exchange from ovarian vein blood to the tubal lumen was found with [85]krypton in sheep (Fig. 8.6). Prostaglandin $F_{2\alpha}$ was found to be transferred from the ovarian vein blood to the ovarian arterial blood (McCracken *et al.* 1972). An increased concentration of endogenous steroid hormones was found in the tubal blood in swine (Hunter *et al.* 1983). Oxytocin was transferred from the ovarian vein blood to the uterine branch of the ovarian artery in sheep (Schramm *et al.* 1986). The testis—epididymis system seems to function in a parallel manner, as ^{133}xenon, ^{85}krypton and ^{3}H-testosterone (Einer-Jensen 1974*b*, *c*; Einer-Jensen and Waites 1977) were exchanged between the testis and epididymis.

The main physiological importance ascribed to the transfer system so far has been the luteolytic action of the uterine $PGF_{2\alpha}$ secretion. Schramm *et al.* (1986) have added to this the theory of a local positive feed back effect of oxytocin on the $PGF_{2\alpha}$ secretion although the distribution of $PGF_{2\alpha}$ production within the uterus (proximal part versus the distal part of the horn) has not been described. Active immunization against oxytocin in the goat suppressed $PGF_{2\alpha}$ and delayed luteolysis thus demonstrating the influence of oxytocin on the uterus (Cooke and Homeida 1984, 1985). $PGF_{2\alpha}$ of uterine origin induces ovarian secretion of oxytocin after local transfer (Heap *et al.* 1986). Oxytocin is produced in the ovary (Flint and Sheldrick 1982; Rodgers, O'Shea, Findlay, Flint, and Sheldrick 1983), including the human ovary (Watches, Swann, Pickering, Porter, Hull, and Drife 1982; Wathes 1984), and transferred locally to the proximal part of the uterine horn in sheep (Schramm *et al.* 1986).

Several luteotrophic substances have been described, for example in man (hCG) (Jaffe 1978), cow (Northey and French 1980; Hickey, Walton, Harper, and Hansel 1985), sheep (trophoblastin, OTP-1: Martal, Lacroix, Loudes, Saunier, and Wintenberger-Torres 1979; protein X: Godkin, Bazer, Moffatt, Sissions, and Roberts 1982) and pig (oestrogens: Bazer, Vallet, Roberts, Sharp, and Thatcher 1986) but others (for example prostaglandins of the E-series: Ellingwood, Nett, and Niswender 1979) may exist even in the human (Batta and Channing 1979). Some of them (but not OTP-1) may be produced by the oviduct and uterus immediately after conception and local transfer of such factors would overcome the problem of dilution in the peripheral circulation and a short half life due to metabolism in the lung, liver and kidney.

Both a Graafian follicle and a young corpus luteum secrete steroid hormones which may be exchanged (Hunter *et al.* 1983) and may have a local effect on the tube. In mono-ovulating animals one may expect structural and morphological differences between the Fallopain tubes during the days immediately following ovulation. A morphological difference has been observed in the bat (Rasweiler 1978) but has to my knowledge not been properly investigated in other animals. McComb and Deibeke (1984) found a unilateral decrease in the number of ovulations in the rabbit after surgical division of the blood vessels between the Fallopian tube and ovary, suggesting an unavoidable surgical trauma or an interaction between the ovary and its oviduct.

Ovarian function in primates including women is said to be independent of the presence of the uterus. The interval between ovulations was not changed by uterine extirpation with the possible exception of the first cycle (Neill,

Johansson, and Knobil 1969; Castracane, Moore, and Shaikh 1979; Beling, Marcus, and Markham 1970; Doyle *et al.* 1971). The effect on this cycle may be due to surgical trauma (Stone, Dickey, and Mickael 1975) causing a transient drop in ovarian blood flow (Janson and Jansson 1977). It may, however, be wise to leave the Fallopian tubes together with the ovaries during uterine extirpation: thus an increased frequency of luteal cysts and irregular cycles has been observed when the uterus and tubes were removed, compared to patients where the tubes were left *in situ* (Arvay, Nyiri, and Buris 1961).

The blood vessels in the ovarian hilus are intimately apposed, thus increasing the area of exchange. The blood borne hormonal environment in the two ovaries may, therefore, be different, for example in the periods when a large follicle or a corpus luteum is found in one of the ovaries. Gougeon and Lefévre (1984) published histological evidence of alternating ovulation in women in 88 per cent of the cycles studied. Ovulations occurring early after parturition in the cow are more likely to be on the previously nonpregnant side than ovulations that occur later (Saiduddin, Riesen, Tyler, and Casida 1967). Follicular development was also significantly depressed unilaterally by the interaction of suckling and the presence of the corpus luteum (Bellin, Hinshelwood, Hauser, and Ax 1984). However, the shift in the side of ovulation is certainly not obligatory.

V   Pharmacological implications

Very little information seems to exist with respect to the possible therapeutic aspects of countercurrent transfer. This section will therefore be short, dealing with suggestions rather than established facts.

Intrauterine administration of [85]krypton to rats (Einer-Jensen 1974*a*) and $PGF_{2\alpha}$ to rhesus monkeys (Einer-Jensen 1973) indicated that compounds instilled into the uterine lumen could reach the ovary and exert a pharmacological effect. At present only one application appears to have been tested.

A norgestrel-releasing IUD can be expected to combine the effect of a conventional IUD with a local gestagen effect on the endometrium caused by slow release of the hormone. The peripheral blood level of norgestrel is presumably too low to induce any effect on any other organ. Countercurrent transfer of norgestrel may occur, however, inducing an increased concentration in the arterial blood to both ovaries. Preliminary results indicate that exchange and a possibly increased concentration in ovarian blood could be found in some, but not all, patients tested (Bendz, Einer-Jensen, and Leidenberger 1986).

Intrauterine administration may also be tested in special cases of ovarian cancer. An intrauterine administration of a depot, lipophilic cytostaticum may, due to local counter current transfer, create high utero-ovarian concentrations. This would improve the therapeutic index. With local administration and the lipophilic nature of the compound, some of it may be removed via the lymph vessels and create high concentrations in the local lymph glands, where metastasis can be suspected.

Acknowledgement

The secretarial assistance from Mrs. Lise Larsen and Mrs. Bente Kuhlmann is appreciated. Dr. Anthony Carter and Dr. Willfried Schramm contributed to the manuscript with valuable suggestions.

References

Abdel Rahim, S. E. A., Bland, K. P., and Poyser, N. L. (1984). Surgical separation of the uterus and ovaries with simultaneous cannulation of the uterine vein extends luteal function in sheep. *J. Reprod. Fert.* **72**, 230–235.

Andersen, D. H. (1927). Lymphatics of the fallopian tube of the sow. *Contrib. Embryol.* **102**, 135–147.

Anderson, D. C. (1976). The role of sex hormone binding globulin in health and disease. In *The Endocrine Function of the Human Ovary* (ed. V. T. H. James, M. Serio, and G. Giusti (pp. 141–158. Academic Press, New York.

Anderson, L. L., Butcher, R. L., and Melampy, R. M. (1961). Subtotal hysterectomy and ovarian function in the gilts. *Endocrinology* **69**, 571–580.

—— Neal, F. C., and Melampy, R. M. (1962). Hysterectomy and ovarian function in beef heifers. *Am. J. vet. Res.* **23**, 794–800.

—— (1962). Effect of uterine distention on the estrous cycle of the gilt. *J. Anim. Sci.* **21**, 597–601.

—— Bland, K. P., and Melampy, R. M. (1969). Comparative aspects of uterine-luteal relationship. *Horm. Res.* **25**, 57–104.

—— Bowerman, A. M., and Melampy, R. M. (1965). Oxytocin on ovarian function in cycling and hysterectomized heifers. *J. Anim. Sci.* **25**, 964–968.

—— Butcher, R. L., and Melampy, R. M. (1963). Uterus and occurrence of oestrus in pigs. *Nature, Lond.* **198**, 311–312.

—— Melampy, R. M. (1962). Effect of uterine denervation and uterine distention on the estrous cycle of the gilt. *Anat, Rec.* **142**, 209.

Andreoli, C. (1965). Corpus luteum activity after hysterectomy in women. *Acta endocr. Copenh.* **50**, 65–69.

Arvay, A., Nyiri, I., and Buris, L. (1961). Die Wirkung der Hysterektomie auf die Morphologie und Funktion des Ovariums. *Acta chir. Acad. Sci. Hung.* **2**, 381–397.

Baker, M. A. (1979). A brain-cooling system in mammals. *Sci. Am.* **240**, 5, 114–122.

Barcroft, J. and Barron, D. H. (1946). Observations upon the form and relations of the maternal and fetal vessels in the placenta of the sheep. *Anat. Rec.* **94**, 569–595.

Barrett, S., Blickey, A. de B., Brown, J. M., Cumming, I. A., Goding, J. R., Mole, B. J., and Obst, J. M. (1971). Initiation of the oestrous cycle in the ewe by infusion of PGF$_2$ to the autotransplanted ovary. *J. Reprod. Fert.* **24**, 136–137.

Batta, S. K. and Channing, C. P. (1979). Preimplantation rhesus monkey blastocyst: secretion of substance capable of stimulating progesterone secretion by granulosa cell cultures. *Life Sci.* **25**, 2057–2063.

Bazer, F. W., Vallet, J. L., Roberts, R. M., Sharp, D. C., and Thatcher, W. W. (1986).

Role of conceptus products in establishment of pregnancy. *J. Reprod. Fert.* **76**, 841–850.

Beling, C. G., Marcus, S. L., and Markham, S. M. (1970). Functional activity of corpus luteum following hysterectomy. *J. clin. Endocr. Metab. B* **30**, 30–39.

Bellin, M. E., Hinshelwood, M. M., Hauser, E. R., and Ax, R. L. (1984). Influence of suckling and side of corpus luteum or pregnancy on folliculogenesis in postpartum Cows. *Biol. Reprod.* **31**, 849–855.

Bendz, A. (1977). The anatomical basis for a possible countercurrent exchange mechanism in the human adnex. *Prostaglandins* **13**, 355–362.

—— (1982*a*). *Countercurrent exchange in the human female reproductive tract.* Thesis, Gothenburg, Sweden.

—— (1982*b*). On the extensive contact between veins and arteries in the human ovarian pedicle. *Acta physiol. scand.* **115**, 179–182.

—— Einer-Jensen, N., Lundgren, O., and Janson, P. O. (1979). Exchange of ⁸⁵Krypton between the blood vessels of the human uterine adnexa. *J. Reprod. Fert.* **57**, 137–142.

—— Lundgren, O., and Hamberger, L. (1982). Countercurrent exchange of progesterone and antipyridine between human utero-ovarian vessels, and of antipyridine between the femoral vessels in the cat. *Acta physiol. scand.* **114**, 611–616.

—— Einer-Jensen, N., and Leidenberger, F. (1986). Überführung von Progesterone und Levonorgestrel durch Gegenstrohm im menschlichen Adnex nach intrauterine Applikationen. *Acta endocr. Suppl.* **274**, 135–136.

Benninghoff, A. and Goerttler, K. (1975). *Lehrbuch der Anatomie des Menschen, Band 1–3,* Urban & Schwarzenberg, München, Berlin, Wien.

Blancardi, S. (1687). *Anatomia Reformata.* Joannem ten Hoorn, 34, 1739. Amstelodami.

Bland, K. P. and Donovan, B. T. (1969). Observations on the time of action and the pathway of the uterine luteolytic effect of the guinea-pig. *J. Endocr.* **43**, 259–264.

Bradbury, J. T. (1937). Prolongation of the life of the corpus luteum by hysterectomy in the rat. *Anat. Rec. Suppl.* **1**, 51.

Bruce, N. W. and Hillier, K. (1974). The effect of prostaglandin $F_{2\alpha}$ on ovarian blood flow and corpora lutea regression in the rabbit. *Nature, Lond.* **249**, 176–177.

Bryant-Greenwood, G. D. (1985). Current concepts on the role of relaxin. *Res. Reprod.* **17**, 1–3.

Burford, T. H. and Diddle, A. W. (1936). Effect of total hysterectomy upon the ovary of the *Macacus* rhesus. *Surg. Gynecol. Obstet.* **62**, 701–707.

Butcher, R. L., Chu, K. Y., and Melampy, R. M. (1962). Utero-ovarian relationships in the guinea pig. *Endocrinology* **71**, 810–815.

—— Barley, D. A., and Inskeep, E. K. (1969). Local relationship between the ovary and uterus of rats and guinea pigs. *Endocrinology* **84**, 476–481.

Calza, L. (1807). Mechanismus der Schwangerschaft. In *Archiv für die Physiologie,* Vol. 7 (ed. C. Reil, J. H. F. Autenrieth) pp. 341–393. Halle.

Carter, A. M. (1975). *Placental Circulation. Comparative Placentation.* Essays in structure and function (ed. D. H. Steven) pp. 108–160. Academic Press, London.

—— Göthlin, J., and Olin, T. (1971). An angiographic study of the structure and function of the uterine and maternal placental vasculature in the rabbit. *J. Reprod. Fert.* **25**, 201–210.

Casida, L. E. and Venzke, W. G. (1936). Observations on reproductive processes in dairy cattle and their relation to breeding efficiency. In abstracts of *The annual meeting of the American society of animal production,* pp. 221–223, abstr. 29.

Castracane, V. D., Moore, G. T., and Shaikh, A. A. (1979). Ovarian function in hysterectomized *Macaca fascicularis. Biol. Reprod.* **20**, 462–472.

Chaichareon, D. P., Rankin, J. H., and Ginther, O. J. (1976). Factors which affect the relative contributions of ovarian and uterine arteries to the blood supply of reproductive organs in guinea pigs. *Biol. Reprod.* **15**, 281–290.

Clara, M. (1956). *Weibliche Geschlechtsorgane. Anatomie die arterio-venösen Anastomosen.* pp. 148–155. Springer Verlag, Wien.

Clemens, J. A., Minaguchi, H., and Meites, J. (1968). Relation of local circulation between ovaries and uterus to lifespan of corpora lutea in rats. *Proc. Soc. exp. Biol. Med.* **127**, 1248–1251.

Cooke, R. G. and Homeida, A. M. (1984). Delayed luteolysis and suppression of the pulsatile release of oxytocin after indomethacin treatment in the goat. *Vet. Sci.* **36**, 48–51.

—— —— (1985). Suppression of prostaglandin $F_{2\alpha}$ release and delay of luteolysis after active immunization against oxytocin in the goat. *J. Reprod. Fert.* **75**, 63–68.

Coudert, S. P., Phillips, G. D., Faiman, C. Chernecki, W., and Palmer, M. (1974a). A study of the utero-ovarian circulation in sheep with reference to local transfer between venous and arterial blood. *J. Reprod. Fert.* **36**, 319–331.

—— —— —— —— Palmer, M. (1974b). Infusion of tritiated prostaglandin $F_{2\alpha}$ into the anterior uterine vein of the ewe: absence of a local venous-arterial transfer. *J. Reprod. Fert.* **36**, 333–343.

Del Campo, C. H. and Ginther, O. J. (1972). Vascular anatomy of the uterus and ovaries and the unilateral luteolytic effect of the uterus: guinea pigs, rats, hamsters and rabbits. *Am. J. vet. Res.* **33**, 2561–2578.

—— —— (1973). Vascular anatomy of the uterus and ovaries and the unilateral luteolytic effect of the uterus: angioarchitecture in sheep. *Am. J. vet. Res.* **34**, 1377–1385.

—— —— (1974). Vascular anatomy of the uterus and ovaries and the unilateral luteolytic effect of the uterus: histological structure of uteroovarian vein and ovarian artery in sheep. *Am. J. vet. Res.* **35**, 397–399.

Dhinesa, D. S. and Dziuk, P. J. (1968). Effect on pregnancy in the pig after killing embryos or fetuses in one uterine horn in early gestation. *J. Anim. Sci.* **27**, 122–126.

Dierschke, D. J., Walsh, S. W., Mapletoft, R. J., Robinson, J. A., and Ginther, O. J. (1975). Functional anatomy of the testicular vascular pedicle in the rhesus monkey: evidence for a local testosterone concentrating mechanism. *Proc. Soc. exp. Biol. Med.* **148**, 236–242.

Doboszynska, T., Janowics, K., Stefanowski, T., and Zamojska, D. (1980). Vascularization of the ovary in the period of estrus in the pig. *Folia Morphol., Warszawa*, **39**, 37–53.

Dunn, J. F., Nisula, B. C., and Rodbard, D. (1981). Transport of steroid hormones: binding of 21 endogenous steroids to both testosterone-binding globulin and corticosteroid-binding globulin in human plasma *J. clin. Endocr. Metab.* **53**, 58–68.

Eddy, C. A. and Pauerstein, C. J. (1980). Anatomy and physiology of the Fallopian tube. *Clin. Obstet. Gynec.* **23**, 1177–1193.

Eglinton, G., Raphael, R. A., Smith, G. N., Hall, W. J., and Pickles, V. R. (1963). Isolation and identification of two smooth muscle stimulants from menstrual fluid. *Nature, Lond.* **200**, 960, 993–995.

Egund, N. and Carter, A. M. (1974). Uterine and placental circulation in the guinea-pig: an angiographic study. *J. Reprod. Fert.* **40**, 401–410.

Einer-Jensen, N. (1973). Decreased endometrial blood flow and plasma progesterone level after instillation of 10 μg prostaglandin $F_{2\alpha}$ into the lumen of uteri of rhesus monkeys. *Prostaglandins* **4**, 517–522.

—— (1974a). Local transfer of [133]Xenon from the uterine horn to the ipsilateral ovary in the mouse, hamster and guinea-pig. *J. Reprod. Fert.* **40**, 479–482.

—— (1974b). Local recirculation of ^{133}xenon and ^{85}krypton to the testes and the caput epididymidis in rats. *J. Reprod. Fert.* **37**, 55–60.

—— (1974c). Local recirculation of injected (H^3) testosterone from the testis to the epidymal fat pad and the corpus epididymidis in the rat. *J. Reprod. Fert.* **37**, 145–148.

—— (1976a). Vascular counter current exchange in the male and female reproductive systems. *Karger Gazette* **33**, 1, 6–7.

—— (1976b). Counter current exchange in reproductive biology. *Acta physiol. scand. Suppl.* **440**, 23.

—— (1984a). Slow binding of progesterone to plasma proteins. *Acta Pharmacol. Toxicol.* **55**, 18–20.

—— (1984b). Binding of steroid hormones to bovine plasma proteins. *Xth Int. Congr. anim. Reprod.* Univ. Ill., U.S.A. Abstr. 511.

—— (1986). Slow binding of steroids to human and bovine plasma proteins. *First Internat. Symp. on binding Proteins: Steroid Hormones.* Lyon. Abstr. 21.

—— McCracken, J. A. (1976). ^{85}Krypton measurement of capillary blood flow in the ovine corpus luteum during PGF$_2$ induced luteolysis. *Adv. Prost. Th.* **2**, 901.

—— —— (1977). The effect of prostaglandin F$_2$ on the counter-current transfer of ^{85}Krypton in the ovarian pedicle of the sheep. *Prostaglandins* **13**, 763–775.

—— —— (1981a). The transfer of progesterone in the ovarian vascular pedicle of the sheep. *Endocrinology* **109**, 310, 471–470.

—— —— (1981b). Physiological aspects of corpus luteum blood flow and of the counter current system in the ovarian pedicle of the sheep. *Acta vet. scand. Suppl.* **77**, 89–101.

—— Waites, G. M. H. (1977). Testicular blood flow and a study of the testicular venous to arterial transfer of radioactive krypton and testosterone in the rhesus monkey. *J. Physiol. Lond.* **267**, 1–15.

Ellinwood, W. E., Nett, T. M., and Niswender, G. D. (1979). Maintenance of the corpus luteum of early pregnancy in the ewe. II. Prostaglandin secretion by endometrium *in vitro* and *in vivo*. *Biol. Reprod.* **21**, 845–856.

Fischer, T. V. (1967). Local uterine regulation of the corpus luteum. *Am. J. Anat.* **121**, 425–442.

Flint, A. P. F. and Sheldrick, E. L. (1982). Ovarian secretion of oxytocin is stimulated by prostaglandin. *Nature Lond.* **297**, 587–588.

Ford, S. P. (1982). Control of uterine and ovarian blood flow throughout the estrous cycle and pregnancy of ewes, sows and cows. *J. Anim. Sci.* **55** Suppl. 2, 32–42.

Ford, S. P. and Chenault, J. R. (1981). Blood flow to the corpus luteum-bearing ovary and ipsilateral uterine horn of cows during the oestrous cycle and early pregnancy. *J. Reprod. Fert.* **62**, 555–562.

Free, M. J., Jaffe, R. A., Jain, S. K., and Gomes, W. R. (1973). Testosterone concentrating mechanism in the reproductive organs of the male rat. *Nature, New Biol.* **244**, 24–26.

Gerrits, R. G. and Hawk, H. W. (1966). Effect of intrauterine devices on fertility in pigs. *J. Anim. Sci.* **25**, p. 1266. Abstr. 99.

Ghazi, R. (1981). Angioarchitectural studies of the utero-ovarian component in the camel. *J. Reprod. Fert.* **61**, 43–46.

Ginther, O. J. (1967). Local utero-ovarian relationships. *J. Anim. Sci.* **26**, 578–585.

—— (1974). Internal regulation of physiological processes through local venoarterial pathways: a review. *J. Anim. Sci.* **39**, 550–564.

—— (1976). Comparative anatomy uteroovarian vasculature. *Vet. Scope* **20**, 1–17.

—— Bisgaard, G. E. (1972). Role of the main uterine vein in local action of an intrauterine device on the corpus luteum in sheep. *Am. J. vet. Res.* **33**, 1783–1787.

—— Dierschke, D. J., Walsh, S. W., and Del Campo, C. H. (1974). Anatomy of arteries and veins of uterus and ovaries in rhesus monkeys. *Biol. Reprod.* **11**, 205–219.

—— Pope, A. L., and Casida, L. E. (1965). Some effects of intra-uterine plastic coils in ewes. *J. Anim. Sci.* **24**, p. 918. Abstr. 301.

—————— (1966). Local effect of an intrauterine plastic coil on the corpus luteum of the ewe. *J. Anim. Sc.* **25**, 472–475.

—— Woody, C. O., Janakiraman, K., and Casida, L. E. (1966). Effect of an intrauterine plastic coil on the oestrous cycle of the heifer. *J. Reprod. Fert.* **12**, 193–198.

Goding, J. R., McCracken, J. A., and Baird, D. T. (1967). The study of ovarian function in the ewe by means of a vascular autotransplantation technique. *J. Endocr.* **38**, 39–52.

Godkin, J. D., Bazer, F. W., Moffatt, J., Sissions, F., and Roberts, R. M. (1982). Purification and properties of a major, low molecular weight protein released by the trophoblast of sheep blastocysts at day 12–21. *J. Reprod. Fert.* **65**, 141–150.

Gourgeon, A. and Lefévre, B. (1984). Histological evidence of alternating ovulation in women. *J. Reprod. Fert.* **70**, 7–13.

Grinsted, J., Blenstrup. K., Andreasen, M. P., and Byskov, A. G. (1980). Temperature measurements of rabbit antral follicles. *J. Reprod. Fert.* **60**, 149–155.

Halket, J. M., Leidenberger, F., Einer-Jensen, N., and Bendz, A. (1985). Measurements of progesterone-13C in plasma with a GCMS-method. *Biomedical Mass Spectrometry* **12**, 429–31.

Hammond, G. L., Nisker, J. A., Jones, L. A., and Siiteri, P. K. (1980). Estimation of the percentage of free steroid in undiluted serum by centrifugal ultrafiltration-dialysis. *J. biol. Chem.* **255**, 5023–5026.

Hansel, W. and Asdell, S. A. (1950). *The Effect of Estrogen and Progesterone on the Arterial System of the Uterus of the Cow.* Ward Scientific Co., Rochester, N. Y.

—— Wagner, W. C. (1960). Luteal inhibition in the bovine as a result of oxytocin injections, uterine dilatation, and intrauterine infusions of seminal and preputial fluids. *J. Dairy Sci.* **43**, 796–805.

—— Dowd, W. (1986). New concepts of the control of corpus luteum function. *J. Reprod. Fert.* **78**, 755–768.

Harrison, R. G. and Weiner, J. S. (1949). Vascular patterns of the mammalian testis and their functional significance. *J. exp. Biol.* **26**, 304–316.

Hartman, C. G. (1932). Studies in the reproduction of the monkey *Macacus* (*Pithecus*) *rhesus*, with special reference to menstruation and pregnancy. *Contrib. Embryol.* **134**, 3–160.

—— Straus, W. L. Jr. (1933). *The Anatomy of the Rhesus Monkey* (*Macaca mulatta*). Williams & Wilkins Company, Baltimore.

Hawk, H. W., Conley, B. S., Brinsfield, T. H., and Righter, B. (1973) Contraceptive effect of plastic devices in cattle uteri. *Intrauterine Contraception* (eds. S. J. Segal, A. L. Southam, and D. K. Shafer) pp. 189–193. University of Wisconsin Press.

Heap, R. B., Fleet, I. R., and Hamon, M. (1985). Prostaglandin F-2 is transferred from the uterus to the ovary in the sheep by lymphatic and blood vascular pathways. *J. Reprod. Fert.* **74**, 645–656.

—— Fleet, I. F., Flint, A. P. F., Sheldrick, E. L., and Goode, J. A. (1986). Ovarian oxytocin release after infusion of prostaglandin $F_{2\alpha}$ into a uterine lymphatic or uterine vein in the sheep. *J. Endocr.* **108** *Suppl.* 238A.

Hickey, G. J., Walton, S. J., Harper, H., and Hansel, W. (1985). Partial purification of a luteotropic substance from allantoic fluids of 28–37 day bovine conceptuses. *Eighteenth ann. meet. Soc. Study Reprod.* McGill Uni. Montreal. Canada, p. 64.

Hixon, J. E. and Hansel, W. (1974). Evidence for preferential transfer of prostaglandin $F_{2\alpha}$ to the ovarian artery following intrauterine administration in cattle. *Biol. Reprod.* **11**, 543–552.

Hoffmann, W., Forbes, T. R., and Westphal, U. (1969). Biological inactivation of progesterone by interaction with corticosteroid-binding globulin and with albumin. *Endocrinology* **85**, 778–781.

Hoggan, G. and Hoggan, F. J. (1882). On the comparative anatomy of the lymphatics of the uterus. *J. Anat. Physiol.* **16**, 50–89.

Hossain, M. I. and O'Shea, J. D. (1983). The vascular anatomy of the ovary and the relative contribution of the ovarian and uterine arteries to the blood supply of the ovary in the guinea-pig. *J. Anat.* **137**, 457–467.

Howe, G. R. (1965). Influence of the uterus upon cyclic ovarian activity in the guinea pig. *Endocrinology* **77**, 412–414.

Hunter, R. H. F., Cook, B., and Poyser, N. L. (1983). Regulation of oviduct function in pigs local transfer of ovarian steroids and prostaglandins: a mechanism to influence sperm transport. *Eur. J. Obstet. repr. Biol.* **14**, 225–231.

—— Poyser, N. L. (1982). Uterine secretion of prostaglandin $F_{2\alpha}$ in anaesthetized pigs during the oestrous cycle and early pregnancy. *Reprod. Nutr. Develop.* **22**, 1013–1023.

Inskeep, E. K., Oloufa, M. M., Howland, B. E., Pope, A. L., and Casida, L. E. (1962). Effect of experimental uterine distention on estrual cycle lengths in ewes. *J. Anim. Sci.* **21**, 331–332.

Ishiama, A. (1959). *Yokohama Med. Jour.* **10**, 89–101 (in Japanese, see *Clin. Obstet, Gynec.* (1964) **7**, 814).

Jaffe, R. B. (1978). The endocrinology of pregnancy; placental protein hormones. In *Reproductive Endocrinology* (eds. S. S. C. Yen and R. B. Jaffe) pp. 525–527. W. B. Saunders Company, Philadelphia.

Janowics, K., Doboszynska, T., Stefanowski, T., Zamojska, D. (1981). Histological structure of the blood vessels of the paraovarian arterio-venous plexus in pig. *Folia Morphol. Warszawa* **40**, 257–270.

Janson, P. O. and Jansson, L. (1977). The acute effect of hysterectomy on ovarian blood flow. *Am. J. Obstet. Gynec.* **127**, 349–352.

Jones, G. E. S. and Telinde, R. W. (1941). The metabolism of progesterone in the hysterectomized woman. *Am. J. Obstet. Gynec.* **41**, 682–687.

Kiracofe, G. H., Menzies, C. S., Gier, H. T., and Spies, H. G. (1966). Effect of uterine extracts and uterine or ovarian blood vessels ligations on ovarian function of ewes. *J. Anim. Sci.* **25**, 1159–1163.

—— Spies, H. G. (1963). Length of corpus luteum maintenance in hysterectomized ewes. *J. Anim. Sci.* **22**, 862.

Kirton, K., Ducan, G., Oesterling, T., and Forbes, A. (1971). Prostaglandins and reproduction in the rhesus monkey. *Ann. N. Y. Acad. Sci.* **180**, 443–455.

Kotwicka, J. (1980). Mechanism of prostaglandin $F_{2\alpha}$ penetration from the horn of the uterus to ovaries in pigs. *J. Reprod. Fertil.* **59**, 237–241.

—— and Krzymowski, T. (1988). Role of mesovarium and mesosalpinx in counter current exchange of hormones. *Acta Physiol. Pol.* **5** *in press.*

—— Krzymowski, T., Stefanczyk, S., Nowicka, R., Debek, J., Czarnocki, J., and Kuznia, S. (1980). Steroid concentrating mechanism in the sow's ovarian vascular pedicle. *Physiol. Sci.* **20**, 149–152.

—————— Koziorowski, M. Czarnocki, J., and Ruszczyk, T. (1983). A new route of prostaglandin $F_{2\alpha}$ transfer from the uterus into the ovary in swine. *Anim. Reprod. Sci.* **5**, 303–309.

—— Williams, G. L., and Marchello, M. J. (1982). Counter-current transfer of testosterone by the ovarian vascular pedicle of the cow: evidence for a relationship to follicular steroidogenesis. *Biol. Reprod.* **27**, 778–789.

Koziorowski, M., Stefanczyk-Krzymowska, S., Czarnocki, J., and Krzymowski, T. (1988). Counter current transfer and back transport of ^{3}H-PGF$_{2\alpha}$ in the cow's broad ligament vasculature ipsilateral and contralateral to the corpus luteum. *Acta Physiol. Pol.* **5**, *in press.*

Krzymowski, T., Stefanczyk-Krzymowska, S., and Koziorowski, M. (1988). Counter current transfer of PGF$_{2\alpha}$ in the mesometrial vessels as a mechanism for prevention of luteal regression in early pregnancy. *Acta Physiol. Pol.* **5**, *in press.*

—— Czarnocki, J., Koziorowski, M., Kotwica, J., Stefanczyk, S., and Krzymowska, S. (1986). Counter current transfer of ^{3}H-PGF$_{2\alpha}$ in the mesometrium: a possible mechanism for prevention of luteal regression. *Anim. Reprod. Sci.* **11**, 259–272.

—— Kotwica, J., Okrasa, S., Doboszynska, T., and Ziecik, A. (1978). Luteal function in sows after unilateral infusion of PGF$_{2\alpha}$ into the anterior uterine vein on different days of the oestrous cycle. *J. Reprod. Fert.* **54**, 21–27.

—— —— Stefanczyk, S. (1979). Venousarterial transfer of testosterone in the sow's ovarian vascular pedicle. In *Proceedings of the 21st World Veterinary Congress*, Moscow, p. 70.

—— —— —— (1981). Venousarterial counter-current transfer of 3H-testosterone in the vascular pedicle of the sow ovary. *J. Reprod. Fert.* **61**, 317–323.

—— —— —— Czarrocki, J., and Koziorowski, M. (1986). Prostaglandin F$_{2\alpha}$ back transfer into the uterus in sow's mesometrium during early pregnancy and estrogen-induced pregnancy. *Exp. Clin. Endocr.* (in press).

—— —— —— Debek, J., and Czarnocki, J. (1981). Venous-arterial counter current exchange of testosterone, estradiol and progesterone in sow's ovarian vascular pedicle. *Adv. Physiol. Sci.* **20**, 153–157.

—— —— —— Czarnocki, J., and Debek, J. (1982). A subovarian exchange mechanism for the countercurrent transfer of ovarian steroid hormones in the pig. *J. Reprod. Fert.* **65**, 457–465.

—— —— —— Debek J., and Czarnocki, J. (1982). Steroid transfer from the ovarian vein to the ovarian artery in the sow. *J. Reprod. Fert.* **65**, 441–456.

—— Stefanczyk, S., Kotwica, J., Czarnocki, J., Glazer, T., Janowske, T., and Chmiel, J. (1981). ^{3}H-oestradiol-17 counter current transfer from the ovarian vein into the ovarian artery in cows. *Anim. Reprod. Sci.* **4**, 199–206.

—— —— Koziorowski, M., Czarnocki, J., Ruszczyk, T., and Nowicka, R. (1982). Role of the mesosalpinx oviduct vasculature in the counter-current transfer of steroid hormones into the ovary. *Anim. Reprod. Sci.* **5**, 25–39.

Land, R. B., Baird, D. T., and Scaramuzzi, R. J. (1976). Dynamic studies of prostaglandin F$_{2\alpha}$ in the uteroovarian circulation of the sheep. *J. Reprod. Fert.* **47**, 209–214.

Lee, C. S. and O'Shea, J. D. (1976). The extrinsic blood vessels of the ovary of the sheep. *J. Morph*, **148**, 287–304.

Lindloff, G., Holtz, W., Elsaisser, F., Kreikenbaum, K., and Schmidt, D. (1976). The effect of prostaglandin F$_{2\alpha}$ on corpus luteum function in the Göttingen miniature pig. *Biol. Reprod.* **15**, 303–310.

Loeb, L. (1908). The production of deciduomata and the relation between the ovaries and the formation of the decidua. *J. Am. med. Ass.* **50**, 1897–1901.

—— (1923). The effect of extirpation of the uterus on the life and function of the corpus luteum in the guinea pig. *Proc. Soc. exp. Biol. Med.* **20**, 441–443.

—— (1927). The effects of hysterectomy on the system of sex organs and the periodicity of the sexual cycle in the guinea pig. *Am. J. Physiol.* **83**, 202–224.

Lundgren, O. (1976). Intestinal countercurrent exchange and its significance for absorption. *Acta physiol. scand. Suppl.* **440**, 22.

Magness, R. P. and Ford, S. P. (1982). Steroid concentrations in uterine lymph and uterine arterial plasma of gilts during the estrous cycle and early pregnancy. *Biol. Reprod.* **27**, 871–877.

——— (1983). Estrone, estradiol-17β and progesterone contractions in uterine lymph and systemic blood throughout the porcine estrous cycle. *J. Anim. Sci.* **57**, 449–455.

Mapletoft, R. J., Del Campo, and Ginther, O. J. (1975). Unilateral luteotropic effect of uterine venous effluent of the gravid uterine horn in sheep. *Proc. Soc. exp. Biol. Med.* **150**, 129–133.

—— Ginther, O. J. (1975). Adequacy of main uterine vein and ovarian artery in the local venoarterial pathway for uterine-induced luteolysis in ewes. *Am. J. vet. Res.* **36**, 957–963.

Lapin, D. R., and Ginther, O. J. (1976). The ovarian artery as the final component of the local luteotropic pathway between a gravid uterine horn and ovary in ewes. *Biol. Reprod.* **15**, 414–421.

Martal, J., Lacroix, M.-C., Loudes, C., Saunier, M., and Wintenberger-Torrés, S. (1979). Trophoblastin, an antiluteolytic protein present in early pregnancy in sheep. *J. Reprod. Fert.* **56**, 63–73.

Mattner, P. E., Brown, B. W., and Hales, J. R. S. (1981). Evidence for functional arterio-venous anastomoses in the ovaries of sheep. *J. Reprod. Fert.* **63**, 279–284.

McComb, P. and Deibeke, L. (1984). Decreasing the number of ovulations in the rabbit with surgical division of the blood vessels between the Fallopian tube and ovary. *J. Reprod. Med.* **29**, 827–829.

McCracken, J. (1971). Prostaglandin $F_{2\alpha}$ and corpus luteum regression. *Ann. N.Y. Acad. Sci.* **180**, 456–472.

McCracken, J. A. (1984). Update on luteosin-receptor of pulsatile secretion of prostaglandin $F_{2\alpha}$ from the uterus. *Reproduction* **16**, 1–2.

—— Baird, D. T. (1969). The study of ovarian functions by means of transplantation of ovary in the ewe. In *The Gonads* (ed. K. W. McKerns), pp. 175–209. Appleton-Century Crofts, New York.

—— Carlson, J. C., Glew, M. E., Goding, J. R., Baird, D. T., Green, K., and Samuelsson, B. (1972). Prostaglandin $F_{2\alpha}$ identified as a luteolytic hormone in sheep. *Nature, New Biol.* **238**, 129–134.

—— Einer-Jensen, N. (1976). The counter current transfer of progesterone in the ovarian vascular pedicle. *The V internat. Congr. Endocr.*, Hamburg, Fed. Rep. Germany, 320.

—— Glew, M. E., and Scaramuzzi, R. J. (1970). Corpus luteum regression induced by prostaglandin F_2. *J. clin. Endocr. Metab.* **30**, 544–546.

—— Schramm. W., Barcikowski, B., and Wilson, L. (1981). The identification of prostaglandin $F_{2\alpha}$ as a uterine luteolytic hormone and the hormonal control of its synthesis. *Acta vet. scand. Suppl.* **77**, 71–88.

——— Einer-Jensen, N. (1984). The structure of steroids and their diffusion through blood vessel walls in a counter-current system. *Steroids* **43**, 293–303.

Mercier, C., Alfsen, A., and Baulieu, E. E. (1966). A testosterone binding globulin. In *Androgens* (ed. A. Vermeulen). *Excerpta Medica Int. Congr. Ser.* **101**, 212.

Milvae, R. A. and Hansel, W. (1980). Concurrent uterine venous and ovarian arterial prostaglandin $F_{2\alpha}$ concentrations in heifers treated with oxytocin. *J. Reprod. Fert.* **60**, 7–15.

Moore, W. W. and Nalbandov, A. V. (1953). Neurogenic effects of distention on the estrous cycle of the ewe. *Endocrinology* **53**, 1–11.

Mossman, H. W. and Duke, K. L. (1973). Mammalian ovary: Ch. 1. Blood vessels. In *Comparative Morphology of the Mammalian Ovary, Gross Anatomy* (ed. H. W. Mossman) University of Wisconsin Press.

Neill, J. D., Johansson, E. D. B., and Knobil, E. (1969). Failure of hysterectomy to influence the normal pattern of cyclic progesterone secretion in the rhesus monkey. *Endocrinology* **84**, 464–465.

Northey, D. L. and French, L. R. (1980). Effect of embryo removal and intrauterine infusion of embryonic homogenates on the lifespan of the bovine corpus luteum. *J. Anim. Sci.* **50**, 298–302.

Oloufa, M. M., Inskeep, E. K., Pope, A. L., and Casida, L. E. (1961). Effect of uterine distention on the estrous cycle of the mature ewe. *J. Anim. Sci.* **20**, 975–76.

Oppenheimer, W. (1959). Prevention of pregnancy by the Grafenberger ring method. *Am. J. Obstet. Gynec.* **78**, 446–454.

Orsine, M. W. (1968). Discussion. *J. Anim. Sci. Suppl.* 1, 131.

Oxenreider, S. L., McClure, R. C., and Day, B. N. (1965). Arteries and veins of the internal genitalia of female swine. *J. Reprod. Fert.* **9**, 19–27.

Paturet, G. (1958). Artéres de la gonade, Artéres spermatiques. In *Traité D'anatomie humaine*. 526–530 (ed. Masson & Cie), Libraires de l'Académie de Médecine, Paris.

Peppler, R. D. (1976). Effect of uterine artery ligation on ovulation in the rat. *Anat. Rec.* **184**, 183–185.

Pexton, J. E., Weems, C. W., and Inskeep, E. K. (1975). Prostaglandins $F_{2\alpha}$ in uterine venous plasma, ovarian arterial and venous plasma and in ovarian and luteal tissue of pregnant and nonpregnant ewes. *J. Anim. Sci.* **41**, 154.

Pharriss, B. B. and Wyngarden, L. J. (1969). The effect of prostaglandin $F_{2\alpha}$ on the progesterone content of ovaries from pseudopregnant rats. *Proc. Soc. exp. Biol. Med.* **130**, 92–94.

——(1970). The possible vascular regulation of luteal function. *Perspect. Biol. Med.* **13**, 434–444.

——(1971). Prostaglandin induction of luteolysis. *Ann. N.Y. Acad. Sci.* **180**. 436–444.

Piper, P. J., Vane, J. R., and Wyllie, J. H. (1970). Inactivation of prostaglandins by the lungs. *Nature, Lond.* **225**, 600–604.

Piper, P. and Vane, J. (1971). The release of prostaglandins from lung and other tissues. *Ann. N.Y. Acad. Sci.* **180**, 383–385.

Poyser, N. L. (1976). Prostaglandin $F_{2\alpha}$ is the uterine luteolytic hormone in the guinea pig: the evidence reviewed. *Prostaglandins and thromboxane research* **2**, 633–643.

Rasweiler, J. J. (1978). Unilateral oviductal and uterine reactions in the little bulldog bat, *Noctilio albiventris*. *Biol. Reprod.* **19**, 467–492.

Rathmacher, R. P., Anderson, L. L., Kawata, K., and Melampy, R. M. (1967). Intra-uterine foreign body and pregnancy in pigs. *J. Reprod. Fert.* **13**, 559–561.

Reynolds, S. R. M. (1973). Blood and lymph vascular systems of the ovary. In *Handbook of Physiology* 7, 2 (ed. R. O. Greep and E. B. Astwood), pp. 261–316. American Physiological Society, U.S.A.

Robinson, B. (1903). *The Utero-ovarian Artery*. E. H. Colegrove, Chicago.

Rodgers, R. J., O'Shea, J. D., Findlay, J. K., Flint, A. P. F., and Sheldrick, E. L. (1983). Large luteal cells the source of luteal oxytocin in the sheep. *Endocrinology* **13**, 2302–1304.

Rowson, L. E. A. and Moor, R. (1965). Effect of partial hysterectomy on the length of the dioestrous interval in sheep. *5th Congr. intern. riprod. anim. fecondaz. artific.* 1964, Torino, Italia. 2, pp. 394–398.

Saiduddin, S., Reisen, J. W., Tyler, W. J., and Casida, L. E. (1967). Some carry-over effects of pregnancy on postpartum ovarian function in the cow. *J. Dairy Sci.* **50**, 1846–1867.

Schmidt-Nielsen, K. (1972). *How Animals Work*. Cambridge University Press, New York.

Schramm, W., Bovaird, L., Glew, M. E., Schramm, G., and McCracken, J. A. (1983). Corpus luteum regression induced by ultra-low pulses of prostaglandin F_2. *Prostaglandins* **26**, 347–364.

—— Einer-Jensen, N., and Schramm, G. (1986). Direct venous-arterial transfer of ^{125}I-radiolabelled relaxin and tyrosine in the ovarian pedicle in sheep. *J. Reprod. Fert.* **77**, 513–521.

—— —— —— McCracken (1986). Local exchange of oxytocin from the ovarian vein to the ovarian arteries in the sheep. *Biol. Reprod.* **34**, 671–680.

Sejersen, P. (1970). Convection and diffusion of inert gases in cutaneous, subcutaneous, and skeletal muscle tissue. *Capillary Permeability*, Alfred Benzon Symp. 2, (ed. Chr. Crone and N. A. Lassen) pp. 586–596. Munksgaard, Copenhagen.

—— (1985). Shunting by diffusion of gas in skeletal muscle and brain. *Cardiovascular shunts*, Alfred Benzon Symposium 21 (eds. K. Johansen & W. W. Burggren) pp. 452–466. Munksgaard, Copenhagen.

Setchell, B. P. and Waites, G. M. H. (1967). Pulse attenuation and countercurrent heat exchange in the internal spermatic artery of some Australian marsupials. *J. Reprod. Fert.* **20**, 165–169.

Southam, A. L. (1964). Historical review of intra-uterine devices. In *Intra-uterine Contraception* (eds. S. J. Segal, A. L. Southam, and K. D. Shafer), pp. 3–5. Excerpta Medica Foundation.

Spies, H. G., Zimmerman, D. R., Self, H. L., and Casida, L. E. (1960). Effect of exogenous progesterone on the corpora lutea of hysterectomized gilts. *J. Anim. Sci.* **19**, 101–107.

Staples, L. D., Fleet, I. R., and Heap, R. B. (1982). Anatomy of the utero-ovarian lymphatic network and the same composition of afferent lymph in relation to the establishment of pregnancy in the sheep and goat. *J. Reprod. Fert.* **64**, 409–420.

Steen, J. B. (1976). General properties of counter current exchange and multiplication. *Acta physiol. scand. Suppl.* **440**, p. 24.

Stone, S. C., Dickey, R. P., and Mickael, A. (1975). The acute effect of hysterectomy on ovarian function. *Am. J. Obstet. Gynec.* **121**, 193–197.

Stormshak, F. and Hawk, H. W. (1966). Effect of intrauterine spirals and HCG on the corpus luteum of the ewe. *J. Anim. Sci.* **25**, 931.

—— Lehmann, R. P., and Hawk, H. W. (1967). Effect of intra-uterine plastic spirals and HCG on the corpus luteum of the ewe. *J. Reprod. Fert.* **14**, 373–378.

Sund, T. (1976). Rate constants and counter current multiplication in the fish swim bladder. *Acta physiol. scand. Suppl.* **440**, 25.

Towers, P. A., Shaw, G., and Renfree, M. B. (1986). Urogenital vasculature and local steroid concentrations in the uterine branch of the ovarian vein of the female tammar wallaby (*Macropus eugenii*). *J. Reprod. Fert.* **78**, 37–47.

Wagenen, G. van and Catchpole, H. R. (1941). Hysterectomy at parturition and ovarian function in the monkey (*M. mulatta*). *Proc. Soc. exp. Biol. Med.* **46**, 580–582.

Walsh, S. W., Yutrzenkan, G. J., and Davis, J. R. (1979). Local steroid concentrating mechanism in the reproductive vasculature of the ewe. *Biol. Reprod.* **20**, 1167–1171.

Walters, D. L., Schams, D., and Schallenberger, E. (1984). Pulsatile secretion of gonadotrophins, ovarian steroids and ovarian oxytocin during the luteal phase of the oestrous cycle in the cow. *J. Reprod. Fert.* **71**, 479–491.

Wathes, D. C. (1984). Possible actions of gonadal oxytocin and vasopression. *J. Reprod. Fert.* **71**, 315–345.

—— Swann, R. W., Pickering, B. T., Porter, D. G., Hull, M. G. R., and Drite, J. O. (1982). Neurohypophysial hormones in the human ovary. *Lancet* 410–412.

Wehrenberg, W. B., Dierschke, D. J., and Wolf, R. C. (1979). Utero-ovarian pathways and maintenance of early pregnancy in rhesus monkeys. *Biol. Reprod.* **20**, 601–605.

Westphal, U. (1978). Bindung von Progesteron und andern Steroid-hormonen an Proteine des Blutserums. *Hoppe-Seyler's Z. physiol. Chem.* **359**, 431–447.

Wilson, L., Cenedella, R. J., Butcher, R. L., and Inskeep, E. K. (1972). Levels of prostaglandins in the uterine endometrium during the ovine estrous cycle. *J. Anim. Sci.* **34**, 93–99.

Wiltbank, J. N. and Casida, L. E. (1956). Alteration of ovarian activity by hysterectomy. *J. Anim. Sci.* **15**, 134–140.

Vollmerhaus, B. (1964). Gefässarchitektonische Untersuchungen am Geschlechtsapparat des weiblichen Hausrindes. *Zentralblatt für Veterinärmedizin, Reihe A.* **11**, 599–646.

Yamauchi, S. and Sasaki, F. (1968). Studies on the vascular supply to the uterus of the cow. I. Morphological studies of arteries in the broad ligament. *Bull. Univ. Osaka Prefecture Serie B*, **20**, 33–47.

Yoffey, J. M. and Courtice, F. C. (eds.) (1970). *Lymphatics, Lymph and the Lymphomyeloid Complex.* pp. 250–266. Academic Press, London, New York.

Ziecik, A., Doboszynska, T., and Dusza, L. (1986). Concentrations of LH, prolactin and progesterone in early-pregnant and oestradiol-treated pigs. *Anim. Reprod. Sci.* **10**, 215–224.

9   Control of transfer across the mature placenta

C. P. SIBLEY AND R. D. H. BOYD

I   Introduction

The rate of movement of solutes and of water across the placenta may be altered by changes in a wide range of factors, and any gross physiological or pathological disturbance in mother or fetus is likely to alter placental transfer. The question of general biological interest is how far and by what mechanism is placental transfer matched to fetal need. Teleologically there might be expected firstly, to be abilities to compensate for acute or chronic disturbance of fetal homeostasis such as bleeding, or amniotic fluid loss for example, or to protect the fetus to some extent from maternal dehydration or malnutrition. Secondly, there is the absolute requirement to match the growth of the conceptus to the net rate of transfer into the fetal compartment bearing in mind that this does not necessarily mean that the rate of transfer is the limiting factor in fetal growth; the exact opposite may be the case. Finally, there is the need for the fetus to begin to accumulate at an appropriate moment in later gestation the stores of macro- and micro-nutrients and,

in some species, of immunoglobulin G required for the early stages of extra-
uterine life. Is there control of what is the appropriate moment?

There is at present, as will become obvious, too little understanding of this
topic to provide a clear analysis of how and to what degree placental transfer
is quantitatively related to these needs. In this review we firstly attempt to
sketch in some of the quantitative background to an understanding of placental transfer rates with particular emphasis on those factors which are most
likely to relate to alterations in transfer rates. In the second part of the review
we have selected a number of individual substances for more detailed consideration on the basis that there is a certain amount of experimental information about physiological modulation of their transfer rates. In each
case, to put the limited evidence in context, it has been necessary to review
generally the transport of the substance in question. However, for no single
permeant including those considered is the control of transfer fully understood at present and for each there are many experimental opportunities
open to the interested investigator.

II   Placental transfer: mechanisms and control

1   QUANTITATIVE CONSIDERATIONS

Movement of nutrients, gases and waste products to and from the fetus is a
quantitative process and analysis of the control of transfer demands at least a
semi-quantitative approach. In this first section we provide a simplified summary of the principles and assumptions which underlie most analyses of placental transfer. A more rigorous consideration is to be found in Faber and
Thornburg (1983). The explanation of symbols as used here is given in Table
9.1.

$$J_{net\,(total)} = J_{net\,(placenta)} \tag{1}$$

This implies that all transport to the fetus is transplacental. Equation (1) is
certainly not strictly true, for example a number of authors have given direct
evidence in the human of solute transport, notably of proteins, from maternal circulation to amniotic fluid without passage through the placenta (Gitlin
and Boesman 1966). Such proteins could be subsequently ingested and their
contained amino acids perhaps absorbed from the fetal gut. For the purpose
of this review we are ignoring this route which is probably, but not certainly,
quantitatively unimportant for most small solutes.

In a quasi steady state

$$J_{net} = J_{mf} - J_{fm} \tag{2}$$

Net transport across all biological membranes is a balance between the
movement of molecules across it in one direction and the movement of the
same molecules in the reverse direction. The placenta is no exception. A

Table 9.1
Abbreviations and symbols in text

[A]	The uterine artery concentration of a substance.
[a]	The umbilical artery concentration of a substance.
[V]	The uterine vein concentration of a substance.
[v]	The umbilical vein concentration of a substance.
$[C_m][C_f]$	The concentration of a substance in maternal or fetal plasma water adjacent to the exchange membrane.
$[\bar{C}_m][\bar{C}_f]$	The mean concentration of a substance in maternal or fetal plasma water along the length of the capillary from which transfer is taking place (see Fig. 9.1).
J_{net} (total)	The net amount of a substance transferred per unit time from mother to fetus by all routes.
J_{net} (placenta)	The net amount of a substance transferred per unit time from mother to fetus across the placenta(e) only (the net transplacental flux).
J_{mf}	Amount of a substance transferred per unit time unidirectionally from mother to fetus (the unidirectional materno-fetal flux).
J_{fm}	Amount of a substance transferred per unit time unidirectionally from fetus to mother (the unidirectional feto-maternal flux).
F_{mf} (F_{fm})	A function relating the unidirectional flux in the materno-fetal (or feto-maternal) direction to the relevant plasma concentration.
k_{mf}, k_{fm}	Proportionality constants relating unidirectional flux to $[C_m]$ or $[C_f]$ respectively for a substance crossing by a first order process.
k	Proportionality coefficient applicable when transfer is first order and $k_{mf} = k_{fm}$
K	Placental clearance; an experimental variable defined in equation 7.
P	Placental permeability for a specified molecule.
V_{max}	Maximal unidirectional transfer rate for a carrier mediated process.
k_m	Affinity constant—the substrate concentration for half maximal unidirectional flux for a carrier mediated process. Here used to mean substrate concentration in appropriate plasma at half maximal J_{mf} or J_{fm}.
D	Diffusion coefficient in placental tissue for a molecule.
D_w	Diffusion coefficient in water.
L	The length of the route through which diffusion is occurring. In the case of diffusion through a uniform barrier, the thickness of the barrier.
S	Surface area of placental tissue lying between maternal and fetal blood streams available for exchange.

change in either J_{mf} or J_{fm} will alter J_{net}. In the succeeding paragraphs we consider in detail factors influencing the magnitude of J_{mf}. Similar principles apply to J_{fm} and the relevant equations for J_{fm} are shown in brackets.

$$J_{mf} = F_{mf}[C_m] \quad (J_{fm} = F_{fm}[C_f]) \tag{3}$$

The function F relates unidirectional transfer rate to the concentration of the solute in question in the maternal plasma water adjacent to the placental exchange area. (For a few solutes, for example iron, the concentration of protein-bound solute rather than plasma water concentration may be the correct variable.) The nature of the function depends on the mechanism of transplacental transfer.

If, and only if, the rate of blood flow past the maternal side of the placental exchange area (that is, that part of uterine blood flow supplying the placenta) is sufficiently high for the concentration fall of solute between artery and vein [A]–[V] to be much less than [A]

$$[C_m] \approx [A] \quad ([C_f] \approx [a]) \tag{4}$$

If this situation does not pertain then $[C_m]$ has to be replaced by $[\bar{C}_m]$, a mean value whose application to transfer kinetics depends on the saturability of transfer mechanisms and the relative geometry of maternal and fetal flow (for example, countercurrent, concurrent, etc.). We do not consider this geometry in detail in this review though it is of great importance in considering placental transfer of molecules to which equation (4) does not apply, that is for molecules whose transfer is "flow limited" (Fig. 9.1). If equation (4) does apply to a substance its transfer is said to be "membrane (or diffusion) limited".

(Meschia, Battaglia, and Bruns 1967; Faber 1973; Faber and Thornburg 1983, discuss this more fully). With certain exceptions discussed below, a change in blood flow may be expected to alter only flow limited and not membrane limited transfer.

If the materno-fetal transfer process is first order, then:

$$J_{mf} = k_{mf}[C_m] \quad (J_{fm} = k_{fm}[C_f]) \tag{5}$$

First order kinetics will apply to those molecules whose movement across the placenta is by free diffusion or by restricted diffusion. It will also apply to carrier-mediated transfer but only over the range of plasma solute concentrations within which the carrier is substantially unsaturated and provided only one molecule need be bound to the carrier for transport to occur.

If, and only if, the transfer process is first order and symmetrical

$$k_{mf} = k_{fm} = k \tag{6}$$

The equal unidirectional proportionality constants being re-defined as a new constant k. If $k_{mf} = k_{fm}$ there can be no transplacental active transport.

$$K = \frac{J_{net}}{([A] - [a])} \tag{7}$$

K is an experimentally measurable value for a given solute transferring across a given placenta. It is usually described as the placental clearance (units ml min^{-1} placenta^{-1} or ml min^{-1} g placenta^{-1}).

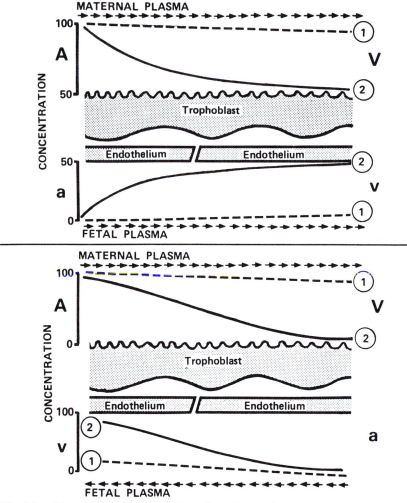

Fig. 9.1.   Diagrammatic representation of membrane limitation and flow limitation. The top diagram shows the expected plasma concentration profiles for two substances [(1) and (2)] in the maternal and fetal circulations of a haemochorial placenta. Plasma flows from the arterial end (A and a) to the venous end (V and v) of the exchange area. Flows in the two compartments are assumed to be of equal magnitude and concurrent. Substance (1) is membrane-limited, ([A]–[V]) is a small proportion of [A] and ([a]–[v]) is a small proportion of [a]; $[C_m] \simeq [A]$. $[C_f] \simeq [a]$. Substance (2) is much more permeable and transfer is flow limited, ([A]–[V]) is a large proportion (half in this case) of [A] and ([a]–[v]) is half of [a]. $[C_m]$ or $[C_f]$ could only be found by integrating the concentration along the length of the exchange area to give $[\bar{C}_m]$ or $[\bar{C}_f]$.

The bottom diagram shows transfer of the same substances across the placental exchange area when the magnitudes of blood flow in the two circulations are again equal but when the flow direction is counter-current. The concentration profile is hardly changed for membrane-limited substance (1) compared to the top diagram but is quite different for flow-limited substance (2).

For a membrane limited permeant (Equation 4 applies) crossing by a symmetrical first order process (Equations 5 and 6 apply)

$$k = K \tag{8}$$

For *diffusional* transfer across a membrane, Fick's equation (Riggs 1972) leads to the equality

$$J_{mf} = [C_m] \, DS/L \quad (J_{fm} = [C_f] \, DS/L) \tag{9}$$

in which D/L is usually written as P so that

$$J_{mf} = [C_m] \, PS \quad (J_{fm} = [C_f] \, PS) \tag{10}$$

and thus for steady state diffusional transfer

$$J_{net} = PS \, ([C_m] - [C_f]) \tag{11}$$

If transfer of an uncharged solute is by membrane-limited diffusion only and the solute is not metabolized within the placenta then

$$PS = K \tag{12}$$

which is useful because of the ease with which K can be measured (see below).

2 MECHANISMS OF TRANSPLACENTAL TRANSFER

The possible mechanisms of transplacental transfer are reviewed by Miller, Kozelka, and Brent 1976; Faber and Thornburg 1983; Morriss and Boyd 1987. They include: *i. Transcellular* lipophilic diffusion.* This is presumed to predominate in the case of small relatively lipid soluble molecules such as respiratory gases. Being lipophilic they can penetrate the lipid bilayer of placental trophoblast and capillary cell membranes, and the entire surface area of tissue between the closely apposed maternal and fetal placental microcirculations is thus available for their diffusion. Because transfer is rapid, velocity of blood flow is normally the rate limiting factor in the transfer which is thus usually flow limited; *ii. Paracellular diffusion.* Hydrophilic substances (for example, mannitol), which, being lipid insoluble, cannot easily penetrate through cell membranes, are presumed to cross through paracellular water-filled channels. In the absence of additional carrier mechanisms their transfer is generally fairly slow and is thus usually found to be membrane limited. If transferring molecules are almost as big as the paracellular channels their passage is restricted by steric effects and *restricted diffusion* is said to be taking place; *iii. Facilitated diffusion* involving a carrier but

---

*Several placentae include a syncytial layer (for example, syncytiotrophoblast of human placenta). For reasons of convenience we refer to its limiting membrane as being its cell membrane and refer to putative extra-cytoplasmic routes across the syncytiotrophoblast as being paracellular.

no active transport (for example, D-glucose); *iv. Carrier mediated active transport* (for example, calcium, most amino acids); *v. Receptor mediated endocytosis* (for example, immunoglobulin G transfer across primate placenta); and perhaps *vi. Fluid phase endocytosis*, and *vii. Ultra-filtration*.

3 PLACENTAL PERMEABILITY

As already mentioned, placental clearance, K, of a substance can be fairly readily measured from equation (7), and for inert hydrophilic substances crossing by unrestricted diffusion in the absence of electrical effects this will, provided there is no flow limitation, equal PS. Measurements of K have been useful in allowing comparison of the diffusional permeability of placentae of different species (see for example, Faber and Thornburg 1983). Furthermore, for a group of such substances of different D_w the ratio of K to D_w for each should be a constant if their transfer meets the above conditions. Thus if K/D_w is calculated for a series of hydrophilic substances then some initial judgments as to the possible mode of transfer of each one may be made.

Values of K/D_w for several hydrophilic substances obtained using *in vivo* preparations of three species with haemochorial placentae (human, guinea-pig, rat) and one, the sheep, which has an epithelio-chorial placenta, are shown in Table 9.2.

The values of K/D_w for each substance are quite similar for each of the three haemochorial types, suggesting that the presence of three trophoblast layers in the rat as compared to one in the human and guinea-pig (see Mossman 1987) has little effect on the rate limiting diffusional barrier. Also in the rat and guinea-pig K/D_w for erythritol, mannitol, CrEDTA and inulin appears to be relatively constant, suggesting that for these transfer is by unrestricted paracellular diffusion, but for Na^+ in both species, and cyanocobalamin in the guinea-pig, K/D_w is somewhat higher, suggesting that for these two substances one or more of the above conditions has not been met. For Na^+ the likely explanation is that either its diffusion is influenced by a transplacental potential difference or that part of its transfer is transcellular utilizing carriers (see **Section III below).** It is rather more difficult to explain the high K/D_w for cyanocobalamin because although there is certainly the possibility of its transcellular carrier-mediated movement this is unlikely under the conditions used by Willis, O'Grady, Faber, and Thornburg (1986). In the human there is as yet too little data to form a clear picture although in this species also the K/D_w for cyanocobalamin appears to be very high.

In the sheep the results are rather different from those for the haemo-chorial placenta in that proceeding from erythritol in the direction of increasing molecular size to mannitol and CrEDTA, K/D_w drops dramatically. This is most simply explained by hypothesizing that the water-filled channels or pores constituting the paracellular route through which diffusion is occurring have in this species a radius only slightly greater than the permeant molecules

Table 9.2

Placental K/D_w values for hydrophilic substances in several species close to term studied with intact placentae and living fetuses

	D_w at 37°C (cm² sec⁻¹)	Epithelio-chorial Sheep[h]	K/D_w (cm g⁻¹ placenta) Haemochorial Human	K/D_w (cm g⁻¹ placenta) Haemochorial Guinea-pig	K/D_w (cm g⁻¹ placenta) Haemochorial Rat
Na⁺	1.7×10^{-5A}	$1.8 \pm 0.6(7)$[i]	32.2[a]	31.3[a] $21.5 \pm 2.6(10)$[d]	44.1[a] $16.7 \pm 1.0(10)$[b]
Erythritol	1.1×10^{-5B}	$5.6 \pm 0.5(12)$		$13.8 \pm 2.3(12)$[d]	
Mannitol	9.9×10^{-6C}	$0.7 \pm 0.1(4)$	20.0[g]	$12.1 \pm 1.5(15)$[e]	$4.9 \pm 0.5(9)$[b] $9.1 \pm 1.0(10)$[c]
Cr-EDTA	7.0×10^{-6D}	$0.03(2)$		$6.4 \pm 0.8(39)$[d]	$7.6 \pm 0.6(17)$[b]
Cyanocobalamin	3.9×10^{-6D}		$61.1 \pm 11.1(10)$[f]	$26.9 \pm 4.3(39)$[f]	
Inulin	2.7×10^{-6B}		8.6[g]	$7.3 \pm 1.2(13)$[e]	$8.0 \pm 1.0(6)$[b] $5.2 \pm 0.5(12)$[c]

Values shown are mean ± s.e. (n) where possible.

References: Diffusion coefficients A—Pappenheimer 1953. B—Faber and Thornburg 1983. C—Štulc, Friedrich, and Jiřička 1969. D—Paaske and Sejrsen 1977. Values for K taken from: a—Flexner, Cowie, Hellman, Wilde, and Vosburgh 1948; b—Robinson, Atkinson, and Sibley unpublished measurements; c—Štulc and Štulcova 1986; d—Hedley and Bradbury 1980, assuming a placental weight of 5 g; e—Thornburg and Faber 1977; f—Willis, O'Grady, Faber, and Thornburg 1986; g—Bain, Copas, Landon, and Stacey 1986; h—data from Boyd, Haworth, Stacey, and Ward 1976 except i—Weedon, Stacey, Ward, and Boyd 1978—assuming sheep placental weight of 500 g.

themselves leading to restricted diffusion, with increasing restriction as molecular size increases. Pores of radius 0.4 nm would be compatible with the sheep data (Boyd, Haworth, Stacey, and Ward 1976). Using larger substances than those shown in Table 9.2 (high molecular weight dextrans) Hedley and Bradbury (1980) found that restriction to diffusion across the guinea-pig placenta is also present. They calculated a pore radius for guinea-pig placenta of 10 nm. The K/D_w for Na⁺ is comparatively too low in sheep rather than too high as in the haemochorial species: K/D_w for Cl⁻ in the sheep has also been shown to be comparatively low (Armentrout, Katz, Thornburg, and Faber 1977), and it has been suggested that the sheep placenta discriminates against the presence but not the sign of an electrical charge (Thornburg, Binder, and Faber 1979*a*) although the basis for such a discrimination is obscure.

Whilst it is easy to define a porous paracellular route physiologically, pores have not been clearly defined morphologically. In placentae such as that of the pig which has a chorionic epithelium with distinct cells, the route probably lies between the cells, but in syncytial placentae such as those of the human, guinea-pig and rat, in which the trophoblast forms a true syncytium, this cannot be, and the nature of the transtrophoblastic channels here is a matter of some debate (see Kaufmann, Schröder, and Leichtweiss 1982; Berhe, Bardsley, Harkes, and Sibley 1987).

4   TRANSPLACENTAL POTENTIAL DIFFERENCE

A knowledge of the value of any transepithelial electrical potential difference (pd) is crucial to the analysis of ion movements. The topic is clearly analysed from first principles in Faber, Binder, and Thornburg (1987). At equilibrium and in the absence of net transport, the passive distribution of charged molecules between two compartments obeys the Nernst equation. The system formed by the placenta as the barrier between the maternal and fetal compartments is no exception, although the assumption of absent net transport is not justified for those ions, such as calcium in some species, whose rate of fetal accumulation is close to the rate of unidirectional materno-fetal transport.

As with other cells there is usually a *transmembrane* potential measurable between the intracellular compartment of human trophoblast and the extracellular fluid (range 0–50 mV interior negative; Carstensen, Leichtweiss, and Schröder 1973; Yano, Okada, Tsuchiya, Kinoshita, Tominaga, and Nishimura 1981; Yano, Okada, Tsuchiya, Kinoshita, and Tominaga 1982). The trophoblast transmembrane pd must have a major influence on ions crossing the placenta unless they are able to by-pass the trophoblast by following a paracellular route along which there is no pd change. Whether there is a substantial transplacental pd between maternal and fetal blood streams as opposed to a trophoblast transmembrane pd and, if so, whether it reflects active transplacental ion transport, is at present uncertain. The available data are apparently contradictory. The experimental problems are considerable. Firstly, a knowledge of the pd drop between two sites in a biological system, for example, between electrodes placed in maternal and fetal blood streams, the usual recording site (see below), is of little value without knowledge, which we do not possess, of both the current flow across and the electrical resistance of the placenta. The placental resistance relative to the resistance of other tissue elements in the electrical circuit, of which the sites of pd measurement are a part, also needs to be known (Thornburg, Binder, and Faber 1979*b*; Faber *et al.* 1987). Secondly, insertion of electrodes directly into maternal and fetal placental capillaries to measure the placental electrical variables directly would be technically difficult because of the solid nature of most placentae (except pig—see below) and has not been done.

The main evidence that a transplacental pd may exist has come from measurement of potential differences between catheters or electrodes placed at various sites in the maternal and fetal extracellular compartments. Reported values for these materno-fetal pd's are shown in Table 9.3 and it is clear that there are major species differences in value and even in sign. Also when Binder, Faber, and Thornburg (1978) and Thornburg *et al.* (1979*b*)

Table 9.3

Directly measured materno-fetal potential differences in various species (maternal electrode is taken to be zero)

Species	P.d. (mV)	Circuit	Reference
Goat	−133 (79 days) −25 (131 days)	Maternal bv—fetal bv	Meschia, Wolkoff, and Barron 1958
	−71 ± 9 (mean of values obtained between 81–141 days; no decline with gestation)	Maternal pc—fetal pc and bv	Mellor 1970
Guinea-pig	−16 ± 2 (35 days) −17 ± 3 (65 days) −21.4 ± 5.6 (52 days–term)	Maternal pc—fetal pc and bv	Mellor 1969
Human	No significant p.d. (term)	Maternal bv—fetal bv	Štulc, Rietveld, Soetman, and Versprille 1972
		Maternal pc—fetal bv	Mellor, Cockburn, Lees, and Blagden 1969
Human	−2.7 ± 1.1 (mid-term)	Maternal bv—fetal bv	Štulc, Svihovec, Drabkova, Stribrny, Kobilkova, Vido, and Dolezal 1978
Pig	−24 ± 18 (92–107 days)	Maternal bv—fetal bv	Boyd, Glazier, and Sibley 1986
Rabbit	No significant p.d. (20–30 days)	Maternal pc—fetal pc and bv	Mellor 1969
Rat	+17 ± 3 (15 days) +17 ± 3 (21 days) +1 ± 2 (22 days)	Maternal pc—fetal bv and pc	Mellor 1969
Sheep	−24 to −40 (140 days)	Maternal pc—fetal bv	Widdas 1961
	−51 ± 7 (mean of values obtained between 67–140 days)	Maternal pc—fetal pc and bv	Mellor 1970
	−25 ± 20 (124–143 days; fall with gestation)	Maternal bv—fetal bv	Weedon et al. 1978

Values given are mean ± S.D. where available. pc = peritoneal cavity, bv = blood vessel.

measured the steady state distribution of radioactive ions (bromide, sulphate, rubidium, lithium) after injection into pregnant guinea-pigs and sheep, they found that the maternal and fetal plasma concentrations approached each other. Application of the Nernst equation (although this may have been inappropriate in the presence of net fluxes) suggested that the transplacental potential difference, unlike the maternofetal potential difference of these species (Table 9.3) was very small. However, pd's have been measured in other preparations where the placenta would seem to be the only likely source of electrical activity. Mounting of the membranous pig placenta in Ussing chambers (Ussing and Zerahn 1951) enabled pd's of 50 mV (fetus negative; Crawford and McCance 1960), 6 mV (fetus negative; Bazer, Goldstein, and Barron 1981), and 5.9 mV (fetus positive; Sibley, Ward, Glazier, Moore, and Boyd 1986) to be recorded; the reason for the variation in the values obtained by the different groups is not clear.

In the case of the guinea-pig placenta, perfused on its fetal side only with the maternal side intact, substantial pd's can again be recorded between electrode catheters located in maternal blood vessels and fetal perfusate. Štulc and Švihovec (1973) found the fetal side to be -20 mV with respect to maternal whereas Bailey, Bradbury, France, Hedley, Naik, and Parry (1979) reported a mean value of fetal side -6.3 ± 0.8 mV (n = 12) with respect to maternal. This pd did not seem to be artefactual as it was reduced both by addition of KCN to and by removal of Na^+ from the perfusate (Štulc and Švihovec 1973; Bailey et al. 1979). Leichtweiss and Schröder (1981) provided evidence, which they regarded as tentative, that the totally isolated guinea-pig placenta perfused through both fetal and maternal circulations with similar solutions also generated a pd ranging from fetal side -4 mV to -17 mV with respect to maternal perfusate. Van Dijk and van Kreel (1982) estimated from the equilibrium distribution of charged molecules ([^{3}H]-choline and [^{14}C] tetra ethyl ammonium bromide) between maternal and fetal perfusate and tissue of the same preparation that the transplacental pd was 6 mV, fetus positive. The only other tissues present in the dually perfused guinea-pig placenta are some uterine muscle and the subplacenta.

The variability in these values obtained for the transplacental pd probably demonstrates the technical difficulties involved, as well as the effects of the different preparations themselves. Nevertheless it seems likely that there is a transplacental pd which may be different in magnitude and even polarity from the maternofetal pd.

There is evidence in the pig that the two pd's are at least part of the same circuit, as adrenaline either injected into instrumented fetuses of conscious sows, or added to the fetal side of pig placenta in Ussing chambers, caused a rapid and marked rise in the measured pd, fetal side becoming more positive (Sibley et al. 1986; Boyd et al. 1987). The effect is mediated by β-adrenergic receptors in both cases (Sibley et al. 1986; Boyd et al. 1987), and the fetal plasma adrenaline concentrations attained in vivo are in the range found during fetal asphyxia (MacDonald, Colenbrander, Versteeg, Heilhecker, and Wensing 1984). It is not possible at present to define this circuit further.

5   MODULATION OF TRANSFER

It follows from equation 3 that the unidirectional transport rate for a substance crossing the placenta from mother to fetus J_{mf} could be altered by either a change in F_{mf} or in $[C_m]$. Conversely, J_{fm} depends on $[C_f]$ and F_{fm}. Thus, from equation 2, a substance's net rate of transport is governed by four factors: $[C_m]$ and $[C_f]$ and F_{mf} and F_{fm}.

$[C_m]$ and $[C_f]$ will depend predominantly on a substance's maternal and fetal arterial concentration which in turn will reflect the balance of its rate of supply, metabolism and excretion within mother and fetus respectively. An example of indirect control of placental transfer through C_m and C_f is seen in the case of glucose. Administration of insulin to mother or fetus markedly affects the placental transfer rate of glucose but does so predominantly or exclusively by an effect on maternal and fetal plasma glucose concentration. Insulin appears to have no direct effect on the placental glucose carrier (see **Section III**). $[C_m]$ and $[C_f]$ also depend on the degree of solute binding in blood, for example a change in plasma pH may lead to a change in the ratio of ionized to bound plasma calcium and thus changes in the concentration of the mineral available for transport (see **Section III**). C_m and C_f will also, in the case of flow limited transport, be greatly affected by the rates of flow in the maternal and fetal placental circulations.

Factors influencing the functions F_{mf} and F_{fm} depend on the type of transport involved. For any individual substance F_{mf} and F_{fm} may be complex, as solutes cross the placenta in many different ways and a particular substance may cross by more than one mechanism.

The factors most likely to alter F_{mf} and F_{fm} for a given solute depend on the mechanism of transfer which predominates for it. For each mechanism we here specify the change likely to increase unidirectional flux for a given solute concentration.

If transfer is by transcellular lipophilic diffusion an increase in placental surface area or a diminution in effective placental thickness ("effective" because thickness is not simple in a non-homogenous multi-layered barrier like the placenta) will have the biggest effect. In the case of paracellular diffusion, an increase in the number of channels, a diminution of their length or an increase in their radius (radius is especially significant in the case of restricted diffusion) will be important. Changes in these parameters might follow from an increase in placental surface area or through a thinning of the placenta, but could also result from increased numbers of channels per unit area or from channel dilatation.

In the case of facilitated diffusion, an increase in V_{max} might be induced by carrier synthesis or recruitment or by an increase in placental surface area if the number of carriers per unit area remained constant. The same is true of carrier-mediated active transport, and of receptor mediated endocytosis, though with these an increased rate constant for the activity of the carrier mechanism could also provide an opportunity for modulation. In the case both of facilitated diffusion and of active transport and receptor mediated endocytosis an alteration in the affinity constant (k_m) between the solute and the carrier may follow either changes in the nature of the carrier or changes in concentration of other solute molecules which also interact with it.

To allow transplacental movement, a facilitated diffusion or active transport system on one plasma membrane must also be accompanied by a carrier in the opposite membrane or there must be a specific plasma membrane pore or "leak" pathway in the latter. Also, as is presumably the case for Ca^{2+}, a system whereby the transported substance is moved across the cell cytoplasm without altering its free cystolic concentration may be involved. The most effective locus of control will be whichever of these three steps (entry, exit, transcytosolic movement) is rate limiting. In many other epithelia it is the rate of leak or of transcytosolic movement rather than of pumping which is controlled (see Ca^{2+} section below and Schultz 1986). "Leaks" and "pumps" are discussed in more detail by Shennan and Boyd (1987).

Two further complicating factors have so far been ignored as far as modulation is concerned. Increased blood flow leading to the recruitment of previously unperfused placental exchange areas might increase the surface area available for membrane limited tansfer. Alteration of flow can thus perhaps alter the transfer of a membrane limited permeant under some circumstances. Also, for charged solutes, transmembrane and any transplacental electrical potential differences will have to be taken into account.

It is not at all clear how far any or all of these possible control strategies (summarized in Fig. 9.2) are deployed during pregnancy to co-ordinate J_{net} with fetal requirements. There are various possibilities.

Firstly, a changing transplacental concentration difference $[C_m] - [C_f]$ will provide some degree of auto-regulation of J_{net} for any solute (and this probably includes all solutes) which crosses at least partly by a diffusional process. Thus, for example, increased uptake of a nutrient by fetal cells will reduce $[C_f]$; J_{net} will rise in response to the increased concentration difference tending to return $[C_f]$ towards its initial value.

Secondly, it is possible that the placenta might be endowed with a programmed sequence of transport functions which wax and wane as gestation proceeds regardless of feedback messages from fetus or mother. For instance, this might be true of the onset of IgG transporting capabilities across the human placenta in the last few weeks of gestation. We are not aware of any direct study of this point; it would be interesting to know if IgG endocytosis develops over time in pre-term trophoblast maintained in culture. In early gestation transport across the trophectoderm changes dramatically over the first few days of life (Benos and Biggers 1983).

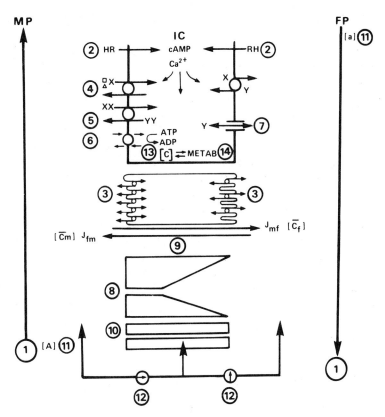

Fig. 9.2.   Schematic representation of those variables which may contribute to the magnitude of unidirectional fluxes across the placenta. (1) Blood flow rate (here shown as counter-current). Increased flow will in the case of flow-limited permeants alter $[C_m]$ and $[C_f]$. Increased flow may be associated with recruitment of unperfused areas, increasing exchange surface area. (2) Receptor (R) activation by hormones (H) may lead to altered concentrations of intracellular second messengers, for example cAMP and Ca^{2+} which may change most remaining variables. (3) Altered number of carriers. (4) Competition for carrier site. (5) Altered carrier affinity. (6) Altered turn-over number of carrier. (7) Altered leak in or out of placental cell. (8) Altered para-cellular path length. (9) Altered width of paracellular path. (10) Altered density of paracellular paths. (11) Alteration in [A] or [a]. (12) Altered potential difference. (13) Altered free intracellular concentration of transferred substance. (14) Altered metabolism of substance by placenta. MP = maternal plasma, FP = fetal plasma, IC = trophoblastic intracellular space.

Thirdly, growth in size of surface area and of vascular beds (and of associated blood flow) of the placenta during gestation will have a profound non-specific effect on J_{net} for both membrane limited and flow limited solutes, and is probably partly inherent and partly a placental response to circulating growth factors.

Finally, the transfer of individual solutes might be influenced by specific endocrine control of individual transport mechanisms. The placenta is extremely well endowed with receptors for a variety of possible effectors, as shown in Table 9.4 for the human. There is no doubt that many of these have a primary role in the control of the endocrine and metabolic functions of the organ itself or they may be the receptor element for transplacental carrier mediated transport of an individual hormone. Nevertheless they might also be related to the control of the placental transfer function for other solutes. As will become clear from the following sections, very few hormones apart from insulin have been widely investigated in this regard. The results of receptor activation may not be limited to the side of the placenta at which the receptor lies. Thus receptor activation in placenta may lead to membrane and intracellular events such as cyclic AMP production (Whitsett *et al.* 1980) and calcium activated phospholipid dependent protein phosphorylation (Moore, Moore, and Cardaman 1986) which may perhaps allow "cross talk" between fetal hormones and maternal side carrier and *vice versa* without the need for the hormone to cross the placenta itself. There appears to be no innervation of the placenta and any CNS control of transfer must be by means of circulating messengers.

III   Placental transfer of selected substances

1   SODIUM

Steady state Na^+ concentrations in maternal plasma are generally close to those in fetal plasma (Table 9.5) though some concentration differences are reported to be significant (for example, human, Faber and Thornburg 1981) and might be great enough in the sheep, when differences in maternal and fetal plasma protein concentration are taken into account, to allow for fetal Na^+ acquisition by diffusion (Canning and Boyd 1984).

Even when net Na^+ flux into the fetus is increased by external urinary drainage in the fetal lamb, relative concentrations are not altered (Gresham, Rankin, Makowski, Meschia, and Battaglia 1972), which implies either a relatively free paracellular diffusion across the organ or alternatively closely controlled transcellular transport.

The foundation for an understanding of placental Na^+ transfer was laid in a series of papers by Flexner and colleagues who measured unidirectional Na^+ fluxes in the materno-fetal direction in several species (Flexner and Gelhorn 1942) including the human (Flexner, Cowie, Hellman, Wilde, and

Table 9.4

Receptors for hormones and neurotransmitters in the human placenta

Receptor	Gestation	Preparation/Location	Reference*
β-Adrenergic	10 wks–term	Enriched in basal (fetal) membrane	Whitsett et al. 1980; Kelley, Smith, and King 1983
Adenosine	Term	Microsomes	Fox and Kurpis 1983
Androgen	Term	Cytosol	Younes, Besch, and Besch 1982
Angiotensin II	Term	Crude membranes	Cooke, Craven, and Symonds 1981
Atrial Natriuretic Peptide	Term	Placental membranes	Sen 1986
Cholinergic (muscarinic)	Term	Microvillous (maternal) membrane	Fant and Harbison 1981
Dopamine D_1	Term	Placental slices	Ferré 1986
Epidermal growth factor	6 weeks to term	Microvillous and basal membrane	Carson et al. 1983; Rao et al. 1985
$GABA_A$	Term	Cell membranes	Erdo et al. 1985
Glucocorticoid	8–11 wks and term	Cytosol	Speeg and Harrison 1979
HCG	Term	Crude particulate fraction (only after neuraminidase treatment)	Paul and Jailkhani 1982
Insulin	26 wks–term	Enriched in microvillous membrane	Potau, Ruidar, and Ballabriga 1981; Whitsett and Lessard 1978
LHRH	10 wks and term	Crude plasma membranes	Currie, Fraser, and Sharpe 1981
Oestrogen	Term	Cytosol	Younes, Besch, and Besch 1981
kappa-opiate	Term	Microvillous membranes	Valette et al. 1980
Progesterone	Term	Cytosol	Younes et al. 1981
Prolactin	Term	Solubilized	Vonderhaar et al. 1985
Basic Somatomedin	Term	Cell membranes	Bhaumick et al. 1982
Somatomedin-A	Term	Plasma membranes	Takano et al. 1975
Somatomedin-C	Term	Cell membranes	Marshall et al. 1974
Somatostatin	Term	Cell membranes	Tsalikian, Foley, and Becker 1984
T_3	Term	Plasma membranes	Alderson, Paston, and Cheng 1985
25-(OH) D_3	Term	Cytosol	Fowler, Williams, and Gray 1978
1,25-$(OH_2) D_3$	Term	Cytosol	Christakos and Norman 1980

* Only a key reference has been given.

Table 9.5

Maternal and fetal plasma [Na$^+$] (mmol l^{-1}) (unless stated)

	Mother	(n)	Fetus	(n)
Human	138.0 ± 0.7	(8)	139.0 ± 1.4[a]	(8)
	135.4 ± 2.4	(7)	137.0 ± 2.8[b]*	(7)
Guinea-pig	142.2 ± 1.5	(19)	138.5 ± 0.9[c]	(45)
Rat	132.8 ± 1.0	(10)	135.4 ± 1.1[d]	(31)
Sheep	145.3 ± 1.2	(9)	143.7 ± 1.9[e]*	(9)
Pig	145.0	(6)	143.0[f]	(12)
	(range 143.2–148.0)		(139.3–146.2)	

Mean ± s.e.m. shown where possible.
a. Mellor *et al.* 1969 (term caesarean section); b. Faber and Thornburg 1981 (mmol kg^{-1} water; term caesarean section); c. Woods, Thornburg, and Faber 1978 (near term); d. Chalon and Garel 1985*b* (19.5 days gestational age); e. Armentrout, Katz, Thornburg, and Faber 1977 (120–145 days gestational age); f. Cummings and Kaiser 1959 (106 ± 1 day gestational age).
* $P < 0.05$ or better v. maternal, where given by authors.

Vosburgh 1948) by injection of ^{24}Na to the mother followed by later ashing and counting of the fetus. They found that in all species studied, except the pig, J_{mf} (mg Na$^+$ g fetus^{-1} hour^{-1}) increased as gestation proceeded until about 9/10 of the way to term when the rate dropped dramatically. These values for J_{mf} were also compared to estimates of J_{net} made from fetal weight gain and proportionate Na$^+$ content and found to be from 3.5 times (pig) to over 1100 times (human) higher, demonstrating the importance of J_{fm} as well as of J_{mf} in determining J_{net}. One of their most important observations was that there appeared to be a relationship between placental morphology and the magnitude of the unidirectional Na$^+$ flux so that, for example in the pig which, as mentioned previously, has an epitheliochorial placenta with a maternal as well as a fetal epithelial layer between maternal and fetal blood (Mossman 1987), J_{mf} was 0.026 mg g^{-1} hour^{-1} at 9/10 of the way through gestation, whereas in the human in which only fetal tissue is interposed between the two circulations it was 6.5 mg g^{-1} hour^{-1} (Flexner *et al.* 1948). Flexner and Gelhorn (1942) were well aware that their study could not address what they termed a fundamental question: "Does the placenta act as an inert membrane or filter placed between the maternal and fetal circulations or does it modify the transmission of substances by contributing energy to the process and so acting as an organ of secretion?"

It is still not possible to answer definitively Flexner's question. The permeabilities of the different placental types to hydrophilic solutes (Table 9.2) as well as the similar Na$^+$ concentrations in maternal and fetal plasma suggest that passive, paracellular diffusion is important and it is widely assumed that this route predominates. Nevertheless there is no doubt that some transmembrane movement of Na$^+$ into and out of trophoblast must occur; all animal cells extrude Na$^+$ actively as part of the "housekeeping" necessary to prevent spontaneous osmotic lysis (Leaf 1981). The question is whether placental cells or syncytium do more than this, transporting Na^+ transcellularly across the trophoblast to a quantitatively important extent either in co-transport with other solutes or alone. The evidence in favour of a transcellular component of transplacental Na^+ movement, a prerequisite for control of active, though not necessarily of passive, movement of Na^+, is as follows:

i   Relative permeability to sodium

As shown in Table 9.2, the K/D_w values for Na^+ for the haemochorial placentae are higher than those for other hydrophilic solutes. Although this may simply reflect the influence of a transplacental pd, fetal side negative, or the existence of heterogenous sized pores (Štulc, Friedrich, and Jiřička 1969), it might also be the result of a substantial transcellular contribution to the totality of Na^+ movement across the placenta.

ii   Short circuit current studies

In vitro investigation of pig placenta in Ussing chambers has provided straight-forward evidence of transcellular Na^+ transfer, although there are some contrasts in the data from different laboratories. Crawford and McCance (1960) found that with the voltage across the placenta clamped to OmV, the short circuit current (scc) was exactly equivalent to the net accumulation of Na^+ (measured by flame photometer) on the maternal side. Under these conditions any net Na^+ transfer must be active and, as the scc provides a measure of total net ion flux, then the equivalence of the two suggested that Na^+ was the only ion actively transported. More recently Sibley et al. (1986) measured Na^+ fluxes under the same conditions using radio-isotopes and found no significant net Na^+ flux under control conditions but addition of ouabain to the fetal side produced a net flux towards the maternal compartment and addition of adrenaline to the fetal side caused a net Na^+ flux towards the fetal compartment. Comparison of scc data with net Na^+ fluxes and the effect of ion substitution showed that although other ions may be transported under control conditions and in the presence of ouabain, the adrenaline effect was solely due to increased net Na^+ transport. Active, transcellular, transport of Na^+ across the pig placenta in vitro can thus be demonstrated in both feto-maternal and materno-fetal directions.

iii Na^+/K^+ ATPase

Firth, Farr, and Koppel (1979) using techniques which would not be expected to demonstrate levels of Na^+/K^+ ATPase activity associated with cell maintenance but which do locate the higher concentrations of enzyme associated with transepithelial Na^+ pumping (Firth 1983) demonstrated Na^+/K^+ ATPase activity histochemically on both the maternal and fetal-facing plasma membranes of the syncytiotrophoblast of the guinea-pig placenta. Similarly, in the pig, Firth, Sibley, and Ward (1986) localized Na^+/K^+ ATPase activity to the fetal-facing chorionic epithelium of the areolar regions of the placental surface.

iv Syncytial cell membrane vesicles

In the human placenta there is evidence for a variety of Na^+ coupled transport systems in the maternal-facing (microvillous) plasma membrane of the syncytiotrophoblast. A Na^+ coupled amino acid uptake system has been demonstrated in vesicles derived from the maternal-facing plasma membrane (Ruzycki, Kelley, and Smith 1978; Boyd and Lund 1981—see **Section III** 5 Amino Acids). Boyd, Chipperfield, and Lund (1980) found evidence of a Na^+Cl^- co-transport system in a similar preparation. Also Balkovetz, Leibach, Mahesh, Devoe, Cragoe, and Ganapathy (1986) have reported a Na^+/H^+ exchanger in vesicles again from the maternal-facing plasma membrane of the syncytiotrophoblast. Finally, it is interesting to note that Kelley, Smith, and King (1983) reported greater purification of ouabain binding sites with fetal-facing plasma membrane than with the microvillous plasma membrane of the syncytiotrophoblast. The necessary elements of a "pump" and "leak" system (see **Section II**) for Na^+ therefore seem to be present at least in the human placenta.

v Na^+ fluxes across the in situ *placenta*

Štulc and Švihovec (1973) suggested that the materno-fetal pd in the guinea-pig arose partly from electrogenic active feto-maternal Na^+ transfer across the placenta, as perfusion of Na^+-free fluid through the fetal side of the guinea-pig reduced but did not abolish the pd. Ouabain (strophanthin $10\,\mu\text{moles l}^{-1}$) in the fetal side perfusion fluid had no effect on the pd but this could have been due to failure of sufficient ouabain to penetrate to sites of Na^+/K^+ ATPase activity on the maternal-facing plasma membrane of the syncytiotrophoblast. In three studies of transplacental radio-isotope Na^+ fluxes (Štulc and Švihovec 1977; Weedon, Stacey, Ward, and Boyd 1978; Bailey *et al.* 1979) it was concluded that it was necesary to postulate an active component of J_{fm} to balance the effect of supposed fetal electronegativity (see **Section II**) on transplacental Na^+ movements. However, it is notable that only Štulc and Švihovec (1973) found Na^+ fluxes to be asymmetrical, providing evidence for active transport in at least one direction even if there is no transplacental pd; the biological significance of net feto-maternal Na^+ transfer, as they found, remains obscure.

There is thus evidence that transplacental Na^+ movement does occur by a transcellular pathway which may be active and electrogenic as well as by passive diffusion through a paracellular pathway. The relative importance of the transcellular route is however not yet clear. It is likely to be more substantial for epitheliochorial than for haemochorial placentae (Faber and Thornburg 1981). The existence of a transcellular route suggests that Na^+ pumping mechanisms may be utilized by the placenta as a source of energy for the active transport of other solutes by means of Na^+ linked co-transport. It may also allow control of active Na^+ transport, as shown by the effect of adrenaline on the pig placenta, which is substantial (Sibley *et al.* 1986). If the stimulation of net Na^+ flux found across the *in vitro* pig placenta in response to physiological increases in adrenaline concentration on its fetal side were replicated *in vivo* and maintained, the increased fetal content of Na^+ would, if accompanied by isotonic flow of water, double the fetal extracellular volume within 24 hours. Hormonal control of passive paracellular Na^+ permeability may also be important and in at least one theoretical analysis of placental transport has been found to be a rate limiting element in fetal growth (Conrad and Faber 1977; Thornburg, Binder, and Faber 1979*a*). It is also interesting to note that Barlet (1985) reported a marked increase in the transplacental Na^+ concentration gradient (fetal plasma higher) following prolactin administration to pregnant ewes.

2 POTASSIUM

There is quite a lot of *in vivo* evidence for control of K^+ transfer across the placenta, although there is little information concerning the mechanism of its transfer.

In general, fetal plasma K^+ concentrations are significantly different from maternal (Table 9.6).

However, because of the known fetal susceptibility to extracellular ion shifts with the hypoxia that is likely to occur during plasma collection (Battaglia, Meschia, Hellegers, and Barron 1958) such values have to be treated with some caution. Indeed both Fantel (1975) and Woods *et al.* (1978), in the rat and guinea-pig respectively, found a correlation between materno-fetal K^+ concentration differences and the time from laparotomy that the fetal plasma samples were taken. Table 9.6 shows values for maternal and fetal plasma concentrations of K^+ taken from studies in which artefact was unlikely because they were chronic preparations or because specific notice was taken of the possible effects of hypoxia.

Indirect evidence of control of transplacental K^+ movements is available from several studies in which fetal K^+ content or fetal plasma K^+ concentrations were shown to be unaffected by maternal hypokalaemia. Stewart and

Table 9.6

Maternal and fetal plasma [K⁺] (mmol l⁻¹) (unless stated)

	Mother	(n)	Fetus	(n)
Human[a]	3.8 ± 0.1	(7)	5.7 ± 0.6*	(7)
Guinea-pig[b]	4.46 ± 0.21	(19)	5.08 ± 0.16	(45)
Rat[c]	4.8 ± 0.2	(10)	2.9 ± 0.6	(89)
Dog[d]	4.1	(5)	5.3*	(21)
Sheep[e]	4.46 ± 0.11	(8)	3.91 ± 0.14*	(9)

Mean ± s.e.m. shown where possible.
a. Faber and Thornburg 1981 (mmol kg⁻¹ water; term caesarean section); b. Woods *et al.* 1978 (near term); c. Fantel 1975 (20 days gestational age); d. Serrano, Talbert, and Welt 1964 (term); e. Armentrout *et al.* 1977 (120–145 days gestational age).
* P < 0.05 or better v. maternal, where given by authors.

Welt (1961) found that a reduction in rat maternal plasma K^+ concentration from 5.2 to 2.3 mmol l⁻¹ had no effect on the overall K^+ concentration in the fetus (mmol 100 g⁻¹), although fetal weight was significantly reduced. In pregnant dogs, Serrano, Talbert, and Welt (1964) found that when the maternal femoral artery K^+ concentration was reduced from 4.1 to 2.7 mmol l⁻¹, the umbilical vein and artery plasma K^+ concentration remained the same. Dancis and Springer (1970) found a similar result in rats. These data must be viewed with some caution because of the possibility of extracellular ion shifts at the time of blood sampling, but the studies of Fantel (1975, 1978) seem particularly well controlled. He found that there was a significant linear correlation between fetal plasma K^+ concentration and the time from initial uterine incision (Fantel 1975). This regression line was used to calculate the fetal plasma K^+ concentration at time zero. As shown in Table 9.6, this value was lower than the maternal plasma K^+ concentration. Fantel noted that the ratio of fetal to maternal plasma K^+ concentrations was close to that predicted from the Nernst equation, using the simultaneously measured materno-fetal pd and assuming it to be transplacental which suggested passive distribution of K^+. This was not the case during maternal hypokalaemia however, when the fetal plasma concentration at time zero was 4.0 mmol l⁻¹ compared to a maternal plasma K^+ of 2.8 mmol l⁻¹ in the absence of a comparable change in pd (Fantel 1978). These results strongly suggest that K^+ transfer is controlled in some way regardless of how materno-fetal pd is interpreted.

Boyd (1982, 1983) found evidence of carrier-mediated K^+ transfer across the microvillous plasma membrane of human placental syncytiotrophoblast. Thus efflux of ⁸⁶Rb (used as a substitute for K^+) from pre-loaded isolated chorionic villi was inhibited by removal of Na^+ and Cl^- from the external bathing solution and by frusemide. A hypertonic bathing solution on the other hand stimulated efflux in a frusemide-inhibitable fashion. A hypotonic bathing solution also stimulated efflux but this was not inhibited by frusemide. The most likely explanation for these results is that the frusemide sensitive, Na^+ and Cl^- dependent, response to hypertonic solution is due to the presence of a $Na^+2Cl^-K^+$ co-transport system, and the frusemide insensitive response is due to the presence of a K^+ channel. Both of these systems have been reported in erythrocytes where they are involved in cell volume regulation (see Geck and Heinz 1986), and this is a likely role in the placenta (Boyd 1983). However, evidence of a Ca^{2+} sensitive K^+ channel in the plasma membrane was also reported by Boyd (1983). This is also common to several other types of cell (Pappone and Cahalan 1986) and raises the possibility of control of placental K^+ permeability by cytoplasmic Ca^{2+}. In the same study adrenaline stimulated ^{86}Rb efflux from villi, and this effect could again have been via a rise in intracellular Ca^{2+}. Stimulation was blocked by the α-adrenergic antagonist phenoxybenzamine. Data presented by Chalon and Garel (1985b), who were studying the effect of fetal hormones on plasma Ca^{2+} in the rat fetus, show a significant rise in fetal plasma K^+ concentrations in decapitated fetuses, without there being any significant change in Ca^{2+}, $PO_4{}^{2-}$ or Na^+ concentrations. Although the usual reservations concerning plasma K^+ concentrations apply here, this might also suggest some neuroendocrine control of K^+ transfer.

There have been few direct measurements of transplacental K^+ transfer. Bailey *et al.* (1979), using the perfused guinea-pig placenta, found that unidirectional fluxes of ^{42}K were similar in both materno-fetal and feto-maternal directions. Nevertheless, ouabain at $10\,\mu\text{mol}\,l^{-1}$ in the perfusion fluid significantly inhibited extraction by the placenta of ^{42}K from the perfusate; transfer to the perfusate from maternal plasma was inhibited only at $50\,\mu\text{mol}\,l^{-1}$ at which concentration there were also effects on ^3H_2O transfer. The permeability of the placenta was some 3.2 times greater for K^+ than for Na^+ while the ratio of D_w for K^+ to Na^+ is approximately 1.5, supporting the idea of a, presumably transcellular, route of K^+ transfer inaccessible to Na^+. In the intact guinea-pig Hedley and Bradbury (1980) found K^+ permeability to be five-fold greater than for Na^+ and the K/D_w value for K^+ calculated from their data is $120\,\text{cm}\,g^{-1}$, suggesting (by comparison with Table 9.2) only a minor role for the paracellular route in K^+ transfer. However, in the human Dancis, Kammerman, Jansen, and Levitz (1983) failed to find any effect of ouabain on ^{86}Rb transfer across the perfused placenta, although there was an effect on accumulation into the tissue. This cautions against putting too much emphasis on the data of Bailey *et al.* (1979) on ^{42}K extraction. In the sheep K^+ permeability measured using Fick's principle and ^{42}K again showed a K^+:Na^+ permeability ratio of 2.5 (Boyd, Canning, Stacey, and Ward 1981).

Although the weight of evidence favours a carrier-mediated mechanism as the most important route of K^+ transfer this has yet to be directly proven. In addition, there is no evidence as to whether the control of transplacental K^+ movement, evidenced by the ability of the placenta to maintain J_{net} of K^+ in the presence of maternal hypokalaemia, reflects a change in F_{mf} or in F_{fm} or in both.

3   CALCIUM

It has been known for many years that the total calcium concentration in fetal plasma is significantly higher than that in maternal plasma. However, it was not until Delivoria-Papadopoulos, Battaglia, Bruns, and Meschia (1967) demonstrated that the ultrafilterable fraction of calcium was also at a higher concentration in fetal plasma that it could be certain that the difference was not primarily due to differences in protein binding capacity in the two plasmas.

Subsequent measurements using Ca^{2+}-selective electrodes also demonstrate a significantly higher ionized calcium (Ca^{2+}) activity in fetal plasma in every species investigated (Table 9.7). One of these species is the rat in which the materno-fetal pd is fetal side positive, an observation incompatible with the idea that materno-fetal electrical gradients rather than active transport are responsible for the higher fetal concentration of the positively charged Ca ion.

There is further evidence of active transport from perfusion experiments. Twardock and Austin (1970) found a significant rise in total calcium concentration of perfusate after a single pass through the *in situ*, umbilically perfused, guinea-pig placenta, despite the fact that the perfusate concentration started higher than that of the maternal plasma. In the sheep, Care and colleagues (Care 1980; Weatherley, Ross, Pickard, and Care 1983; Care, Caple, Abbas, and Pickard 1986) have found that the concentration of calcium (total and Ca^{2+}) in autologous fetal blood perfused in closed circuit through the fetal circulation of the placenta, gradually increases despite the initial level being higher than maternal plasma. In the isolated, dually perfused, human placental lobule, Abramovich, Dacke, Elcock, and Page (1987a) have reported that closed circuit perfusion of the fetal circulation resulted in an increase in calcium concentration in the fetal perfusate reservoir. Abramovich *et al.* (1987b) also provided further evidence of active calcium transfer in that dinitrophenol (DNA) reduced the ratio of clearances of ^{45}Ca compared to that of ^3H_2O in the materno-fetal direction but not the feto-maternal direction. However, the results might alternatively be attributable to a DNP induced alteration in flow pattern and thus ^3H_2O clearance. In the rat, Štulc and Štulcova (1986) showed, using an *in situ* umbilically perfused preparation, that both $0.5\,mmol\,l^{-1}$ DNP and $1.0\,mmol\,l^{-1}$ NaCN in the perfusate reduced the materno-fetal calcium flux.

Flux measurements have, in general, supported the idea of active calcium transfer although, surprisingly for a molecule in which there has been so much interest, there are few reports which pay particular attention to accurate measurement of both materno-fetal and feto-maternal unidirectional

Table 9.7
Maternal and fetal plasma [calcium] (mmol l⁻¹)

	Mother				Fetus			
	Total	(n)	Ultrafilterable/ ionized	(n)	Total	(n)	Ultrafilterable/ ionized	(n)
Human[a]	2.13±0.15	(115)	1.12±0.06[i]	(115)	2.65±0.19*	(115)	1.41±0.09[i]*	(115)
Monkey[b]	2.05±0.008	(7)	1.05±0.006[i]	(5)	2.10±0.008	(7)	1.33±0.005[i]*	(5)
Rat[c]	2.23±0.06	(9)	1.10±0.03[i]	(17)	2.75±0.002	(49)*	1.42±0.03[i]*	(18)
Guinea-pig[d]	1.70±0.06	(13)	1.23±0.05[u]	(13)	2.30±0.03	(13)*	1.53±0.04[u]*	(13)
Sheep[e]	2.12±0.05	(40)	0.99±0.09[i]	(5)	3.30±0.05	(42)	1.59±0.16[i]	(6)
Pig[f]	2.53±0.02	(19)	0.79±0.03[i]	(5)	3.10±0.04*	(22)	1.12±0.14*	(5)

Values are mean ± s.e. u = ultrafilterable, i = ionized.

a. Schauberger and Pitkin 1979 (36–42 weeks gestational age); b. Northrop, Misenhimer, and Becker 1977 (150 ± 5 days gestational age); c. Total calcium from Chalon and Garel 1985b (19.5 days gestational age). Ionized calcium provided by N. Robinson in our laboratory (21 days gestational age); d. Twardock, Kuo, Austin, and Hopkins 1971 (56–65 days gestational age); e. Care, Ross, Pickard, Weatherley, and Robinson 1981 (near term); f. Care, Ross, Pickard, Weatherley, Garel, Manning, Allgrove, Papapoulos, and O'Riordan 1982 (near term).

* P<0.05 or better, v. maternal, where given by authors.

fluxes, and in only one (Štulc and Štulcova 1986) were the flux values given confidence limits. They reported the unidirectional materno-fetal flux of Ca^{2+} in non-anaesthetized rats to be 100.2 ± 7 nmol min^{-1} fetus^{-1}. Anaesthesia had no effect in intact animals but umbilical perfusion was associated with a slightly lower flux. The feto-maternal flux was estimated from the steady state extraction of ^{45}Ca from fetal perfusate to be only about 20 per cent of materno-fetal transfer. The calculated net flux was thus approximately 80 nmol min^{-1} fetus^{-1}, slightly but significantly less than that estimated from the increase in fetal calcium content between 20 and 21 days of pregnancy of 100 ± 4 nmol min^{-1} fetus^{-1}. Twardock's guinea-pig study (1967) disclosed a similar discrepancy in that the unidirectional materno-fetal flux was 30 mg day^{-1} fetus^{-1} as compared to a net fetal calcium accretion in late gestation of 40 mg day^{-1}.

Calculation of unidirectional fluxes from tracer studies requires knowledge of which plasma fraction of calcium is involved in the transport process and this is not clear. The discrepancy between net flux calculated from unidirectional flux asymmetry and that measured by growth in calcium content could have a methodological explanation if, following injection of ^{45}Ca as radio tracer, the specific activity of different plasma calcium fractions is not identical within the time course of the studies. The studies described above made the assumption that ^{45}Ca was distributed between the different plasma pools proportionate to their concentration. In support of this assumption ultrafiltration of plasma containing ^{45}Ca has generally revealed no difference in specific activity between the ultrafilterable and the non-filterable fractions (Wiester, Whitla, and Goldsmith 1963; Giese and Comar 1964). However, Twardock *et al.* (1971) did note the specific activity of ultrafilterable calcium to be slightly lower than that of protein bound calcium in maternal and fetal blood from pregnant guinea-pigs which could, for example, lead to an underestimate of materno-fetal flux and thus of net flux if ultrafilterable calcium is the fraction mainly transported and if its equilibration is faster in the fetal perfusate. Specific activities of ionized and chelated forms of calcium can also be different, particularly in situations where plasma calcium is lower than normal (Wiester *et al.* 1963; Giese and Comar 1964). This uncertainty requires clarification before calcium flux studies can be securely interpreted. An alternative explanation of the discrepancy between calculated net flux and that measured from growth could be a slower time course component of materno-fetal flux (perhaps in rodents *via* the yolk sac) that was not identified in isotope studies of relatively short duration, such as the above.

A K/D_w ratio for unidirectional materno-fetal calcium flux can be calculated from the data of Twardock and Austin (1970) for the guinea-pig of 72 cm g^{-1}, and from the data of Štulc and Štulcova (1986) for the rat of 154 cm g^{-1}. Both values are considerably higher than the ratios for extracellular tracers (Table 9.2) providing further support for the importance of transcellular transfer of calcium across the placenta.

By analogy with other calcium transporting epithelia such as the duodenal mucosa it seems likely that three steps are required in transtrophoblastic calcium movement. Firstly, transfer across the (in the human microvillous*) maternal-facing plasma membrane. In the duodenum this step occurs down an electrochemical gradient with a channel or carrier involved (Bronner, Pansu, and Stein 1986). Secondly, Ca^{2+} (presuming ionized calcium to be the transported fraction) must diffuse through the intracellular compartment. If, as in other cell types, the placental intracellular Ca^{2+} concentration is in the sub-micromolar range, this must be maintained, despite the high flux of Ca^{2+} through the cell. In the duodenum this is achieved by a calcium binding protein (CaBP) in association with which Ca^{2+} moves across the cell (Bronner *et al.* 1986). Total placental calcium content is high (Cittadini, Paparella, Costaldo, Romar, Polsinelli, Carelli, Bompiani, and Terranova 1977) and it would be interesting to confirm that cytoplasmic Ca^{2+} concentration within the trophoblastic syncytium is indeed very low. Thirdly, Ca^{2+} must be extruded from the cell against a highly adverse electrochemical gradient, again presuming intracellular Ca^{2+} concentration to be sub-micromolar. In duodenum, Ca^{2+} pumping at the basolateral* membrane is by a $Ca^{2+}Mg^{2+}$-ATPase (Bronner *et al.* 1986).

There is evidence for calcium carriers located at both surfaces of placental trophoblast as well as for CaBP within at least haemochorial and yolk sac placentae, and for a calcium pump at the basal (fetal) face of trophoblast. Thus Sweiry and Yudilevich (1984) and Sweiry, Page, Dacke, Abramovich, and Yudilevich (1986) compared the unidirectional uptake of Ca^{2+}, calculated from the delay in Ca^{2+} transit time through the perfused maternal or fetal circulations compared to that of Na^+ used as a reference molecule (the single pass indicator dilution technique), and found evidence for specific saturable (that is, carrier mediated) uptake of Ca^{2+} into both the maternal and fetal sides of the placenta in both human and guinea-pig.

The unidirectional influx was higher on the fetal than on the maternal side in the guinea-pig (Sweiry and Yudilevich 1984) but equal on the two sides in the human placenta (Sweiry *et al.* 1986). Backflux, that is Ca^{2+} re-appearing in perfusate, presumably from the tissue, was also found at both maternal and fetal sides, and in both placental types the backflux on the fetal side was significantly higher than on the maternal side. These backflux data can be interpreted as being the result of the higher capacity of the fetal-facing plasma membrane of the syncytiotrophoblast for calcium extrusion, entirely appropriate for the placental transfer of Ca^{2+} and perhaps representing the

---

* The microvillous surface of human syncytiotrophoblast is in contact with maternal blood in the intervillous space and microvillous membrane vesicles are thus maternal side trophoblast cell membrane vesicles. By analogy with the gut the non-microvillous face of syncytiotrophoblast is referred to as basal in the preparation of membrane vesicles. This must not be confused with the basal plate of the placenta which refers to the maternal side in contradistinction to the chorionic plate on the fetal side.

Ca^{2+} pump. However, Štulc and Štulcova (1985) found significant extra-cellular binding of Ca^{2+} in the guinea-pig placenta which could influence esti-mates of Ca^{2+} fluxes using non-steady state techniques such as these. The possibility of carrier-mediated uptake of Na^+ may also lead to some further inaccuracy.

The extrusion step from syncytiotrophoblast on its fetal side has been investigated by Fisher, Kelley, and Smith (1987) who measured Ca^{2+} uptake into plasma membrane vesicles prepared from the basal face of the human placental syncytiotrophoblast. The uptake was ATP-dependent, required Mg^{2+}, had a high affinity and was not inhibited by azide (which did block uptake into purified mitochondria). It is implied in the conclusions drawn by Fisher *et al.* (1987) that the uptake they measured *into* the vesicles was repre-sentative of what would normally be extrustion of Ca^{2+} *from* the syncytio-trophoblast which would be the case if basal membrane vesicles were of inside out orientation, as in similar preparations from, for example, duo-denum (Ghijsen, van Os, Heizmann, and Murer 1986). As Fisher *et al.* (1987) also found that the intracellular calcium-dependent regulatory protein cal-modulin, when added to a suspension of preformed vesicles, stimulated Ca^{2+} uptake, it seems likely that the orientation was indeed inside out but this is not yet certain. The Ca^{2+} transporter is presumably a Ca^{2+}-ATPase, although no such enzyme has yet been characterized in the fetal-facing plasma membrane. McKercher, Derewlany, and Radde (1983) questioned the role of Ca^{2+}-ATPase in Ca^{2+} transfer, as ethacrynic acid, an inhibitor of the enzyme (McKercher, Derewlany, and Radde 1984), had no effect on Ca^{2+} transfer across the *in situ* perfused guinea-pig placenta when injected into the mother or when added to the fetal perfusate.

However, the failure of these authors (McKercher *et al.* 1983) to provide evidence of active Ca^{2+} transfer across their preparation, together with the known effect of perfusion in increasing the passive permeability of the guinea-pig placenta (Hedley and Bradbury 1980) sheds doubt on this study. In excitable tissues calcium is often extruded *via* a plasma membrane $Na^+/$ Ca^{2+} exchanger (Nicholls 1986) and such an exchanger could also con-ceivably function in uptake, but there is at present no evidence for this in the placenta.

As mentioned above, placental tissue calcium content is high. The pre-sumed submicromolar concentration of Ca^{2+} within cytoplasm could be achieved by a combination of sequestration into intracellular organelles (mito-chondria, endoplasmic reticulum, nucleus) and binding to specific cytoplasmic proteins. In the guinea-pig placenta subcellular fractionation after perfusion with ^{45}Ca revealed that most of the radioactivity as well as most stable Ca^{2+} was in the nucleus (van Kreel and van Dijk 1983), though Croley (1973), who used pyroantimonate to complex calcium into a granular form in the human placenta and which could be observed under the electron microscope, found granular deposits in membrane associated with endoplasmic reticulum, in mitochondria and in vesicles in the syncytiotrophoblast. An ATP-dependent Ca^{2+} uptake into a microsomal fraction of human placenta has been previously reported by Whitsett and Tsang (1980) and both low and high affinity Ca^{2+} ATPase activity has been found to co-purify with the microvillous plasma membrane of the human syncytiotrophoblast (Whitsett and Lessard 1978; Whitsett and Tsang 1980; Treinen and Kulkarni 1986). Treinen and Kulkarni suggest that the high affinity enzyme might function in the maintenance of a low intracellular Ca^{2+} concentration rather than in transplacental transfer, but this matter is not yet well understood.

Bruns, Fausto, and Avioli (1978) isolated a CaBP from the rat placenta which was immunologically identical to the vitamin D-dependent CaBP of intestine; it had a molecular weight of 10,500, a binding constant for Ca^{2+} (Kd) of $0.12 \pm 0.03 \, \mu mol \, l^{-1}$ and bound approximately 2.4 molecules of Ca^{2+} per molecule of the protein. The similarity of rat placental and intestinal CaBP has been confirmed by Warembourg, Perret, and Thomasset (1986) using DNA/RNA hybridization assays, although they suggest the molecular weight to be 9000. Warembourg *et al.* (1986) report the concentration of CaBP RNA and of CaBP itself to be much less in placenta than in intestine, confirming an earlier report of Marche, Delorme, and Cuisinier-Gleizes (1978). The content of CaBP in the rat placenta does, however, increase dramatically during the last quarter of gestation, mirroring the increase in fetal and placental weight, as well as the increase in the rate of fetal bone mineralization (Bruns *et al.* 1978; Delorme, Marche, and Garel 1979) and thus J_{net} for calcium. CaBP's have also been reported in mouse (Bruns, Wallstein, and Bruns 1982; Tuan and Cavanaugh 1985) and in human placenta (Fowler, Williams and Gray 1978; Tuan 1982, 1985). In the latter there is some discrepancy in their reported molecular weights. Fowler *et al.* (1978) found a 12,000 MW CaBP, whereas Tuan (1982) found a 150,000 MW dimer, which was different from human intestinal CaBP. The CaBP of Tuan (1982, 1985) had a Kd for Ca^{2+} of $5 \, \mu mol \, l^{-1}$, and had 10 binding sites for Ca^{2+} per molecule of protein, so was also different from that of the rat placenta (Bruns *et al.* 1978). It is possible that several different CaBP's may be present in human placenta. With the exception of the effect of calmodulin (one type of CaBP) on Ca^{2+} uptake by human basal membrane vesicles (Fisher *et al.* 1987) and perhaps an effect on microsomes (Tuan 1985) there is no direct evidence at present for a role of CaBP's in placental calcium transport.

Duodenal CaBP is vitamin D dependent and this is probably an important means by which 1,25 $(OH)_2$ $VitD_3$ stimulates intestinal calcium transfer (Bronner *et al.* 1986). As there are receptors for 1,25 $(OH)_2D_3$ in both human and rat placentae (Fowler *et al.* 1978; Christakos and Norman 1980; Pike, Gooze, and Haussler 1980; see Table 9.4), 1,25 $(OH)_2D_3$ might also be expected to influence placental CaBP (van Dijk 1981). However, Marche *et al.* (1978) showed that vitamin D deprivation of the pregnant rat, whilst lowering maternal plasma calcium levels and decreasing concentrations of intestinal CaBP had no effect on placental CaBP. In order to investigate this issue more rigorously Garel, Delorme, Marche, Nguyen, and Garabedian (1981) thyroparathyroidectomized (TPTX) pregnant rats to remove the stimulus normally provided by parathyroid hormone (PTH) to renal production of $1,25 (OH)_2 D_3$. The concentrations of CaBP in maternal duodenal mucosa and in entire fetal intestine (21.5 days' gestation) were markedly reduced by TPTX; this effect was reversed by injection of $1,25 (OH)_2 D_3$ but not $24,25$-$(OH)_2 D_3$. However, in the placenta CaBP concentrations were not significantly different in TPTX rats versus control. Therefore it is clear both from this study and that of Marche *et al.* (1978) that placental CaBP in the rat is, at the very least, far less sensitive to maternal $1,25 (OH)_2 D_3$ than is its intestinal counterpart.

Whether associated with alteration in CaBP control or not, the possibility of control of placental transfer has been more widely investigated for calcium than for any other ion. Much of the available data has previously been reviewed by Care (1980) and Care and Ross (1984). However, many studies which attempt to infer placental transfer effects have depended only on measurement of the fetal plasma calcium concentrations which is influenced by many other factors in addition to placental transfer rate. Recent examples include studies by Chalon and Garel (1985a,b,c), and by Brommage and DeLuca (1984). As pointed out by Care (1980) these cannot differentiate between effects on fetal bone and effects on a putative placental calcium pump. Direct measurements of placental calcium transfer in animals with altered hormone status have however been made in guinea-pigs, rats and sheep. Durand, Barlet, and Braithwaite (1983) measured fetal calcium content and ^{45}Ca transfer from mother to fetus in guinea-pigs treated with $1,25 (OH)_2 D_3$ and non-injected controls at days 40, 50, and 66 of gestation. The total calcium content of the fetuses (that is, J_{net}) was not affected by the hormone treatment at 40 or 66 days' gestation, although there was a significant increase at 50 days. J_{mf} for calcium flux as measured by ^{45}Ca content was not significantly altered by $1,25 (OH)_2 D_3$ at any of three gestational times, although interestingly the calcium transfer to the placental tissue measured in the same way was significantly increased by the hormone at all stages of gestation. In a preliminary report, Mughal, Robinson, and Sibley (1987) also failed to find any effect of maternal vitamin D status on materno-fetal calcium flux across the perfused rat placenta. In their study both J_{mf} and J_{net} of calcium were similar in TPTX rats and TPTX rats injected with $1,25(OH)_2 D_3$, paralleling the lack of any marked effect of $1,25(OH)_2 D_3$ on placental CaBP (Garel *et al.* 1981).

In sheep, Barlet (1985a) found that calcitonin replacement in thyroidectomized (THX) ewes led to a significant reduction in fetal accumulation of ^{45}Ca over 7 days near term, as compared to control ewes and to ewes subjected to THX without replacement. Barlet (1985b) also showed that prolactin injection to pregnant ewes increased both intestinal and placental calcium transfer as measured by ^{45}Ca accretion. This effect was reversed by bromocriptine and calcitonin, and potentiated by 1,25(OH)$_2$D$_3$ injection. The role, if any, of calcitonin and prolactin in placental calcium transfer in other species is unknown. As the fetal radiocalcium concentration was not reported in Durand's and Barlet's studies it is not clear whether ^{45}Ca concentration in the fetal plasma became sufficiently high for feto-maternal flux of ^{45}Ca as well as materno-fetal flux to be of relevance in interpreting their results. As materno-fetal flux is much greater than feto-maternal flux in the sheep, the calculated value would not be very much in error but it is important to remember that the experimental design did not rule out an increase in feto-maternal rather than a decrease in materno-fetal flux as the mechanism of the diminished net flux of ^{45}Ca into the fetus seen, for example, with calcitonin.

Weatherley et al. (1983) used the sheep placenta perfused through the fetal circulation with autologous fetal blood to investigate the effect of maternal thyroparathyroidectomy. The maternal calcium concentration in TPTX animals was maintained normal by calcium infusion during the three days between initial surgery and perfusion. During closed circuit perfusion the Ca^{2+} concentration of the perfusate increased in both the normal and TPTX animals despite starting levels being higher than maternal plasma. The rate of increase was slower in the TPTX animals, although only significantly so at two time points. The authors concluded that maternal PTH did not have a major effect on placental calcium transfer and suggest that the slightly slower rate of increase in TPTX animals was more likely to be due to the TPTX associated lowering of maternal plasma 1,25(OH)$_2$D$_3$ concentration leading to a lower fetal plasma concentration of this hormone in the days before placental perfusion. The effect of fetal hormones has also been studied directly by Care's group. The ratio of fetal to maternal plasma Ca^{2+} concentration was significantly reduced in TPTX sheep fetuses despite plasma 1,25(OH)$_2$D$_3$ concentrations being normal, suggesting a PTH effect alone (Care et al. 1986). However, the net clearance of calcium across the placenta to autologous fetal blood during subsequent closed circuit perfusion (fetus removed) was not different, whether TPTX or normal, an apparent discrepancy the authors explained as being due to a lower diffusional feto-maternal backflux in the TPTX situation associated with the lower fetal plasma Ca^{2+} concentration found at the start of perfusion in these fetuses. Studies of feto-maternal unidirectional calcium fluxes at different perfusate Ca^{2+} concentrations would be required to establish if this explanation is correct. More recently Abbas, Caple, Care, Loveridge, Martin, Pickard, and Rodda (1987) have provided preliminary evidence that fetal parathyroid glands in the sheep produce a substance which is not PTH and which may be capable of stimulating calcium transport across the perfused sheep placenta.

There are therefore strong but inconclusive hints from these direct studies of control of placental calcium transfer by both maternal and fetal hormones, particularly in the sheep, but much remains to be done, particularly in relation to investigating hormonal effects on both unidirectional and net calcium fluxes before it will be clear if there is definite specific endocrine control of transplacental Ca^{2+} transfer and whether the locus of control is alteration in J_{mf} or in J_{fm}.

4   GLUCOSE

In normoglycaemia, the major source of glucose for the developing fetus is that transferred across the placenta from maternal blood (Kalhan, D'Angelo, Savin, and Adam 1979; Hay, Sparks, Wilkening, Battaglia, and Meschia 1984; Jones and Rolph 1985). Fetal gluconeogenesis may increase during maternal starvation (Jones and Rolph 1985). Fetal whole blood or plasma D-glucose concentrations are below those of the mother (Huggett, Warren, and Warren 1951; Kalhan *et al*. 1979; Armentrout *et al*. 1977; Faber and Thornburg 1981) and it was clear from early studies that glucose passes easily across both epitheliochorial (Huggett *et al*. 1951) and haemochorial placentae (Chinard, Danesino, Hartmann, Huggett, Pauls, and Reynolds 1956). More recent work has confirmed that, over a physiological concentration range, net transplacental glucose flux is directly proportional to the maternal arterial concentration (Hay *et al*. 1984; Hauguel, Desmaizieres, and Challier 1986) and to the maternofetal concentration difference (Simmons, Battaglia, and Meschia 1979).

However, the placental transfer of D-glucose is not simply related to its diffusion coefficient. Huggett *et al*. (1951) demonstrated in the sheep that the transfer of glucose was much more rapid than fructose, which is of similar size, and on the basis of these experiments Widdas (1952) postulated the mechanism of glucose transfer which has become known as facilitated diffusion. The existence of a facilitated transplacental transport system for D-glucose and certain other monosaccharides has since been amply confirmed by demonstration of stereo and chemical specificity, saturability, competition in transfer and independence from energy sources in a variety of species including guinea-pig (Ely 1966; Schröder, Leichtweiss, and Madee 1975; Yudilevich, Eaton, Short, and Leichtweiss 1979; Bissonnette, Hohimer, Cronan, and Black 1979), sheep (Stacey, Weedon, Haworth, Ward, and Boyd 1978), and human (Carstensen, Leichtweiss, Molsen, and Schröder 1977; Rice, Rourke, and Nesbitt 1976, 1979; Challier, Nandakumaran, and Mondon 1985; Hauguel, Desmaizieres, and Challier 1986).

The D-glucose carrier of the human placenta has been investigated in both microvillous membrane vesicles (Johnson and Smith 1980; Bissonnette, Black, Wickham, and Acott 1981), and in vesicles prepared from the basal membrane of the syncytiotrophoblast (Johnson and Smith 1985). In both preparations glucose uptake (measured using [^{3}H] or [^{14}C] D-glucose or the non-metabolisable analogue, 3-O-[1-^{3}H] methyl-D-glucose) was rapid, stereospecific, Na^+-independent and inhibitable by cytocholasin B, phloretin and phloridzin. The k_m for the carrier was also similar, being 31 mmol l^{-1} in the microvillous vesicles (Johnson and Smith 1980), and 23 mmol l^{-1} in the basal vesicles (Johnson and Smith 1985). Both these k_m values are sufficiently high for it to be likely that transport would be linearly related to concentration in the physiological range, as was indeed found in the perfused human placenta (Hauguel *et al.* 1986).

The properties of the D-glucose carrier demonstrated using vesicles thus accords well with the nature of D-glucose transfer across the intact organ. This renders the isolation of the carrier (or its components) from microvillous membrane vesicles of some interest. Johnson and Smith (1982) found that the binding of cytocholasin B to a protein of 52,000 molecular weight was sensitive to D-glucose, whereas Ingermann, Bissonnette, and Koch (1983) found a protein of 60,000 molecular weight to show D-glucose sensitive cytocholasin B binding. It would therefore seem that the D-glucose carrier protein of the microvillous plasma membrane of the human placenta is, or contains, a protein of between 52,000 to 60,000 molecular weight. The isolated carrier retains its function when reconstituted into lipid vesicles (Bissonnette, Black, Thornburg, Acott, and Koch 1982). The glucose transport carrier also appears to have an affinity for dehydro-ascorbic acid both in the human (Ingermann, Stankova, and Bigley 1986; Ingermann 1988) and guinea-pig (Leichtweiss, Lisboa, and Steinborn 1986) placenta, although the significance of this observation is unclear.

Detailed analysis of the kinetics of transplacental glucose transfer is complicated by the rapid sugar metabolism demonstrated by placenta and uterine tissue. Simmons *et al.* (1979) estimated from the measurement of V-A glucose concentration differences and blood flows on the maternal and fetal sides of the sheep placenta that the net flux of glucose from the placenta to fetus was only about one-third of that from the uterine circulation to the uterus as a whole. Similarly, in the perfused human placental cotyledon, Hauguel *et al.* (1986) found that only about 40 per cent of the glucose taken up from the maternal circulation was transferred to the fetal circulation. Simmons *et al.* (1979) empirically derived an equation for sheep relating net transplacental glucose flux to maternal and fetal arterial D-glucose concentrations, which included a term for glucose metabolism by the placenta, and calculated a k_m for the sheep carrier of 3.9 mmol l^{-1}, much lower than the human but above the mean plasma glucose concentrations found in the ewe. Tracer studies by Hay, Sparks, Battaglia, and Meschia (1984) suggest that the relationship between uteroplacental metabolism and D-glucose transfer is even more complex in that they calculated that 40 per cent of the D-glucose utilized by the placenta is actually derived from the fetal glucose pool.

The dependence of net transplacental D-glucose transfer on maternal and fetal plasma D-glucose concentrations that follows from the non-active nature of D-glucose transfer makes it certain that net glucose flux to the fetus is directly influenced by those factors which alter maternal and fetal plasma glucose concentrations. However, the demonstration of insulin receptors on the maternal surface of the placenta in a variety of species (Table 9.4; Posner 1974; Owens, Brinsmead, Waters, and Thorburn 1980) suggested that insulin might also have a direct effect on glucose transfer rates and this appeared to be the case in studies with the sheep in which insulin was infused into the uterine artery (Paxson, Morriss, and Adcock 1978; Crandell, Palma, and Morriss 1982). However, several recent studies have failed to confirm any such effect. In sheep, Hay, Sparks, Gilbert, Battaglia, and Meschia (1984) infused insulin into near term pregnant ewes at a rate which produced a 9-fold increase in its plasma concentration. Arterial plasma glucose concentrations were maintained constant by the glucose "clamp" technique. Glucose extraction by a control tissue, the ewes hind limb, was increased 4.9-fold by the insulin infusion but neither uptake into the uterus as measured from the uterine arterio-venous concentration difference nor release into the fetus (umbilical arterio-venous concentration difference) was altered. In a further similar study, Hay, Meznarich, Sparks, Battaglia, and Meschia (1985) also failed to find any effect of fetal hyperinsulinaemia on fetal glucose uptake from the placenta. Rankin and colleagues (Jodarski, Shanahan, and Rankin 1985; Rankin, Jodarski, and Shanahan 1986) investigated the possibility of an insulin effect on transfer independent of any metabolic effect by measuring the clearance of 3-0-methyl-D-glucose (3MG) from the fetus, again using the glucose clamp technique. Insulin infusion to the fetus sufficient to cause a 30-fold increase in its plasma concentration had no effect (Jodarski *et al.* 1985). Infusion of insulin into the mother's circulation sufficient to increase insulin concentration over 200-fold also had no effect and prior fasting of the ewe did not alter the result (Rankin *et al.* 1986). In the perfused human placental cotyledon insulin concentrations in the maternal perfusate ranging from 50 to $1200 \mu U \, ml^{-1}$ had no effect on glucose uptake or transfer (Challier, Hauguel, and Desmaizieres 1986). Similarly, Johnson and Smith (1980) failed to find any increase in 3MG uptake by microvillous membrane vesicles prepared from human placental villous tissue incubated with $10 \, mU \, ml^{-1}$ insulin as compared to that into vesicles isolated from tissue incubated without insulin. These experiments were carried out under conditions in which exogenous insulin degradation was minimized and in which biologically active insulin was likely to be present throughout in the experimental system (Smith 1981).

This lack of effect of insulin on glucose uptake or transfer might seem surprising but, as pointed out by Johnson and Smith (1980), the capacity of the placenta to transfer glucose, at least in man, greatly exceeds fetal requirements. The insulin receptor could of course have other functions. In recent work a striking effect of insulin on rat placental glucose utilization has been demonstrated before, but not at, term (Leturque, Hauguel, Kande, and Girard 1987), and Fletcher and Bassett (1986) noted that insulin injection into rabbit fetuses caused a significant increase in placental weight and RNA content in keeping with its mitogenic activity in other tissues (Hill and Milner

1985). In addition, the effect of lowering insulin concentrations below normal *in vivo* has not been investigated and therefore a permissive effect on the glucose transport carrier cannot be ruled out.

The high capacity of the human placenta for glucose transfer in relation to fetal requirements, whilst possibly explaining the lack of an insulin effect, might also suggest that direct regulation by inhibition of transfer is a possibility (Smith 1981). Such an effect would reduce the close parallel relationship between maternal and fetal plasma glucose concentrations normally seen. It is therefore interesting that 3MG uptake into microvillous and basal membrane vesicles from human placenta was inhibited by oestrogens and progesterone (Johnson and Smith 1980, 1985). Although the concentrations required for inhibition ($1.0\,\mu\mathrm{mol\,l^{-1}}$ and upward) were supraphysiological, compared to normal circulating steroid concentrations, it could be that placental production would lead to similarly high local concentrations (Johnson and Smith 1985). The mechanism and function of this steroid inhibitory effect is unclear at present but it would be interesting to know whether there is a similar effect on glucose transfer across the whole organ.

Thus on present evidence transplacental glucose movement does not appear to be specifically controlled in the short run. Whether there is any feedback mechanism stimulating the necessary increase in placental transport capacity for glucose as gestation proceeds is unknown.

5   AMINO ACIDS

The extensive literature on placental amino acid transfer has been well reviewed by Yudilevich and Sweiry (1985). We here present a summary of the more important aspects, with particular attention to the literature on control.

The total fetal plasma amino nitrogen concentration has been known for many years to exceed that of maternal plasma in women, guinea-pigs and rabbits leading to the assumption that diffusion alone could not explain placental transfer (Morse 1917; Christensen and Streicher 1948). More recent investigations in a variety of species near term, including sheep (Young and McFadyen 1973), guinea-pig (Hill and Young 1973), and human (Young and Prenton 1969) have shown that the fetal plasma concentrations of most, but not all, individual amino acids are also higher than their maternal counterpart. This relationship also exists in the human at mid-gestation (18–29 weeks) (McIntosh, Rodeck, and Heath 1984; Soltesz, Harris, Mackenzie, and Aynsley-Green 1985).

Placental intracellular free amino acid concentrations are generally much higher than either maternal or fetal plasma concentrations in sheep (Young, Stern, Horn, and Noakes 1982), guinea-pig (Hill and Young 1973), and human (Pearse and Sornson 1969; Yudilevich and Sweiry 1985). It is therefore likely that the concentration difference between fetal and maternal plasma is achieved by concentrative energy-requiring uptake from maternal plasma into an intracellular placental compartment, followed by a passive release into fetal plasma (Hill and Young 1973). The former step undoubtedly requires plasma membrane carriers on the maternal side and these have now been partially characterized as detailed below. Release into fetal plasma also probably requires carriers and these have been identified for neutral and basic amino acids on the basis of single pass indicator dilution studies in the guinea-pig (Yudilevich and Eaton 1980; Eaton and Yudilevich 1981; Eaton, Mann, and Yudilevich 1982).

An energy requirement for placental intracellular accumulation has been clearly shown. Thus, amino acid uptake *in vitro* into both human and guinea-pig placenta was inhibited in the absence of oxygen or the presence of metabolic inhibitors (Sybulski and Tremblay 1981; Litonjua, Canlas, Soliman, and Paulino 1967; Dancis, Money, Springer, and Levitz 1968). However, Longo, Yuen, and Gusseck (1973) found that glycogenolysis could support some ^{14}C-aminoisobutyric acid (AIB, a non-metabolizable amino acid) uptake under anaerobic conditions. Interestingly, the first order rate constant for ^{14}C-AIB efflux from human placental slices was also diminished by DNP (Miller and Berndt 1974), raising the possibility of some energy consumption during efflux.

There are carrier systems with six distinct specificity patterns involved in tissue uptake of neutral amino acids by animal tissues but only three of these, the A, ASC, and L Systems, have been clearly demonstrated to be present in the placenta (Yudilevich and Sweiry 1985). It is not yet clear whether there are separate carriers for acidic and basic amino acids. Enders, Judd, Donohue, and Smith (1976) studied the three neutral systems in villous fragments from term human placenta. System A was inhibited by DL-N-methylalanine (NMA) and mediated uptake of AIB, glycine, proline, alanine, serine, threonine and glutamine. System L was inhibited by DL-b-2-aminobicyclo-[2,2,1] heptane-2-carboxylic acid (BCH) and mediated isoleucine, valine, phenylalanine, BCH, alanine, serine, threonine and glutamine uptake. System ASC, which was partially inhibited by both NMA and BCH, mediated alanine, serine, threonine and glutamine uptake. Most of the amino acids are thus carried by more than one system. Ruzycki, Kelley, and Smith (1978) investigated AIB uptake by microvillous plasma membrane vesicles from human placenta and found it to be via System A. It was temperature dependent, saturable and was also markedly stimulated by an inward Na^+ gradient (that is, inside concentration lower than outside), providing evidence for secondary active transport of amino acids. It is presumed that in the intact organ, energy-requiring Na^+/K^+ ATPase maintains an inwardly directed Na^+ gradient thus energising concentrative uptake by System A carriers co-transporting amino acids with Na^+. Boyd and Lund (1981) found that an inward Na^+ gradient also stimulated L-[5-^{3}H] proline uptake by human placental microvillous membrane vesicles. L-proline uptake was not electrically neutral in that it was influenced by a K^+ diffusion potential; competitive in-hibition studies again suggested that the uptake was mainly via System A. More recently Ganapathy, Leibach, Mahesh, Howard, Devoe, and Ganapathy (1986) found that the uptake of tryptophan by microvillous membrane vesicles was Na^+-independent and appeared to be via System L. It is not clear what, in the absence of Na^+ co-transport, energises concentrative movement of amino acids using this carrier.

The indicator dilution method has been used to study amino acid uptake into and efflux from both maternal and fetal sides of the isolated, dually per-fused, guinea-pig placenta (Yudilevich and Eaton 1980; Eaton and Yudile-vich 1981; Eaton *et al.* 1982; Yudilevich and Sweiry 1985). As far as the neutral amino acids were concerned, there was evidence for the presence of System L (Yudilevich and Eaton 1980; Eaton *et al.* 1982) and System ASC (Eaton *et al.* 1982) transport on both maternal and fetal sides but, interest-ingly, unlike the human, System A was absent (Eaton *et al.* 1982). Whether or not any of the other systems for neutral amino acids are present in the guinea-pig placenta remains to be seen (Yudilevich and Sweiry 1985). These studies in the guinea-pig are the only ones which directly address the question of how release of amino acid from the fetal side is mediated. While influx on the two sides was similar except for differences in Na^+-dependency (Eaton *et al.* 1982), efflux (backflux) from tissue to perfusate appeared to be greater on the fetal side (Eaton and Yudilevich 1981). They propose that net transfer from mother to fetus is due to this asymmetric efflux (Eaton and Yudilevich 1981). The development of methods to prepare vesicles from the basal, fetal-facing plasma membrane of the human placental trophoblast, used success-fully for studies on glucose (Johnson and Smith 1985) and calcium (Fisher *et al.* 1986) carriers should, when applied to amino acids, help to confirm or refute this hypothesis.

There is good evidence that basic amino acids are also readily transferred across the placenta. In the sheep, Young and McFadyen (1973) found that lysine, histidine, ornithine, and arginine showed a rise in umbilical vein con-centration after injection into the maternal jugular vein within the range seen for neutral amino acids. Placental uptake of basic amino acids from the uter-ine circulation and release into the umbilical circulation was also seen, as for neutral amino acids, under physiological conditions (Holzman, Lemons, Meschia, and Battaglia 1979). Transfer of lysine and arginine across the per-fused human placenta again resembled that of the neutral amino acids (Schneider, Mohlen, and Dancis 1979) and on recirculation lysine concentra-tion in fetal perfusate increased to about 1.5 times maternal (Schneider, Mohlen, Challier, and Dancis 1979). In the dually perfused guinea-pig pla-centa studied by the indicator dilution technique lysine, arginine and histi-dine behaved similarly to most neutral amino acids (Eaton and Yudilevich 1981). Although some interaction between lysine and alanine was noted in the guinea-pig (Eaton *et al.* 1982; Yudilevich and Sweiry 1985) in which lysine uptake may be Na^+ dependent (Eaton *et al.* 1982), there is little detailed information available about basic amino acid carriers.

Placental handling of acidic amino acids (glutamate, aspartate, cystine) appears to be quite different, as in the sheep and human there appears to be uptake from the fetal circulation (as shown by negative umbilical [a]–[v] differences) and little materno-fetal transfer *in vivo* (Lemons, Adcock, Jones, Naughton, Meschia, and Battaglia 1976; Holzman *et al.* 1979; Morriss, Adcock, Paxson, and Greeley 1979; Hayashi, Sanada, Sagawa, Yamada, and Kido 1978), and in perfused human placental preparations (Schneider, Mohlen, Challier, and Dancis 1979; Schneider, Mohlen, and Dancis 1979). Glutamate does appear to be concentrated by human placental slices (Schneider and Dancis 1974) although this could have occurred from the fetal side. Similar absence of glutamate and aspartate transfer in the maternal to fetal direction is found in rhesus monkeys following bolus maternal injection (Stegink, Pitkin, Reynolds, Filer, Boaz, and Brummel 1975; Stegink, Pitkin, Reynolds, Brummel, and Filer 1979). In the guinea-pig uptake of glutamate and aspartate by either maternal or fetal faces was low or absent according to indicator dilution studies (Eaton and Yudilevich 1981) although some unidirectional flux of glutamate into placenta from the fetal circulation was seen during steady state perfusion (Bloxham, Tyler, and Young 1981). A net flux from placenta to fetus was evidenced by a positive umbilical [a]–[v] difference (Hill and Young 1973; Bloxham *et al.* 1981).

Placental uptake of dipeptides might provide another source of amino-nitrogen for the fetus (Crandell, Adcock, and Morriss 1981). There is evidence for an affinity of System A for the dipeptide glycyl-L-proline in that it inhibited L-proline uptake by human microvillous plasma membrane vesicles (Boyd and Lund 1981). Ganapathy, Mahesh, Devoe, Leibach, and Ganapathy (1985) also demonstrated uptake of [1-^{14}C] glycylsarcosine by a process which was inhibitable by other dipeptides and which was Na^+-independent. Whether such uptake can result in complete transfer of the dipeptide to the fetus or whether any uptake is followed by hydrolysis within the placenta remains to be demonstrated.

Control of placental amino acid transfer has been investigated at several levels and some possible mechanisms are shown in Figure 9.3.

In the intact rat the overall plasma amino acid concentration in fetal plasma was maintained when total maternal plasma amino acid concentration was reduced by 50 per cent through dietary manipulation and glucagon infusion (Domenech, Gruppuso, Nishino, Susa, and Schwartz 1986), and AIB transfer per gram of placenta is enhanced in growth retarded fetal guinea-pigs from protein-restricted mothers as compared to normal control animals (Young and Widdowson 1975). It appears from these findings that amino acid transfer to the fetus can be modulated to compensate for defects in maternal supply. There are a number of partially understood mechanisms which might allow this. One possibility might be a change in blood flow rate

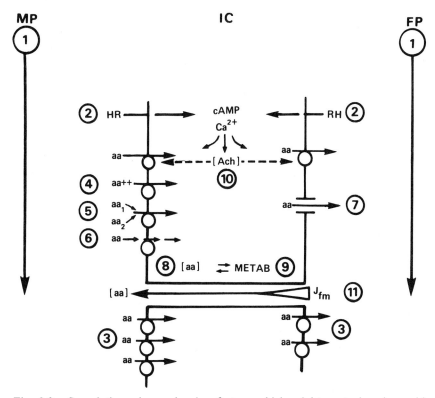

Fig. 9.3. Speculative scheme showing factors which might control amino acid transfer. (1) Blood flow rate; under conditions of low flow an increase might lead to perfusion of previously unperfused areas of placenta. (2) Hormone (H) Receptors (R) on either side; activation leads to altered intracellular concentrations of second messenger, for example cAMP and Ca^{2+} which might in turn alter any of (3) to (11). (3) Altered V_{max} of carrier; will increase with increased number of carrier sites. (4) Altered affinity of carrier, for example through change in its configuration. (5) Competition for carrier. (6) Altered turnover number of carrier. (7) Altered leak out of (or into) cell. (8) Altered intracellular free amino acid concentration. (9) Altered amino acid metabolism by placenta. (10) Permissive effect of acetylcholine. (11) Altered paracellular permeability leading to increased or decreased diffusional flux from fetus to mother. An increase in permeability of this route will reduce the overall net flux from mother to fetus (There will be unidirectional fluxes in both directions through this pathway but, as fetal amino acid concentrations are higher, J_{fm} through the paracellular pathway will be greater than J_{mf}; for clarity only J_{fm} is shown).

MP = maternal plasma. IC = trophoblast intracellular compartment. FP = fetal plasma (qv Fig. 9.2). Blood flows are arbitrarily shown as concurrent.

in that amino acid transfer across the perfused guinea-pig placenta was reduced when blood flow diminished (Reynolds and Young 1971), but flows were probably unphysiologically low in this preparation and there might have been surface area effects as discussed in **Section II**. It seems unlikely that amino acid transfer is substantially affected by changes in blood flow rate within a physiological range in that, at any rate in the sheep, the proportionate extraction of amino acid from uterine arterial blood is lower than that for glucose whose transport is only moderately flow-limited (Holzman *et al.* 1979; Simmons *et al.* 1979). Cellular mechanisms seem more likely to be important.

Simple incubation of placental tissue leads to an increase in its ability to accumulate amino acids. This effect was first reported by Longo *et al.* (1973) for AIB uptake by human placental slices and by Smith, Adcock, Teasdale, Meschia, and Battaglia (1973) for AIB uptake by human villous fragments. The increased uptake was directly proportional to the length of pre-incubation and resulted from an increase in V_{max} and a decrease in k_m of the transport system (Smith *et al.* 1973). It was later found that the effect was abolished by inhibition of protein synthesis during pre-incubation (Smith and Depper 1974; Gusseck, Yuen, and Longo 1975), suggesting that synthesis of new transport sites was involved. Although addition of AIB or alanine to the pre-incubation medium initially appeared to block the stimulation of uptake, this was found to be due to intracellular accumulation (trans-inhibition, see below) and the pre-incubation effect is not due to absence of amino acids during pre-incubation. In fact the stimulus for the increased uptake capacity is still unclear (Smith 1981) although it is tempting to suppose it might be related to physiological control mechanisms. It is restricted to System A carriers (Enders *et al.* 1976).

Trans-inhibition of existing carriers provides an alternative potential control mechanism with intracellular accumulation of amino-acid leading to decreased uptake via System A (Smith and Depper 1974; Steel, Smith, and Kelley 1982). Trans-inhibition occurs at physiological intracellular amino acid concentrations and appears to be due to a direct interaction between intracellular amino acid and carrier (Steel *et al.* 1982). Both pre-incubation effects and trans-inhibition have also been observed in non-placental tissue (Guidotti, Borghetti, and Gazzola 1978).

A proportion of the amino acid taken up is used by placenta itself both for protein synthesis and for intermediary metabolism. Young (1981) and Carroll and Young (1983) have suggested that there might be a relationship between the ratio of free to bound (that is, incorporated into protein) pools of amino acid in the placenta and transport to the fetal circulation. In support of this, they found that when cyclohexamide, an inhibitor of protein synthesis, was added to the fetal perfusate of the *in situ* perfused guinea-pig placenta (maternal circulation intact), both incorporation of ^{14}C-lysine, -leucine, -glycine and -aspartate into protein and their transfer from mother to fetus were reduced. On the other hand, the transfer of ^{14}C-AIB, which cannot be used for protein synthesis, was unaffected. Inhibitors of protein synthesis were found to have no effect on the "normal" (as opposed to the pre-incubation stimulated) uptake of amino acid by human placental slices (Sybulsky and Tremblay 1967; Gusseck et al. 1975). This suggests that any relationship between placental protein synthesis and amino acid transfer may be mediated by an effect on efflux into the fetal circulation rather than on uptake.

Neither steroid hormones, such as progesterone, testosterone, oestradiol and cortisone (Litonjua et al. 1967; Sybulsky and Tremblay 1967), nor protein hormones such as chorionic gonadotrophin and insulin (Litonjua et al. 1967) have been found to have any consistent effect on amino acid uptake by human placental slices. Dancis et al. (1968) did note stimulation of ^{14}C-AIB uptake into human placental slices by insulin but there was no effect on uptake into guinea-pig placental slices. More recently Steel, Mosley, and Smith (1979) rigorously investigated effects of insulin on uptake into human placental microvillous vesicles of proline, glycine and AIB as System A substrates, of phenylalanine as a System L substrate and of alanine as a System ASC substrate. Under conditions during which bio-active insulin was shown to be present throughout, there was no stimulation in the uptake of any of the amino acids compared to controls. Despite the lack of demonstration of any clear hormonal effects on amino acid transfer so far, there are a wide range of other hormones which have effects on amino acid transfer in other tissues (Guidotti et al. 1978) which have yet to be studied in placental systems. In view of its properties as a growth factor (Hill and Milner 1985), the lack of an acute influence of insulin does not rule out the possibility of a chronic effect over the long term.

As in some other non-nervous tissues, there are components of a cholinergic system in the human placenta (reviewed by Sastry and Sadavongvivad 1979). There is good evidence for the presence of acetylcholine and choline acetyltransferase (Sastry and Sadavongvivad 1979) as well as of muscarinic receptors (Table 9.4). Cholinesterase activity is found in microvillous plasma membrane from human placenta (Fant and Harbison 1981). There is also a rapid uptake system for choline in both guinea-pig (Sweiry and Yudilevich 1985) and human placenta (Sweiry et al. 1986). Because of an association between maternal smoking and poor fetal growth, a possible relationship between the cholinergic system and amino acid transport has been looked for. Although cholinergic receptor blockers such as atropine have been found to reduce ^{14}C-AIB accumulation in microvillous membrane vesicles from human placenta (Fant and Harbison 1981), the concentrations required (1 mmol l^{-1}) suggest a non-specific effect. Cholinergic agonists or raised acetylcholine concentrations appear to have little effect on AIB uptake by human placenta (Sastry and Sadavongvivad 1979; Fant and Harbison 1981) but there is strong evidence for a permissive effect of acetylcholine in that inhibitors of choline acetyltransferase (required for acetylcholine synthesis) reduce ^{14}C-AIB uptake by isolated human placental villi (Rowell and Sastry 1981). Potency of inhibition of choline acetyltransferase by a series of drugs closely paralleled inhibition of AIB uptake in the absence of any apparent effect on cell viability. Similar results were reported by Welsch, Wenger, and Stedman (1981) who found that 50 per cent inhibition of choline acetyltransferase activity was produced by $0.4 \, \text{mmol} \, l^{-1}$ and of AIB uptake by human placental fragments by $0.1 \, \text{mmol} \, l^{-1}$ 2-benzoylethyl trimethylammonium (BETA). BETA also reduced tissue acetylcholine concentrations (Welsch *et al.* 1981). The function of Ach dependent amino acid uptake is unclear and may indeed not be physiologically important in that Jarmer, Shoaf, and Harbison (1985) found carnitine rather than choline to be the preferred substrate of the acetylase of human microvillous membrane vesicles. Its activity may relate to transfer of fatty acid residues, as in mitochondria, rather than to amino acid transport.

Carrier mechanisms for active amino acid transport into and across the placenta are thus present and their activity can be modulated under a number of circumstances in *in vitro* systems. However, yet again a coherent picture of how this relates to transport across the intact organ or to fetal and placental needs is not yet visible. It is also unclear how far the carrier mechanisms studied in non-vectorial *in vitro* systems such as slices and vesicles are related to the amino acid needs of the placenta both for "housekeeping" and for synthesis of protein hormones, and how far they contribute to its role as an epithelial transporting organ. Finally, the possibility that modulation of paracellular diffusional backflux of amino acids from fetus to mother (bearing in mind the higher fetal plasma amino acid concentration) is important in controlling net fetal amino acid acquisition has not been systematically addressed. The above remarks can, with even greater strength, be applied to the control of transfer of all the other solutes we have considered.

Acknowledgements

A symposium on the topic of this review organized by Professor Tony Care was held in conjunction with the Physiological Society in Leeds, England, in December 1986. Review papers on different aspects of control were given by Dr. H. Schröder, Dr. C. H. Smith, Dr. E. Marelyn Wintour, Dr. J. Štulc (1988) Dr. J. P. van Dijk (1988) and Dr. R. F. Ingermann (1988), and these contributed considerably to our thinking about control. We would like to express our indebtedness to these speakers and to other contributors to the meeting. We are also most grateful to Dr. W. G. Bardsley and Dr. C. A. R. Boyd for many helpful discussions on the theoretical formulation of placental transport, and to Dr. J. Štulc for critical advice. The authors' work reported in this chapter was supported by Action Research for the Crippled Child, The Wellcome Trust and the North Western Regional Health Authority. Mrs. Vera Green typed and re-typed the manuscript without flagging or complaint, and Mrs. Karen Irving obtained many references.

References

Abbas, S. K., Caple, I. W., Care, A. D., Loveridge, N., Martin, T. J., Pickard, D. W., and Rodda, C. (1987). The role of the parathyroid glands in fetal calcium homeostasis in the sheep. *J. Physiol., Lond.* **386**, 27P.

Abramovich, D. R., Dacke, C. G., Elcock, C., and Page, K. R. (1987*a*). Calcium transport across the isolated dually perfused human placental lobule. *J. Physiol., Lond.* **382**, 397–410.

—— —— —— —— (1987*b*). Effect of dinitrophenol on calcium transfer across the dually perfused human placental lobule. *J. Physiol., Lond.* **386**, 21P.

Alderson, R., Pastan, I., and Cheng, S.-Y. (1985). Characterization of the 3, 3′, 5-triiodo-L-thyronine-binding site on plasma membranes from human placenta. *Endocrinology* **116**, 2621–2630.

Armentrout, T., Katz, S., Thornburg, K. L., and Faber, J. J. (1977). Osmotic flow through the placental barrier of chronically prepared sheep. *Am. J. Physiol.* **233**, H466–H474.

Bailey, D. J., Bradbury, M. W. B., France, V. M., Hedley, R., Naik, S., and Parry, H. (1979). Cation transport across the guinea-pig placenta perfused *in situ*. *J. Physiol., Lond.* **287**, 45–56.

Bain, M., Copas, D. K., Landon, M. J., and Stacey, T. E. (1986). Passive permeability in the human placenta *in vivo*. *Early hum. Dev.* **14**, 137–138.

Balkovetz, D. F., Leibach, F. H., Mahesh, V. B., Devoe, L. D., Cragoe, E. J., and Ganapathy, V. (1986). $Na^+ - H^+$ exchanger of human placental brush-border membrane: identification and characterization. *Am. J. Physiol.* **251**, C852–C860.

Barlet, J.-P. (1985*a*). Calcitonin may modulate placental transfer of calcium in ewes. *J. Endocr.* **104**, 17–21.

—— (1985*b*). Prolactin and calcium metabolism in pregnant ewes. *J. Endocr.* **107**, 171–175.

Battaglia, F. C., Meschia, G., Hellegers, A., and Barron, D. H. (1958). The effects of acute hypoxia on the osmotic pressure of the plasma. *Q. Jl. exp. Physiol.* **43**, 197–208.

Bazer, F. W., Goldstein, M. H., and Barron, D. H. (1981). Water and electrolyte transport by pig chorio-allantois. In *Fertilization and embryonic development in vitro* (ed. L. Mastroianni and J. D. Biggers) pp. 299–321. Plenum, New York.

Benos, D. J. and Biggers, J. D. (1983). Sodium and chloride co-transport by preimplantation rabbit blastocysts. *J. Physiol., Lond.* **342**, 23–33.

Berhe, A., Bardsley, W. G., Harkes, A., and Sibley, C. P. (1987). Molecular charge effects on the protein permeability of the guinea-pig placenta. *Placenta* **8**, 365–380.

Bhaumick, B., Armstrong, G. D., Hollenberg, M. D., and Bala, R. M. (1982). Characterization of the human placental receptor for basic somatomedin. *Can. J. Biochem.* **60**, 923–932.

Binder, N. D., Faber, J. J., and Thornburg, K. L. (1978). The transplacental potential difference as distinguished from the materno-fetal potential difference of the guinea-pig. *J. Physiol., Lond.* **282**, 561–570.

Bissonnette, J. M., Black, J. A., Thornburg, K. L., Acott, K. M., and Koch, P. L. (1982). Reconstitution of D-glucose transporter from human placental microvillous plasma membranes. *Am. J. Physiol.* **242**, C166–C171.

—— —— Wickham, W. K., and Acott, K. M. (1981). Glucose uptake into plasma membrane vesicles from the maternal surface of human placenta. *J. Membrane Biol.* **58**, 75–80.

—— Hohimer, A. R., Cronan, J. Z., and Black, J. A. (1979). Glucose transfer across the intact guinea-pig placenta. *J. dev. Physiol.* **1**, 415–426.

Bloxam, D. L., Tyler, C. F., and Young, M. (1981). Foetal glutamate as a possible precursor of placental glutamine in the guinea pig. *Biochem. J.* **198**, 397–401.

Boyd, C. A. R. (1982). The regulation of K$^+$ permeability of human placental trophoblast *in vitro*. *J. Physiol., Lond.* **325**, 64P.

—— (1983). Co-transport systems in the brush border membrane of the human placenta. In *Brush Border Membranes. Ciba Fdn Symp.* No. 95, (eds. R. Porter and G. M. Collins), pp. 300–314. Pitman Publishing Ltd., London.

—— Chipperfield, A. R., and Lund, E. K. (1980). NaCl co-transport by human placental plasma membrane vesicles. *J. Physiol., Lond.* **307**, 86P.

—— and Lund, E. K. (1981). L-proline transport by brush border membrane vesicles prepared from human placenta. *J. Physiol., Lond.* **315**, 9–19.

Boyd, R. D. H., Canning, J. F., Stacey, T. E., and Ward, R. H. T. (1981). Steady state ion distribution and feto-maternal ion fluxes across the sheep placenta. *Placenta Suppl.* **2**, 229–234.

—— Glazier, J. D., and Sibley, C. P. (1986). Stimulation by adrenaline of the *in vivo* materno-fetal potential difference in the pig. *J. Physiol., Lond.* **378**, 84P.

—— —— —— and Ward, B. S. (1987). Effects of adrenergic agonists and a β-antagonist on the maternofetal potential difference (p.d.) fetal heart rate and fetal blood pressure (B.P.) in the unanaesthetized pig. *J. Physiol., Lond.* **386**, 25P.

—— Haworth, C., Stacey, T. E., and Ward, R. H. T. (1976). Permeability of the sheep placenta to unmetabolized polar non-electrolytes. *J. Physiol., Lond.* **256**, 617–634.

Brommage, R. and Deluca, H. F. (1984). Placental transport of calcium and phosphorus is not regulated by vitamin D. *Am. J. Physiol.* **246**, F526–F529.

Bronner, F., Pansu, D., and Stein, W. D. (1986). An analysis of intestinal calcium transport across the rat intestine. *Am. J. Physiol.* **250**, G561–G569.

Bruns, M. E. H., Fausto, A., and Avioli, L. V. (1978). Placental calcium binding protein in rats. *J. biol. Chem.* **253**, 3186–3190.

—— Wallstein, V., and Bruns, D. E. (1982). Regulation of calcium-binding protein in mouse placenta and intestine. *Am. J. Physiol.* **242**, E47–E52.

Canning, J. F. and Boyd, R. D. H. (1984). Mineral and water exchange between mother and fetus. In *Fetal physiology and medicine.* (eds. R. W. Beard and P. W. Nathanielsz) pp. 481–509. Marcel Dekker Inc., New York.

Care, A. D. (1980). Calcium homeostasis in the fetus. *J. dev. Physiol.* **2**, 85–99.

—— Caple, I. W., Abbas, S. K., and Pickard, D. W. (1986). The effect of fetal thyroparathyroidectomy on the transport of calcium across the ovine placenta to the fetus. *Placenta* **7**, 417–424.

—— and Ross, R. (1984). Fetal calcium homeostasis. *J. dev. Physiol.* **6**, 59–66.

—— —— Pickard, D. W., Weatherley, A. J., Garel, J.-M., Manning, R. M., Angrove, J., Papapoulos, S., and O'Riordan, J. L. H. (1982). Calcium homeostasis in the fetal pig. *J. dev. Physiol.* **4**, 85–106.

—— —— —— and Robinson, J. S. (1981). The role of the kidney in ovine foetal calcium and phosphate homeostasis. *Adv. Physiol. Sci.* **20**, 45–51.

Carroll, M. J. and Young, M. (1983). The relationship between placental protein synthesis and transfer of amino acids. *Biochem. J.* **210**, 99–105.

Carson, S. A., Chase, R., Ulep, E., Scommegna, A., and Benveniste, R. (1983). Ontogenesis and characteristics of epidermal growth factor receptors in human placenta. *Am. J. Obstet. Gynec.* **147**, 932–939.

Carstensen, M., Leichtweiss, H.-P., Molsen, G., and Schröder, H. (1977). Evidence for a specific transport of D-hexoses across the human term placenta *in vitro. Arch. Gynaek.* **222**, 187–196.

—— —— and Schröder, H. (1973). Zellpotentiale in der Placenta des Menschen. *Arch. Gynaek.* **215**, 299–303.

Challier, J. C., Hauguel, S., and Desmaizieres, V. (1986). Effect of insulin on glucose uptake and metabolism in the human placenta. *J. clin. Endocr. Metab.* **62**, 803–807.

—— Nandakumaran, M., and Mondon, F. (1985). Placental transport of hexoses: a comparative study with antipyrine and amino acids. *Placenta* **6**, 497–504.

Chalon, S. and Garel, J.-M. (1985a). Plasma calcium control in the rat fetus. I. Influence of maternal hormones. *Biol. Neonate* **48**, 313–322.

—— —— (1985b). Plasma calcium control in the rat fetus. II. Influence of fetal hormones. *Biol. Neonate* **48**, 323–328.

—— —— (1985c). Plasma calcium control in the rat fetus. III. Influence of alterations in maternal plasma calcium on fetal calcium level. *Biol. Neonate* **48**, 329–335.

Chan, S. T. H. and Wong, P. Y. D. (1978). Evidence of active sodium transport in the visceral yolk sac of the rat. *J. Physiol., Lond.* **279**, 385–394.

Chinard, F. P., Danesino, V., Hartmann, W. L., Huggett, A. St. G., Pauls, W., and Reynolds, S. R. M. (1956). The transmission of hexoses across the placenta in the human and the rhesus monkey. *J. Physiol., Lond.* **132**, 289–303.

Christakos, S. and Norman, A. W. (1980). Specific receptors/binding proteins for $1,25(OH)_2$-vitamin D_3 in rat and human placenta. *Fedn Proc.* **39**, 560.

Christensen, H. N. and Streicher, J. A. (1948). Association between rapid growth and elevated cell concentrations of amino acids. *J. biol. Chem.* **175**, 95–100.

Cittadini, P., Paparella, P., Castaldo, F., Romar, R., Polsinelli, F., Carelli, G., Bompiani, A., and Terranova, T. (1977). Water and ion metabolism in placenta. *Acta obstet. gynec. scand.* **56**, 233–238.

Conrad, E. E. and Faber, J. J. (1977). Water and electrolyte acquisition across the placenta of the sheep. *Am. J. Physiol.* **233**, H475–H487.

Cooke, S. F., Cronen, D. J., and Symonds, E. M. (1981). A study of angiotensin II binding sites in human placenta, chorion and amnion. *Am. J. Obstet. Gynec.* **140**, 689–692.

Crandell, S. S., Adcock, E. W., and Morriss, F. H. (1981). Hydrolysis of dipeptides by human and ovine placentae. *Pediat. Res.* **15**, 357–361.

—— Palma, P. A., and Morriss, F. H. (1982). Effect of maternal serum insulin on umbilical extraction of glucose and lactate in fed and fasted sheep. *Am. J. Obstet. Gynec.* **142**, 219–224.

Crawford, J. D. and McCance, R. A. (1960). Sodium transport by the chorio-allantoic membrane of the pig. *J. Physiol., Lond.* **151**, 458–471.

Croley, T. E. (1973). The intracellular localization of calcium within the mature human placental barrier. *Am. J. Obstet. Gynec.* **117**, 926–932.

Cummings, J. N. and Kaiser, I. H. (1959). The blood gases, pH, and plasma electrolytes of the sow and fetal pig at 106 days of pregnancy. *Am. J. Obstet. Gynec.* **77**, 10–17.

Currie, A. J., Fraser, H. M., and Sharpe, R. M. (1981). Human placental receptors for luteinizing hormone releasing hormone. *Biochem. Biophys. Res. Comm.* **99**, 332–338.

Dancis, J., Kammerman, S., Jansen, V., and Levitz, M. (1983). The effect of ouabain on placental transport of ^{86}Rb. *Placenta* **4**, 351–360.

—— Money, W. L., Springer, D., and Levitz, M. (1968). Transport of amino acids by placenta. *Am. J. Obstet. Gynec.* **101**, 820–829.

—— and Springer, D. (1970). Fetal homeostasis in maternal malnutrition: potassium and sodium deficiency in rats. *Pediat. Res.* **4**, 345–351.

Delivoria-Papadopoulos, M., Battaglia, F. C., Bruno, P. D., and Meschia, G. (1967). Total, protein-bound, and ultrafilterable calcium in maternal and fetal plasmas. *Am. J. Physiol.* **213**, 363–366.

Delorme, A. C., Marche, P., and Garel, J.-M. (1979). Vitamin D-dependent calcium-binding protein. Changes during gestation, prenatal and postnatal development in rats. *J. dev. Physiol.* **1**, 181–194.

Domenech, M., Gruppuso, P. A., Nishino, V. T., Susa, J. B., and Schwartz, R. (1986). Preserved fetal plasma amino acid concentrations in the presence of maternal hypoaminoacidema. *Pediat. Res.* **20**, 1071–1076.

Durand, D., Barlet, J.-P., and Braithwaite, G. D. (1983). The influence of 1,25-dihydroxycholecalciferol on the mineral content of foetal guinea pigs. *Reprod. Nutr. Develop.* **23**, 235–244.

Eaton, B. M., Mann, G. E., and Yudilevich, D. L. (1982). Transport specificity for neutral and basic amino acids at maternal and fetal interfaces of the guinea-pig placenta. *J. Physiol., Lond.* **328**, 245–258.

—— and Yudilevich, D. L. (1981). Uptake and asymmetric efflux of amino acids at maternal and fetal sides of placenta. *Am. J. Physiol.* **241**, C106–C112.

Ely, P. A. (1966). The placental transfer of hexoses and polyols in the guinea-pig, as shown by umbilical perfusion of the placenta. *J. Physiol., Lond.* **184**, 255–271.

Enders, R. H., Judd, R. M., Donohue, T. M., and Smith, C. H. (1976). Placental amino acid uptake. III. Transport systems for neutral amino acids. *Am. J. Physiol.* **230**, 706–710.

Erdo, S. L., Laszlo, A., Kiss, B., and Zsolnai, B. (1985). Presence of gamma-aminobutyric acid and its specific receptor binding sites in the human term placenta. *Gynecol. obstet. Invest.* **20**, 199–203.

Faber, J. J. (1973). Diffusional exchange between foetus and mother as a function of the physical properties of the diffusing materials. In *Foetal and Neonatal Physiology* (eds. R. S. Comline, K. W. Cross, G. S. Dawes, and P. W. Nathanielsz) pp. 306–327. Cambridge University Press, Cambridge.

—— Binder, N. D., and Thornburg, K. L. (1987). Electrophysiology of extrafetal membranes. *Placenta* **8**, 89–108.

—— and Thornburg, K. L. (1981). The forces that drive inert solutes and water across the epitheliochorial placentae of the sheep and the goat and the haemochorial placentae of the rabbit and the guinea pig. *Placenta Suppl.* **2**, 203–214.

—— —— (1983). *Placental Physiology*. Raven Press, New York.

Fant, M. E. and Harbison, R. D. (1981). Syncytiotrophoblast membrane vesicles: a model for examining the human placental cholinergic system. *Teratology* **24**, 181–199.

Fantel, A. G. (1975). Fetomaternal potassium relations in the fetal rat on the twentieth day of gestation. *Pediat. Res.* **9**, 527–530.

—— (1978). Fetomaternal potassium relations in the rat on the twentieth day of gestation. II. Effects of maternal hypokalemia. *Pediat. Res.* **12**, 977–979.

Ferré, F. (1986). Dopamine-stimulated adenylate cyclase in human term placenta. *Life Sci.* **39**, 1893–1900.

Firth, J. A. (1983). Microscopical analysis of electrolyte secretion. In *Progress in Anatomy* Vol. 3 (eds. V. Navaratnam and R. J. Harrison) pp. 33–55. Cambridge University Press, Cambridge.

—— Farr, A., and Koppel, H. (1979). The localization and properties of membrane adenosine triphosphatases in the guinea-pig placenta. *Histochemistry* **61**, 157–165.

—— Sibley, C. P., and Ward, B. S. (1986). Histochemical localization of phosphatases in the pig placenta: II potassium-dependent and potassium-independent p-nitrophenyl phosphatases at high pH; relation to sodium-potassium-dependent adenosine triophosphatase. *Placenta* **7**, 27–35.

Fisher, G. J., Kelley, L. K., and Smith, C. H. (1987). ATP-dependent calcium transport across basal plasma membranes of human placental trophoblast. *Am. J. Physiol.* **252**, C38–C46.

Fletcher, J. M. and Bassett, J. M. (1986). Increased placental growth and raised plasma glucocorticoid concentrations in fetal rabbits injected with insulin *in utero*. *Horm. metabol. Res.* **18**, 441–445.

Flexner, L. B., Cowie, D. B., Hellman, M. D., Wilde, W. S., and Vosburgh, G. J. (1948). The permeability of the human placenta to sodium in normal and abnormal pregnancies and the supply of sodium to the human fetus as determined with radioactive sodium. *Am. J. Obstet. Gynec.* **55**, 469–480.

——and Gellhorn, A. (1942). The comparative physiology of placental transfer. *Am. J. Obstet. Gynec.* **43**, 965–974.

Fowler, S. A., Williams, M. E., and Gray, T. K. (1978). Calcium and 25-hydroxyvitamin D_3 binding proteins in human placenta. *Biol. Neonate* **33**, 8–12.

Fox, I. H. and Kurpis, L. (1983). Binding characteristics of an adenosine receptor in human placenta. *J. biol. Chem.* **258**, 6952–6955.

France, V. M. (1976). Active sodium uptake by the skin of foetal sheep and pigs. *J. Physiol., Lond.* **258**, 377–392.

Ganapathy, M. E., Leibach, F. H., Mahesh, V. B., Howard, J. C., Devoe, L. D., and Ganapathy, V. (1986). Characterization of tryptophan transport in human placental brush-border membrane vesicles. *Biochem. J.* **238**, 201–208.

——Mahesh, V. B., Devoe, L. D., Leibach, F. H., and Ganapathy, V. (1985). Dipeptide transport in brush-border membrane vesicles isolated from normal term human placenta. *Am. J. Obstet. Gynec.* **153**, 83–86.

Garel, J.-M., Delorme, A. C., Marche, P., Nguyen, T. M., and Garabedian, M. (1981). Vitamin D_3 metabolite injections to thyroparathyroidectomized pregnant rats: effects on calcium-binding proteins of maternal duodenum and of fetoplacental unit. *Endocrinology* **109**, 284–289.

Geck, P. and Heinz, E. (1986). The Na-K-2Cl co-transport system. *J. Membrane Biol.* **91**, 97–106.

Ghijsen, W. E. J. M., van Os, C. H., Heizmann, C. W., and Murer, H. (1986). Regulation of duodenal Ca^{2+} pump by calmodulin and vitamin D-dependent Ca^{2+}-binding protein. *Am. J. Physiol.* **251**, G223–G229.

Gibson, J. S. and Ellory, J. C. (1984). Effect of amphotericin B and gestational age on sodium transport across the rat visceral yolk sac placenta *in vitro*. *J. Reprod. Fert.* **72**, 529–535.

Giese, W. and Comar, C. L. (1964). Existence of non-exchangeable calcium compartments in plasma. *Nature, Lond.* **202**, 31–33.

Gitlin, D. and Boesman, M. (1966). Serum α-fetoprotein, albumin and γ-globulin in the human conceptus. *J. clin. Invest.* **45**, 1826–1838.

Gresham, E. L., Rankin, J. H. G., Makowski, E. L., Meschia, G., and Battaglia, F. C. (1972). An evaluation of fetal renal function in a chronic sheep preparation. *J. clin. Invest.* **51**, 149–156.

Guidotti, G. G., Borghetti, A. F., and Gazzola, G. C. (1978). The regulation of amino acid transport in animal cells. *Biochim. Biophys. Acta* **515**, 329–366.

Gusseck, D. J., Yuen, P., and Longo, L. D. (1975). Amino acid transport in placental slices: mechanism of increased accumulation by prolonged incubation. *Biochim. Biophys. Acta.* **401**, 278–284.

Hauguel, S., Desmaizieres, V., and Challier, J. C. (1986). Glucose uptake, utilization, and transfer by the human placenta as functions of maternal glucose concentration. *Pediat. Res.* **20**, 269–273.

Hay, W. W., Jr., Meznarich, H. K., Sparks, J. W., Battaglia, F. C., and Meschia, G. (1985). Effect of insulin on glucose uptake in near-term fetal lambs. *Proc. Soc. exp. Biol. Med.* **178**, 557–564.

—— Sparks, J. W., Battaglia, F. C., and Meschia, G. (1984). Maternal-fetal glucose exchange: necessity of a three-pool model. *Am. J. Physiol.* **246**, E528–E534.

—— —— Gilbert, M., Battaglia, F. C., and Meschia, G. (1984). Effect of insulin on glucose uptake by the maternal hindlimb and uterus, and by the fetus in conscious pregnant sheep. *J. Endocr.* **100**, 119–124.

—— —— Wilkening, R. B., Battaglia, F. C., and Meschia, G. (1984). Fetal glucose uptake and utilization as functions of maternal glucose concentration. *Am. J. Physiol.* **246**, E237–E242.

Hayashi, S., Sanada, K., Sagawal, N., Yamada, N., and Kido, K. (1978). Umbilical vein-artery differences of plasma amino acids in fetal tissues. *Biol. Neonate* **34**, 11–18.

Hedley, R. and Bradbury, M. W. B. (1980). Transport of polar non-electrolytes across the intact and perfused guinea-pig placenta. *Placenta* **1**, 277–285.

Hill, D. J. and Milner, R. D. G. (1985). Insulin as a growth factor. *Pediat. Res.* **19**, 879–886.

Hill, P. M. M. and Young, M. (1973). Net placental transfer of free amino acids against varying concentrations. *J. Physiol., Lond.* **235**, 409–422.

Holzman, I. R., Lemons, J. A., Meschia, G., and Battaglia, F. C. (1979). Uterine uptake of amino acids and placental glutamine-glutamate balance in the pregnant ewe. *J. dev. Physiol.* **1**, 137–149.

Huggett, A. St. G., Warren, F. L., and Warren, N. V. (1951). The origin of the blood fructose of the foetal sheep. *J. Physiol., Lond.* **113**, 258–275.

Ingermann, R. L. (1988). Control of placental glucose transfer. *Placenta.* **8**, 557–571.

—— Bissonnette, J. M., and Koch, P. L. (1983). D-glucose-sensitive and -insensitive cytochalasin B binding proteins from microvillous plasma membranes of human placenta. *Biochim. Biophys. Acta* **730**, 57–63.

—— Stankova, L., and Bigley, R. H. (1986). Role of monosaccharide transporter in vitamin C uptake by placental membrane vesicles. *Am. J. Physiol.* **250**, C637–C641.

Jarmer, S., Shoaf, A. R., and Harbison, R. D. (1985). Comparative enzymatic acetylation of carnitine and choline by human placental syncytiotrophoblast membrane vesicles. *Terat. Carcin. Mut.* **5**, 445–461.

Jodarski, G. D., Shanahan, M. F., and Rankin, J. H. G. (1985). Fetal insulin and placental 3-O-methyl glucose clearance in the near-term sheep. *J. dev. Physiol.* **7**, 251–258.

Johnson, L. W. and Smith, C. H. (1980). Monosaccharide transport across microvillous membrane of human placenta. *Am. J. Physiol.* **238**, C160–C168.

—— —— (1982). Identification of the glucose transport protein of the microvillous membrane of human placenta by photoaffinity labelling. *Biochem. Biophys. Res. Commun.* **109**, 408–413.

—— —— (1985). Glucose transport across the basal plasma membrane of human placental syncytiotrophoblast. *Biochim. Biophys. Acta* **815**, 44–50.

Jones, C. T. and Rolph, T. P. (1985). Metabolism during fetal life: a functional assessment of metabolic development. *Physiol. Rev.* **65**, 357–430.

Kalham, S. C., D'Angelo, L. J., Savin, S. M., and Adam, P. A. J. (1979). Glucose production in pregnant women at term gestation. *J. clin. Invest.* **63**, 388–394.

Kaufmann, P., Schröder, H., and Leichtweiss, H.-P. (1982). Fluid shift across the placenta: II. Fetomaternal transfer of horseradish peroxidase in the guinea-pig. *Placenta* **3**, 339–348.

Keeley, V. L. (1984). *Physiology of the Fetal Lamb Allantois.* PhD thesis, University of Cambridge.

Kelley, L. K., Smith, C. H., and King, B. F. (1983). Isolation and partial characterization of the basal cell membrane of human placental trophoblast. *Biochim. Biophys. Acta.* **734**, 91–98.

Leaf, A. (1981). Transport properties of cell membranes. *Placenta Suppl.* **2**, 79–88.

Leichtweiss, H.-P., Lisboa, B., and Steinborn, C. (1986). Transport of ^{14}C-dehydro-ascorbic acid in the isolated guinea-pig placenta. *Placenta* **7**, 454.

—— and Schröder, H. (1981). Dual perfusion of the isolated guinea-pig placenta. *Placenta Suppl.* **2**, 119–128.

Lemons, J. A., Adcock, E. W., Jones, M. D., Naughton, M. A., Meschia, G., and Battaglia, F. C. (1976). Umbilical uptake of amino acids in the unstressed fetal lamb. *J. clin. Invest.* **58**, 1428–1434.

Leturque, A., Hauguel, S., Kande, J., and Girard, J. (1987). Glucose utilisation by the placenta of anaesthetised rats: effect of insulin, glucose and ketone bodies. *Pediat. Res.* **22**, 483–487.

Litonjua, A. D., Canlas, M., Soliman, J., and Paulino, D. Q. (1967). Uptake of α-aminoisobutyric acid in placental slices at term. *Am. J. Obstet. Gynec.* **99**, 242–246.

Longo, L. D., Yuen, P., and Gusseck, D. J. (1973). Anaerobic, glycogen-dependent transport of amino acids by the placenta. *Nature, Lond.* **243**, 531–533.

McIntosh, N., Rodeck, C. H., and Heath, R. (1984). Plasma amino acids of the mid-trimester human fetus. *Biol. Neonate* **45**, 218–224.

McKercher, H. G., Derewlany, L. O., and Radde, I. C. (1983). Placental calcium and phospherus transfer in the guinea pig: lack of effect of modulators of Ca-ATPase and alkaline phosphatase activity. *Can. J. Physiol. Pharmacol.* **61**, 1354–1360.

—————— (1984). Effects of pharmacological and physiological modulators on Ca-ATPase and alkaline phosphatase activities from guinea-pig placenta *in vitro*. *Placenta* **5**, 281–292.

MacDonald, A. A., Colenbrander, B., Versteeg, D. H. G., Heilhecker, A., and Wensing, C. J. G. (1984). Catecholamines in fetal pig plasma and the response to acute hypoxia and chronic fetal decapitation. *Roux Arch. dev. Biol.* **193**, 19–23.

Marche, P., Delorme, A., and Cuisinier-Gleizes, P. (1978). Intestinal and placental calcium-binding proteins in vitamin D-deprived or -supplemented rats. *Life Sci.* **23**, 2555–2562.

Marshall, R. N., Underwood, L. E., Voina, S. J., Foushee, D. B., and van Wyk, J. J. (1974). Characterization of the insulin and somatomedin-C receptors in human placental cell membranes. *J. clin. Endocr. Metab.* **39**, 283–292.

Mellor, D. J. (1969). Potential differences between mother and foetus at different gestational ages in the rat, rabbit and guinea-pig. *J. Physiol., Lond.* **204**, 395–405.

— (1970). Distribution of ions and electrical potential differences between mother and foetus at different gestational ages in goats and sheep. *J. Physiol., Lond.* **207**, 133–150.

—— Cockburn, F., Lees, M. M., and Blagden, A. (1969). Distribution of ions and electrical potential differences between mother and fetus in the human at term. *J. Obstet. Gynaec. Br. Commonw.* **76**, 993–998.

Meschia, G., Battaglia, F. C., and Bruns, P. D. (1967). Theoretical and experimental study of transplacental diffusion. *J. appl. Physiol.* **22**, 1171–1178.

—— Wolkoff, A. S., and Barron, D. H. (1958). Difference in electric potential across the placenta of goats. *Proc. natn. Acad. Sci. U.S.A.* **44**, 483–485.

Miller, R. K. and Berndt, W. O. (1974). Characterization of neutral amino acid accumulation by human term placental slices. *Am. J. Physiol.* **227**, 1236–1242.

——Kozelka, T. R., and Brent, R. L. (1976). The transport of molecules across placental membranes. In *The Cell Surface in Animal Embryogenesis and Development* (eds. G. Porte and G. L. Nicholson) pp. 145–223. North Holland Publishing Co., Amsterdam.

Moore, J. J., Moore, R., and Cardaman, R. C. (1986). Calcium-phospholipid enhanced protein phosphorylation in human placenta. *Proc. Soc. exp. Biol. Med.* **182**, 364–371.

Morriss, F. H., Adcock, E. W., Paxson, C. L., and Greeley, W. J. (1979). Uterine uptake of amino acids throughout gestation in the unstressed ewe. *Am. J. Obstet. Gynec.* **135**, 601–608.

——and Boyd, R. D. H. (1987). Placental transport. In *The Physiology of Reproduction* (eds. E. Knobil and J. D. Neill) pp. 2043–2083. Raven Press, New York.

Morse, A. (1917). The amino-acid nitrogen of the blood in cases of normal and complicated pregnancy and also in the new-born infant. *Bull. John Hopkins Hospital* **28**, 199–204.

Mossman, H. W. (1987). *Vertebrate Fetal Membranes*. MacMillan Press, London.

Mughal, M. Z., Robinson, N. R., and Sibley, C. P. (1987). The role of maternal vitamin D status in maternofetal calcium exchange across the rat placenta. *J. Physiol., Lond.* **386**, 22P.

Nicholls, D. G. (1986). Intracellular calcium homeostasis. *Br. med. Bull.* **42**, 353–358.

Northrop, G., Misenhimer, H. R., and Becker, F. O. (1977). Failure of parathyroid hormone to cross the nonhuman primate placenta. *Am. J. Obstet. Gynec.* **129**, 449–453.

Olver, R. E. and Strang, L. B. (1974). Ion fluxes across the pulmonary epithelium and the secretion of lung liquid in the foetal lamb. *J. Physiol., Lond.* **241**, 327–357.

Owens, P. C., Brinsmead, M. W., Waters, M. J., and Thorburn, G. D. (1980). Ontogenic changes in multiplication-stimulating activity binding to tissues and serum somatomedin-like receptor activity in the ovine fetus. *Biochem. Biophys. Res. Commun.* **96**, 1812–1820.

Paske, W. P. and Sejrsen, P. (1977). Transcapillary exchange of ^{14}C-inulin by free diffusion in channels of fused vesicles. *Acta physiol. scand.* **100**, 437–445.

Pappenheimer, J. R. (1953). Passage of molecules through capillary walls. *Physiol. Rev.* **33**, 387–423.

Pappone, P. A. and Cahalan, M. D. (1986). Ion permeation in cell membranes. In *Physiology of membrane disorders* 2nd Edn. (eds. T. E. Andreoli, J. F. Hoffman, D. D. Fanestil, and S. G. Schultz) pp. 249–272. Plenum, New York.

Paul, S. and Jailkhani, B. L. (1982). Unmasking by neuraminidase of specific chorionic gonadotrophin binding activity of human placental syncytiotrophoblast. *J. Reprod. Fert.* **66**, 445–450.

Paxson, C. L., Morriss, F. H., and Adcock, E. W. (1978). Effect of uterine artery insulin infusions on umbilical glucose uptake in sheep. *Pediat. Res.* **12**, 864–867.

Pearse, W. H. and Sornson, H. (1969). Free amino acids of normal and abnormal human placenta. *Am. J. Obstet. Gynec.* **105**, 696–701.

Philipps, A. F., Porte, P. J., Stabinsky, S., Rosenkrantz, T. S., and Raye, J. R. (1984). Effects of chronic fetal hyperglycaemia upon oxygen consumption in the ovine uterus and conceptus. *J. clin. Invest.* **74**, 279–286.

Pike, J. W., Gooze, L. L., and Haussler, M. R. (1980). Biochemical evidence for 1,25-dihydroxyvitamin D receptor macromolecules in parathyroid, pancreatic, pituitary, and placental tissues. *Life Sci.* **26**, 407–414.

Posner, B. I. (1974). Insulin receptors in human and animal placental tissue. *Diabetes* **23**, 209–217.

Potau, N., Ruidar, E., and Ballabriga, A. (1981). Insulin receptors in human placenta in relation to fetal weight and gestational age. *Pediat. Res.* **15**, 798–802.

Rankin, J. H. G., Jodarski, G., and Shanahan, M. R. (1986). Maternal insulin and placental 3-O-methyl glucose transport. *J. dev. Physiol.* **8**, 247–253.

——and McLaughlin, M. K. (1979). The regulation of the placental blood flow. *J. dev. Physiol.* **1**, 3–30.

Rao, C. V., Ramani, N., Chegini, N., Stadig, B. K., Carman, F. R., Woost, P. G., Schultz, G. S., and Cook, C. L. (1985). Topography of human placental receptors for epidermal growth factor. *J. biol. Chem.* **260**, 1705–1710.

Reynolds, M. L. and Young, M. (1971). The transfer of free α-amino nitrogen across the placental membrane in the guinea-pig. *J. Physiol., Lond.* **214**, 583–597.

Rice, P. A., Rourke, J. E., and Nesbitt, R. E. L. (1976). *In vitro* perfusion studies of the human placenta. IV. Some characteristics of the glucose transport system in the human placenta. *Gynec. Invest.* **7**, 213–221.

——————(1979). *In vitro* perfusion studies of the human placenta. *Am. J. Obstet. Gynec.* **133**, 649–655.

Riggs, D. S. (1972). *The Mathematical Approach to Physiological Problems.* MIT Press, Cambridge p. 183.

Rowell, P. P. and Sastry, B. V. R. (1981). Human placental cholinergic system: depression of the uptake of α-aminoisobutyric acid in isolated human placental villi by choline acetyltransferase inhibitors. *J. Pharmac. exp. Ther.* **216**, 232–238.

Ruzycki, S. M., Kelley, L. K., and Smith, C. H. (1978). Placental amino acid uptake. IV. Transport by microvillous membrane vesicles. *Am. J. Physiol.* **234**, C27–C35.

Sastry, B. V. R. and Sadavongvivad, C. (1979). Cholinergic systems in non-nervous tissues. *Pharmac. Rev.* **30**, 65–120.

Schauberger, C. W. and Pitkin, R. M. (1979). Maternal-perinatal calcium relationships. *Obstet. Gynecol.* **53**, 74–76.

Schneider, H. and Dancis, J. (1974). Amino acid transport in human placental slices. *Am. J. Obstet. Gynec.* **120**, 1092–1098.

——Möhlen, K.-H., Challier, J.-C., and Dancis, J. (1979). Transfer of glutamic acid across the human placenta perfused *in vitro*. *Br. J. Obstet. Gynaec.* **86**, 299–306.

———— and Dancis, J. (1979). Transfer of amino acids across the *in vitro* perfused human placenta. *Pediat. Res.* **13**, 236–240.

Schröder, H., Leichtweiss, H.-P., and Madee, W. (1975). The transport of D-glucose, L-glucose and D-mannose across the isolated guinea pig placenta. *Pflügers Arch.* **356**, 267–275.

Schultz, S. G. (1986). Cellular models of epithelial ion transport. In *Physiology of membrane disorders* 2nd Edn. (eds. T. E. Andreoli, J. F. Hoffman, D. D. Fanestil, and S. G. Schultz) pp. 519–534. Plenum, New York and London.

Sen, I. (1986). Identification and solubilization of atrial natriuretic factor receptors in human placenta. *Biochem. Biophys. Res. Commun.* **135**, 480–486.

Serrano, C. V., Talbert, L. M., and Welt, L. G. (1964). Potassium deficiency in the pregnant dog. *J. clin. Invest.* **48**, 27–31.

Shennan, D. B. and Boyd, C. A. R. (1987). Ion transport by the placenta: a review of membrane transport systems. *Biochim. Biophys. Acta.* **917**, 1–30.

Sibley, C. P., Ward, B. S., Glazier, J. D., Moore, W. M. O., and Boyd, R. D. H. (1986). Electrical activity and sodium transfer across *in vitro* pig placenta. *Am. J. Physiol.* **250**, R474–R484.

Simmons, M. A., Battaglia, F. C., and Meschia, G. (1979). Placental transfer of glucose. *J. dev. Physiol.* **1**, 227–243.

Smith, C. H. (1981). Incubation techniques and investigation of placental transport mechanisms *in vitro*. *Placenta Suppl.* **2**, 163–176.

—— Adcock, E. W., Teasdale, F., Meschia, G., and Battaglia, F. C. (1973). Placental amino acid uptake: tissue preparation, kinetics, and pre-incubation effect. *Am. J. Physiol.* **224**, 558–564.

—— and Depper, R. (1974). Placental amino acid uptake. II. Tissue pre-incubation, fluid distribution, and mechanisms of regulation. *Pediat. Res.* **8**, 697–703.

Soltesz, G., Harris, D., Mackenzie, I. Z., and Aynsley-Green, A. (1985). The metabolic and endocrine milieu of the human fetus and mother at 18–21 weeks of gestation. I. Plasma amino acid concentrations. *Pediat. Res.* **19**, 91–93.

Speeg, K. V. and Harrison, R. W. (1979). The ontogony of the human placental glucocorticoid receptor and inducibility of heat-stable alkaline phosphatase. *Endocrinology* **104**, 1364–1368.

Stacey, T. E., Weedon, A. P., Haworth, C., Ward, R. H. T., and Boyd, R. D. H. (1978). Fetomaternal transfer of glucose analogues by sheep placenta. *Am. J. Physiol.* **3**, E32–E37.

Steel, R. B., Mosley, J. D., and Smith, C. (1979). Insulin and placenta: degradation and stabilization, binding to microvillous membrane receptors, and amino acid uptake. *Am. J. Obstet. Gynec.* **135**, 522–529.

—— Smith, C. H., and Kelley, L. K. (1982). Placental amino acid uptake. VI. Regulation by intracellular substrate. *Am. J. Physiol.* **243**, C46–C51.

Stegink, L. D., Pitkin, R. M., Reynolds, W. A., Filo, L. J., Boaz, D. P., and Brummel, M. G. (1975). Placental transfer of glutamate and its metabolites in the primate. *Am. J. Obstet. Gynec.* **122**, 70–78.

—— —— —— Brummel, M. C., and Filer, L. J. (1979). Placental transfer of aspartate and its metabolites in the primate. *Metabolism* **28**, 669–676.

Stewart, E. L. and Welt, L. G. (1961). Protection of the fetus in experimental potassium depletion. *Am. J. Physiol.* **200**, 824–826.

Štulc, J. (1988). Is there control of solute transport at placental level? *Placenta.* **9**, 19–26.

—— Friedrich, R., and Jiřička, Z. (1969). Estimation of the equivalent pore dimensions in the placenta of the rabbit. *Life Sci.* **8**, 167–180.

—— Rietveld, W. J., Soeteman, D. W., and Versprille, A. (1972). The transplacental potential differences in guinea-pigs. *Biol. Neonate* **21**, 130–147.

—— and Štulcová, B. (1985). Extracellular binding of calcium in the guinea-pig placenta. *Placenta* **6**, 259–264.

—— —— (1986). Transport of calcium by the placenta of the rat. *J. Physiol., Lond.* **371**, 1–16.

—— and Švihovec, J. (1973). Effects of potassium cyanide, strophantin or sodium-free perfusion fluid on the electrical potential difference across the guinea-pig placenta perfused *in situ*. *J. Physiol., Lond.* **231**, 403–415.

—— —— (1977). Placental transport of sodium in the guinea-pig. *J. Physiol., Lond.* **265**, 691–703.

—— —— Drábková, J., Střibrný, J., Kobilková, J., Vido, I., and Doležal, A. (1978). Electrical potential difference across the mid-term human placenta. *Acta obstet. gynec. scand.* **57**, 125–126.

Sweiry, J. H., Page, K. R., Dacke, C. G., Abramovich, D. R., and Yudilevich, D. L. (1986). Evidence of saturable uptake mechanisms at maternal and fetal sides of the perfused human placenta by rapid paired-tracer dilution: studies with calcium and choline. *J. dev. Physiol.* **8**, 435–445.

—— and Yudilevich, D. L. (1984). Asymmetric calcium influx and efflux at maternal and fetal sides of the guinea-pig placenta: kinetics and specificity. *J. Physiol., Lond.* **355**, 295–311.

—— —— (1985). Characterization of choline transport at maternal and fetal interfaces of the perfused guinea-pig placenta. *J. Physiol., Lond.* **366**, 251–266.

Sybulski, S., and Tremblay, P. C. (1967). Uptake and incorporation into protein of radioactive glycine by human placentae *in vitro. Am. J. Obstet. Gynec.* **97**, 1111–1118.

Takano, K., Hall, K., Fryklund, L., Hobmgren, A., Sievertsson, H., and Uthne, K. (1975). The binding of insulin and somatomedin-A to human placental membrane. *Acta endocr. Copenh.* **80**, 14–31.

Thornburg, K. L., Binder, W. D., and Faber, J. J. (1979*a*). Diffusion permeability and ultrafiltration-reflection-coefficients of Na⁺ and Cl⁻ in the near-term placenta of the sheep. *J. dev. Physiol.* **1**, 47–60.

—— —— —— (1979*b*). Distribution of ionic sulphate, lithium and bromide across the sheep placenta. *Am. J. Physiol.* **236**, C58–C65.

—— and Faber, J. J. (1977). Transfer of hydrophilic molecules by placenta and yolk sac of the guinea pig. *Am. J. Physiol.* **233**, C111–C124.

Treinen, K. A. and Kulkarni, A. P. (1986). High-affinity, calcium-stimulated ATPase in brush border membranes of the human term placenta. *Placenta* **7**, 365–373.

Tsalikian, E., Foley, T. P., and Becher, D. J. (1984). Characterization of somatostatin specific binding in plasma cell membranes of human placenta. *Pediat. Res.* **18**, 953–957.

Tuan, R. S. (1982). Identification and characterization of a calcium-binding protein from human placenta. *Placenta* **3**, 145–158.

—— (1985). Ca²⁺-binding protein of the human placenta. *Biochem. J.* **227**, 317–326.

—— and Cavanaugh, S. T. (1985). Identification and characterization of a calcium-binding protein in the mouse chorioallantoic placenta. *Biochem. J.* **233**, 41–49.

Twardock, A. R. (1967). Placental transfer of calcium and strontium in the guinea pig. *Am. J. Physiol.* **213**, 837–842.

—— and Austin, M. K. (1970). Calcium transfer in perfused guinea pig placenta. *Am. J. Physiol.* **219**, 540–545.

—— Kuo, E. Y. H., Austin, M. K., and Hopkins, J. R. (1971). Protein binding of calcium and strontium in guinea pig maternal and fetal blood plasma. *Am. J. Obstet. Gynec.* **110**, 1008–1014.

Ussing, H. H. and Zerahn, K. (1951). Active transport of sodium as the source of electric current in the short-circuited isolated frog skin. *Acta physiol. scand.* **23**, 110–127.

Valette, A., Reme, J. M., Pontannier, G., and Cros, J. (1980). Specific binding for opiate-like drugs in the placenta. *Biochem, Pharmac.* **29**, 2657–2661.

van Dijk, H. P. (1981). Active transfer of the plasma bound compounds calcium and iron across the placenta. *Placenta Suppl.* **1**, 139–164.

—— (1988). Regulatory aspects of placental iron transfer: a comparative study. *Placenta.* In press.

—— and van Kreel, B. K. (1982). Electric potential difference in the isolated guinea-pig placenta. *J. dev. Physiol.* **4**, 23–38.

van Kreel, B. K. and van Dijk, J. P. (1983). Mechanisms involved in the transfer of calcium across the isolated guinea pig placenta. *J. dev. Physiol.* **5**, 155–165.

Vonderhaar, B. K., Bhattacharya, A., Alhadi, T., Liscia, D. S., Andrew, E. M., Young, J. K., Ginsburg, E., Bhattacharjee, M., and Horn, T. M. (1985). Isolation, characterization and regulation of the prolactin receptor. *J. Dairy Sci.* **68**, 466–488.

Warembourg, M., Pernet, C., and Thomasset, M. (1986). Distribution of vitamin

D-dependent calcium-binding protein messenger ribonucleic acid in rat placenta and duodenum. *Endocrinology* 119, 176–184.

Weatherley, A. J., Ross, R., Pickard, D. W., and Care, A. D. (1983). The transfer of calcium during perfusion of the placenta in intact and thyroparathyroidectomised sheep. *Placenta* 4, 271–278.

Weedon, A. P., Stacey, T. E., Canning, J. F., Ward, R. H. T., and Boyd, R. D. H. (1980). Maternofetal electrical potential difference in conscious sheep: effect of fetal death or acidosis. *Am. J. Obstet. Gynec.* 138, 422–428.

—— —— Ward, R. H. T., and Boyd, R. D. H. (1978). Bidirectional sodium fluxes across the placenta of conscious sheep. *Am. J. Physiol.* 235, F536–F541.

Welsch, F., Wenger, W. C., and Stedman, D. B. (1981). Acetylcholine in human term placenta: tissue levels in intact fragments after inhibition *in vitro* of choline acetyltransferase and relationship to [^{14}C] α-aminoisobutyric acid uptake. *Placenta Suppl.* 3, 339–351.

Whitsett, J. A., Johnson, C. L., Noguchi, A., Darovec-Beckerman, C., and Costello, M. (1980). Adrenergic receptors and catecholamine-sensitive adenylate cyclase of the human placenta. *J. clin. Endocr. Metab.* 50, 27–32.

—— and Lessard, J. L. (1978). Characteristics of the microvillus brush border of human placenta: insulin receptor localization in brush border membranes. *Endocrinology* 103, 1458–1468.

—— and Tsang, R. C. (1980). Calcium uptake and binding by membrane fractions of human placenta: ATP-dependent calcium accumulation. *Pediat. Res.* 14, 769–775.

Widdas, W. F. (1952). Inability of diffusion to account for placental glucose transfer in the sheep and consideration of the kinetics of a possible carrier transfer. *J. Physiol., Lond.* 118, 23–29.

—— (1961). Transport mechanisms in the foetus. *Br. med. Bull.* 17, 107–111.

Wiester, M. J., Whitla, S. H., and Goldsmith, R. (1963). Evidence for multiple calcium pools in the blood of dogs. *Nature, Lond.* 197, 1170–1171.

Willis, D. M., O'Grady, J. P., Faber, J. J., and Thornburg, K. L. (1986). Diffusion permeability of cyanocobalamin in human placenta. *Am. J. Physiol.* 250, R459–R464.

Woods, L. L., Thornburg, K. L., and Faber, J. J. (1978). Transplacental gradients in the guinea-pig. *Am. J. Physiol.* 235, H200–H207.

Yano, J., Okada, Y., Tsuchiya, W., Kinoshita, M., and Tominaga, T. (1982). Dependence of membrane potential on Ca^{2+} transport in cultured cytotrophoblasts of human immature placentae. *Biochim. Biophys. Acta* 685, 162–168.

—— —— —— —— —— and Nishimura, T. (1981). Oscillation of membrane potential in syncytiotrophoblast of human term placenta in culture. *Acta obst. gynaec. jpn.* 33, 137–141.

Younes, M. A., Besch, N. F., and Besch, P. K. (1981). Estradiol and progesterone binding in human term placental cytosol. *Am. J. Obstet. Gynec.* 141, 170–174.

—— —— —— (1982). Evidence for an androgen binding component in human placental cytosol. *J. steroid Biochem.* 16, 311–315.

Young, M. (1981). Placental amino acid transfer and metabolism. *Placenta Suppl.* 2, 177–184.

—— and McFadyen, I. R. (1973). Placental transfer and fetal uptake of amino acids in the pregnant ewe. *J. Perinat. Med.* 1, 174–182.

—— and Prenton, M. A. (1969). Maternal and fetal plasma amino acid concentrations during gestation and in retarded fetal growth. *J. Obstet. Gynaec. Br. Commonw.* 76, 333–344.

—— Stern, M. D. R., Horn, J., and Noakes, D. E. (1982). Protein synthetic rate in the sheep placenta *in vivo*: the influence of insulin. *Placenta* 3, 159–164.

—— and Widdowson, E. M. (1975) The influence of diets deficient in energy, or in protein, on conceptus weight, and the placental transfer of a non-metabolisable amino acid in the guinea pig. *Biol. Neonate* **27**, 184–191.

Yudilevich, D. L. and Eaton, B. M. (1980). Amino acid carriers at maternal and fetal surfaces of the placenta by single circulation paired tracer dilution. Kinetics of phenylalanine transport. *Biochim. Biophys. Acta* **596**, 315–319.

—— —— Short, A. H., and Leichtweiss, H.-P. (1979). Glucose carriers at maternal and fetal sides of the trophoblast in guinea pig placenta. *Am. J. Physiol.* **237**, C205–C212.

—— and Sweiry, J. H. (1985). Transport of amino acids in the placenta. *Biochim. Biophys. Acta* **822**, 169–201.

10 Control of myometrial contractility: Role and regulation of gap junctions

R. E. GARFIELD, M. G. BLENNERHASSETT AND S. M. MILLER

I Introduction

II Basis of contractility of the uterus
   1 Structure of the uterus and myometrium
      i Anatomy of the uterus
     ii Cell-to-cell contacts between myometrial cells
    iii Innervation of the myometrium
   2 Electrophysiology
      i Resting membrane potential
     ii Spontaneous electrical activity
    iii Pacemaker activity
    iv Longitudinal and circular muscle layer interaction
     v Transmission of electrical activity
- Conduction velocity
- Conduction distance
- Cell-to-cell coupling during pregnancy
- Structural basis of cell-to-cell coupling
    vi Electromechanical coupling

III Modulation of uterine contactility
   1 Hormonal control
      i Steroid hormones
     ii Prostaglandins
    iii Oxytocin
    iv Relaxin
   2 Neurogenic control
   3 Myogenic control

IV Gap junctions as sites for electrical and metabolic coupling
   1 Description and definition
      i Spectroscopic analysis
     ii Biochemistry
   2 Junctional protein of smooth muscle
   3 Functional studies
      i Molecular permeability

I Introduction

The mechanisms which regulate the transformation of uterine motility from inactive to active at the end of gestation are complex. Normally, at the end of pregnancy in all species, the uterus becomes increasingly active and reactive to contractile agents eventually to reach a state in which it contracts frequently and forcefully (labour). These contractions aid in dilating the cervix and succeed in evacuating the uterine contents. Uterine muscle, compared to other smooth muscles, is unique in that it displays this highly active and synchronous contractility only during parturition. Indeed, the coordination of contractions occurring in different regions of the uterine wall are only associated with and are believed to be necessary for normal labour and delivery (Reynolds 1949; Fuchs 1969, 1978; Liggins 1979; Thorburn and Challis 1979; Csapo 1981). Furthermore, the absence of this activity and the prevailing state of relative inactivity, quiesence and asynchrony are believed to be indispensable for implantation of the fertilized ovum and appropriate nourishment of the developing fetus during pregnancy (Reynolds 1949; Liggins 1979; Csapo 1981).

It has long been recognized that, because the myometrium is composed of billions of small muscle cells, some specialized mechanism of conduction must be present between the cells to coordinate their activity (see review by Csapo 1981). Bozler (1938) concluded that: "uncoordinated activity of the small muscle cells could never produce the regular movements which are observed; rhythmic contractions can only be understood by postulating some mechanism of conduction which coordinates the activity of the numerous elements . . .". The innervation of the uterus was not considered to be of importance in regulating myometrial activity associated with the birth process by Reynolds (1949), because procedures which interfere with nerve impulses fail to change the outcome. Furthermore, there is no recognized specialized conduction pathway comparable to the Purkinje fibre system of the heart present in the myometrium. Additionally, the effects of stimulants on contractility cannot account for synchronous patterns of contractility during labour. For these reasons, it is though that electrical activity must propagate from cell-to-cell to coordinate mechanical events among the individual myometrial muscle cells (Abe and Tomita 1968; Abe 1970; Finn and Porter 1975; Csapo 1981). Thus, it is evident that the manner by which this activity propagates between the smooth mucle cells and the factors which regulate this process during gestation and labour are of singular importance to an understanding of the mechanisms which maintain pregnancy and initiate parturition.

The recent observations that gap junctions (see **Section IV**) are only present in large numbers and sizes between uterine smooth muscle cells at parturition is thought to be significant with regard to propagation and the development of coordinated contractility during labour (Garfield *et al.* 1977, 1978). These specialized sites of cell-to-cell contact are believed to provide pathways (that is, cell-to-cell channels) for direct intercellular communication and to permit the synchronization of electrical and metabolic activities in other multicellular tissues (Peracchia 1980; Loewenstein 1981; Hooper and Subak-Sharpe 1981; Spray *et al.* 1984). Thus, the observation that gap junctions are only present during labour was, we believe, a major breakthrough in understanding uterine contractility as it explains how the myometrium behaves as a functional syncytium during labour. More recent studies show that the steroid hormones and prostaglandins, long recognized as controlling labour, regulate the development of the junctions (Garfield *et al.* 1980*a,b*; 1982; Puri and Garfield 1985, 1982). Moreover, functional studies indicate that the myometrial cells, like other cells, are better coupled electrically and metabolically when the junctions are present (Sims *et al.* 1982; Cole *et al.* 1985; Cole and Garfield 1986).

In this review, we will first describe the anatomy and electrophysiology of the myometrium because these are fundamental to understanding uterine contractility. We will then discuss studies of the presence, the function and the control of gap junctions in other tissues and the myometrium. Studies of these elements suggest that the presence and permeability of the myometrium gap junctions may be regulated physiologically and that either pathological conditions or pharmacological intervention might alter their normal course and function to modify labour and delivery.

II   Basis of contractility of the uterus

1   STRUCTURE OF THE UTERUS AND MYOMETRIUM

i   *Anatomy of the uterus*

The uterus is a tubular organ which opens from the oviducts on one end and

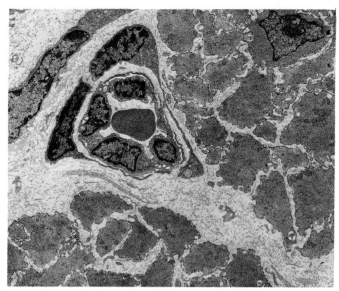

Fig. 10.1   Low magnification electron micrograph of the uterus from a rabbit show-ing muscle cells cut in cross-section. X4,800.

into the vagina on the other end. The wall of the uterus is composed of three coats: (1) an inner mucosa, the endometrium, lining the uterine cavity; (2) a surrounding muscle coat, the myometrium; (3) a thin external serous coat, the serosa (Ham and Cormack 1979). The myometrium of most species con-sists of two distinct smooth muscle layers with vascular zones in between: an outer longitudinal layer and inner circular layer, whose muscle fibres are arranged respectively, parallel to or concentrically around the long axis of the uterus (Finn and Porter 1975).

The longitudinal and circular layers of the myometrium are composed of smooth muscle cells embedded in a connective tissue matrix (Fig. 10.1) and arranged into bundles of approximately 10 to 50 partially ovarlapping cells, their orientation being in the direction of the bundle's long axis (Melton and Saldivar 1964; Burnstock 1970; Garfield and Daniel 1974; Garfield and Somlyo 1985). Muscle bundles are inter-connected: those of the longitudinal myometrium interlace to form a network while those of the circular myo-metrium appear less clearly arranged (Silva 1967; Garfield 1984; Garfield and Somlyo 1985). Transmission of force along the length of the muscle results from transmission of tension (generated by individual muscle cells) from cell to cell (within a bundle and/or between connecting bundles) and from cell to the connective tissue stroma (Gabella 1979, 1984).

The individual uterine smooth muscle cells are the physiological units of contractile function (Csapo 1971; Marshall 1974). They are small, spindle-shaped cells with dimensions 2–10 µm wide and 50–800 µm long, with these dimensions varying with the functional state of the uterus (Marshall 1974; Finn and Porter 1975). During pregnancy the myometrial muscle cells increase both in number and size, the largest cells being observed at term when they may become 3–5 times as long as at the time of implantation (Carsten 1968; Abe 1970; Shoenberg 1977; Aftin and Elce 1978).

Ultrastructurally, uterine smooth muscle cells are similar to other smooth muscle types (Finn and Porter 1975; Shoenberg 1977) and contain a single elongated central nucleus as well as myofilaments, microtubules and various organelles in the cytoplasm. The myofilaments are arranged into bundles (fibrils) running parallel (Gabella 1979) or obliquely (Shoenberg 1977) to the cell axis and include thick (myosin) and thin (actin) contractile filaments as well as intermediate (desmin) filaments, the latter perhaps provides a supporting framework for the contractile filaments (Shoenberg 1977; Gabella 1981, 1984).

ii   Cell-to-cell contracts between myometrial cells

Smooth muscle cells within the muscle bundles that comprise the contractile coat of the uterus and most hollow organs are separated from their neighbours over most of their surface by a space of greater than 50 to 100 nm (Burnstock 1970; Gabella 1981). However, intermittently neighbouring smooth muscle cells come into closer apposition forming specialized contacts. In rat uterine smooth muscle four different types of cell-to-cell contacts have been described on the basis of their appearance in thin section electron micrographs: (1) intermediate contacts; (2) close contacts or simple appositions; (3) interdigitations; and (4) nexuses or gap junctions (Bergman 1968; Garfield and Daniel 1974; Garfield *et al.* 1977, 1978; Garfield and Somlyo 1985). These contacts have distinct characteristic structure but only the gap junctions are thought to provide the basis for intercellular communication (see below).

iii   Innervation of the myometrium

In most species, the uterus is thought to be innervated primarily by post-ganglionic noradrenergic fibres from the sympathetic nervous system which originate from cell bodies in the lumbar and mesenteric ganglia and from short adrenergic axons (Marshall 1981). However, there appear to be species differences. The rat uterus receives a dual adrenergic and cholinergic innervation entering the uterine wall either directly or along the uterine arteries and veins to positions in the connective tissue between the inner and outer muscle layers (Adham and Schenk 1969; Marshall 1970; Bell 1972; Papka *et al.* 1985; Garfield 1986). Branches from the hypogastric and pelvic nerve trunks give rise to fibres that innervate bundles of uterine smooth muscle as well as smooth muscle in the walls of blood vessels (Silva 1967; Adham and Schenk

1969; Papka *et al.* 1985; Garfield 1986). With histochemical techniques, it has been shown that in pregnant and nonpregnant rat uterus the adrenergic fibres predominantly innervate blood vessels while the cholinergic fibres innervate both uterine smooth muscle and blood vessels (Bell 1972; Hervonen *et al.* 1973; Singh *et al.* 1985; Garfield 1986). In addition to adrenergic and cholinergic nerve fibres, peptidergic (namely, vasoactive intestinal polypeptide, substance P, neuropeptide Y and gastrin-releasing peptide) nerves have recently been found in the nonpregnant rat uterus (Larsson *et al.* 1977; Alm *et al.* 1980; Ottesen *et al.* 1981; Ottesen 1981; Stjernquist *et al.* 1983, 1986; Gu *et al.* 1984; Papka *et al.* 1985). Their distribution, like that of the adrenergic and cholinergic nerves, is to the vascular and nonvascular smooth musculature of the myometrium, except for gastrin-releasing peptide (GRP) nerve fibres which are exclusively associated with the nonvascular smooth muscle. They (with the exception of GRP) are also similarly more numerous in the cervix than in the uterine horns.

Concerning the nerve-muscle relationship, there is a low density of nerves to myometrial smooth muscle cells as compared to richly innervated smooth muscle such as rat vas deferens and urinary bladder. In the latter, each muscle cell is closely related to a nerve axon whereas in the myometrium nerve fibres are associated with groups or bundles of muscle cells (Silva 1967; Marshall 1970, 1973; Adham and Schenk 1979; Garfield 1986). Below, we consider neural regulation of the myometrium.

2   ELECTROPHYSIOLOGY

The contractile activity of the uterus which is associated with parturition is a direct consequence of the underlying electrical activity in the myometrial cells (Abe 1970; Kao 1977*a*; Anderson *et al.* 1981; Csapo 1981). Kao (1977*a*) reviewed uterine electrophysiology but dealt exclusively with the longitudinal muscle whose electrical properties were also assumed for the circular muscle. Since then, detailed studies of the circular muscle have revealed some marked differences in electrical properties compared to the longitudinal muscle and also species differences (Osa and Fujino 1978; Anderson *et al.* 1981; Bengtsson *et al.* 1984*b*; Parkington *et al.* 1985). The following is a brief account of the electrical properties of the myometrium with emphasis on studies of the rat myometrium.

i   Resting membrane potential

Uterine smooth muscle from the longitudinal or circular layer spontaneously generates electrical discharges between periods of quiescence. During the silent period, the membrane potential reaches a value ("resting" membrane potential) that varies from about 40 to 70 mV (inside negative) (Abe 1970). In the longitudinal muscle during pregnancy, the membrane potential increases from a nonpregnant level of 30 to 40 mV, reaching a maximum (60 to 70 mV) at midpregnancy, and then declines to approximately 50 mV at term (Casteels and Kuriyama 1965; Kuriyama and Suzuki 1976; Kanda and Kuriyama 1980). Permeability changes of the membrane to K^+ ions during pregnancy (caused by changes in the levels of oestrogen and progesterone, and stretch) are the likely cause of the membrane potential changes (Casteels and Kuriyama 1965; Marshall 1974; Kao 1977a; Kanda and Kuriyama 1980; Mironneau et al. 1981). The membrane potential of the circular muscle (42 mV, day 17), while consistently lower than that of the longitudinal muscle (52 mV, day 17), follows a similar decline during the middle to late stages of gestation, except at term when there is an abrupt transient increase in membrane potential which subsequently declines during delivery (35–40 mV; Anderson et al. 1981; Kishikawa 1981; Bengtsson et al. 1984b). There are also regional differences in membrane potential with muscle cells at placental sites having a consistently higher membrane potential (up to 7 mV on gestation days 17 to 19) than those from nonplacental regions; however they are the same during delivery (Thiersch et al. 1959; Kuriyama 1961; Casteels and Kuriyama 1965; Abe 1970; Kanda and Kuriyama 1980). Thus, the membrane potential varies at different sites and at different times during pregnancy, but all locations show similar values (35–40 mV) when the muscle becomes optimally coupled (see below) during delivery. The membrane potential is an important consideration because changes in the resting potential affect the spontaneous electrical activity of the muscle cells which is, in turn, responsible for the contractile events of the muscle.

ii   Spontaneous electrical activity

The sequence of contraction and relaxation of the myometrium results from the cyclic depolarization and repolarization of the membranes of the muscle cells. The spontaneous electrical discharges in longitudinal muscle from the uteri consist of intermittent bursts of spike-action potentials (Kuriyama 1961; Csapo 1962; Marshall 1962; Ohkawa 1975; Kuriyama and Suzuki 1976; Kanda and Kuriyama 1980). In contrast, discharges of the circular muscle consist of single plateau-type (that is, spike followed by a slow, sustained depolarization) action potentials in nonpregnant, early and midpregnant uteri, changing to repetitive spike-shaped action potentials superimposed on a plateau that gradually diminishes in amplitude towards term, to spike bursts at delivery (Osa and Fujino 1978; Anderson et al. 1981; Osa et al. 1983; Bengtsson et al. 1984b). Uterine volume (chronic stretch) and ovarian hormones (principally oestrogen), contribute to the change in action potential shape through their effect on the resting membrane potential (Kawarabayashi and Marshall 1981). A single spike can initiate a contraction but multiple spikes are needed for forceful, maintained contractions (Marshall 1984).

During the progress of gestation, the spontaneous discharge frequencies of the longitudinal versus circular muscle change in a reciprocal fashion. Thus, the discharge frequency in preterm longitudinal muscle is less than in the circular muscle, with activity increasing in the former but decreasing in the latter until immediately before delivery when they are the same (Anderson *et al.* 1981; Bengtsson *et al.* 1984*a*, 1984*b*). Throughout gestation, except at term, placental regions are electrically less active than nonplacental regions (Kanda and Kuriyama 1980). In longitudinal muscle, the spike discharge is highest at midterm and then declines towards term (Thiersch *et al.* 1959; Kuriyama and Csapo 1961; Kuriyama and Suzuki 1976; Anderson *et al.* 1981; Kishikawa 1981). When spikes appear in the circular muscle at term they are discharged at a frequency lower than that in the longitudinal layer (Anderson *et al.* 1981).

As in other excitable tissues, the action potential in uterine smooth muscle results from voltage- and time-dependent changes in membrane ionic permeabilities (Anderson 1978). In longitudinal (Abe 1971; Anderson *et al.* 1971; Mironneau 1973; Vassort 1981; Mironneau *et al.* 1981, 1984*b*) and circular muscles (Kawarabayashi and Osa 1976; Osa and Kawarabayashi 1977; Bengtsson *et al.* 1984*a*; Chamley and Parkington 1984) the depolarizing phase of the spike is due to an inward current carried mainly by Ca^{2+} ions but also by Na^+ ions (Anderson 1969; Anderson *et al.* 1971; Kao and McCullough 1975) although their exact contributions remain uncertain (Abe 1970; Kao 1977*b*; Anderson 1978; Kuriyama 1981; Tomita 1981). The outward current causing repolarization (studied in detail in longitudinal and assumed for circular muscles) is carried by K^+ ions and consists of a fast (voltage-dependent) and slow (Ca^{2+}-activated) component (Kao and McCullough 1975; Mironneau and Savineau 1980; Mironneau *et al.* 1981, 1984*b*). In preterm circular muscle, the plateau-type action potential is likely due to a combined effect of a sustained inward Ca^{2+} or Na^+ current and a decrease in the voltage-sensitive outward current (Osa and Kawarabayashi 1977; Kawarabayashi 1978; Bengtsson *et al.* 1984*b*; Chamley and Parkington 1984).

iii   Pacemaker activity

Spontaneous electrical activity in uterine smooth muscle arises from pacemaker cells (pacesetters) which can be identified from nonpacemakers (pacefollowers or propagating cells) by the occurrence or absence, respectively, of characteristic pacemaker potential (Kuriyama 1961; Kuriyama and Csapo 1961; Marshall 1962; Kao 1977*a*). The resting membrane potential of pacemaker cells is significantly lower than that of nonpacemakers and in contrast, remains constant throughout pregnancy (Lodge and Sproat 1981). Pacemaker regions are about 2 mm by 4 mm in size (Marshall 1959; Daniel and Lodge 1973; Lodge and Sproat 1981) and therefore greater than one cell.

They have no discrete anatomical location, that is, each myometrial cell is capable of becoming a pacemaker and thus pacemaker sites can shift from one area to another (Daniel 1960; Kuriyama and Csapo 1961; Marshall 1962; Osa *et al.* 1983). The development of new pacemaker activities is associated with membrane depolarization, although depolarization alone is not sufficient to initiate pacemaker activity (Tomita 1970; Kao 1977*a*; Lodge and Sproat 1981).

It has not been definitely resolved as to whether the uterus *in situ* has a single localized pacemaker (Marshall 1974; Osa *et al.* 1983). The ovarian and cervical ends of the late pregnant rat uterus were found to contract independently and asynchronously, while at delivery contractions were claimed to originate at the ovarian end (Fuchs 1973, 1978). Similar claims regarding the location of pacemaker activity at the uterine tubal end during delivery have been made for the uteri of the cat, rabbit, monkey and human (Daniel 1960; see Marshall 1974; Wolfs and van Leeuwen 1979; Osa *et al.* 1983). However, detailed *in vivo* recordings of electrical activity at multiple sites along the uterus during pregnancy or at delivery in several species have not revealed a constant, localized pacemaker from which excitation spreads (rabbit: Kao 1959; Csapo and Takeda 1965; rat: de Paiva and Csapo 1973; human: Wolfs and van Leeuwen 1979; sheep: Thorburn *et al.* 1984).

Pacemaker activity in longitudinal rat myometrium is characterized by subthreshold oscillations in membrane potential of two types (Kuriyama 1961; Reiner and Marshall 1975; Anderson and Ramon 1976; Kao 1977*a*; Tomita 1981): (1) pacesetter potential (also called slow pacemaker potential, oscillatory local potential or slow wave depolarization), varying in amplitude from 3 to 20 mV with a duration period of 10 seconds to minutes, generates a burst of spike discharges; and (2) prepotential, having an amplitude of 2 to 7 mV and a period of 800 to 1200 msec, triggers an individual action potential within a spike burst. Thus, the frequencies of the pacesetter and prepotential set respectively the burst discharge frequency and spike frequency. The pacesetter and prepotential pacemaker potentials are functionally related in that the slow pacemaker depolarizes the membrane to threshold for prepotential pacemaker activity (the latter triggering a spike), maintains this depolarization to allow repetitive spiking and modulates the rate (slope) of prepotential depolarization and hence spike discharge frequency (Marshall 1959; Kuriyama and Csapo 1961; Reiner and Marshall 1975; Anderson and Ramon 1976). Conversely, spiking is terminated when the slow pacemaker potential repolarizes the membrane below the threshold for prepotential pacemaker activity.

iv   Longitudinal and circular muscle layer interaction

Several *in vitro* studies provide evidence for electrical coupling between the longitudinal and circular muscles in late pregnant mouse (Osa 1974) and midpregnant rat (Ohkawa 1975; Osa and Katase 1975) uteri: (1) electrical or mechanical activity of one muscle layer occurs in synchrony with that of the other layer; and (2) electrical stimulation of the long axis of the longitudinal layer produces a membrane response in the circular layer and *vice versa*. The marked differences in electrical properties (for example, membrane potential, action potential and strength-duration relationship) as well as frequency of spontaneous electrical activity (see above) mitigates against an extensive electrotonic interaction between the muscle layers during most of gestation, except just before and during delivery (Osa and Katase 1975; Bengtsson *et al.* 1984*b*).

v   Transmission of electrical activity

Propagation of electrical activity in uterine as well as other visceral smooth muscles is brought about by local circuit current flow between active and resting regions of the tissue (Prosser 1962; Barr *et al.* 1968; Tomita 1970; Daniel and Lodge 1973). Whereas the individual smooth muscle cell is the electrical unit of excitation, the functional unit of conduction is believed to be a bundle of smooth muscle cells of approximately 100–200 μm diameter and several millimeters in length (Burnstock and Prosser 1960; Csapo 1961; Melton and Saldivar 1964; Marshall 1974). It is assumed that the individual cells within the bundle, and at points where bundles overlap, are electrically coupled through low-resistance pathways to form a multicellular electrical syncytium (Tomita 1970, 1975; Anderson 1978). This assumption is based on the many studies showing that small strips or bundles of uterine (rat: Abe 1970;, 1971; Daniel and Lodge 1973; Kuriyama and Suzuki 1976*a*; Kawarabayshi 1978; Kanda and Kuriyama 1980; Sims *et al.* 1982; Chamley and Parkington 1984; mouse: Osa 1974; sheep: Thorburn *et al.* 1984; Parkington 1985) as well as other smooth muscle types (Tomita 1970, 1975; Creed 1979) show cable-like (core-conductor) properties, analogous to those of nerve and skeletal muscle fibres (Tomita 1970, 1975; Marshall 1974), when part of the tissue is polarized with large external electrodes (Abe and Tomita 1968). In other words, the spatial decay of the electrotonic potential recorded along the muscle bundles is exponential and the value of the length constant is much larger than the length of a single muscle cell (Tomita 1975; Creed 1979).

Electrical activity (for example, action potentials triggered in pacemaker cells) propagates away from its site of origin by current flow (its spatial distribution governed partly by the tissue's cable properties) between the active source (source) and adjacent electrically-coupled membrane regions (sinks). The organization of uterine smooth muscle into branching bundles of electrically coupled cells therefore provides the basis for considering the myometrium a functional syncytium (Anderson 1978). However, despite its superficial resemblance to core conductors, the current flow through such a syncytium with its three dimensional intercellular connections is likely complex (Nagai and Prosser 1963; Tomita 1970; 1975; Daniel and Lodge 1973), and the geometrical factor of branching may produce some complications in spike propagation. Furthermore, impulses may travel at different velocities over distinct pathways (Melton and Saldivar 1964, 1965, 1967; Tomita 1970) like the Purkinje fibres of the heart. The *linea uteri* (a thick region of bundles of longitudinal muscle located along the uterine midline) may be the prime conduction pathway allowing rapid spread of action potentials along the uterine horn and perhaps also between the two uterine horns (Melton and Saldivar 1967), although this aspect has never been fully studied.

• *Conduction velocity*. Conduction velocity of action potentials (measured in the longitudinal direction) in uterine smooth muscle studied *in vitro* varies between species and as a function of gestational state within a given species (Abe 1970; Daniel and Lodge 1973). In the rat uterus, mean conduction velocities of 1.25 to 10.5 cm/sec for nonpregnant and hormone-treated ovariectomized animals (Daniel 1960; Melton and Saldivar 1964, 1965, 1967, 1970; Melton 1965; Saldivar and Melton 1966) and 7 to 33 cm/sec for pregnant animals (Daniel 1960; Kanda and Kuriyama 1980) have been obtained for spontaneous and evoked action potentials. In the longitudinal myometrium of rat, the conduction velocity of evoked spikes increases during gestation in both nonplacental and placental regions (Kanda and Kuriyama 1980). However, the conduction velocity of the uterus is different depending on the direction. In the pregnant mouse uterus for example, longitudinal conduction velocity is higher in the cervical than in the ovarian direction (Kuriyama 1961). It is claimed that longitudinal conduction velocity is the same in both directions in the nonpregnant rat myometrium (Melton and Saldivar 1964), although this has not been extensively studied. In addition, conduction velocity between two cells in the same or in different bundles is higher in the longitudinal than in the transverse direction in the uterus (Lander *et al.* 1959; Goto *et al.* 1961; Kuriyama 1961; Osa 1974; Zelcer and Daniel 1979). These higher propagation velocities and length constants in the longitudinal as compared to the tangential direction of the cell axis are probably directly related to the much lower internal resistance in the long axis of the cells that arises from the smaller number of cells and cell junctions per unit length.

• *Conduction distance*. The extent of conduction (that is, the distance travelled by the same action potential) in the uterus has been studied in some detail in nonpregnant and pregnant rat uterus. Conduction distance of less than 1 to 2 mm for ovariectomized, nonpregnant (Melton and Saldivar 1964), 6 to 37 mm for nonpregnant oestrogen and oestrogen-plus progesterone-treated (Melton 1956; Daniel 1960; Melton and Saldivar 1964, 1965, 1967) and 14.4 mm for early pregnant (Daniel 1960) animals have been reported. However, the velocity and extent of uterine conduction during pregnancy has not been adequately studied in any animal species despite the importance of such information in understanding the mechanisms underlying the synchronization of uterine smooth muscle cell activity at parturition (Csapo 1971, 1981).

- *Cell-to-cell coupling during pregnancy*. Gestational changes in the degree of current spread (cell-to-cell coupling) in uterine smooth muscle have been assessed by measurement of passive electrical (that is, cable) properties. In circular (Thorburn *et al.* 1984; Parkington 1985) and longitudinal (Sims *et al.* 1982) smooth muscle from sheep and rat uteri, respectively, the length constants and membrane time constants were higher at term than at preterm indicating an enhanced cell-to-cell coupling at term. Others (Abe 1971; Daniel and Lodge 1973; Kuriyama and Suzuki 1976; Daniel and Zelcer 1979) studying longitudinal rat myometrium found no such gestational changes. However, with the exception of the study of Sims *et al.* (1982) the tissues used for electrical coupling measurements in rat myometrium were not examined for the presence of gap junctions which may have developed *in vitro* in the preterm tissues before or during electrical recording (Garfield *et al.* 1980*a,b*) and contributed to the electrical coupling (see below). The degree of stretch can also influence the measured length constant (Tomita 1975; Sims *et al.* 1982). Using a different method (impedance measurements) to assess electrical coupling in the rat myometrium, Sims *et al.* (1982) found a lower junctional resistance (that is, increased coupling) during delivery compared to before term or postpartum.

Action potential conduction velocity (v) in a core-conductor such as the myometrium can be calculated from the relation $v = S\lambda/\tau$ where S is a safety factor depending on both the excitability of the muscle cell membrane and the action potential amplitude, λ is the length constant and τ is the membrane time constant (Holman 1968). Thus, changes in passive or active membrane properties during gestation may affect action potential conduction velocity.

- *Structural basis of cell-to-cell coupling*. Electrical interaction between uterine smooth muscle cells is usually assumed to imply the existence of intercellular low resistance pathways for current. It is generally believed that the gap junction represents the structural basis for electrical coupling in smooth muscle (Barr *et al.* 1968; Dewey and Barr 1968; Gabella 1981; Garfield 1985*b*) and other tissue (Loewenstein 1981; Bennett *et al.* 1984). The role and regulation of myometrial gap junctions is considered in greater detail below.

v   Electromechanical coupling

Contractile activity in uterine smooth muscle is initiated by a rise in the intracellular concentration of free ionized Ca^{2+} to approximately 10^{-5} M from a resting level of about 10^{-7} M (Daniel *et al.* 1983). The source of this activator Ca^{2+} is extracellular (that is, Ca^{2+} ions which flow into the cell down their electrochemical gradient in response to a change in membrane permeability) or intracellular (that is, Ca^{2+} ions released from intracellular storage sites) or a combination of both (Mironneau 1973, 1976*a*; Marshall 1974; Grosset and Mironneau 1977; Kuriyama 1981; Daniel *et al.* 1983; Mironneay *et al.* 1984*a,b*). Conversely, a reduction of intracellular free Ca^{2+} (either as a result of efflux into the extracellular space or re-uptake into intracellular storage sites) terminates contraction (Marshall 1974; Kuriyama 1981; Daniel *et al.* 1983).

In longitudinal and circular muscles of pregnant rat myometrium, the inward Ca^{2+} current (through voltage-dependent transmembrane Ca^{2+} channels) during a single action potential initiates a twitch contraction (Mironneau 1973; Kawarabayashi and Osa 1976; Bengtsson *et al.* 1984*b*). When action potentials are repetitively discharged (for example, in spike bursts) the contraction amplitude is increased because: (1) the intracellular level of free Ca^{2+} is increased; and (2) the increments in tension triggered by individual spikes can summate (Kuriyama and Csapo 1961; Mironneau 1973; Reiner and Marshall 1975; Kao 1977*a*). When action potentials are discharged at a rate higher than about 1 c/s, a fused tetanic type contraction is generated (Kuriyama and Csapo 1961; Mironneau 1973). Thus, the frequency and duration of spike discharge can control the contraction height and duration, respectively (Marshall 1974). In this way, the frequency of pacemaker discharge (pacesetter and prepotential) determines, respectively, the rate and intensity of uterine contractions (Marshall 1973; Kao 1977*a*). However, increases in contractile force in the longitudinal muscle may be achieved by a more or less synchronous stimulation of large areas of the myometrium (that is, increased cell-to-cell coupling), as opposed to a faster rate of stimulation of individual cells (Kao 1977*a*). This is supported by the recent observation that between late pregnancy and delivery the amplitude of spontaneous contractions increases in the longitudinal myometrium despite a decrease in spike frequency and lack of change in the ability of the muscle to contract (Anderson *et al.* 1981; Bengtsson *et al.* 1984*a,b*; Izumi 1985).

In the circular muscle at midterm, contractions are weaker than in the longitudinal muscle, but as pregnancy progresses they increase in strength until at term when they are of similar strength (or just slightly weaker) than contractions in the longitudinal layer (Osa and Katase 1975; Anderson *et al.* 1981; Bengtsson *et al.* 1984*b*; Izumi 1985). These changes are not due to changes in the ability of the circular muscle to contract (see above) but rather mainly to changes in membrane electrical events (that is, stronger contractions generated by spike bursts at term than by single, plateau-type action potentials at midterm) and cell-to-cell coupling (Osa and Katase 1975; Anderson *et al.* 1981; Bengtsson *et al.* 1984*a,b*; Thorburn *et al.* 1984). The differences in magnitude of contraction between longitudinal and circular

MECHANISMS FOR
CONTROL OF SMOOTH MUSCLE

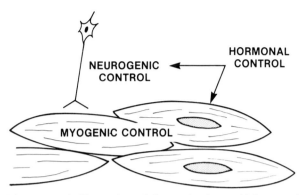

Fig. 10.2.   Diagrammatic illustration of the mechanisms which control the contractility of smooth muscle.

muscles at midterm may also be explained by differences in membrane electrical events, cell-to-cell coupling and perhaps intracellular Ca^{2+} release (Kawarabayashi 1978; Bengtsson *et al.* 1984*a,b*; Izumi 1985).

It should be noted that some drugs can cause contraction or relaxation of smooth muscle without any change in the resting membrane potential or action potential frequency. This process has been termed pharmacomechanical coupling (Somlyo and Somlyo 1968). It is probable that in a variety of experimental and physiological conditions electromechanical and pharmacomechanical processes contribute to activation (or inhibition) simultaneously (Johansson and Somlyo 1980).

III   Modulation of uterine contractility

The electrical activity of any smooth muscle tissue may be modulated by either myogenic, neurogenic or hormonal mechanisms (Fig. 10.2). Myogenic activity refers to spontaneous activity which occurs in the absence of any neural or hormonal input. Myogenic properties are those active and passive electrical characteristics of the muscle which include the basic intrinsic excitability, the ability to contract spontaneously, and the mechanisms for conduction to produce rhythmic contractions (see above). Neurogenic and hormonal activity are attributed to the neural or humoral control systems which are superimposed on the myogenic mechanisms to modify myometrial activity. Physiologically, the stimulation or inhibition of myometrial contractility occurs through the influence or manipulation of these three mechanisms. These systems can also be altered pharmacologically or changed by abnormal pathological conditions. The three systems will be considered in more detail below.

1   HORMONAL CONTROL

There is a wealth of information that myometrial contractility is under very dominant control by humoral mechanisms (Fuchs 1978; Csapo 1981; Challis and Lye 1986). Progesterone, oestrogens, prostaglandins, oxytocin and perhaps relaxin have tremendous influence on the myogenic properties and control of the myometrium as well as on neurogenic mechanisms (Marshall 1962, 1981; Carsten 1968; Finn and Porter 1975; Thorburn and Challis 1979; Challis 1984; Challis and Lye 1986). It is evident from the results of many studies that the maintenance of pregnancy and the initiation of term or preterm labour is accomplished through the action of the hormones on the myometrium. The possible mechanisms of action of these substances are considered below.

i   Steroid hormones

Changes in the synthesis of the steroid hormones, as reflected in tissue and plasma levels, occur before normal or premature labour in most species of animals (Fuchs 1978; Csapo 1981; Challis 1984). Csapo (see review, 1981) in many elegant studies of uterine contractile patterns and hormonal profiles showed that when progesterone levels decline either at term or preterm the myometrium is converted from inactive to active and reactive muscle. These observations led Csapo first to propose the *progesterone block theory* (Csapo 1961), and later the *seesaw theory* (Csapo 1977). These theories predicted that the maintenance of pregnancy involved a balance in the myometrium of opposing forces, between stimulants and suppressors, whereas the onset and progression of labour depended upon a regulatory imbalance (decrease in progesterone) in favour of the stimulants (increase in oestrogen and prostaglandins). Csapo's work has been overwhelmingly supported by other studies although the interpretation and conclusions have not always been unanimous.

It is now known that antiprogesterone agents are efficient abortifacents at any time during pregnancy in all species used including humans (Baulieu 1985; Garfield and Baulieu 1987). These studies would support the role of progesterone in the maintenance of pregnancy, since anti-progesterone agents bind to the steroid receptor and prevent its action, effectively producing the same effect as progesterone withdrawal (Baulieu 1985).

The basis for the effects of the steroid hormones on myometrial function lies in the fact that they regulate myogenic mechanisms in muscle cells. Generally, oestrogen and progesterone have opposing effects on the myogenic properties of the myometrial cells: oestrogen stimulates and progesterone inhibits. The steroid hormones have both delayed effects as well as short onset or acute effects, and they can have different effects *in vitro* from those *in vivo*. An explanation for this may be that both oestrogen and progesterone have immediate effects either directly on Ca^{2+} binding and transport or indirectly on prostaglandin production (Thorburn and Challis 1979), and delayed effects dependent on the steroid-receptor system. This latter action probably involves the stimulation (oestrogen) or inhibition (progesterone) of new proteins, including those for excitation-contraction coupling, receptors, contractile filaments, high energy phosphates and gap junctions (see below). The increase in excitability and conduction may be explained on the basis of the classical steroid receptor and the control of protein synthesis by the steroid hormones.

Progesterone treatment *in situ* may affect uterine excitability and conduction by increasing (hyperpolarizing) the resting membrane potential and thus the threshold for excitation (Marshall 1959, 1962; Daniel 1960; Csapo 1961, 1962; Kuriyama 1961; Kuriyama and Csapo, 1961; Talo and Csapo 1970; Kuriyama and Suzuki 1976). However, it is not certain whether progesterone actually affects the resting membrane potential in uterine smooth muscle, because factors such as stretch, time after removal of uterus from the animal before measurement of the membrane potential, solvent used for dissolving the hormone *in vitro* and cell sampling bias may not have been adequately controlled (Kao and Nishiyama 1964; Carsten 1968; Marshall 1974; Kao 1977a). On the other hand, the higher resting membrane potential, decreased spontaneous electrical activity and slower conduction velocity in placental than in nonplacental regions of the myometrium (see **Section II**, 2 Electrophysiology) support the idea of a local blocking effect of progesterone (Csapo 1961, 1969). However, it cannot be definitely concluded that those properties of the placental region are due solely to the action of progesterone (Abe 1970; Finn and Porter 1975; Kanda and Kuriyama 1980).

We have suggested that progesterone blocks conduction by decreasing electrical coupling between myometrial cells (Garfield *et al.* 1977, 1978, 1980a,b). Progesterone treatment increases junctional resistance (and hence electrical coupling) in the myometrium (Ichikawa and Bortoff 1970; Bortoff and Gilloteaux 1980) and we have demonstrated that progesterone inhibits gap junctions (see below). The suppression of gap junctions by progesterone may be the molecular mechanism responsible for the "block" described by Csapo (1961, 1981).

The levels of oestrogens increase prior to parturition in many species (Thorburn and Challis 1979). Oestrogens have well known excitatory effects on uterine contractility (Bozler 1941) and it seems probable that oestrogens play a role in the initiation of parturition (Csapo 1971). Oestrogens: (1) decrease the uterine smooth muscle cell resting membrane potential into a range where excitability is improved and action potential discharge (spontaneous or drug-induced) is readily initiated, and impulse propagation enhanced (Bozler 1941; Melton 1956; Marshall 1959; Kuriyama 1961; Melton and Saldivar 1964; Casteels and Kuriyama 1965; Marshall 1973; Kuriyama and Suzuki 1976; Kao 1977a; Fuchs 1978; Ogasawara *et al.* 1980);

(2) decrease resting membrane potential and transition from plateau-type action potential to repetitive spiking in the circular muscle of the late pregnant rat myometrium (Kawarabayashi and Marshall 1981); (3) induce formation of gap junctions in the myometrium (see below), which may be the basis for the increased cell-to-cell coupling (length constant) in oestrogen-dominated (Kuriyama and Suzuki 1976) or parturient rat myometrium (Garfield 1977; Sims *et al.* 1982); (4) enhance uterine contractility through their stimulation of uterine prostaglandin production (see Thorburn and Challis 1979; Williams 1983) and actomyosin formation (Michael and Schofield 1969; Csapo 1971, 1981); and (5) enhance the pharmacologic response of the uterus (Csapo 1971; Kuriyama and Suzuki 1976; Kao 1977*a*; Fuchs 1978) by modulating both the resting membrane potential and the number of agonist receptors in the myometrium (oxytocin: Alexandrova and Soloff 1980; Fuchs 1983; relaxin: Mercado-Simmen *et al.* 1982; Marshall 1973).

ii Prostaglandins

The evidence suggesting a role for prostaglandins in parturition in all species is extensive (Aiken 1972; Csapo 1977, 1981; Fowler 1977; Fuchs 1978; Thorburn and Challis 1979; Novy and Liggins 1980; Anderson *et al.* 1981; Phillips and Poyser 1981; Dubin *et al.* 1982; Puri and Garfield 1982; Chan 1983; Mitchell 1984). In brief, the grounds for this contention are: (1) prostaglandins (administered *in vivo* or *in vitro*) markedly stimulate uterine contractility and enhance uterine sensitivity to other uterotonic agents (for example, oxytocin); (2) prostaglandin production by uterine and intraluminal tissues increases as pregnancy proceeds (but which tissue is the major prostaglandin source is controversial); (3) parturition is delayed or accelerated by agents which block or stimulate prostaglandin synthesis or metabolism, respectively; and (4) *in vivo* administration of prostaglandins induces parturition. However, it is not clear whether prostaglandins enhance uterine contractility or excitability *in situ* by either their direct action on the myometrium or indirectly.

Concerning their direct action on the myometrium, prostaglandins ($PGF_{2\alpha}$, PGE_1 and PGE_2) increase the electrical and mechanical activity of uterine smooth muscle *in vitro*. Depending on the amount, prostaglandins cause an increase in frequency of phasic contractions or an increase in tonic tension upon which phasic contractions are superimposed (Reiner and Marshall 1976; Chamley and Parkington 1984). Underlying the phasic contractile change is a slow membrane depolarization (caused by an inward Na^+ current: Reiner and Marshall 1976; Ca^{2+} current unchanged, Grossett and Mironneau 1977) that initiates or increases the frequency of spike burst discharges, whereas the increased resting (tonic) tension is likely to be due to prostaglandin-stimulated intracellular Ca^{2+} release (Reiner and Marshall 1976; Grosset and Mironneau 1977; Anderson 1978; Villar *et al.* 1985) and perhaps inhibition of active Ca^{2+} extrusion from the cell (see Popescu *et al.*

1985). The sensitivity of the isolated longitudinal rat myometrium to exogenous prostaglandins increases during the last stage of gestation (Reiner and Marshall 1976; Kanda and Kuriyama 1980).

Prostaglandins also enhance uterine contractility (that is, synchronized activity) by indirect effects. In this way, prostaglandins may: (1) modulate oxytocin release and sensitivity; (2) inhibit progesterone synthesis in the corpus luteum or placenta; (3) mediate cervical ripening; (4) regulate uteroplacental blood flow (Novy and Liggins 1980; Fuchs 1983); (5) modify steroid receptors in muscle cells (Sanfillipo *et al.* 1983); and (6) control of cell-to-cell coupling (see below).

iii   Oxytocin

Oxytocin is a potent uterine stimulant which initiates contractions in quiescent uteri or increases contractile frequency and force in active uteri (Fuchs 1973, 1978, 1983, 1985; Marshall 1974; Kao 1977a). When administered *in vivo* (Csapo and Takeda 1965; Csapo 1969, 1971; Fuchs 1973, 1978) or *in vitro* (Marshall 1974; Ohkawa 1975; Kao 1977a; Anderson 1978; Chamley and Parkington 1984), oxytocin produces these effects primarily by increasing the duration and repetition frequency of spike burst discharges, increasing the spike discharge frequency within individual bursts and perhaps also by improving electrical conduction.

The ionic basis of oxytocin enhancement of membrane electrical activity is disputed, that is, increase in Ca^{2+} (Marshall 1974) as opposed to Na^+ (Kao 1977a) current. However, voltage-clamp studies on pregnant rat longitudinal myometrium show that oxytocin selectively potentiates the inward Ca^{2+} current (Mironneau 1976b). As well as stimulating Ca^{2+} entry into the cell at the plasma membrane, oxytocin might raise intracellular Ca^{2+} (and thus enhance uterine contractility) by mobilizing Ca^{2+} from intracellular stores (Schild 1964; Sakai *et al.* 1981; Batra 1985) or by inhibiting active Ca^{2+} extrusion from the cell (Soloff and Sweet 1982; Popescu *et al.* 1985; Soloff 1985).

In term pregnant rat, serum oxytocin levels (as detected with sensitive and specific radioimmunoassay of serial samples taken during parturition) remain unchanged until just before expulsion of the first fetus when they increase markedly, remain elevated during delivery and then decrease after expulsion of the last fetus (Higuchi *et al.* 1985). Peak concentrations of oxytocin, coincident with delivery of the first or second fetus, have also recently been shown in serial blood samples obtained just before and during parturition in the rabbit (O'Byrne *et al.* 1986). Uterine sensitivity to oxytocin increases as gestation progresses and reaches maximal levels shortly before or at parturition (Csapo 1971; Fuchs 1973, 1978, 1985; Chan 1983). There is some evidence that the circular muscle of the rat myometrium behaves differently from the longitudinal muscle with respect to gestational changes in oxytocin sensitivity (Ohkawa 1975; Crankshaw 1985).

Marshall (1974) has described the relationship between uterine resting membrane potential and oxytocin sensitivity. When membrane potentials are highest (at midterm) the uterus exhibits a reduced sensitivity to oxytocin. The fall in membrane potential towards the end of gestation lowers the threshold for spike discharge and hence facilities oxytocin's stimulatory effect. Paradoxically, however, sensitivity of the muscle cell membrane to oxytocin is higher in placental as compared to that in nonplacental regions at any stage of gestation (Kanda and Kuriyama 1980) even when the membrane potential of the former region is higher (see above).

Oxytocin binding sites increase in number (but not affinity for oxytocin) just prior to parturition in the myometrium of rats, rabbits, sheep and humans (Soloff *et al.* 1979; Fuchs *et al.* 1982; Sheldrick and Flint 1985; Reiner *et al.* 1986). It is assumed that these sites represent oxytocin receptors on the plasma membrane. In rats, when gestation is terminated prematurely or is prolonged, myometrial oxytocin receptor concentration rises prematurely or is delayed, respectively (Soloff 1985). While the uterine oxytocin sensitivity is well correlated with the change in receptor concentration (Soloff *et al.* 1979; Alexandrova and Soloff 1980, Fuchs 1983), additional factors such as the level of the membrane potential (see above) or electrical coupling between smooth muscle cells (see below) may also be important in oxytocin action. Ovarian hormones, stretch, or a combination of both are involved in the regulation of rat uterine oxytocin receptors, for example, oestrogens or oestrogens in combination with uterine stretch, increase receptor number while progesterone inhibits the oestrogen-induced increase (Soloff *et al.* 1979; Alexandrova and Soloff 1980; Fuchs 1983, 1985).

Whether oxytocin and also prostaglandins play obligatory or facilitory roles in parturition has not been resolved. Also unresolved is the role gap junctions or cell coupling might have on the contractile response of the myometrium to oxytocin or any uterotonic drug. One would expect that the presence of the junctions would amplify the contractile response by either changing the sensitivity or maximal effect of any agent. However, little attention has been paid to this aspect.

iv *Relaxin*

Recent data suggest that relaxin may also participate in the maintenance of pregnancy (Porter *et al.* 1979; Downing and Sherwood 1985a,b). The levels of this hormone in the rat are elevated during pregnancy, reach a peak during the 24–36 hours before parturition, but decline around the time of labour (Sherwood *et al.* 1980; Downing and Sherwood 1985a). Relaxin inhibits spontaneous activity in the myometrium of the rat (*in vitro* and *in vivo*) (Sanborn *et al.* 1980; Downing and Sherwood 1985b) and it is postulated that this hormone may be responsible for the marked uterine quiescence and maintenance of pregnancy in the rat over the final 24–36 hours of pregnancy after the levels of progesterone have fallen considerably (Porter *et al.* 1979; Downing and Sherwood 1985*b*). It is also evident that this hormone partici-pates in the control of cervical maturation to permit the passage of the fetus(es) (Downing and Sherwood 1985*c*). Our studies show that relaxin does not change the presence of myometrial gap junctions, but it may regulate their permeability (see below).

2 NEUROGENIC CONTROL

Neurogenic activity of smooth muscle is that component of contractility which occurs in response to the discharge of nerves and the effects of trans-mitters on the muscle. In spontaneously active smooth muscle, such as the uterus, neurogenic activity is that element which is superimposed upon myo-genic mechanisms to influence the quality and quantity of contractility. Release of neurotransmitters from nerves of the autonomic nervous system in other smooth muscle tissues results in stimulation, inhibition or initiation of contractility. However, it is generally believed that uterine contractility is controlled by myogenic and hormonal mechanisms and that modulation by nerves is not important. This opinion is based upon observations that show that transection of the spinal cord or peripheral nerves has no consistent effect on implantation, pregnancy or parturition and the fact that the myo-metrium contracts spontaneously *in vitro*. This misconception has been per-petuated without a complete understanding of the neural pathways to the uterus or an appreciation for the intrinsic nerves.

The uterus is innervated by postganglionic adrenergic and cholinergic fibres from the autonomic nervous system (see above). Thorbert and col-leagues (1977) have recently described the origin and distribution of adrener-gic nerves in the guinea pig. These studies show that about 35 per cent of the fibres to the uterus enter the uterus at the tubal end and are present in the sus-pensory ligament. The remaining 65 per cent of the nerves which enter the uterus are present in the hypogastric nerve and about 35 per cent of these nerves are from short adrenergic fibres from paracervical ganglia present at or near the wall of the cervix. These findings are highly significant since cutting the hypogastric nerves may not eliminate supply in the suspensory ligament. Furthermore, transection of the cord or cutting the pelvic nerves and nerves in the suspensory ligament will not exclude local control by post-ganglionic nerves which originate in the paracervical ganglia. In addition, the axons from the nerves may continue to release transmitters in tissues studied *in vitro*.

There is an abundance of biochemical, histochemical and electron micro-scopic evidence which support the observations that adrenergic nerves in the uterine wall disappear near term during pregnancy in humans and other mammals (Marshall 1981). The only normal adrenergic nerves still present at term are those in the cervix and tubal ends of the uterus. We have demonstrated that the cholinergic nerves also disappear near term in the rat (Garfield 1986). These observations have led to the hypothesis that the maintenance of pregnancy and the initiation of labour may be related to the disappearance of nerves (Thorbert 1979; Marshall 1981). Myogenic activity could be normally suppressed by the adrenergic nerves throughout pregnancy and with their disappearance near term myogenic contractions may increase. However, Thorbert (1979) has shown that implantation, pregnancy and parturition occur with subsequent pregnancies and prior to the regeneration of the adrenergic nerves. These studies would tend to argue against involvement of the adrenergic nerves but because the nervous supply to the uterus and nerve-muscle interactions are complex, we cannot simply dismiss the possible involvement of neurogenic mechanisms in the control and initiation of labour (see Marshall 1981).

The adrenergic neurotransmitter, noradrenaline, is inhibitory at midgestation and at term in the longitudinal muscle of the rat uterus (Kawarabayashi and Osa 1976; Chow and Marshall 1981). The circular muscle, in contrast, shows a different response, that is, excitation at preterm and inhibition at term (Kawarabayashi and Osa 1976; Osa and Watanabe 1978; Chow and Marshall 1981; Kishikawa 1981). It is thus possible that the adrenergic nerves, until they degenerate in late pregnancy, have a tonic inhibitory effect on longitudinal muscle and excitatory effect on circular muscle contractions. However, the uterine body of the rat has a notably poor adrenergic innervation (Marshall 1970; Hervonen *et al.* 1973; Garfield 1986). On the other hand, acetylcholine is excitatory on the myometrium (both longitudinal and circular muscle layers) at all stages of gestation (Marshall 1973; Kao 1977*a*; Bengtsson *et al.* 1984*b*; Izumi 1985). It is therefore perplexing why the cholinergic nerves should degenerate in the uterine body in late pregnancy if they played a role in initiating parturition. It has been suggested that adrenergic or cholinergic nerves may function to initiate pacemaker activity at the utero-tubal junction (where they remain intact) during the late stages of pregnancy and at parturition (Thorbert 1979; Singh *et al.* 1985). However, adrenergic nerves may not be involved in the initiation of parturition, at least in the rat, since sympathetic denervation of the utero-tubal junctions and upper uterine horn at midgestation does not affect the time of onset or duration of parturition (Roche *et al.* 1985).

The motor effects of peptidergic nerves, substance P (SP) and gastrin-releasing peptide (GRP) increase both the frequency and amplitude of uterine smooth muscle contractions, while vasoactive intestinal polypeptide (VIP) has an inhibitory effect on both the spontaneous and SP-stimulated contractions of the nonpregnant uteri of rats and other species (Ottesen 1981; Ottesen *et al.* 1981, 1983; Stjernquist *et al.* 1986). Neuropeptide Y, on the other hand, inhibits neurally (for example, cholinergic nerves) evoked but not spontaneous contractions of the nonpregnant rat uterus (Stjernquist *et al.* 1983). The effects of peptidergic nerves on the motility of the pregnant rat uterus have not been studied. However, it is likely that their effects are weak, especially in late pregnancy when the nerve distribution is sparse (Ottesen 1981; Ottesen *et al.* 1981). In the human uterus, as the number of VIP-containing nerve fibres decreases during pregnancy, their inhibitory effect on spontaneous uterine contractions diminishes (Ottesen *et al.* 1982).

3   MYOGENIC CONTROL

Myogenic control of contractility refers to the basic intrinsic properties of the muscle cells including the active and passive electrical properties as described above (**Section II**, 2 Electrophysiology). Myogenic activity of smooth or cardiac muscle is considered that component of contractility which is observed in the complete absence of neural or humoral influences. The signal for myogenic contractions originates in the spontaneous oscillation of the membrane potential which is provided by pacemaker cells. We have concentrated on the process that controls propagation or conduction in uterine smooth muscle because this property appears to be one of the key myogenic mechanisms of uterine smooth muscle which is absent during most of pregnancy, that develops in response to the steroid changes and may be responsible for synchronous or coordinated contractility of the myometrium. Below we consider the general concept of conduction between cells and then review studies of the uterus and the role and regulation of myometrial gap junctions.

IV   Gap junctions as sites for electrical and metabolic coupling

The initial observations that started research into intercellular coupling were that electrical current injected into one cell could alter the electrical potential of an adjacent cell (Furshpan and Potter 1959), and that certain membrane specializations in the intercellular region (Revel and Karnovsky 1967), now termed gap junctions, could be responsible. By now, sufficient evidence has correlated the physical presence of gap junctions with functional aspects of intercellular coupling that the gap junction is indeed taken as the morphological equivalent of cell-cell coupling (for review see Peracchia 1980; Loewenstein 1981; Larsen 1983; Pitts and Finbow 1986).

1   DESCRIPTION AND DEFINITION

Gap junctions are intercellular channels that link cells to their neighbours by allowing the passage of inorganic ions and small molecules. They have been found between cells in every tissue and organ examined, and are essentially ubiquitous in the animal kingdom. In the electron microscope they appear in regions of close apposition between cells as zones of paired, parallel mem-

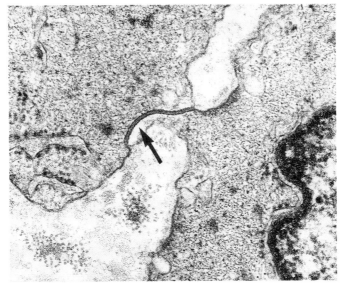

Fig. 10.3   Intermediate magnification micrograph of portions of two myometrial cells from rabbit during parturition showing gap junction (arrow). X71,000.

branes, of unusually smooth outline, separated by a narrow space of constant width—the "gap" (Figs 10.3 and 10.4). Both negative staining and freeze-fracture show that these regions have characteristic membrane particles present in each contributing membrane, that meet in register at regularly spaced intervals along the gap (Chalcroft and Bullivant 1970; McNutt and Weinstein 1970). Each cell of the pair contributes a set of particles, with one particle termed a hemi-channel or connexon, and two together forming one intercellular channel. While connexons, like other membrane proteins, can move laterally in the fluid membrane, unpaired connexons are not functional and gap junctions seem to be formed only when these particles aggregate into characteristic, roughly circular plaques. The mechanism for this is not known, nor whether functional channels can occur between single pairs of connexons.

i   Spectroscopic analysis

Analysis of electron and x-ray diffraction patterns from isolated gap junctions has provided valuable information about molecular organization within the connexon. With standard electron microscopy, these preparations show large numbers of intact gap junctions, with connexons in hexagonal arrays. Electron diffraction (Unwin and Zampighi 1980) showed that each connexon was composed of 6 subunits arranged around a central pore. Unwin and Ennis (1983, 1984) analysed x-ray diffraction patterns to deduce a

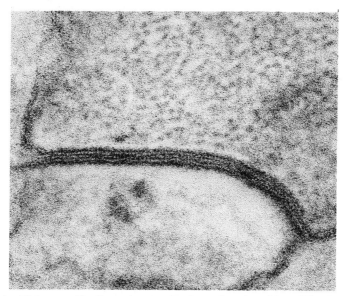

Fig. 10.4   High magnification micrograph of gap junction. Note 7-lined appearance. X319,000.

3-dimensional map of the channel. Essentially, a connexon is a hexameric structure with a 6.4 nm diameter in the plane of the membrane, and is 7.0 nm in height, with less than 1.0 nm protruding into the extracellular space, and more (1.5–2.0 nm) into the cytoplasmic space. The six subunit proteins stand at an angle to the long axis and surround a central channel, which Makowski *et al.* (1984) showed to have a distinct funnel shape, decreasing from approximately 5.0 nm diameter at the cytoplasmic entrance to 2.0 nm at a level within the membrane. Unwin and Ennis (1983) experimentally increased Ca^{2+} to levels associated with loss of intercellular coupling, and found that the subunits tilted further, reducing the channel diameter by 1.8 nm, suggesting channel closure may have occurred. Thus, intercellular passage of a molecule requires entry into and passage through two complex intramembrane proteins in series, each capable of conformational change in response to (at least) cytoplasmic Ca^{2+} levels.

ii   Biochemistry

The widespread occurrence and uniform morphology of gap junctions in different tissues and different species suggest there is phylogenetic conservation of the junctional protein. However, there is also considerable diversity in apparent molecular weight of the junction protein. For example, the hepatocyte gap junction, historically the focus of structural and biochemical research of the gap junction, has been isolated and purified with biochemical analysis showing co-purification of a 27 kD membrane protein. However, such analysis in other systems has found values ranging between 16 kD and 54 kD (Revel *et al.* 1985). Examination shows, that this variation occurs mainly between tissues and not between species, and sequence homology is usually retained. Thus, the molecular weight of the cardiac gap junction proteins is 47 kD (Manjunath *et al.* 1987), but there is sequence homology with the liver 26 kD protein (Nicholson *et al.* 1987; Hertzberg and Skibbins 1984). The 17 kD sequence of the cardiac protein not found in the liver protein (Paul 1986) is thought to lie in the cytoplasmic domain, and may confer the unique properties of regulation of cardiac junctional permeability (White *et al.* 1985).

Gap junctions in the eye lens represent another example of tissue specific variation in the junctional protein. The avascular crystalline lens appears to rely upon extensive cell communication to transfer metabolites and ions between central and superficial cells, reflected in the extremely large proportion (30–60 per cent) of the surface membrane occupied by gap junctions (Benedetti *et al.* 1974). The amino acid sequence of the 26 kD junctional protein (major intrinsic protein MIP 26) has been determined (Gorin *et al.* 1984; Revel *et al.* 1985). While there is no homology between MIP 26 and either liver or cardiac junctional proteins (Nicholson *et al.* 1985), monoclonal antibodies to MIP 26 have been shown to bind to lens junctional membrane (Sas *et al.* 1985), and incorporation of MIP 26 into an artificial lipid system results in formation of functional channels (Zamphighi *et al.* 1985).

2   JUNCTIONAL PROTEIN OF SMOOTH MUSCLE

Current information about the junctional protein of smooth muscle is drawn from the rat uterus, where a 28 kD protein has been isolated (Zervos *et al.* 1985). Peptide mapping studies have shown that the 28 kD polypeptides of uterus, liver and heart are homologous, although on the basis of a 10 kD polypeptide component (Zervos *et al.* 1985). Further, antibodies to an N-terminal fraction of the liver proteins cross-reacted with all three tissues. As is the case of the heart and liver junctional proteins (see above), the uterine (smooth muscle) junctional protein may possess peptide sequences that confer tissue specific properties. However, Zervos *et al.* (1985) used whole uterine tissues containing endometrium vascular tissues, etc. and thus their results may not be truly representative of the myometrium.

3   FUNCTIONAL STUDIES

In excitable cells, electrical coupling allows transmission of action potentials by local circuit current flow. Therefore, the net junctional resistance is determined by the number of channels, the mean open time of each channel, and perhaps the predominant conductance state. Regulation of cell coupling in any of these ways will determine the rate of propagation and the extent of coordination of contraction. In these tissues, and non-excitable tissues, a second consequence will be variation in metabolic coupling, that is, the number and size of molecules passing between cells.

In both non-excitable cells (rat lacrimal gland: Neyton and Trautmann 1985) and excitable cells (chick heart: Veenstra and DeHaan 1986) gap junctions appear to be composed of ionic channels with a large unitary conductance of 120–165 pS, that undergo spontaneous openings and closings, and discriminate poorly between cations and anions. The rate of transition between states is slow, which may relate to the intermediate states of conductance seen in both studies, and is suggested to be due to gradual conformational change in the channel macromolecule.

i Molecular permeability

Gap junctions act as molecular filters, allowing only ions and hydrophilic molecules that are smaller than the channel diameter to pass from cell to cell. As judged by passage or obstruction of fluorescently labelled molecules, the upper limit for passage is between 1.6 nm and 2.0 nm in mammalian cells (Schwarzmann et al. 1981), which would allow spherical molecules of roughly 1000 MW to pass. Thus, monovalent and divalent ions, small peptides and cyclic nucleotides may pass, which could serve in homeostasis and interestingly, to convey signals intercellularly.

Analysis of the rates of intercellular transfer via gap junctions clearly indicates a functionally important level of communication can occur. Kam et al. (1986) have shown that the fluorescent probe Lucifer Yellow (MW 440) when injected into one cell of the mouse dermis passes into more than 500 surrounding cells in less than 5 minutes. In a cell culture system combining two mutant fibroblast strains, one incapable of synthesizing purine nucleotides and the other incapable of synthesizing thymidine nucleotides, growth occurs at the normal rate of the wild type strain as a result of nucleotide exchange occurring at a rate greater than 10^7 nucleotides per second per cell pair (Pitts and Finbow 1986). The consequences of increased gap junctional area in smooth muscle of the uterus at parturition are therefore immense, allowing coordination of metabolism and facilitation of homeostasis in a severely stressed tissue, as well as mechanical coordination via increased electrical coupling.

ii Regulation of junctional permeability

In tissues where gap junction number is modulated in response to extracellular stimuli, for example, steroid hormones and the uterus (this review) or glucose stimulation of pancreatic beta-cells (Meda et al. 1979), the increase probably occurs by insertion of single particles, which then aggregate (Lane and Swales 1980; Yancey *et al.* 1979; Shivers *et al.* 1985). When gap junction number decreases, either particles disperse into surrounding membrane (Lane and Swales 1980; Yancey *et al.* 1979) or intact junctional membrane is taken into the cytoplasm by one cell, as an annular gap junction (Larsen 1981). In general, modulation of gap junction area occurs over time measurable in hours, which corresponds with the approximately 5-hour half-life of liver gap junction protein (Fallon and Goodenough 1981).

Rapid closure of gap junctions is necessary to isolate healthy cells from damaged neighbours, to prevent the general loss of the small cytoplasmic components, and also apparently to allow rapid adjustment of coupling levels to changing circumstances. Experimentally, cell coupling can be rapidly and reversibly blocked or raised with modulation of permeability occurring within seconds to minutes, according to the following, apparently independent mechanisms: (1) intracellular Ca^{2+} and intracellular pH; (2) intracellular cAMP and/or phosphorylation; (3) transjunctional voltage.

In addition, there are a number of pharmacological agents which act either through at least one of the above mechanisms, directly on gap junction structure, or else have an unknown mode of action (Spray and Bennett 1985).

• *Regulation by Ca and pH.* According to the Ca^{2+} hypothesis for regulation of cell coupling (Loewenstein 1966) elevation of Ca_i from normally low (approx. 10^{-7} M) levels results in junctional uncoupling. All systems tested for such a Ca effect can respond by reduced coupling (see Loewenstein 1981; Spray and Bennett 1985). However, this occurs at levels of 10^{-5} to 10^{-4} M Ca_i, which is felt to be unphysiological except in the case of irreversible cell injury (Spray *et al.* 1982). Nonetheless, Ca_i in excitable cells such as cardiac myocytes can normally reach 10^{-5} M (Fabiato 1982).

pH sensitivity of gap junctions has been demonstrated by Spray *et al.* (1981, 1986) who used embryonic cells of fish and amphibians to show that intercellular conductance decreases rapidly when pH_i (independently of Ca_i^{2+}) is displaced from value. Junctional conductance between mammalian cardiac cells is also very sensitive to slight shifts in pH_i. Although in general, cytoplasmic pH is well buffered and unlikely to fluctuate, such acidification of cardiac cytoplasm can accompany brief ischaemic episodes (Jacobus *et al.* 1982). In contrast, hepatocytes from rat liver are insensitive to alterations in pH_i up to 1 unit, which supports the idea of tissue specificity of gap junctions.

In addition, cytoplasmic components may be involved in determining the Ca sensitivity of coupling: Johnston and Ramon (1981) showed that internal perfusion of the septate axon caused loss of sensitivity to both Ca^{2+} and H^+, and Peracchia *et al.* (1985) have shown that lens gap junctions reconstituted in liposomes are initially Ca^{2+} insensitive, but that incorporation of calmodulin confers Ca^{2+} sensitivity and the ability to uncouple fully. Further, both Welsh *et al.* (1982) and Hertzbeg and Gilula (1981) have shown that calmodulin binds to lens and liver gap junctions. Therefore, some experiments may underestimate the role of Ca^{2+} due to an incomplete system. Ultimately, Ca^{2+} vs H^+ sensitivity *in vivo* rests on the relative stability of these ion levels in the cytosolic micro-environment immediately adjacent to the gap junction.

• *cAMP and junctional phosphorylation.* The intracellular second messenger, cAMP, is a recently recognized important agent in the regulation of cell coupling. An increased level of cAMP appears to alter the level of coupling in a variety of systems, according to two different mechanisms. First, a slow increase in coupling, involving hours to days, is seen in cultured mammalian cells treated with substances increasing cAMP (for example, Flag-Newton *et al.* 1981; Azarnia *et al.* 1981 and Radu *et al.* 1982). This generally involves increased numbers of gap junctions, and is dependent upon protein synthesis, which suggests that control is exerted at the level of junction formation.

A rapid response to cAMP or agents causing increased cAMP is seen in other systems such as insect salivary gland (Hax *et al.* 1974), vertebrate heart muscle (DeMello 1984) and vertebrate retinal horizontal cells (Lasater and Dowing 1985; Teranishi *et al.* 1984; Piccolino 1984). Recently, both direct elevation of cAMP and glucagon induced elevation of cAMP were shown to increase junctional conductance in liver hepatocytes (Saez *et al.* 1986).

In the case of the retina, and in rat uterine smooth muscle (Cole and Garfield 1986), increased cAMP decreases permeability. In the uterus, Cole and Garfield (1986) found that both direct elevation of cAMP (through stable analogues, theophyline and forskolin) and physiological agents elevating cAMP in the uterus (such as relaxin or isoproterenol) resulted in decreased junctional permeability, without change in junctional area. These fast permeability changes are presumably due to direct gating effects which, interestingly, can act either to increase *or* decrease coupling.

A possible mechanism for cAMP regulation of junctional permeability is by phosphorylation of the junctional protein, through cAMP dependent protein kinase. This mechanism is inferred from findings that both lens junctional protein (Louis *et al.* 1985) and liver gap junctions (Saez *et al.* 1986) are phosphorylated *in vivo* and *in vitro*. Further, cAMP causes phosphorylation of isolated liver junction protein, and this is correlated with increased junctional conductance in intact hepatocytes (Saez *et al.* 1986).

The dichotomy of responses to cAMP reflects tissue specialization, that is not, however, uniform among non-excitable or excitable cells. Conceivably, a stimulation of the exocrine function of the liver that also causes increased junctional permeability will tend to equalize the distribution of small molecules and second messengers, as well as involving uniformity of response among coupled cells by overcoming deficits in exposure to the signal, or in receptor number. In the uterus, cAMP control of junctional permeability represents a level of control additional to the hormonal regulation of junctional number (see text). Physiological agents such as relaxin that maintain a quiescent uterus in the latter stages of pregnancy may do so in part by decreasing permeability of the gap junctions as they develop. Ultimately, the agents that terminate pregnancy will counteract this tendency. Such a scheme for coordination of gap junction number and permeability emphasizes the increasingly complex events that appear to occur, as continuing research shows the importance of transmembrane signalling and the involvement of soluble cytosolic intermediates in regulation of coupling.

• *Transjunctional voltage.* In some tissue, a transjunctional voltage of the order of 10's of millivolts reversibly decreases junctional conductance (reviewed in Spray *et al.* 1984). While there is no voltage dependence of coupling between cell pairs from liver (Spray and Bennett 1985) or heart (White *et al.* 1985), this form of regulation in rectifying electrotonic synapses allows action potentials arriving in one direction only to be conducted (Auerbach and Bennett 1969; Margiotta and Walcott 1983). In the early embryo, transjunctional voltage sensitivity may isolate groups of cells that develop different membrane potential, and so support developmental autonomy. There is currently no information on voltage sensitivity in smooth muscle cells, but such regulation of coupling, while technically very difficult to demonstrate, could be significant in coordination of contraction by oscillations in membrane potential (slow wave activity), or in determining the extent of entrainment by pacemaker cells.

4   DIRECT EVALUATION OF CELL-CELL COUPLING IN SMOOTH MUSCLE

The study of cell-cell coupling in smooth muscle is made difficult by cellular heterogeneity and tissue geometry. Detection and measurement of electrical coupling by current injection is virtually impossible due to the rapid spatial decay of voltage, as current flows to neighbouring cells in 3-dimensions. Extracellular polarization of an area of a strip of muscle produces measurable electrotonic potentials, but the current strength is unknown, and changes in spatial decay cannot be interpreted as due to junctional effects alone (Tomita 1970). One method that circumvents these problems is to use tissue culture techniques to provide an effectively 2-dimensional monolayer of cultured smooth muscle, where cell type and extracellular fluid composition are fully defined, with excellent visibility of cells.

In the only available quantitative study of junctional communication in smooth muscle, we examined directly cell-cell coupling in cultured vascular smooth muscle (Blennerhassett *et al.* 1987a). Cultured aortic smooth muscle cells, with membrane potentials matching those *in vivo*, were grown to confluence, and a quantitative study of electrical coupling was performed using a 2-dimensional model of electrical coupling that allows separation of junctional from nonjunctional resistance. The intercellular resistance was 900–1400 ohm cm, much lower than the nonjunctional or lateral membrane resistance of $> 10^4$ ohm cm^2. These values and the calculated space constants were comparable to those estimated by cable analysis in intact smooth muscle.

Figure 10.5 A & B illustrates the intercellular spread of the fluorescent dye 6-carboxy fluorescein (CF; MW 376), between the cultured aortic smooth muscle cells at an early stage of their growth and shows the capability for rapid molecular transfer between cells. When confluent cultures were exposed to Ca^{2+} ionophores and retested for CF passage, CF passage persisted in conditions that uncoupled fibroblasts in equivalent cultures. However, CF passage was blocked by the calcium selective ionophore ionomycin, at 10^{-5} M (Fig. 10.5 C & D). In addition to allowing direct study of both molecular transfer and electrical coupling, such methods can be used to determine the agents that regulate coupling in smooth muscle.

V   Myometrial gap junctions

Labour and parturition in humans and other mammals is characterized by the development of intense, synchronous, and coordinated contractile activity of the myometrium of the uterus from a relatively inactive state during most of pregnancy (Csapo 1981). In all species studied, the onset and progression of this contractile activity during term or preterm labour is invariably associated with the presence of large numbers of gap junctions between the myometrial cells (Garfield et al. 1977, 1978). Moreover, improved electrical (Sims et al. 1982) and metabolic (Cole et al. 1985) communication between uterine smooth muscle cells is observed concomitant with the formation of junctions, supporting the hypothesis that the gap junctions permit the myometrium to behave as a functional syncytium during parturition. These alterations in the extent of structural and functional coupling are significant, therefore, in that the presence of gap junctions and cell-to-cell communication probably represents the biophysical basic for synchronous and effective uterine contractile activity during labour.

Evidence from our studies and that of others suggests the presence of specific physiological mechanisms for regulating and producing alterations in structural and functional coupling in the myometrium during pregnancy and parturition. Below we describe the possible mechanisms involved in the control of coupling in uterine smooth muscle: (1) the presence of the gap junctions and hence the extent of structural coupling; and (2) the permeability of the gap junctions, and hence the extent of functional coupling in the myometrium (Fig. 10.6). The integrated function of these control mechanisms presumably operates to ensure the appropriate activation and maintenance of synchronous activity in the myometrium and effective delivery of the fetus(es).

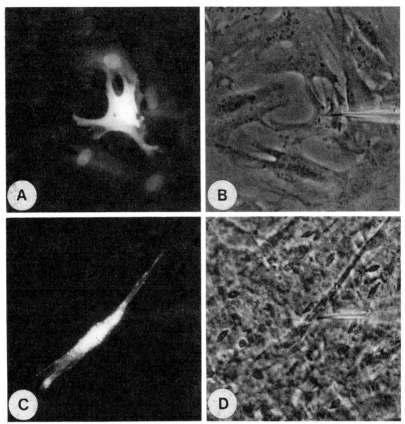

Fig. 10.5 Intracellular injection via microelectrodes of a fluorescent probe can directly show the passage of small molecules between cells. Here, the fluorescent tracer 6-carboxy fluorescein (CF; MW376) passes between cultured aortic smooth muscle cells, but passage is blocked by the application of a calcium ionophore (A-D, x 280). A. Fluorescence is detectable in five adjacent cells at 28 sec. of injection of CF into one cell of a low density culture. B. Phase contrast view of cell field in A. C. The Ca ionophore ionomycin (10^{-5} M) blocks intercellular spread of CF in a high density culture (45 sec of injection into the source cell, with constant membrane potential). D. Phase contrast view of C. Cell shape changes with increasing culture density (compare C with A,B).

1 PRESENCE OF GAP JUNCTION DURING LABOUR

It is now established that gap junctions occupy a significant percentage of the area of the uterine smooth muscle cell plasma membrane (ca. 0.1–0.4 per cent) only during term or preterm labour, that is, when the muscle is functionally active (Fig. 10.7). Quantitative transmission electron microscopic morphometric analysis was employed to document alterations in structural coupling in the uterine muscle of a variety of mammalian species throughout

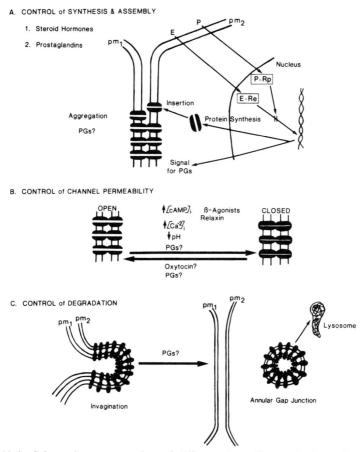

MODEL of REGULATION of GAP JUNCTIONS

BETWEEN MYOMETRIAL CELLS

Fig. 10.6   Schematic representation of different sites of control of gap junctions between plasma membranes (pm$_1$ and pm$_2$) of two myometrial cells. Shown are: A) possible roles for oestrogens (E) and progesterone (P) in controlling the synthesis and assembly of the junctions; B) regulation of permeability of the junctions by cAMP and Ca^{2+} and agonists which influence their levels; and ·C) degradation of junctions.

pregnancy (Garfield *et al.* 1972, 1986). Gap junctions are consistently absent, or present in low frequency and small size in nonpregnant, as well as preterm and postpartum animals (Garfield *et al.* 1977, 1978, 1979). In pregnant animals, the junctions begin to form about one day prior to the onset of labour. Gap junctions are always present in large numbers (ca. 1000 per cell) and increased size (ca. 250 nm) during normal delivery of the fetuses, but disappear within 24 hours after parturition (Garfield *et al.* 1977). This pattern of altered

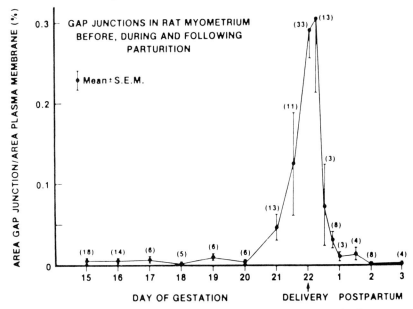

Fig. 10.7. Area of gap junctions (% of plasma membrane area) in the rat myo-metrium from day 15 of gestation to 3 days postpartum. Note that area begins to rise on day 21, is at its highest peak during delivery, and then declines sharply 24 hours following delivery.

structural coupling in the myometrium is particularly prominent in rats and rabbits (Garfield *et al.* 1978; Demianczuk *et al.* 1984). Guinea pigs and sheep (Garfield *et al.* 1982; Garfield *et al.* 1979) appear to differ slightly in that they demonstrate higher numbers of gap junctions prior to term. The gap junction profile in women during pregnancy and normal, spontaneous vaginal delivery is not known, but junctions are present in greater numbers in tissues from women undergoing Caesarean section and in labour compared with those not in labour (Garfield and Hayashi, 1981). It is also significant that gap junctions are invariably present in myometrial tissues from animals undergoing premature labour, either as a result of experimental manipulation or pathology (Garfield *et al.* 1978; Garfield *et al.* 1980; Garfield *et al.* 1982; MacKenzie and Garfield 1985, 1986). Moreover, if the development of gap junctions is delayed, then pregnancy is prolonged (Garfield *et al.* 1978). Thus, myometrial gap junctions are dynamic and transitory structures whose presence is inextricably associated with the conversion of the uterus into an active organ just prior to parturition. There is no known exception to this phenomenon and, for this reason, gap junctions appear to be necessary for effective labour. Gap junctions also develop spontaneously when a tissue is incubated *in vitro* (Garfield *et al.* 1978, 1979). The junctions begin to appear in significant numbers after about two hours, and after three days *in vitro* attain values approximately three times those present during labour. These results demonstrate that there is some mechanism suppressing the development of the junctions *in vivo* prior to term and should be borne in mind with *in vitro* investigations of electrical properties.

2   REGULATION OF THE PRESENCE OF THE JUNCTION

In a variety of species there are changes in the synthesis and/or breakdown of several circulating and local hormones immediately prior to term or preterm labour (Thorburn and Challis 1979; Challis and Lye 1986). In particular, in species such as rats and sheep, oestrogens increase, whereas progesterone decreases (progesterone withdrawal), at term. Some of the cycloxygenase breakdown products of arachidonic acid, such as prostaglandins $F_{2\alpha}$ and E_2, increase whereas others, such as prostacyclin may decrease immediately prior to parturition (Thorburn and Challis 1979). The role(s) of the steroid hormones and prostaglandins play in regulating the presence of myometrial gap junctions has been studied both *in vivo* and *in vitro*. These studies show that oestrogen stimulates, whereas progesterone inhibits gap junction formation in the myometrium (Garfield 1977; Garfield *et al.* 1978; Puri and Garfield 1982; MacKenzie and Garfield 1985; MacKenzie and Garfield 1986). The role of the prostaglandins remains unclear: there is evidence for both stimulatory and inhibitory prostanoids (MacKenzie *et al.* 1983).

i   Steroids

The changes in the synthesis of steroid hormones, as reflected in tissue and plasma levels that occur before labour are thought to be responsible for activating the synthesis of myometrial gap junctions, probably through a genomic mechanism. Evidence that progesterone suppresses myometrial gap junctions is suggested from the following studies. Ovariectomy of pregnant rats subsequent to day 15 following conception leads to premature formation of gap junctions and labour within 24–48 hours (Garfield *et al.* 1977). The fact that this premature alteration in gap junctions results from an experimentally induced progesterone withdrawal is suggested by a decline in progesterone levels and the ability of progesterone injections to prevent the descent, prevent the formation of gap junctions, and prolong pregnancy for as long as treatment is continued. Administration of progesterone to intact rats over the last few days of pregnancy (that is, day 19 onward) will similarly prevent normal progesterone withdrawal, the appearance of gap junctions, and parturition. This hormone will also inhibit gap junctions development in tissues in nonpregnant animals *in vivo*, or incubated *in vitro* (Garfield *et al.* 1978; MacKenzie and Garfield 1985). These observations prompted us (Garfield *et al.* 1977, 1978; Garfield, 1984) to suggest that progesterone inhibits labour and maintains pregnancy by suppressing the formation of gap junctions and limiting electrical cell-to-cell coupling in the myometrium.

Moreover, we proposed that the development of gap junctions and their control by steroids may be the basis of the progesterone block hypothesis advanced by Csapo (1981).

However, it is not entirely evident how progesterone manifests its influence on myometrial gap junction formation. We have contended that this steroid could act at one or more of the following: inhibition of protein (connexon) synthesis either directly or indirectly through inhibition of the oestrogen nuclear receptor-genome interaction, and/or an indirect effect through manipulation of the synthesis of prostaglandins. More recently, we have demonstrated that the antiprogesterone compound RU 38486 stimulates the premature development of gap junctions and that this is accompanied by premature delivery (Garfield and Baulieu, 1987). These studies would tend to support the hypothesis that progesterone suppresses the synthesis of gap junctions through a receptor-genome mechanism.

Injection of relatively high doses of oestrogen (either oestradiol or diethylstilbesterol, a synthetic oestrogen) into immature (Merk *et al*. 1980; Mackenzie *et al*. 1983), mature ovariectomized nonpregnant (Mackenzie and Garfield 1985), and intact, day 18–20 pregnancy (MacKenzie and Garfield 1986) rats will induce the formation of myometrial gap junctions, and in the latter animals, it will provoke premature labour. Similarly, oestrogen will potentiate the formation of junctions in tissues incubated *in vitro* (Garfield *et al*. 1978). This stimulatory influence of oestrogen on gap junction synthesis is not unique to the myometrium and has been observed in other reproductive tissues, such as the ovary (Merk *et al*. 1972), where steroid receptors are present.

It is thought that the oestrogens give rise to the development of gap junctions by stimulating the synthesis of connexon proteins (Garfield *et al*. 1978). The ability of the steroid to stimulate uterine protein synthesis is well recognized and is believed to result from a direct stimulation of transcription through interaction of oestrogen-nuclear receptor complexes with the genome (Gorski and Gannon 1976). The ability of an antioestrogen, tamoxifen (MacKenzie and Garfield 1985), as well as actinomycin D and cyclohexamine (Garfield *et al*. 1980b), to inhibit gap junction development in nonpregnant animals in response to oestrogens is consistent with a nuclear control of connexon synthesis. Moreover, mRNA obtained from the myometria of oestrogen-primed rats is thought to induce the formation of gap junctions when incorporated into mutant cells that normally lack the capability to form junctions (Dahl *et al*. 1980). However antioestrogens will not prevent the development of gap junctions or delivery at term in rats (Garfield, unpublished observations).

ii   Prostaglandins

The ability of metabolites of arachidonic acid liberated from the plasma membrane of smooth muscle cells, endometrial cells, or the fetal membranes to influence the development of gap junctions in the myometrium is suggested by several lines of experimentation. However, the exact species of prostaglandins, their role and the mechanism by which they operate remain to be identified.

Gap junctions were significantly less numerous in the nondistended horns of unilaterally ovariectomized, parturient rats (Garfield *et al.* 1977). Similarly, gap junction frequency in the myometrium of postpartum rats following oestrogen injections was the greatest, and more closely approached that observed in parturient animals, in uterine horns distended by intrauterine balloons (Wathes and Porter 1982). Distention of the myometrium is a well-documented stimulus of uterine prostaglandin synthesis (Thorburn and Challis 1979), and on the basis of these observations it seems probable that prostanoids and stretch are required to attain the extent of structural coupling observed in the distended horns of parturient animals.

Evidence for stimulation and inhibition of myometrial gap junction development by prostaglandins has been obtained through the use of two inhibitors of cycloxygenase (the initial enzyme in the prostaglandin pathway), indomethacin, and sodium meclofenamate. When myometrial tissues from midterm pregnant rats were incubated *in vitro* in the presence of indomethacin, the normal development of gap junctions was dramatically suppressed (Garfield *et al.* 1980*a,b*). This would suggest that a stimulatory prostaglandin is involved in the *in vitro* formation of gap junctions. In contrast, however, treatment of immature or ovariectomized mature rats (with nondistended uteri) with a combination of oestrogen and either indomethacin or sodium meclofenamate potentiates the stimulatory effect of the steroid (MacKenzie and Garfield 1985). Neither cycloxygenase inhibitor exerts an effect when administered alone.

In vitro experiments show that some prostaglandins, for example thromboxanes and endoperoxides, may stimulate and others inhibit the development of gap junctions (Garfield *et al.* 1980*a,b*). These data exemplify the complex nature of the prostaglandin influence of myometrial gap junctions. The *in vivo* actions of the cycloxygenase inhibitors have been interpreted to indicate that some product of the cycloxgenase pathway, perhaps prostacyclin, inhibits the stimulus provided by oestrogen treatment or, alternatively, that a product of the lipoxygenase pathway of arachidonic acid metabolism may potentiate the oestrogenic stimulation of gap junction formation (MacKenzie *et al.* 1983). In the presence of the cycloxygenase inhibitors, the activity of lipoxygenase may be enhanced and the effect of oestrogen potentiated. It should be noted, however, that the effects of the lipoxygenase products or their inhibitors on gap junction formation are as yet untested.

How the arachidonic acid metabolites may influence gap junction synthesis remains to be clarified. They could influence the synthesis of connexons directly, either at the level of the steroid receptors or translation (MacKenzie *et al.* 1983). Alternatively, it was previously suggested that they may be involved in the regulation of connexon aggregation by either altering membrane fluidity or influencing protein-protein cross-linking in the plasma membrane leading to a stabilization of developing gap junctions (Garfield *et al.* 1980*a,b*). Prostaglandins can influence the interaction between steroids and their specific receptors and can inhibit the binding of these hormones by uterine tissues (Sanfilippo *et al.* 1983). Thus, an interaction between, and down regulation of, the oestrogen receptor by some inhibitory prostanoid is a distinct possibility. Clearly, considerably more experimentation is required before we fully understand the role of the prostaglandins in the regulation of myometrial gap junction development.

It should be noted that many of our studies have implied that changes in the synthesis of gap junctions are important determinants in regulating their presence in the myometrium. However, the destruction or degradation of the gap junctions may also play a significant role in this process. Gap junction degradation previously was suggested to involve an aggregation of small gap junctions into very large contacts, prior to the formation of annular junctions and a withdrawal of these into one of the adjacent cells by an endocytotic mechanism (Fig. 10.6; Garfield *et al.* 1980*b*). It remains to be shown what regulates gap junction degradation; perhaps prostaglandins or other substances are responsible for either stimulating or inhibiting this process. Clearly, this is an important aspect since an inhibition of the breakdown of gap junctions may have the same effect as stimulating their formation, and a stimulation of degradation should have the opposite effect.

3   FUNCTIONAL STUDIES

In 1977 we proposed that the role of the gap junctions was to couple uterine smooth muscle cells into a functional syncytium, facilitating synchronized electrical and contractile activity in the uterine wall (Garfield *et al.* 1977). That the junctions should function in this capacity is consistent with their role as sites for cell-to-cell communication in other tissues (see above). However, attempts to evaluate whether there is a change in electrical coupling in the myometrium at term have produced conflicting results. This is most likely due to the inherent difficulties in the study of electrical coupling between smooth muscle cells (see Holman and Neild 1979).

This difficulty has led to the development of techniques that are thought to evaluate the extent of coupling indirectly. Four studies on coupling in the myometrium during pregnancy and parturition have used the so-called Abe-Tomita method (see Sims *et al.* 1982). Sims *et al.* (1982) observed a significant increase in the space constant in myometrial tissues from parturient rats (3.68 ± 1.0 mm) compared with day 17–22 pregnant, nondelivering rats (2.59 ± 0.84 mm), suggestive of improved cell-cell flow of current during labour. However, the inability of three previous studies (Daniel and Lodge 1973; Kuriyama and Suzuki 1976; Zelcer and Daniel 1979) to identify similar changes in tissues at term and the consistently large magnitude of this parameter in nondelivering tissues, which were thought to possess very few junctions, remain to be adequately explained. This may be the result of gap junction formation *in vitro* during recording, a failure to document the extent of structural coupling in the tissues employed for the functional measurements, and/or that the Abe-Tomita recording technique does not accurately reflect the extent of cell-to-cell coupling. In addition, it may be that the level of coupling prior to term as measured *in vitro* overestimates that *in vivo* because of the lack of regulation of functional coupling (that is, permeability) by circulating hormones (see below).

In light of some of these problems, we recently chose to evaluate alterations in coupling by comparing cell-to-cell diffusion of a radiolabelled metabolite, [^{3}H]2-deoxyglucose (2DG), in small strips of longitudinal myometrium from day 17–20 pregnant rats (few gap junctions) and parturient rats (many gap junctions) (Cole *et al.* 1985). A two-compartment bathing chamber technique similar to that used by Weidmann (1966) was employed. One portion of strips of longitudinal myometrium was exposed to the tracer and its longitudinal distribution determined following 5 hours for diffusion. 2DG enters the cells by substituting for glucose on the facilitative carrier; once inside it is phosphorylated and in this form is not metabolized further and it is not able to recross the plasma membrane. Thus, the tracer remains in a diffusible pool in the cytoplasmic compartment of the muscle cells and, given appropriate cell-to-cell pathways, it can diffuse through the muscle strip.

The spatial distribution of 2DG was found to be considerably greater in strips from parturient compared with the nonparturient rats (Cole *et al.* 1985; Cole and Garfield 1986). This implies an increased diffusivity of 2DG in the myometrium of delivering rats which is reflected, in quantitative terms, by an almost 10-fold change in the apparent diffusion coefficient for the tracer at term. Similar experiments using tritiated sucrose and mannitol indicated that diffusion through the extracellular space could not account for either the distribution of 2DG in the different tissues or its altered diffusivity in those from parturient animals. We concluded that the increased rate of tracer redistribution in the parturient tissues was the result of the larger area of gap junction between the smooth muscle cells. The results of these diffusion experiments suggest that there is a dramatic increase in the extent of functional coupling in the myometrium at term concomitant with the formation of gap junctions.

The diffusion of radiolabelled deoxyglucose is an indirect measure of molecular coupling in the uterus. Direct study is possible using *ex situ* preparations of longitudinal muscle from rat myometrium. These are viewed with a standard microscope with phase contrast/fluorescent optics, while a fluorescent dye such as carboxy fluorescein (CF) is injected with an intracellular

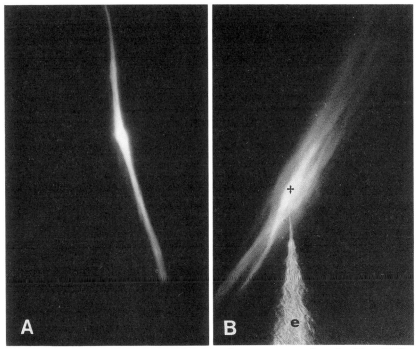

Fig. 10.8. Using microelectrodes, the fluorescent probe Lucifer Yellow (LY, MW440) can be injected into single cells of an intact preparation of uterine longitudinal smooth muscle. The rate of spread of LY from the injected cell to its neighbours, as seen by the development of fluorescence within adjacent cells, is a direct measure of the extent of intracellular coupling. Intracellular spread of LY in very slow (A) in preterm (Day 17) myometrium, where gap junctions are rare, but is relatively rapid (B) in the parturient myometrium with abundant gap junctions (A, B; x 112). A). Fluorescence micrograph of the result of injection of LY for 70 sec into longitudinal muscle from preterm myometrium, showing a single brightly fluorescent cell and failure of spread of fluorescence to adjacent cells; B). Fluorescence micrograph of the result of injection of LY for 60 sec from a microelectrode (e) into one cell of parturient longitudinal muscle, showing fluorescence filling cells adjacent to the source cell (+).

microelectrode. Electrical parameters can be recorded, and the intercellular passage of CF evaluated (Blennerhassett *et al.* 1987*b*). Using this technique, we have examined the effect of gap junction number and the rate of spread of the fluorescent probe Lucifer Yellow (LY, MW440), by taking longitudinal muscle from the myometrium of rats at day 17 (D17) of gestation or during normal delivery (D22). Figure 10.8A shows the result of LY injection for 70 sec into one cell in D17 myometrium: the injected cell is brightly fluorescent, but no dye transfer to adjacent cells is detectable. In contrast, Figure 10.8B shows that injection of LY into a cell of D22 myometrium results in spread to adjacent cells. The dye injection in Figure 10.8A was continued for a further 500 sec before the membrane potential depolarized, by which time fluorescence was detectable in two adjacent cells, proving that the injected cell was not artefactually uncoupled, but rather lacked the ability for rapid transfer of dye. This correlates with the presence of the few, small gap junctions that can be seen in D17 tissue.

Therefore, direct studies show that extensive molecular coupling (and hence metabolic cooperation *in vivo*) becomes possible in delivering myometrium as a consequence of increased gap junction area. This upswing in molecular coupling contrasts somewhat with the moderately effective electrical coupling already present before term (see text), and leads to the suggestion that increased metabolic cooperation is at least as important a consequence as improved electrical coordination. Possibly, endogenous down-regulation of cell coupling is present *in vivo* but is lost when the tissue is isolated.

We have also recently completed studies of the spontaneous electrical activity of the myometrium on tissues from preterm (day 17 of gestation) and delivering animals (Miller *et al.* 1987). Electrical events were recorded at six separate sites along a 15 mm distance in the longitudinal and circular direction with extracellular electrodes. The results of this study revealed that burst discharges propagated over the entire recording distance in the longitudinal axis of the myometrium at preterm and at parturition. However, individual spikes within the bursts propagated further and with higher velocity into the long axis, at parturition. In the transverse axis of the myometrium, both bursts, and individual spikes within bursts, propagated over longer distances at parturition than before. Furthermore, analysis of the propagation of spikes evoked by electrical stimulation confirmed that spike propagation was improved (that is, higher velocity and longer distance of spread) in both the longitudinal and transverse axis of the myometrium during delivery when the junctions were present. Thus, improved propagation of electrical discharges was associated with an increase in gap junction contact. These results are consistent with the hypothesis that the presence of gap junction at term improves electrical coupling of myometrial cells and coordinates contractility.

4 PERMEABILITY OF THE JUNCTIONS

Alterations in the extent of functional coupling are thought to occur in the absence of a change in structural coupling in a wide variety of cell types (see above). These observations suggest that changes in junctional permeability may be obtained through alterations in the gap junctional connexons, such as an all-or-none closure or dilation of the cell-cell channel, leading to a state of either decreased or enhanced coupling, respectively. Examples of a modulation of junctional permeability by hormones (Hermsmeyer 1982) or neurotransmitters (Laufer and Salas 1981; Piccolino *et al.* 1982) have been described in the literature. That the permeability of the junctions and, therefore, the extent of functional coupling in the myometrium, may be regulated by endogenous mechanisms is an intriguing possibility. Improved functional coupling would be expected to promote greater electrical and contractile synchrony in the uterine wall and lead to an enhanced rate of intrauterine pressure development and more effective labour. Alternatively, gap junction closure would reduce synchrony and lead to ineffective labour and a prolongation of pregnancy.

5   CONTROL OF PERMEABILITY

We have used the 2DG diffusion technique described above to study the influence of several agents on coupling in the myometrium of parturient rats. Our studies suggests that the permeability of the junctions in the parturient myometrium is influenced by intracellular Ca^{2+}, pH and cAMP (Cole and Garfield 1986). Elevated intracellular Ca^{2+} and lowered pH produced by the calcium ionophore A23187 and the substituted benzyl ester, o-nitrobenzyl acetate (see Spray $et\ al.$ 1984), respectively, reduced the longitudinal distribution and apparent diffusion coefficient of 2DG in the parturient myometrium: however the latter drug is rather ineffective compared with A23187. Smooth muscle cells may possess a very effective mechanism for regulating interacellular pH and this might explain the inability of the substituted ester to reduce coupling in the myometrium as markedly as that seen in other cell types (Spray $et\ al.$ 1984). It appears that calmodulin may be required to confer calcium sensitivity to the junctions because a normal distribution of 2DG is observed in tissues treated with A23187 in the presence of calmodulin antagonists, such as chloropromazine and calmidazolium. However, the role played by calmodulin in the regulation of junctional permeability is complex in that the calmodulin antagonists are able to produce a dose-dependent uncoupling when administered alone. This latter result would seem to suggest that calmodulin may be involved indirectly in the regulation of coupling as well, perhaps by participating in the regulation of intracellular cAMP. Morphometric analysis of the tissues used in these uncoupling experiments failed to show any differences in the extent of structural coupling in the treatment groups, consistent with uncoupling by a modulation of channel permeability.

Elevated cAMP produced by treatment with dibutyryl- or 8-bromo-cAMP also reduces cell-cell diffusion of 2DG in the myometrium, and this can be mimicked by inhibiting phosphodiesterase activity with theophylline or by stimulating adenylate cyclase with forskolin. That cAMP may play a role in regulating functional coupling in the myometrium is significant in that relaxin, prostacyclin (PI_2), and β_2-adrenoceptor agonists appear to influence labour and parturition (Thorburn and Challis 1979) and exert inhibitory effects on the myometrium by elevating intracellular cAMP (Vesin $et\ al.$

1979; Sanborn *et al.* 1980). Moreover, in preliminary experiments with porcine relaxin, carbacyclin (a stable PGI_2 analog), and isoproterenol (a nonspecific β-adrenoceptor agonist), the diffusivity of 2DG was found to be significantly lower in treated compared with control tissues. Evidently, there may be specific receptor and secondary messenger-mediated physiological mechanisms for controlling cell-cell communication in the myometrium independent of the systems controlling structural coupling.

It is tempting to speculate that a cAMP-mediated uncoupling mechanism may be involved in maintaining pregnancy in instances of premature junction formation and/or in species such as the guinea pig, sheep, and possibly human, in which low, but significant numbers of gap junctions are present throughout pregnancy. Perhaps the high levels of relaxin and prostacyclin observed in preterm pregnant animals act to elevate intracellular cAMP and, alternatively, prevent synchronous activity in the myometrium. A decline in these hormones or their receptors, or an antagonism of their action by oxytocin or stimulatory prostaglandins (for example PGF_2 or PGE_2), may facilitate a shift to patent gap junction channels and the development of syncytial behaviour.

Whether junctional permeability in the myometrium can be enhanced by circulating hormones is unresolved at this time. Oxytocin or a stimulatory prostaglandin, such as $PGF_{2\alpha}$, PGE_1, or PGE_2, could increase cell-cell communication between uterine smooth muscle cells either through a direct interaction with the gap junctions or, as noted above, by reversing the inhibition of coupling produced by relaxin and/or prostacyclin. These agonists all promote labour and there is evidence that their action is in part to increase the rate of intrauterine pressure development (Csapo 1981). As noted previously, the rate of pressure development is dependent on the rate of activation of individual cells in the myometrium and hence influenced by the extent of functional coupling.

VI Summary

In this review we have attempted to describe the morphology and electrophysiology of the myometrium and also the myogenic, neurogenic and hormonal systems which control its activity. The synthesis of myometrial gap junctions appears to play a significant role in the gradual evolution of uterine contractility during labour. We suggest that the development of myometrial gap junctions is physiologically regulated by the steroid hormones, their receptors and prostaglandins. Furthermore, the permeability or conductance of the gap junction channels as well as their degradation seems to be hormonally regulated (Figure 10.6). Thus, there are several sites for the control of the presence and function of the gap junctions and the management of uterine contractility.

Acknowledgement

The work reported here by the authors has been supported by a grant from the Medical Research Council of Canada to R. E. Garfield.

References

Abe, Y. (1968). Cable properties of smooth muscle. *J. Physiol., Lond.* **196**, 7–100.

—— (1970). The hormonal control and the effects of drugs and ions on the electrical and mechanical activity of the uterus. In *Smooth Muscle.* (eds E. Bulbring, A. F. Brading, A. W. Jones, and T. Tomita) pp. 396–417. Edward Arnold, London.

—— (1971). Effects of changing the ionic environment on passive and active membrane properties of pregnant rat uterus. *J. Physiol., London.* **214**, 173–190.

Adham, N. and Schenk, E. A. (1969). Autonomic innervation of the rat vagina, cervix and uterus and its cyclic variation. *Am. J. Obstet. Gynec.* **104**, 508–516.

Afting, E.-G. and Elce, J. S. (1978). DNA in rat uterus myometrium during pregnancy and postpartum involution. *Annals. Biochem.* **86**, 90–99.

Aiken, J. W. (1972). Aspirin and indomethacin prolong parturition in rats. Evidence that prostaglandins contribute to the expulsion of the fetus. *Nature., Lond.* **24**, 21–25.

Alexandrova, M. and Soloff, M. S. (1980). Oxytocin receptors and parturition. I. Control of oxytocin receptor concentration in the rat myometrium at term. *Endocrinology,* **106**, 730–735.

Alm, P., Alumets, J., Hakanson, R., Owman, Ch., Sjoberg, N.-O., Sundler, F., and Walles, B. (1980). Origin and distribution of VIP (Vasoactive Intestinal Polypeptide)-nerves in the genito-urinary tract. *Cell Tissue Res.* **205**, 337–347.

Anderson, G. F., Kawarabayashi, T., and Marshall, J. M. (1981). Effect of indomethacin and aspirin on uterine activity in pregnant rats: comparison of circular and longitudinal muscle. *Biol. Reprod.* **24**, 359–372.

Anderson, N. C. (1969). Voltage-clamp studies on uterine smooth muscle. *J. gen. Physiol.* **54**, 145–165.

—— Ramon, F., and Snyder, A. (1971). Studies on calcium and sodium in uterine smooth muscle excitation under current-clamp and voltage-clamp conditions. *J. gen. Physiol.* **58**, 322–339.

—— —— (1976). Interaction between pacemaker electrical behaviour and action potential mechanism in uterine smooth muscle. In *Physiology of Smooth Muscle.* (eds E. Bülbring and M. F. Shuba) pp. 53–63. Raven Press, New York.

—— (1978). Physiological basis of myometrial function. *Semin. Perinatology,* **2**, 211–222.

Auerbach, A. A. and Bennett, M. V. L. (1969). A rectifying synapse in the central nervous system of a vertebrate. *J. gen. Physiol.* **53**, 211–237.

Azarnia, R., Dahl, G., and Loewenstein, W. R. (1981). Cell junctions and cyclic AMP: III. Promotion of junctional membrane permeability and junctional membrane particles in a junction-deficient cell type. *J. Membr. Biol.* **63**, 133–146.

Barr, L., Berger, W., and Dewey, M. M. (1968). Electrical transmission at the nexus between smooth muscle cells. *J. gen. Physiol.* **51**, 347–368.

Batra, S. (1985). Effect of oxytocin on calcium movements in uterine smooth muscle. *Regul. Peptides Suppl.* **4**, 8–81.

Baulieu, E. E. (1985). RU486: an antiprogestin with contragestive activity in women. In *The Antiprogestin Steroid RU486 and Human Fertility Control.* (eds E. E. Baulieu and S. J. Segal) pp. 1–25. Plenum Press, New York.

Bell, C. (1972). Autonomic nervous control of reproduction: circulatory and other factors. *Pharmac. Rev.* **24**, 657–736.

Benedetti, E. L., Dunia, I., and Bloemendahl, H. (1974). Development of junctions during differentiation of lens fibres. *Proc. natn. Acad. Sci. U.S.A.* **71**, 5073–5077.

Bengtsson, B., Chow, E. M. H., and Marshall, J. M. (1984*a*). Calcium dependency of pregnant rat myometrium: comparison of circular and longitudinal muscle. *Biol. Reprod.* **30**, 869–878.

——— ——— ——— (1984*b*). Activity of circular muscle of rat uterus at different times in pregnancy. *Am. J. Physiol.* **246**, C216–C223.

Bennett, M. V. L., Spray, D. C., Harris, A. L., Ginzberg, R. D., Campos de Cavalho, A., and White, R. L. (1984). Control of intracellular communication by way of gap junctions. In *The Harvey Lectures*, **Series 78**, pp. 23–57. Academic Press, Orlando.

Bergman, R. A. (1968). Uterine smooth muscle fibres in castrate and estrogen-treated rats. *J. Cell Biol.* **36**, 639–648.

Blennerhassett, M. G., Kannan, M. S., and Garfield, R. E. (1987*a*). Functional characterization of cell-to-cell coupling in cultured rat aortic smooth muscle. *Am. J. Physiol.* **252**, C555–C569.

——— ——— ——— (1987*b*). Regulation of gap junctions in myometrial smooth muscle. *Ann. N.Y. Acad. Sci.* **484**, 196–198.

Bortoff, A. and Gilloteaux, J. (1980). Specific tissue impedances of estrogen- and progesterone-treated rabbit myometrium. *Am. J. Physiol.* **238**, C34–C42.

Bozler, E. (1938). Electrical stimulation and conduction of excitation in smooth muscle. *Am. J. Physiol.* **122**, 616–623.

——— (1941). Influence of estrone on the electric characteristics and motility of uterine muscle. *Endocrinology*, **29**, 225–227.

Burnstock, G. and Prosser, C. L. (1960). Conduction in smooth muscles: comparative electrical properties. *Am. J. Physiol.* **199**, 553–559.

——— (1970). Structure of smooth muscle and its innervation. In *Smooth Muscle.* (eds E. Bulbring, A. F. Brading, A. W. Jones, A. W., and T. Tomita) pp. 1–69. Edward Arnold, London.

Carsten, M. E. (1968). Regulation of myometrial composition, growth and activity. In *Biology of the Uterus.* **Vol. 1** (ed. N. S. Assali) pp. 355–423. Academic Press, New York.

Casteels, R. and Kuriyama, H. (1965). Membrane potential and ionic content in pregnant and nonpregnant rat myometrium. *J. Physiol., Lond.* **177**, 263–287.

Chalcroft, J. P. and Bullivant, S. (1970). An interpretation of liver cell membrane and junction structure based on observation of freeze-fracture replicas of both sides of the fracture. *J. Cell Biol.* **47**, 49–60.

Challis, J. R. G. (1984). Characteristics of parturition. In *Maternal-Fetal Medicine.* (eds R. K. Creasy and R. Resnik) pp. 401–414. W. B. Saunders, Philadelphia.

——— Lye, S. J. (1986). Parturition. In *Oxford Rev. Reprod. Biol.* Vol. 8 (ed. J. R. Clarke) pp. 61–129. Clarendon Press, Oxford.

Chamley, W. A. and Parkington, H. C. (1984). Relaxin inhibits the plateau component of the action potential in the circular myometrium of the rat. *J. Physiol., Lond.* **353**, 51–65.

Chan. W. Y. (1983). Uterine and placental prostaglandins and their modulation of oxytocin sensitivity and contractility in the parturient uterus. *Biol. Reprod.* **29**, 680–688.

Chow, E. H. M. and Marshall, J. M. (1981). Effects of catecholamines on circular and longitudinal uterine muscle. *Eur. J. Pharmac.* **76**, 157–165.

Cole, W. C., Garfield, R. E., and Kirkaldy, J. S. (1985). Gap junctions and direct intercellular communication between rat uterine smooth muscle cells. *Am. J. Physiol.* **249**, C20–C31.

—— —— (1986). Evidence for physiological regulation of gap junction permeability. *Am. J. Physiol.* **251**, C411–C420.

Crankshaw, D. J. (1985). Oxytocin receptors: comments on their regulation. In *Oxytocin: Clinical and Laboratory Studies.* (eds J. A. Amico and A. G. Robinson) pp. 277–283. Elsevier, Amsterdam.

Creed, K. E. (1979). Functional diversity of smooth muscle. *Br. med. Bull.* **35**, 243–247.

Csapo, A. I. (1961). Defence mechanism of pregnancy. In *Progesterone and the Defence Mechanism of Pregnancy.* Ciba Foundation Study Group No. **9**, pp. 1–27. Little Brown, Boston.

—— (1962). Smooth muscle as a contractile unit. *Physiol. Rev.* **42**, 7–33.

—— (1969). The luteo-placental shift, the guardian of prenatal life. *Postgrad. med. J.* **45**, 57–64.

—— (1971). The uterus: a model for medical considerations. In *Contractile Proteins and Muscle* (ed. K. Laki) pp. 413–482. Marcel Dekker, New York.

—— (1977). The "see-saw" theory of parturition. In *The Fetus and Birth* Ciba Fdn. Symp. **Vol 47** (eds J. Knight and M. O'Connor) pp. 159–195. Elsevier, Amsterdam.

—— (1981). Force of labour. In *Principles and Practice of Obstetrics and Perinatology.* (eds L. Iffy and H. A. Kamientzky) pp. 761–799. John Wiley and Sons, New York.

—— and Kuriyama, H. (1963). Effects of ions and drugs on cell membrane activity and tension in the postpartum rat myometrium. *J. Physiol., Lond.* **165**, 575–592.

—— and Takeda, H. (1965). Effect of progesterone on the electric activity and intra-uterine pressure of pregnant and parturient rabbits. *Am. J. Obstet. Gynec.* **91**, 221–231.

Dahl, G., Azarnia, R., Werner, R. (1980). *De novo* construction of cell-to-cell channels. *In vitro* **16**, 1068–1075.

Daniel, E. E. (1960). The activation of various types of uterine muscle during stretch-induced contraction. *Can. J. Biochem. Physiol.* **38**, 1327–1362.

—— Grover, A. K., and Kwan, C. Y. (1983). Control of intracellular calcium in smooth muscle. In *Calcium Regulation by Calcium Antagonists.* (eds R. G. Rahwan, and D. T. Witiak) pp. 73–88. ACS Symposium Series, No. 201.

—— and Lodge, S. (1973). Electrophysiology of myometrium. In *Uterine Contraction: Side Effects of Steroidal Contraceptives.* (ed. J. B. Josimovich) pp. 19–64. John Wiley and Sons, New York.

DeMello, W. (1984). Effect of intracellular injection of cAMP on the electrical coupling of mammalian cardiac cells. *Biochem. Biophys. Res. Commun.* **119**, 1001–1007.

Demianczuk, N., Towell, M., and Garfield, R. E. (1984). Myometrial electrophysiologic activity and gap junctions in the pregnant rabbit. *Am. J. Obstet. Gynec.* **149**, 485–491.

Dewey, M. M. and Barr, L. (1968). Structure of vertebrate intestinal smooth muscle. In *Handbook of Physiology*, **Section 6**: *Alimentary Canal*, **Vol. IV**: *Motility*. (ed. C. F. Code) pp. 1629–1654. American Physiological Society, Washington, D. C.

Downing, S. J. and Sherwood, O. D. (1985a). The physiological role of relaxin in the pregnant rat. I. The influence of relaxin on parturition. *Endocrinology* **116**, 1200–1205.

—— —— (1985b). The physiological role of relaxin in the pregnant rat. II. The influence of relaxin on uterine contractile activity. *Endocrinology* **116**, 1206–1214.

—— (1985c). The physiological role of relaxin in the pregnant rat. III. The influence of relaxin on cervical extensibility. *Endocrinology* **116**, 1215–1220.

Dubin, N. H., Blake, D. A., Ghodgaonkar, R. B., and Egner, P. G. (1982). Thromboxane B$_2$, 6-keto-prostaglandin F$_1$ and prostaglandin F$_2$ by contracting pregnant rat uteri *in vitro*. *Biol. Reprod.* **26**, 281–288.

Fabiato, A. (1982). Fluorescence and differential light absorption recordings with calcium probes and potential-sensitive dyes in skinned cardiac cells. *Can. J. Physiol. Pharmacol* **60**, 556–567.

Fallon, R. F. and Goodenough, D. A. (1981). Five hour half-life of mouse liver gap junction protein. *J. Cell Biol.* **90**, 521–526.

Finn, C. A. and Porter, D. G. (1975). *The Uterus*. Elek Science, London.

Flagg-Newton, J. L., Dahl, G., and Loewenstein, W. R. (1981). Cell junction and cyclic AMP: I. Upregulation of junctional membrane permeability and junctional membrane particles by administration of cyclic nucleotide or phosphodiesterase inhibitor. *J. Membr. Biol.* **63**, 105–121.

Fowler, R. J. (1977). The role of prostaglandins in parturition with special reference to the rat. In *The Fetus and Birth*. Ciba Fdn. Symp. **Vol. 47** (eds J. Knight and M. O'Connor) pp. 297–312. Elsevier, Amsterdam.

Fuchs, A.-R. (1973). Parturition in rabbits and rats. In *Endocrine Factors in Labour*, Mem. Soc. Endocr. **Vol 20** (eds A. Klopper and J. Gradner) pp. 163–185. Cambridge University Press, London.

—— (1978). Hormonal control of myometrial function during pregnancy and parturition. *Acta endocr. Suppl.* 221, 1–70.

—— Fuchs, F., Hurstein, P., Soloff, M. S., and Fernstrom, M. J. (1982). Oxytocin receptors and human parturition: a dual role for oxytocin in the initiation of labor. *Science, N.Y.* 215, 1396–1398.

—— (1983). The role of oxytocin in parturition. *Curr. Top. exp. Endocr.* **4**, 231–265.

—— (1985). Oxytocin in animal parturition. In *Oxytocin: Clinical and Laboratory Studies*. (eds J. A. Amico and A. G. Robinson) pp. 207–235. Elsevier, Amsterdam.

Furshpan, E. J., and Potter, D. D. (1959). Transmission at the giant motor synapses of the crayfish. *J. Physiol., Lond.* **145**, 289–325.

Gabella, G. (1979). Smooth muscle cell junctions and structural aspects of contraction. *Br. med. Bull.* **35**, 213–218.

—— (1981). Structure of smooth muscles. In *Smooth Muscle: an Assessment of Current Knowledge* (eds E. Bulbring, A. F. Brading, A. W. Jones, and T. Tomita) pp. 1–46. Edward Arnold, London.

—— (1984). Structural apparatus for force transmission in smooth muscles. *Physiol. Rev.* **64**, 455–477.

Garfield, R. E. and Daniel, E. E. (1974). The structural basis of electrical coupling (cell-to-cell contacts) in rat myometrium. *Gynec. Invest.* **5**, 284–300.

—— Sims, S., and Daniel, E. E. (1977). Gap junctions: their presence and necessity in myometrium during gestation. *Science, N.Y.* **198**, 958–960.

—— Sims, S. M., Kannan, M. S., and Daniel, E. E. (1978). Possible role of gap junctions in activation of myometrium during parturition. *Am. J. Physiol.* **235**, C168–179.

—— Rabideau, S., Challis, J. R. G., and Daniel, E. E. (1979). Hormonal control of gap junction formation in sheep myometrium during parturition. *Biol. Reprod.* **21**, 999–1007.

—— Merrett, D., and Grover, A. K. (1980b). Gap junction formation and regulation in myometrium. *Am. J. Physiol.* **239**, C217–C228.

—— Kannan, M. S., and Daniel, E. E. (1980a). Gap junction formation in myo-

482   R. E. Garfield, M. G. Blennerhassett and S. M. Miller

metrium: control by estrogens, progesterone and prostaglandins. *Am. J. Physiol.* **238**, C81–C89.

—— and Hayashi, R. H. (1981). Appearance of gap junctions in the myometrium of women during labor. *Am. J. Obstet. Gynec.* **140**, 254–260.

—— Daniel, E. E., Dukes, M., and Fitzgerald, J. D. (1982). Changes in gap junctions in myometrium of guinea pig at parturition and abortion. *Can. J. Physiol. Pharmacol.* **60**, 335–41.

—— Puri, C. P., and Csapo, A. I. (1982). Endocrine, structural and functional changes in the uterus during premature labor. *Am. J. Obstet. Gynec.* **142**, 21–27.

—— (1984). Myometrial ultrastructure and uterine contractility. In *Uterine Contractility* (eds S. Bottari, J. P. Thomas, A. Vokaer, and R. Vokaer) pp. 81–109. Masson Publ., New York.

—— and Somlyo, A. P. (1985*a*). Structure of smooth muscle. In *Calcium and Contractility.* (eds A. K. Grover and E. E. Daniel) pp. 1–36. Humana Press, Clifton.

—— (1985*b*). Cell-to-cell communication in smooth muscle. In *Calcium and Contractility* (eds A. K. Grover and E. E. Daniel) pp. 143–173. Humana Press, Clifton.

—— (1986). Structural studies of innervation on nonpregnant rat uterus. *Am J. Physiol.* **251**, C41–C56.

—— and Baulieu, E. E. (1987). The antiprogesterone steroid RU486: a short pharmacological and clinical review with emphasis on the interruption of pregnancy. *Baillieres clin. Endocr. Metab.* **1**, 207–221.

Gorin, M. B., Yancey, S. B., Cline, J., Revely, J.-P., and Horowitz, J. (1984). The major intrinsic proteins (MIP) of the bovine lens fibre membrane: characterization and structure based upon cDNA cloning. *Cell* **39**, 49–59.

Gorski, J. and Gannon, F. (1976). Current models of steroid hormone action: a critique. *A. Rev. Physiol.* **38**, 425–450.

Goto, M., Kuriyama, H., and Abe, Y. (1961). Refractory period and conduction of excitation in the uterine muscle cells of the mouse. *Jpn. J. Physiol.* **11**, 369–377.

Grosset, A. and Mironneau, J. (1977). An analysis of the actions of prostaglandin E_1 on membrane currents and contraction in uterine smooth muscle. *J. Physiol., Lond.* **270**, 765–784.

Gu, J., Polak, J. M., Su, H. C., Blank, M. A., Morrison, J. F. B., and Bloom, S. R. (1984). Demonstration of paracervical ganglion origin for the vasoactive intestinal peptide-containing nerves of the rat uterus using retrograde tracing techniques combined with immunocytochemistry and denervation procedures. *Neurosci. Lett.* **51**, 377–382.

Ham, A. W. and Cormack, D. H. (1979). *Histology.* J. B. Lippincott, Philadelphia.

Hax, W. M. A., Van Venrooj, G. E. P., and Vossenberg, J. B. (1974). Cell communication: a cyclic-AMP mediated phenomenon. *J. Membr. Biol.* **19**, 253–266.

Hertzberg, E. L. and Gilula, N. B. (1981). Liver gap junctions and lens fiber junctions: comparative analysis and calmodulin interaction. *Cold Spring Harb. Symp. quant. Biol.* **46**, 639–645.

—— and Skibbens, R. V. (1984). A protein homologous to the 27,000-dalton liver gap junction protein is present in a wide variety of species and tissues. *Cell* **39**, 61–69.

Hermsmeyer, K. (1982). Angiotensin II increases electrical coupling in mammalian ventricular myocardium *Circ. Res.* **47**, 524–529.

Hervonen, A., Kanerva, L., and Lietzen, R. (1973). Histochemically demonstrable catecholamines and cholinesterases of the rat uterus during estrus cycle, pregnancy and after estrogen treatment. *Acta physiol. scand.* **87**, 283–288.

Higuchi, T., Honda, K., Fukuoka, T., Negoro, H., and Wakabayashi, K. (1985).

Release of oxytocin during suckling and parturition in the rat. *J. Endocr.* **105**, 339–346.

Holman, M. E. (1968). Introduction to electrophysiology of visceral smooth muscle. In *Handbook of Physiology, Section 6: Alimentary Canal, Vol. IV: Motility* (ed. C. F. Code) pp. 1665–1708. American Physiological Society, Washington, D. C.

Hooper, M. L. and Subak-Sharpe, J. H. (1981). Metabolic cooperation between cells. *Int. Rev. Cytol.* **69**, 45–104.

Ischikawa, S. and Bortoff, A. (1970). Tissue resistance of the progesterone-dominated rabbit myometrium. *Am. J. Physiol.* **219**, 1763–1767.

Izumi, H. (1985). Changes in the mechanical properties of the longitudinal and circular muscles of the rat myometrium during gestation. *Br. J. Pharmac.* **86**, 247–257.

Jacobus, W. E., Pores, I. H., Lucas, S. K., Kallman, C. H., Weisfeldt, M. L., and Flaherty, J. T. (1982). The role of intracellular pH in the control of normal and ischaemic myocardial contractility. In *Intracellular pH* (eds. R. Nuccitelli and D. Deamer) pp. 537–565. Liss, New York.

Johnston, M. F. and Ramon, F. (1981). Electrotonic coupling in internally perfused crayfish segmented axons. *J. Physiol., Lond.* **317**, 509–518.

Kam, E., Melville, L., and Pitts, J. D. (1986). Patterns of junctional communication in skin. *J. Invest. Derm.* **87**, 748–763.

Kanda, S. and Kuriyama, H. (1980). Specific features of smooth muscle cells recorded from the placental region of the myometrium of pregnant rats. *J. Physiol., Lond.* **299**, 127–144.

Kao, C. Y. (1959). Long-term observations of spontaneous electrical activity of the uterine smooth muscle. *Am. J. Physiol.* **196**, 343–350.

—— (1977*a*). Electrical properties of uterine smooth muscle. In *Biology of the Uterus.* (ed. R. M. Wynn) pp. 423–496. Plenum Press, New York.

—— (1977*b*). Recent experiments in voltage-clamp studies on smooth muscles. In *Excitation-Contraction Coupling in Smooth Muscle* (eds R. Casteels, R. Godfraind, and J. C. Ruegg) pp. 91–96. Elsevier/North-Holand, Amsterdam.

—— and McCullough, J. R. (1975). Ionic currents in the uterine smooth muscle. *J. Physiol., Lond.* **246**, 1–36.

—— and Nishiyama, A. (1964). Ovarian hormones and resting potentials of uterine smooth muscle. *Am. J. Physiol.* **207**, 793–799.

Kawarabayashi, T. (1978). The effects of phenylephrine in various ionic environments on the circular muscle of midpregnant rat myometrium. *Jpn. J. Physiol.* **28**, 627–645.

—— and Marshall, J. M. (1981). Factors influencing circular muscle activity in the pregnant rat uterus. *Biol. Reprod.* **24**, 373–379.

—— and Osa, T. (1976). Comparative investigations of alpha- and beta-effects on the longitudinal and circular muscles of the pregnant rat myometrium. *Jpn. J. Physiol.* **26**, 403–416.

Kishikawa, T. (1981). Alterations in the properties of the rat myometrium during gestation and postpartum. *Jpn. J. Physiol.* **31**, 515–536.

Kuriyama, H. (1961). Recent studies of the electrophysiology of the uterus. In *Progesterone and the Defence Mechanism of Pregnancy.* Ciba Fndn Study Group **Vol. 9**. pp. 51–70. Little Brown, Boston.

—— (1981). Excitation-contraction coupling in various visceral smooth muscles. In *Smooth Muscle: an Assessment of Current Knowledge* (eds E. Bülbring, A. F. Brading, A. W. Jones, and T. Tomita) Edward Arnold, London.

—— and Csapo, A. I. (1961). A study of the parturient uterus with the microelectrode technique. *Endocrinology* **68**, 1010–1025.

—— and Suzuki, H. (1976). Changes in electrical properties of rat myometrium during gestation and following hormonal treatments. *J. Physiol., Lond.* **260**, 315–333.

Landa, J. F., West, T. C., and Thiersch, J. B. (1959). Relationships between contraction and membrane electrical activity in the uterus of the pregnant rat. *Am. J. Physiol.* **196**, 905–909.

Lane, N. J. and Swales, L. S. (1980). Dispersal of junctional particles with internalization during *in vivo* disappearance of gap junctions. *Cell* **19**, 579–589.

Larsen, W. J. (1983). Biological implications of gap junction structure, distribution and composition: a review. *Tissue and Cell* **15**, 645–671.

Larsson, L.-I., Fahrenkrug, J., and Schaffalitzky de Muckadell, O. B. (1977). Vasoactive intestinal polypeptide occurs in nerves of the female genito-urinary tract. *Science, N.Y.* **196**, 1374–1375.

Lasater, E. M. and Dowling, J. E. (1985). Dopamine decreases conductance of the electrical junctions between cultured retinal horizontal cells. *Proc. natn. Acad. Sci. U.S.A.* **82**, 3025–3029.

Laufer, W. and Salas, R. (1981). Intercellular coupling and retinal horizontal cell receptive field. *Neurosci. Lett.* **7**, 5339.

Liggins, G. C., (1979). Initiation of parturition. *Br. med. Bull.* **35**, 145–150.

Lodge, S. and Sproat, J. E. (1981). Resting membrane potentials of pacemaker and nonpacemaker areas in rat uterus. *Life Sci.* **28**, 2251–2256.

Loewenstein, W. R. (1965). Permeability of membrane junctions. *Ann. N.Y. Acad. Sci.* **137**, 441–472.

—— (1981). Junctional intercellular communication in the cell-to-cell membrane channel. *Physiol. Rev.* **61**, 829–913.

Louis, C. F., Johnson, R. Johnson, K., and Turnquist, J. (1985). Characterization of the bovine lens plasma membrane substrates for cAMP-dependent protein kinase. *Eur. J. Biochem.* **150**, 279–286.

MacKenzie, L. W. and Garfield, R. E. (1985). Hormonal control of gap junctions in the myometrium. *Am. J. Physiol.* **248**, C296–C308.

—— —— (1986). Effects of estradiol-17β on myometrial gap junctions and pregnancy in the rat. *Can. J. Physiol. Pharmacol.* **64**, 462–466.

—— Puri, C. P., and Garfield, R. E. (1983). Effect of estradiol-17β and prostaglandins on rat myometrial gap junctions. *Prostaglandins* **26** 925–941.

Makowski, I., Caspar, D., Philips, W., and Baker, T. (1984). Gap junction structures. VI. Variation and conservation in connexon conformation and packing. *Biophys. J.* **45**, 208–218.

Manjunath, C. D., Nicholson, B. J., Teplow, D., Hood, L., Page, E., and Revel, J. P. (1987). The cardiac gap junction protein (M_r 47 000) has a tissue specific cytoplasmic domain of M_r 17 000 at its carboxy-terminus. *Biochem. Biophys. Res. Commun.* **142**, 228.

Margiotta, J. F. and Walcott, B. (1983). Conductance and dye permeability of a rectifying electrical synapse. *Nature, Lond.* **305**, 2–55.

Marshall, J. M. (1959). Effects of estrogen and progesterone on single uterine muscle fibers in the rat. *Am. J. Physiol.* **197**, 935–942.

—— (1962). Regulation of activity in uterine smooth muscle. *Physiol. Rev.* **42**, 213–227.

—— (1970). Adrenergic innervation of the female reproductive tract: anatomy, physiology and pharmacology. *Rev. Physiol. Biochem. Pharmacol.* **62**, 6–67.

—— (1973). The physiology of the uterus. In *The Uterus* (eds H. J. Norris, A. T. Hertig, and M. R. Abell) pp. 89–109. Williams and Wilkins, Baltimore.

—— (1974). Effects of neurohypophysial hormones on the myometrium. In *Handbook of Physiology*, **Section 7**: *Endocrinology* **Vol. IV**: *The pituitary gland*, **Part 1**.

(eds R. O. Greep and E. B. Astwood) pp. 469–492. American Physiological Society, Washington.

——(1981). Effects of ovarian steroids and pregnancy on adrenergic nerves of uterus and oviduct. *Am. J. Physiol.* **240**, C165–C174.

McNutt, N. S. and Weinstein, R. S. (1970). The ultrastructures of the nexus. A correlated thin section and freeze-cleave study. *J. Cell Biol.* **47**, 666–688.

Meda, P., Merrelet, A., and Orci, L. (1979). Increase of gap junctions between pancreatic B-cells during stimulation of insulin secretion. *J. Cell. Biol.* **82**, 441–448.

Melton, C. E. (1956). Electrical activity in the uterus of the rat. *Endocrinology* **58**, 139–149.

—— and Saldivar, J. T. (1964). Impulse velocity and conduction pathways in rat myometrium. *Am. J. Physiol.* **207**, 279–285.

—— ——(1965). Estrous cycle and electrical activity of rat myometrium. *Life Sci.* **4**, 593–602.

—— ——(1967). The *linea uteri*, a conduction pathway in rat myometrium. *Life Sci.* **6**, 297–304.

—— ——(1970). Activity of the rat's uterine ligament. *Am. J. Physiol.* **219**, 122–125.

Mercado-Simmen, R. C., Bryant-Greenwood, G. D., and Greenwood, F. C. (1982). Relaxin receptor in the rat myometrium: regulation by estrogen and relaxin. *Endocrinology*, **110**, 220–226.

Merk, F. B., Botticelli, C. R., and Albright, J. T. (1972). An intercellular response to estrogen by granulosa in the rat ovary: an electron microscopic study. *Endocrinology*. **90**, 992–1007.

—— Kwan, P. W. L., and Lear, I. (1980). Gap junctions in the myometrium of hypophysectomized estrogen-treated rats. *Cell Biol. Intl. Rep.* **4**, 287–294.

Michael, C. A. and Schofield, B. M. (1969). The influence of the ovarian hormones on the actomyosin content and the development of tension in uterine muscle. *J. Endocr.* **44**, 501–511.

Miller, S. M. (1986). *Changes in burst and spike propagation associated with gap junction formation in myometrium at parturition*. Ph. D. Thesis, McMaster University.

Mironneau, J. (1973). Excitation-contraction coupling in voltage-clamped uterine smooth muscle. *J. Physiol., Lond.* **233**, 127–141.

——(1976*a*). Relationship between contraction and transmembrane ionic current in voltage-clamped uterine smooth muscle. In *Physiology of Smooth Muscle* (eds E. Bülbring and M. F. Shuba) pp. 175–183. Raven Press, New York.

——(1976*b*). Effects of oxytocin on ionic currents underlying rhythmic activity and contraction in uterine smooth muscle. *Pflügers Arch.* **363**, 113–118.

—— and Savineau, J.-P. (1980). Effects of calcium ions on outward membrane currents in rat uterine smooth muscle. *J. Physiol., Lond.* **302**, 411–425.

—— —— and Mironneau, C. (1981). Fast outward current controlling electrical activity in rat uterine smooth muscle during gestation. *J. Physiol., Paris* **77**, 851–859.

—— Mironneau, J., and Savineau, J.-P. (1984*a*). Maintained contractions of rat uterine smooth muscle incubated in a Ca^{2+}-free solution. *Br. J. Pharmac.* **82**. 735–743.

—— Lalanne, C., Mironneau, C., Savineau, J.-P., and Lavie, J. L. (1984*b*). Comparison of pinaverium bromide, manganese chloride and D600 effects on electrical and mechanical activities in rat uterine smooth muscle. *Eur. J. Pharmac.* **98**, 99–107.

Mitchell, M. D. (1984). The mechanism(s) of human parturition. *J. dev. Physiol.* **6**, 107–118.

Nagai, T. and Prosser, C. L. (1963). Patterns of conduction in smooth muscle. *Am. J. Physiol.* **204**, 910–914.

Neyton, J. and Trautmann, A. (1985). Single-channel currents of an intercellular junction. *Nature, Lond.* **317**, 331–385.

Nicholson, B. J., Gros, D. E., Kent, S. B., Hood, L. E., and Revel, J.-P. (1985). The M_r 28 000 gap junction proteins from rat heart and liver are different but related. *J. biol. Chem.* **260**, 6514–6517.

Novy, J. and Liggins, G. C. (1980). Role of prostaglandins and thromboxanes in the physiological control of the uterus and in parturition. *Semin. Perinatol.* **4**, 45–66.

O'Byrne, K. T., Ring, J. P. G., and Summerlee, A. J. S. (1986). Plasma oxytocin and oxytocin neurone activity during delivery in rabbits. *J. Physiol., Lond.* **370**, 501–513.

Ogasawara, T., Kato, S., and Osa, T. (1980). Effects of estradiol-17β on the membrane response and K-contracture in the uterine longitudinal muscle of ovariectomized rats studied in combination with the Mn action. *Jpn. J. Physiol.* **30**, 271–285.

Ohkawa, H. (1975). Electrical and mechanical interaction between the muscle layers of rat uterus and different sensitivities to oxytocin. *Bull. Yamaguchi Med. School* **22**, 197–210.

Osa, T. (1974). An interaction between the electrical activities of longitudinal and circular smooth muscles of pregnant mouse uterus. *Jpn. J. Physiol.* **24**, 189–203.

—— and Katase, T. (1975). Physiological comparison of the longitudinal and circular muscles of the pregnant rat uterus. *Jpn. J. Physiol.* **25**, 153–164.

—— and Kawarabayashi, T. (1977). Effects of ions and drugs on the plateau potential in the circular muscle of pregnant rat myometrium. *Jpn. J. Physiol.* **27**, 111–121.

—— and Fujino, T. (1978). Electrophysiological comparison between the longitudinal and circular muscles of the rat uterus during the estrous cycle and pregnancy. *Jpn. J. Physiol.* **28**, 197–209.

—— and Watanabe, M. (1978). Effects of catecholamines on the circular muscle of rat myometria at term during pregnancy. *Jpn. J. Physiol.* **28**, 647–658.

—— Ogasawara, T., and Kato, S. (1983). Effects of magnesium, oxytocin and prostaglandin F_2 on the generation and propagation of excitation in the longitudinal muscle of rat myometrium during late pregnancy. *Jpn. J. Physiol.* **33**, 51–67.

Ottesen, B. (1981). Vasoactive intestinal polypeptide (VIP): effect on rabbit uterine smooth muscle *in vivo* and *in vitro*. *Acta physiol. scand.* **113**, 193–199.

—— Larsen, J.-J, Fahrenkrug, J., Stjernquist, M., and Sundler, F. (1981). Distribution and motor effect of VIP in female genital tract. *Am. J. Physiol.* **240**, E32–E36.

—— Gram, B. R., and Fahrenkrug, J. (1983). Neuropeptides in the female genital tract: effect on vascular and non vascular smooth muscle. *Peptides* **4**, 387–392.

Ottesen, G., Ulrichsen, H., Fahrenkrug, J., Larsen, J.-J., Wagner, G., Schierup, L., and Sondergaard, F. (1982). Vasoactive intestinal polypeptide and the female genital tract: relationship to reproductive phase and delivery. *Am. J. Obstet. Gynec.* **143**, 414–420.

Paiva, C. E. N. de and Csapo, A. I. (1973). The effect of prostaglandin on the electric activity of the pregnant uterus. *Prostaglandins* **4**, 177–188.

Papka, R. E., Cotton, J. P., and Traurig, H. H. (1985). Comparative distribution of neuropeptide tyrosine-, vasoactive intestinal polypeptide-, substance P-immunoreactive, acetylcholinesterase-positive and noradrenergic nerves in the reproductive tract of the female rat. *Cell Tissue Res.* **242**, 475–490.

Parkington, H. C. (1985). Some properties of the circular myometrium of the sheep throughout pregnancy and during labour. *J. Physiol., Lond.* **359**, 1–15.

Paul, D. L. (1986). Molecular cloning of cDNA for rat liver gap junction protein. *J. Cell Biol.* **103**, 123–134.

Peracchia, C. (1980). Structural correlates of gap junction permeation. *Int. Rev. Cytol.* **66**: 81–146.

—— Girsch, S. J., Bernardini, G. and Peracchia, L. L. (1985). Lens junctions are communicating junctions. *Current Eye Research* **4**, 1155–1169.

Piccolino, M., Neyton, M. J., Witkovsky, P., and Gerschenfeld, H. M. (1982). Gamma-aminobutyric acid antagonists decrease junctional communication between L-horizontal cells in retina. *Proc. natn. Acad. Sci. U.S.A.* **79**, 671–3675.

—— Gerschenfeld, H. M., and Neyton, J. (1984). Decrease of gap junction permeability induced by dopamine and cyclic adenosine 3′–5′ monophosphate. *J. Neurosci.* **4**, 477–2488.

Pitts, J. D. and Finbow, M. E. (1986). The gap junction. *J. Cell Sci. Suppl.* **4**, 239–266.

Phillips, C. A. and Poyser, N. L. (1981). Prostaglandins, thromboxanes and the pregnant rat uterus at term. *Br. J. Pharmac.* **73**, 75–80.

Popescu, L. M., Nuto, O., and Panoui, C. (1965). Oxytocin contracts the human uterus at term by inhibiting the myometrial Ca^{2+}-extrusion pump. *Bioscience Rep.* **5**, 21–28.

Porter, D. G., Downing, S. J., and Bradshaw, J. M. (1979). Relaxin inhibits spontaneous and prostaglandin-driven myometrial activity in anaesthetized rats. *J. Endocr.* **83**, 183–192.

Prosser, C. L. (1962). Conduction in nonstriated muscles. *Physiol. Rev.* **42**, 193–206.

Puri, C. P. and Garfield, R. E. (1982). Changes in hormone levels and gap junctions in the rat uterus during pregnancy and parturition. *Biol. Reprod.* **27**, 967–975.

Radu, A., Dahl, G., and Loewenstein, W. R. (1982). Hormonal regulation of cell junction permeability upregulation by catecholamine and prostaglandin E_1. *J. Membr. Biol.* **70**, 239–251.

Reimer, R. K., Goldfien, A. C., Goldfien, A., and Roberts, J. M. (1986). Rabbit uterine oxytocin receptors and *in vitro* contractile response: abrupt changes at term and the role of eicosanoids. *Endocrinology*, **119**, 699–709.

Reiner, O. and Marshall, J. M. (1975). Action of D-600 on spontaneous and electrically stimulated activity of the parturient rat uterus. *Naunyn-Schmiedeberg's Arch. Pharmacol.* **290**, 21–28.

Revel, J.-P. and Karnovsky, M. (1967). Hexagonal array of subunits in intercellular junctions of the mouse heart and liver. *J. Cell Biol.* **33**, C7–C12.

Revel, J.-P., Nicholson, B. J., and Yancey, S. B. (1985). Chemistry of gap junctions. *A. Rev. Physiol.* **47**, 263–279.

Reynolds, S. R. M. (1949). *Physiology of the Uterus*, 2nd Edition, P. B. Hoeber Inc., New York.

Roche, P. J., Parkington, H. C., and Gibson, W. R. (1985). Pregnancy and parturition in rats after sympathetic denervation of the ovary, oviduct and utero-tubal junction. *J. Reprod. Fert.* **75**, 653–661.

Saez, J. C., Spray, D. C., Nairn, A. C., Hertzberg, E., Greengard, P., and Bennett, M. V. L. (1986). cAMP increases junctional conductance and stimulates phosphorylation of the 27 kD principal gap junction polypeptide. *Proc. natn. Acad. Sci. U.S.A.* **83**, 2473–2477.

Sas, D. F., Sas, M. J., Johnson, K. R., Menko, A. S., and Johnson, R. G. (1985). Junctions between lens fibre cells are labelled with a monoclonal antibody shown to be specific for MP 26. *J. Cell Biol.* **100**, 216–225.

Sakai, K., Yamaguchi, T., and Uchida, M. (1981). Oxytocin-induced Ca-free contraction of rat uterine smooth muscle: effects of divalent cations and drugs. *Arch. Int. Pharmacodyn.* **250**, 40–54.

Saldivar, J. T. and Melton, C. E. (1966). Effects *in vivo* and *in vitro* of sex steroids on rat myometrium. *Am. J. Physiol.* **211**, 835–843.

Sanborn, B. M., Kuo, H. S., Weisbrodt, N. W., and Sherwood, O. D., (1980). The interaction of relaxin with the rat uterus. 1. Effects on cyclic nucleotide levels and spontaneous contractile activity. *Endocrinology* **106**, 1210–1215.

Sanfillipo, J. S., Teichman, J., Melvin, T. R., Osyamkpe, J. L., and Wittliff, J. H. (1983). The influence of certain prostaglandin synthetase inhibitors on cytoplasmic estrogen receptors in the uterus. *Am. J. Obstet. Gynec.* **145**, 100–104.

Schild, H. O. (1964). Calcium and the effect of drugs on depolarized smooth muscle. In *Pharmacology of Smooth Muscle*. (ed. E. Bülbring) pp. 95–104. Pergamon Press, Oxford.

Schwartzmann, G., Wiegandt, H., Rose, B., Zimmerman, A., Ben-Haim, D., and Loewenstein, W. R. (1981). Diameter of the cell-to-cell junctional membrane channels as probed with neutral molecules. *Science, N.Y.* **231**, 551–553.

Sheldrick, E. L. and Flint, A. P. F. (1985). Endocrine control of uterine oxytocin receptors in the ewe. *J. Endocr.* **106**, 249–258.

Sherwood, O. D., Crnekovic, V. E., Gordon, W. L., and Rutherford, J. E. (1980). Radioimmunoassay of relaxin throughout pregnancy and during parturition in the rat. *Endocrinology* **107**, 691–698.

Shivers, R. R. and Bowman, P. D. (1985). A freeze-fracture paradigm of the delivery and insertion of gap junction particles into the plasma membrane. *J. submicrosc. Cytol.* **17**, 199–203.

Shoenberg, C. F. (1977). The contractile mechanism and ultrastructure of the myometrium. In *Biology of the Uterus*. (ed. R. M. Wynn) pp. 497–544. Plenum Press, New York.

Silva, D. G. (1967). The ultrastructure of the myometrium of the rat with special reference to the innervation. *Anat. Rec.* **158**, 21–34.

Sims, S. M., Daniel, E. E., and Garfield, R. G. (1982). Improved electrical coupling in uterine smooth muscle is associated with increased gap junctions at parturition. *J. gen. Physiol.* **80** 353–375.

Singh, I., Sankaranarayanan, A., and Bawa, S. R. (1985). Automatic innervation of pregnant and nonpregnant rat uterus. *IRCS Med. Sci.* **13**, 324–325.

Soloff, M. S. (1985). Oxytocin receptors and mechanisms of oxytocin action. In *Oxytocin: Clinical and Laboratory Studies*. (eds J. A. Amico and A. G. Robinson) pp. 259–276. Elsevier, Amsterdam.

—— Alexandrova, M., and Fernstrom, M. J. (1979). Oxytocin receptors: triggers for parturition and lactation? *Science, N. Y.* **204**, 1313–1315.

—— and Sweet, P. (1982). Oxytocin inhibition of $(Ca^{2+} + Mg^{2+})$-ATPase activity in rat myometrial plasma membranes. *J. biol. Chem.* **257**, 10687–10693.

Somlyo, A. V. and Somlyo, A. P. (1968). Electromechanical and pharmacomechanical coupling in vascular smooth muscle. *J. Pharmac. exp. Ther.* **159**, 129–145.

Spray, D. C., Harris, A. C., and Bennett, M. V. L. (1981). Gap junction conductance is a simple and sensitive function of intracellular pH. *Science, N.Y.* **211**, 712–715.

—— Stern, J. H., Harris, A. L., and Bennett, M. V. L. (1982). Comparison of sensitivity of gap junctional conductance to H and Ca ions. *Proc. natn. Acad. Sci. U.S.A.* **79**, 441–445.

—— White, R. L., Campos de Carvalho, A., Harris, A. L., and Bennett, M. V. L. (1984). Gating of gap junctional channels. *Biophys. J.* **45**, 219–230.

—— and Bennett, M. V. L. (1985). Physiology and pharmacology of gap junctions. *Am. J. Physiol.* **47**, 281–303.

—— de Carvalho, A. C., and Bennett, M. V. L. (1986). Sensitivity of gap junctional conductance to H ions in amphibian embryonic cells is independent of voltage sensitivity. *Proc. natn. Acad. Sci. U.S.A.* **83**, 3533–3536.

Stjernquist, M., Emson, P., Owman, Ch., Sjoberg, N.-O., Sundler, F., and Tatemoto,

K. (1983). Neuropeptide Y in the female reproductive tract of the rat. Distribution of nerve fibres and motor effects. *Neurosci. Lett.* **39**, 279–284.

—— Ekblad, E., Owman, Ch., and Sundler, F. (1986). Neuronal localization and motor effects of gastrin-releasing peptide (GRP) in rat uterus. *Regul. Peptides*, **13**, 197–205.

Talo, A. and Csapo, A. I. (1970). Conduction of electric activity in late pregnant and parturient rabbit uteri. *Physiol Chem. Phys.* **2**, 489–494.

Teranishi, T., Negishi, K., and Kato, S. (1984). Regulatory effects of dopamine on spatial properties of horizontal cells of the carp retina. *J. Neurosci.* **4**, 1271–1280.

Thiersch, J. B., Landa, J. F., and West, T. C. (1959). Transmembrane potentials in the rat myometrium during pregnancy. *Am. J. Physiol.* **196**, 901–904.

Thorbert, G., Alm, P. Owman, Ch., and Sjoberg, N.-O. (1977). Regional distribution of autonomic nerves in the guinea pig uterus. *Am. J. Physiol.* **239**, C25–C34.

—— (1979). Regional changes in structure and function of adrenergic nerves in guinea-pig uterus during pregnancy. *Acta obstet. gynecol. scand. Suppl.* **79**, 5–32.

Thorburn, G. D. and Challis, J. R. G. (1979). Endocrine control of parturition. *Physiol. Rev.* **59**, 863–918.

—— Harding, R., Jenkin, G., Parkington, H., and Sigger, J. N. (1984). Control of uterine activity in the sheep. *J. dev. Physiol.* **6**. 31–43.

Tomita, T. (1970). Electrical properties of mammalian smooth muscle. In *Smooth Muscle* (eds E. Bülbring, A. F. Brading, A. W. Jones, and T. Tomita) pp. 197–243. Edward Arnold, London.

—— (1975). Electrophysiology of mammalian smooth muscle. *Prog. Biophys. Molec. Biol.* **30**, 185–203.

—— (1981). Electrical activity (spikes and slow waves) in gastrointestinal smooth muscles. In *Smooth Muscle: an Assessment of Current Knowledge.* (eds E. Bülbring, A. F. Brading, A. W. Jones, and T. Tomita) pp. 127–156. Edward Arnold, London.

Unwin, P. N. T. and Zampighi, G. (1980). Structure of the gap junction between communicating cells. *Nature, Lond.* **283**, 545–549.

—— and Ennis, P. D. (1983). Calcium-mediated changes in gap junction structure. Evidence from the low-angle X-ray pattern. *J. Cell Biol.* **97**, 1459–1466.

—— and Ennis, P. D. (1984). Two configurations of a channel forming membrane protein. *Nature, Lond.* **307**, 609–613.

Vassort, G. (1981). Ionic currents in longitudinal muscle of the uterus. In *Smooth Muscle: an Assessment of Current Knowledge.* (eds E. Bulbring, A. F. Brading, A. W. Jones, and T. Tomita) pp. 353–366. Edward Arnold, London.

Veenstra, R. D. and DeHaan, R. L. (1986). Measurement of single channel currents from cardiac gap junctions. *Science, N.Y.* **233**, 972–974.

Vesin, M. F., Dokhac, L., and Harbon, S. (1979). Prostacyclin as an endogenous modulator of adenosine cyclic 3'–5' monophosphate in the myometrium and endometrium. *Mol. Pharmac.* **16**, 823–840.

Villar, A., D'Ocon, M. P., and Anselmi, E. (1985). Calcium requirement of uterine contraction induced by PGE_1: importance of intracellular calcium stores. *Prostaglandins* **30**, 491–496.

Wathes, D. C. and Porter, D. G. (1982). Effect of uterine distention and oestrogen treatment on gap junction formation in the myometrium of the rat. *J. Reprod. Fert.* **65**, 497–505.

Weidmann, S. (1966). Diffusion of radiopotassium across intercalated disks of mammalian cardiac muscle. *J. Physiol., Lond.* **187**, 323–342.

Welsh, M. J., Aster, J. C., Ireland, M., Alcala, J., and Maisel, H. (1982). Calmodulin binds to chick lens gap junction proteins in a calcium independent manner. *Science, N.Y.* **216**, 642–644.

White, R. L., Spray, D. C., de Carvalho, A. C., Wittenberg, B. A., and Bennett, M. V. L. (1985). Some electrical and pharmacological properties of gap junctions between adult ventricular myocytes. *Am. J. Physiol.* **249**, C444–C455.

Williams, L. (1983). Effects of estradiol and progesterone on uterine prostaglandin levels in pregnant rat. *Prostaglandins* **26**, 47–54.

Wolfs, G. M. J. A. and van Leeuwen, M. (1979). Electromyographic observations on the human uterus during labour. *Acta obstet. gynec. scand. Suppl.* **90**, 1–61.

Yancey, S., Easter, D., and Revel, J.-P. (1979). Cytological changes in gap junctions during liver regeneration. *J. Ultrastruc. Res.* **67**, 229–242.

Zampighi, G. A., Hall, J. E., and Kreman, M. (1985). Purified lens junctional protein forms channels in planar lipid films. *Proc. natn. Acad. Sci. U.S.A.* **82**, 8468–8472.

Zelcer, E. and Daniel, E. E. (1979). Electrical coupling in rat myometrium during pregnancy. *Can. J. Physiol. Pharmacol.* **57**, 490–495.

Zervos, A. S. (1985). Preparation of a gap junction fraction from uteri of pregnant rats: The 28 kD polypeptides of uterus liver and heart gap junctions are homologous. *J. Cell Biol.* **101**, 1363–1370.

11 Softening and dilation of the uterine cervix
HILARY DOBSON

I Introduction

The cervix is a firm, thick-walled tube lined with a mucus-secreting epithelium, which connects the uterus and the vagina. It consists largely of connective tissue and, in the non-pregnant female, acts as a barrier protecting the upper genital tract. Throughout pregnancy the cervix prevents expulsion of the fetus and yet at parturition it must yield to permit passage of the offspring to an independent existence.

For management or clinical reasons it is sometimes desirable to alter the properties of the cervix. For example, during artificial insemination in species such as the sheep, the almost impenetrable structure of the cervix prevents insertion of an insemination rod (intra-uterine insemination results in far superior conception rates). This and the techniques of uterine lavage, or non-surgical embryo transfer, in many species would be easier if penetration of the cervix was less difficult. At the end of pregnancy, failure of part or all the cervix to dilate poses clinical problems, especially in primigravid subjects, and during conditions such as "ringwomb" in sheep. An opposite problem can occur in mid-pregnancy with cervical incompetence (premature dilation) which can result in an increased risk of uterine infection, as well as early delivery of the fetus.

A greater understanding of the regulation of functional changes in the cervix will hopefully lead to an ability to successfully manipulate the physical properties of this important organ. The following review does not attempt to cite every publication on the cervix, but rather to present established facts and highlight areas where further clarification is required. Throughout, the term "softening" refers to changes which take place usually during the last quarter of pregnancy to permit "dilation", which occurs with considerable changes in structure during parturition with the passage of the fetus. A "ripened" cervix is one which has fully softened immediately prior to the onset of first stage labour.

II   Gross structure

The gross morphology of the cervix varies markedly between species (El-Bana and Hafez 1972), no doubt reflecting the different physiological mechanisms of sperm transport as much as the mechanics of parturition.

In non-pregnant sheep the cervix is 5–6 cm long and 2–3 cm wide (Ward 1968). The lumen is practically occluded by 5–8 asymmetrical annular folds (Fig. 11.1). The wall consists of an inner, tough, dense layer of connective tissue with an outer smooth muscle layer covered by peritoneum dorsally and ventrocranially. The major blood vessels run between the outer and inner layers (Ward 1968).

The bovine cervix is 6–13 cm long and of similar appearance, with only 3 or 4 annular projections (Mullins and Saacke 1982). In the pig, the projections are arranged in corkscrew fashion, corresponding to the spiral twisting of the boar's penis. Apart from ruminants and the pig, domestic animals have a straight cervical canal. In the rat and the mouse the cervix is much smaller (1.3×0.7 mm) and the former has two canals.

In the human female, the cervix is 3–4 cm long, with a straight, narrow lumen containing extensively branched glands and a mucus-secreting epithelium. The portion of the cervix continuous with the uterus is the isthmus which merges with the uterus in late pregnancy during the process called "effacement" (Danforth 1983).

III   Microscopic structure

The ovine cervix is lined by a highly folded epithelium consisting of non-secretory tall columnar cells and mucus-secreting goblet cells. The cervical stroma is made up mainly of thick collagen fibre bundles running in all directions, with blood vessels and smooth muscle cells embedded between the col-

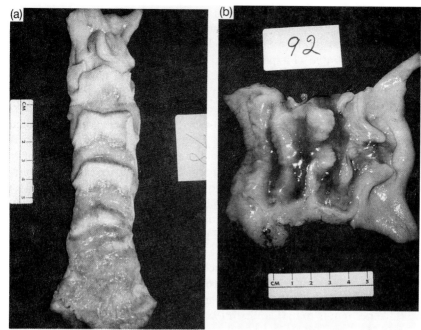

Fig. 11.1.   Gross structure of the ovine cervix a) from a sheep in mid pregnancy, and b) from a sheep during the ripening process just prior to parturition. In each case the tubular structure was sliced in half longitudinally. Note the pronounced luminal folds in both cervices, and the loose mucus in the ripening cervix.

lagen. The matrix is sparse and few cells can be seen (Ward 1968). Studies in man (Danforth 1947; 1983; Hughesdon 1952) and other species (rats: Harkness and Harkness 1959; Bryant, Greenwell, and Weeks 1968; mouse: Leppi 1964; pig: Endell, Holtz, and Schmidt 1976*a,b*; rabbit: Conrad and Hoover 1982; MacLennan, Katz, and Creasy 1985) indicate basically the same microscopic structure as in the ovine cervix. Very small amounts of elastin (< 1 per cent of the total fibrous tissue) are haphazardly distributed in human cervical tissue (Danforth 1947). A higher proportion (2–10 per cent) is present in the ovine cervix (Ward 1968). Elastin may account for the rapid dilatability of the cervix in labour and the return to normal shape afterwards (Leppert, Keller, Cerreta, and Mandl 1982).

The smooth muscle is unevenly distributed throughout the cervix: the upper, middle and lower segments of the human cervix contain about 25, 16 and 6 per cent smooth muscle, respectively. This contrasts with 65–70 per cent in the myometrium (Rorie and Newton 1967). The rat cervix contains a higher proportion of smooth muscle than the human (25–38 per cent, Harkness and Harkness 1959), again with a greater amount at the internal os.

Whether collagen, or the included smooth muscle, is the more important factor in cervical function at parturition has been examined at length (Hughesdon 1952; Buckingham, Beuthe, and Danforth 1965). The ability of cervical muscle to contract and respond to pharmacological agents has been clearly demonstrated (Fitzpatrick 1957; 1977; Conrad and Ueland 1976; Edqvist, Einarsson, Gustafsson, Linde, and Lindell 1975). The active tension developed by the smooth muscle within the collagen layer, however, does not appear to influence the tensile properties of the cervix (Harkness and Harkness 1959; Hollingsworth and Gallimore 1981). The present consensus is that collagen plays the more important role in regulation of cervical function.

Apart from smooth muscle, the fibroblast is the most common cell type in cervical tissue. In cervices from non-pregnant animals these cells are poorly developed, with elongated nuclei and long thread like cytoplasmic processes projecting along the collagen fibre bundles (Junqueira, Zugaib, Montes, Toldeo, Krisztian, and Shigihara 1980; Parry and Ellwood 1981). During pregnancy there is a proliferation and migration of fibroblasts, white cells, macrophages and eosinophils into the cervix (Bassett 1962; Buckingham, Selden, and Danforth 1962), all of which are important in determining form and function (see later). Extracellular collagen forms the bulk of the stroma: 45–80 per cent dry matter in non-pregnant sheep (Ward 1968; Fitzpatrick 1977), 75 per cent in man (Danforth 1980), and 25–55 per cent in rats (Harkness and Harkness 1959). The densely packed collagen bundles are randomly arranged throughout the tissue. With advancing gestation there is an increase in the separation and a decrease in the size of bundles. An increased separation of individual fibrils within the bundles has been observed within the last two weeks of pregnancy in the ewe but there is no apparent change in the nature of each fibril as revealed by collagen banding on transmission electron micrographs (Fitzpatrick 1977; Parry and Ellwood 1981). Scanning electron microscopy of the ovine cervix shows a similar time scale of collagen fibril separation (Owiny, Fitzpatrick, Spiller, and Appleton 1987).

The inter-fibrillar spaces are composed of proteoglycans and glycoproteins in varying states of hydration. Differential staining by alcian blue, in the presence of varying molarities of magnesium chloride (Scott and Dorling 1965) has revealed changes in the proportions of the types of glycosamino-glycans present throughout pregnancy and parturition (Table 11.1).

IV Physical properties

Investigations of tissue compliance and strength have been limited by available techniques. *In vivo* measurements may permit temporal changes to be recorded, assuming that the test itself does not interfere with subsequent events, but ethical considerations in women usually limit this to very short intervals. *In vitro* recording has to be carried out on tissue obtained immediately after death of the animal, and this precludes precise timing in relation to parturition.

Table 11.1

Relative staining intensities (at 620 nm) of histological sections obtained from cervices of sheep. The sections were stained with alcian blue (pH. 5.8) in the presence of different magnesium chloride molarities. (From Fitzpatrick and Dobson 1979, with permission)

MgCl$_2$	0.06M	0.3M	0.5M	0.7M	0.9M
Non-pregnant n = 5	6.96 ± 0.63	1.62 ± 0.54	0.10 ± 0.01	0.06 ± 0.01	0.04 ± 0.01
Late pregnant, non-dilated, n = 5	8.16 ± 0.51	1.12 ± 0.28	0.09 ± 0.01	0.02 ± 0.01	0.01 ± 0.01
Pregnant, dilated n = 4	4.92 ± 0.12	0.22 ± 0.12	0.07 ± 0.01	0.02 ± 0.01	0.02 ± 0.01
	Hyaluronic Acid	Weakly Sulphated	Chondroitins	Heparan Sulphate	Keratan Sulphate

An *in vivo* scoring system for the human cervix and pelvis has been described (Bishop 1964) in which the degree of dilation and effacement, the consistency and position of the cervix and the level of the presenting fetal parts, are assessed. The system has been widely applied to predict the favourability of labour induction in women. Cervical consistency has also been assessed by temporarily inserting rigid intracervical instruments (Johnstone, Boyde, McCarthy, and McClure-Brown 1974; Calder 1981). In sheep digital palpation has been applied to assess cervical state (Ward 1968; Mitchell and Flint 1978), while in cattle palpation *per rectum* is preferred (Putnam, Rice, Wetteman, Lusky, and Pratt 1985; Varner, Hinnichs, Garcia, Osborne, Blanchard, and Kenney 1985). It must be remembered, however, that vaginal distension results in oxytocin and prostaglandin $F_{2\alpha}$ release. As both these hormones can cause uterine contractions the methodology used to investigate vaginal distention may alter rates of spontaneous cervical dilation.

A small but steady increase in lumenal diameter occurs in the early weeks of pregnancy both in nulliparous and parous women; few changes take place in the second trimester but during the last 5–6 weeks, the cervix changes shape, effacement occurs and there is a marked increase in compliance in the absence of obvious uterine contractions. Cervical softening is more advanced in multiparous than nulliparous women, while the converse is true of cervical effacement (Hendricks, Brenner, and Kraus 1970), emphasising the need in all species to take into account parity of the dam. The onset of cervical dilation is often coincident with the onset of labour (Anderson and Turnbull 1969).

Long-term indwelling rubber intracervical balloons have been used in sheep and goats in the last third of pregnancy (Fitzpatrick 1977; Fitzpatrick and Dobson 1979; Stys, Clark, Clewell, and Meschia 1980), allowing repeated measurements of compliance over several days in relation to cervical softening, myometrial activity and administration of different drugs. However, artefactual problems can arise from maintaining balloons in place over long periods. Introduction of the balloons is associated with a transient increase in cervical compliance, indicating that the procedure, at least temporarily, influences cervical function, possibly *via* local inflammation. Nevertheless, dramatic changes in compliance have been recorded during parturition (Fitzpatrick 1977; Stys *et al.* 1980) (Fig. 11.2).

Harkness and co-workers (Harkness and Harkness 1959; Cullen and Harkness 1960) developed an *in vitro* system for measuring changes in the cervical diameter. A static load was applied to rat cervices by a rod suspended through each cervical canal. The extension per unit length per unit time (creep) was used as a measure of softness. Modifications of this method have revealed in rats a gradual increase in extensibility during the last quarter of pregnancy, with a further large rise immediately pre-partum (Hollingsworth 1981). Using this method, similar changes have also been observed in

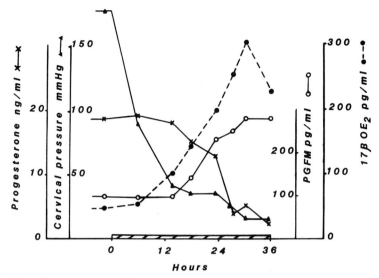

Fig. 11.2.   Concentrations of hormones in maternal vena caval plasma in relation to changes in compliance of the cervix of a ewe pregnant 133 days. Dexamethasone was infused i.v. into the fetus at 1 mg/day for the period indicated by the bar. Cervical compliance (triangles) is shown as the pressure recorded from the balloon *in situ*, less the corresponding pressure in air, when distended with 4 ml water. (From Fitzpatrick 1977, with permission).

sheep (Ward 1968; Ledger Ellwood and Taylor 1983). However, in the sheep, the whole cervix cannot be used and it is difficult to cut rings of standard or regular width. A modification involving cervical strips (10 mm lengths of 3 mm wide, full thickness) and a static load has been applied to ovine cervices (Owiny *et al.* 1987).

Another measure of the strength of cervical tissue is the stretch modulus, where strips of tissue are subjected to a dynamic load and the tension developed is plotted against extension. The stretch modulus is calculated from the linear portion of the resultant J-shaped curve (Fitzpatrick and Dobson 1981; Conrad and Ueland 1983). This method allows alternate extension and relaxation to be applied, simulating forces from uterine contractions.

Both these *in vitro* methods have the disadvantage that measurements can be made only once after killing the animal. However, due to the quantity of tissue available in large animals, for example, sheep, repeated measurements can be made and changes along the length of the cervix can be examined. Creep and extensibility tests usually, although not invariably, give similar results. Care must be taken to restrict tension loads to low limits, avoiding rupture, in order to simulate conditions which may occur spontaneously *in vivo*.

On application of a load, strips are very extensible, that is, small increases in stress produce large extension (strain), with little resistance to deformation. As more load is applied, tissues become stiffer and less extensible (Owiny 1986). Conrad and Ueland (1983) have suggested that highly extensible elastic fibres are called into extension first, with stiffer collagen fibres taking over as extension proceeds. Most of the resistance to deformation occurs after the straightening of the crimp in the collagen fibres (Gathercole and Keller 1975).

During the second half of pregnancy in the sheep, there is a slow and progressive softening of the cervix, reflected in the increase in extensibility and reduction of stretch modulus (Owiny 1986), but even at term the tissue feels "hard" on manual examination. Ripening occurs between full term and the onset of labour (Fig. 11.1), coinciding with gross macroscopic changes in the cervix which becomes markedly enlarged, oedematous and floppy. Similar changes have been recorded in other species (human: Calder 1981; rat: Hollingsworth 1981). The large increase in extensibility (ripening) prior to the onset of labour no doubt represents the culmination of cervical softening, which will accommodate dilation by uterine contractions and the passage of the fetus. The fact that there is considerable cervical ripening before the onset of regular contractions (Owiny 1986) supports the observations that uterine contractions are not essential to the process of ripening (Stys *et al.* 1980; Hollingsworth 1981; Ledger, Webster, Harrison, Anderson, and Turnbull 1985).

Passage of the fetus through the cervix almost overstretches the tissue, so that very little creep will be found in measurements made immediately post partum. This renders dubious any hypotheses based on this measurement at a time when fetal parts have been presented at the vulva. Post partum cervical involution is rapid and the tissue remodelling taking place at this time is as remarkable as that prior to parturition, although it has been studied far less.

During early phases of labour in the sheep, extensibility of the uterine end of the cervix is only half that of the middle zone, suggesting that the former area is already stretched quite early by uterine contractions before the onset of labour (Owiny 1986). A major implication of variable ripening along the ovine cervix is manifest in the clinical condition of ringwomb in which the middle zone of the cervix fails to dilate before the expulsive contractions of the uterus. A similar differential ripening occurs in women during effacement when only a ring of fibrous tissue at the internal os may remain to retain the fetus (Danforth 1983). Indeed, the stretch modulus not only decreases progressively along the cervix, the internal os being stiffer than the external os, but also outwards within the wall, that is, inner > middle > outer (Conrad and Ueland 1983). This indicates that in studies using strips of tissue, care must be taken to define their source and studies may be biased if relying too heavily on results from one zone.

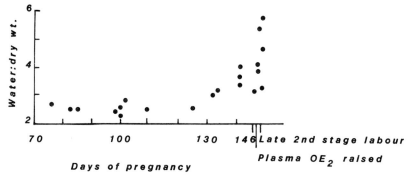

Fig. 11.3. The water: dry weight ratio in cervices from individual ewes at different stages of pregnancy and labour. (From Fitzpatrick and Dobson 1981, with permission).

V   Biochemistry of cervical connective tissue

The extracellular stroma of the cervix consists of collagen fibres and elastin, held together by an amorphous matrix composed principally of proteoglycans, glycoproteins and water. The collagen gives tensile strength, elastin imparts elasticity, while the matrix contributes to the integrity of the tissue.

1   WATER

Studies in several species have shown that cervical water content and dry matter increase disproportionately during pregnancy, the water : dry weight ratio more than doubling around parturition (sheep: Fitzpatrick and Dobson 1981; Fosang, Handley, Santer, Lowther, and Thorburn 1984; woman: Kleissl, van der Rest, Naftolin, Glorieux, and de Leon 1978; Ito, Kitamura, Mori, and Hirakawa 1979; Uldbjerg, Ulmsten, and Ekman 1983b; rat: Harkness and Harkness 1959; Cabrol, Huszar, Romero, and Naftolin 1981; mouse: Rimmer 1973). The water content increases slowly during the latter part of pregnancy but more markedly just prior to and immediately after parturition (Fig. 11.3).

The separation of collagen bundles referred to in histological studies could be due to the presence of increased amounts of water at the end of pregnancy. This, undoubtedly, will affect the tensile strength of the tissue, as mechanical behaviour of tissues is known to be sensitive to water content (Soden and Kershaw 1974). The presence of an increased population of leucocytes at term could influence hydrodynamics through effects on vascular permeability, and changes in the ionic components of the extracellular matrix may result in imbibition of water.

Water content complicates the interpretation of some of the following biochemical analyses. Obviously, certain changes in dry matter constituents lead to increased hydration which is essential for changes in physical properties but equally, increased weight of water cannot itself be responsible for the enzyme changes. Even expression of results per mg DNA as a reduction to a cellular basis can be misleading, due to the increased infiltration of leucocytes at the end of pregnancy.

2   PROTEOGLYCANS AND GLYCOSAMINOGLYCANS (GAGs)

Proteoglycans are complex molecules, composed of a number of highly charged (anionic) linear GAG chains, which extend out from a protein core, allowing each intact macromolecule to occupy large domains in the tissue matrix. Reversible extension can occur, due to solvent accumulation in the molecular domain and altered intermolecular interactions between the GAG chains. This ability of the proteoglycans to control solvent domains gives connective tissues such as the cervix the properties of compliance and yielding to stretch forces. In early studies of proteoglycans and GAGs in cervical tissue, Danforth, Veis, Breen, Weinstein, Buckingham, and Manalo (1974) produced preliminary evidence for alterations in the constituents after cervical dilation in women. It became clear that more detailed studies were required.

The GAGs are long chains of repeating disaccharides, containing one hexosamine (glucosamine or galactosamine) and one uronic acid (glucuronic or iduronic). Individual GAGs vary in the exact combinations of these residues (Fig. 11.4): chondroitin sulphate consists of galactosamine and sulphated glucuronic acid, whereas dermatan sulphate contains chondroitin-sulphate-like segments in parts of the molecule, and other parts in which glucuronic acid has been converted to iduronic acid after incorporation into the chain (leading to heterogeneity and resulting in difficulties in precise biochemical identification). Non-sulphated hyaluronic acid is composed of glucosamine and glucuronic acid and, upon hydration, assumes an extended structure. Heparan sulphate is characterized by glucosamine and a mixture of sulphated glucuronic and iduronic acids. Keratan sulphate is the only GAG which does not contain uronic acid, but has repeating galactose and glucosamine units.

The number and length of GAG chains in proteoglycan molecules is dependent on type of tissue, stage of development and metabolic activity. Thus the overall molecular nature and water structuring properties of proteoglycans are different between tissues and even within tissues at different times. Many of the early studies on cervical tissue involved treatment with proteolytic enzymes and separation of the GAG residues by electrophoresis or column chromatography. Closer identification of individual GAGs was carried out by differential enzyme digestion techniques.

In general, from work in several species, dermatan sulphate appears to be the most abundant GAG present (sheep: Fitzpatrick and Dobson 1981;

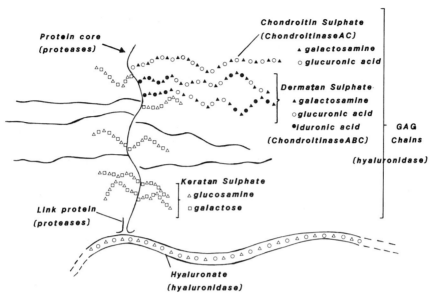

Fig. 11.4. A diagram indicating the relationships between different glycosamino-glycans in proteoglycans of the ovine cervix. Catabolic enzymes are indicated in brackets beneath each constituent.

Fosang *et al.* 1984; woman: von Maillot, Stuhlsatz, Mohanaradhkrishnan, and Greiling 1979; Kitamura, Ito, Mori, and Hirakawa 1980; Uldbjerg *et al.* 1983*b*; rat: Golichowski, King, and Mascaro 1980; Huszar 1981). Estimates of dermatan sulphate, as a percentage of GAGs extracted, vary between 50–80 per cent in cervices from non pregnant subjects, but most studies report a decline in proportion during pregnancy. Conversely, hyaluronic acid, while not a major constituent, does increase significantly, especially during the last quarter of pregnancy.

Differences do occur between different reports in respect of other GAGs. For example, Danforth *et al.* (1974) and von Maillot *et al.* (1979) both referred to a new GAG component which from hexosamine/uronic acid ratios was thought to be keratan sulphate, but others were unable to isolate keratan sulphate (Uldbjerg *et al.* 1983; Fosang *et al.* 1984). Similarly, Kita-mura *et al.* (1980) placed emphasis on heparan sulphate fraction, whereas others (Golichowski *et al.* 1980; Fitzpatrick and Dobson 1981) concentrated on the relative proportions of two dermatan sulphate fractions identified by differential enzyme digestion. Changes in the proportions of GAGs have also been shown using histological staining methods (Table 11.1). The cervical content of hyaluronic acid was greater in late pregnant ewes than in non-pregnant ewes but the reason for the unexpected marked decline in dilated cervices is unknown. The concentrations of the other GAGs declined as preg-nancy proceeded. It is most likely that all these differences between reports are, to a great extent, the sum of different and inadequate technical procedures as well as possible species differences.

Few workers have investigated intact proteoglycans. Briefly, this involves extraction from tissue in the presence of protease inhibitors and purification by ion-exchange chromatography. However, a small molecular weight (approximately 25,000 Daltons) dermatan sulphate proteoglycan has been isolated from human and ovine cervices (Uldbjerg, Malmstrom, Ekman, Sheehan, Ulmsten, and Wingerup 1983*a*; Fosang *et al*. 1984). A larger proteoglycan which may represent a new proteoglycan species, or a proteoglycan monomer with larger GAG side chains, was isolated from preparturient cervices. The rate of synthesis of GAGs is enhanced during pregnancy, especially during the last few weeks, that is, the period of cervical softening and dilation. This increased rate of synthesis in the presence of a diminishing total uronate concentration (Fosang *et al*. 1984) is indicative of enhanced proteoglycan turnover. Dermatan sulphate proteoglycans are closely associated with collagen fibres (Flint 1972) and organize extracellular formation and arrangement of collagen fibrils (Toole and Lowther 1968). Increased turnover and formation of new species with different characteristics will lead to altered mechanical properties. The considerable accumulation of hyaluronic acid, and hence water, between the collagen fibrils during softening and ripening will also disperse the fibrils and increase distensibility of the tissue (Fig. 11.5).

3   COLLAGEN

Collagen which forms the bulk of cervical tissue, has a molecular weight of approximately 300,000 Daltons and includes high concentrations of the uncommon amino acid, hydroxyproline, which is thus often used for collagen estimations. The collagen molecule has a stiff, rod-like shape (300 nm × 15 nm) and consists of three parallel chains wrapped around each other in a helix. A collagen fibril is made up of collagen molecules orientated parallel and staggered, so creating the typical light and dark bands on electron microscopic examination. Cross links between collagen molecules, along with binding of collagen fibrils to the ground substance, provide mechanical strength for the cervical tissue. Different types of collagen exist, depending largely on amino acid sequences, but there is no clear evidence that cervical collagen types change either during pregnancy, softening or dilation (Ito *et al*. 1979).

As the cervix grows in size during pregnancy the total collagen content per cervix increases. However, expressed on a dry matter basis, the collagen concentration declines gradually throughout pregnancy in several species (sheep: Ward 1968; woman: Danforth *et al*. 1974; rat: Harkness and Harkness 1959; mouse: Rimmer 1973; rabbit: Koob, Stubblefield, Eyre, and Ryan 1980). Nevertheless, marked changes in *total* collagen concentration have only once

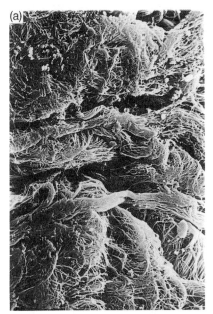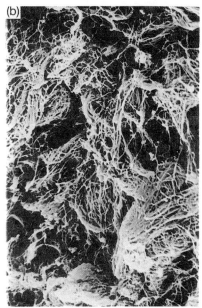

Fig. 11.5.   Scanning electron micrograph of the stroma of cervical tissue obtained a) at full term (× 9600) and b) during early labour (× 9000) in sheep. At full term the appearance is similar to the non pregnant state; during early labour the bundles are disintegrating into fine, widely separated fibrils. (From Owiny 1986, with permission).

been suggested *during* dilation (Fosang *et al.* 1984) and that was on a wet weight basis.

Perhaps more importantly, changes in collagen solubility have been correlated with softening and dilation: the proportion of salt- or acid-soluble collagen increased from 2–5 per cent in cervices from non pregnant subjects to 11–30 per cent at full dilation, whereas insoluble collagen decreased from 68–75 per cent to 30–40 per cent at the same time (sheep: Fitzpatrick 1977; woman: von Maillot and Zimmerman 1976; Kleissl *et al.* 1978; rat: Hillier and Wallis 1982). Detailed analysis of the soluble fractions indicated a prevalence of collagen breakdown products in the samples obtained during dilation. The least cross-linked and most soluble collagen is more easily digested than the insoluble collagen, but removal of the partly degraded collagen is slow (Kleissl *et al.* 1978). This could explain why total collagen concentrations do not change significantly during dilation: the collagen may be degraded during parturition but retained in the cervix for some time. Junqueira *et al.* (1980), using a histological procedure for the detection of collagen assembled in fibres, suggested that a much lower proportion existed in intact fibres in dilated cervices compared with those from non pregnant women.

However, it must not be forgotten that new collagen is being synthesised throughout pregnancy as the cervix grows, so it is possible that degradation of existing collagen is not the sole explanation of the role of collagen in cervical softening. The newly synthesised collagen could form weaker links between fibres and with the proteoglycan matrix to produce a more pliable tissue.

4   COLLAGENASE

Native collagen is cleaved by activated collagenase into three-quarters ($-NH_2$ terminal) and one-quarter (-COOH terminal) fragments which undergo spontaneous denaturation with loss of helical configuration and hence loss of physical strength (van der Rest 1980). The resultant polypeptides are further hydrolysed by non-specific proteases (Gay and Miller 1978).

Increased collagenase activity has been demonstrated in cervices from pregnant women and sheep at term, compared to non pregnant subjects (Kitamura, Ito, Mori, and Hirakawa 1979; Ellwood, Anderson, Mitchell, Murphy, and Turnbull 1981). Indeed, by the 10th week of human pregnancy, increased collagenase activity has been suggested (Uldbjerg *et al.* 1983*b*) which supports the hypothesis that collagen turnover increases in early pregnancy.

Collagenase activity can be regulated in a variety of ways: principally, by the rate of synthesis and secretion of the latent proenzyme, procollagenase; by changing conversion rates from the latent form to the active form; and by the concentration of inhibitors (that is, $\beta1$ anti collagenase and $\alpha2$ macroglobulin). However, Raynes, Clarke, Anderson, Fitzpatrick, and Dobson (1988) have shown that collagenase inhibitor is present in increased amounts in cultured cervical extracts taken from sheep near term. Pretreatment of tissue with trypsin can increase hydroxyproline release, either by unmasking a latent form of the enzyme, or by increasing tissue susceptibility to degradation (Fitzpatrick and Dobson 1981), although there is evidence in human cervices that the proportion of active and inactive forms of collagenase do not change in late pregnancy (Kitamura *et al.* 1979). The latter authors also showed that the enzyme exists complexed with macroglobulin, but this may have been a technical artefact (Uldbjerg *et al.* 1983*b*). As it appears that collagenases are crucial to dilation, more precise work is required, especially by taking care to determine the molecular sizes of the hydroxyproline-containing molecules produced (Fig. 11.6).

Non-specific proteolytic enzymes have been described in cervix tissue: neutral and acid proteinases (Mori, Ito, Hirakawa, and Kitamura 1981) and PZ-peptidase (Ito, Naganeo, Mori, Hirakawa, and Hayashi 1977). The latter displays increased activity immediately post partum, as does leucocyte elastase, a serine protease localized in granules of polymorphonuclear leucocytes (Uldbjerg *et al.* 1983*b*). The physiological significance of these enzymes is unknown, but they may prepare collagen fibrils for collagenase digestion

a)

b)

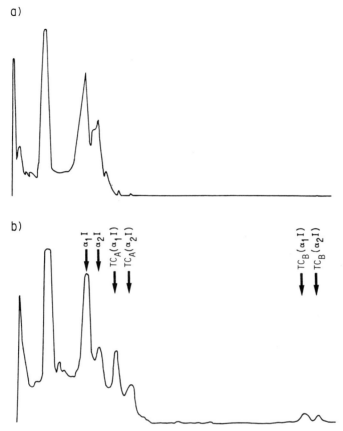

Fig. 11.6.   Laser densitometer trace of electrophoresis on a 7.5% polyacrylamide gel of collagen substrate incubated a) without and b) with concentrated explant culture medium for 18 h at 25°C. Specific cleavage products TC_A and TC_B are indicated. (From Raynes *et al*. 1988, with permission).

(Raynes, Anderson, Fitzpatrick, and Dobson 1988), degrade collagen already cleaved by collagenase, or may be of importance in the catabolism of proteoglycans and other components in the extracellular matrix.

VI   Cells of cervix connective tissue

Although a large proportion of the stromal connective tissue of the cervix is comprised of extracellular matrix, changes in its constituents must be governed by the cellular component. The most common cell types, apart from smooth muscle cells, are fibroblasts which become much enlarged and increasingly active as pregnancy progresses (Bassett 1962; Junqueira *et al*. 1980; Parry and Ellwood 1981). Fibroblasts are involved in both collagen and glycosaminoglycan metabolism (Buckingham *et al.* 1962; Chang, Abe, and Murota 1977) and, no doubt, production of the required enzyme systems. In late pregnancy there is also an infiltration of leucocytes, around which a zone of connective tissue degradation has been reported, suggesting secretion of collagenase. Leucocytes have also been shown to influence vascular permeability (Wedmore and Williams 1981) which may regulate tissue hydration. The factors stimulating leucocytic infiltration at term are unknown; an inflammatory response to the trauma of parturition could be involved (Liggins 1981). The presence of collagen fragments, attributed to increased collagen turnover, may provide a further stimulus through a chemotactic effect (Postlethwaite and Kang 1976).

VII   Hormonal regulation of cervical connective tissue

Whatever dictates the role of fibroblasts and leucocytes, it is obvious that their presence, activities and, consequently, the mechanical strength of the cervix, must be under close control. During pregnancy and, especially at parturition, a complex sequence of events has to be finely synchronized: for example, the cervix must soften to allow increased uterine activity to coincide with cervical dilation and expulsion of the fetus. It is understandable that all these events should be integrated by the reproductive hormones, especially progesterone, oestradiol, prostaglandins and relaxin. Receptors for these hormones are expected to exist on selected cell types in order to exert influences on the expression of specific cellular activities (although this still requires proof for cervical tissue). There is evidence that prostaglandins can be produced within the cervical tissue; the other hormones are thought to originate elsewhere.

In spite of many experiments conducted to elucidate precise mechanisms for hormonal control of cervical dilation, there is currently little conclusive evidence for a direct effect of any hormone. Any *in vivo* experiment to alter the hormonal environment may affect uterine activity, which could cause prostaglandin release; similarly, administration of oestradiol may directly influence prostaglandin synthesis and release. Hormonal actions will be mediated via fibroblasts, leucocytes and smooth muscle cells, and it is known that all these cell types from different tissues and species respond in different ways; consequently, extrapolation across tissue types may be very misleading. Much of the following discussion will therefore be speculative.

1   OESTRADIOL

Clinical application of different oestrogen preparations to women have improved cervical scores in late pregnancy, no doubt partially through an action on uterine motility (MacKenzie 1981). Stys *et al.* (1980) prevented measurable uterine contractions in sheep by simultaneous administration of progesterone with diethylstilboestrol, and showed a reversible increase in compliance measured *in vivo*. Fitzpatrick and Dobson (1981) confirmed for sheep the oestradiol-induced change in mechanical strength; a release of prostaglandins of the E and F series was also found, but the physical changes were only partially inhibited by simultaneous infusion of a prostaglandin synthetase inhibitor, meclofenamic acid. Parallel analyses of $^{35}SO_4$ incorporation into GAGs revealed an alteration in total GAG synthesis caused by oestradiol, which may in turn have accounted for changes in mechanical strength.

In vivo application of an oestradiol precursor, dehydroepiandrosterone sulphate, resulted in increased cervical compliance in women (Mochizuki and Tojo 1980), along with an increase in collagenase activity. However, others (Ellwood *et al*. 1981; Wallis and Hillier 1981) could only show an oestradiol-induced inhibition of *in vitro* collagenase activity. Oestrogens may be responsible for the influx of collagenase-active leucocytes at the end of pregnancy (Bassett 1962), an event that would not be reproduced *in vitro*.

2 PROGESTERONE

The role of progesterone in cervical dilation is also far from clear. It has been used to suppress uterine activity and permit softening during oestrogen administration in sheep (Stys *et al*. 1980) and yet, at the end of pregnancy, results implied an inhibitory role for progesterone (Fitzpatrick and Dobson 1979). Progesterone also reduces collagenase activity and leucocytic migration, albeit in the post partum uterus (Jeffrey and Koob 1980).

3 PROSTAGLANDINS

Clinically, prostaglandins have been used to improve cervical scores, but this may be a pharmacological rather than a physiological effect. Any reported action of prostaglandin $F_{2\alpha}$ in the rat must be clearly distinguished from its luteolytic effects (Hollingsworth 1981). Intra-vaginal administration of prostaglandin E_2 has been used extensively in human medicine (MacKenzie 1981) and administration in late pregnant sheep appears to increase compliance by a local mechanism (Fitzpatrick 1977; Stys *et al*. 1980).

Ellwood *et al*. (1981) were unable to detect an effect of prostaglandin E_2 on collagenase activity, possibly because tissue was incubated and assessed for small hydroxyproline-containing peptides released into the medium. This would not detect larger cleavage fragments left trapped in the tissue, although structural strength had diminished. Other studies (Uldbjerg *et al*. 1983*b*), using smaller peptide fragments as substrate, have shown an increase in collagenase after prostaglandin E_2. Caution in interpretation must be urged here as many experiments compare tissue from non pregnant subjects with that obtained after the start of labour or, even worse, after delivery, then regard the results as an explanation of events during softening. Passive mechanical stretching of tissue (as in labour) will increase tissue prostaglandin concentration (Hillier and Coad 1982) and this, in turn, may augment collagenase activity: this is an indirect effect of increased uterine activity, with no relation to the direct effects on cervical connective tissue metabolism.

In vivo administration of prostaglandin E_2 to women has also been shown to alter the proportions of GAGs in cervical tissue (Uldbjerg *et al*. 1983*a*; Norstrom 1984). Although not directly comparable, prostaglandins increased hyaluronic acid production in cultured rat granuloma fibroblasts (Murota, Abe, and Otsuka 1977) and this may be important in cervical softening.

4   RELAXIN

Only crude preparations of this hormone have been available for clinical use. However, application has led to softening of the cervix in a variety of species, especially if the cervix has been primed with oestradiol (woman: MacLennan 1981; Evans, Dougman, Moawad, Evans, Bryant-Greenwood, and Greenwood 1983; rat: Cullen and Harkness 1960; Steinetz, O'Byrne, and Kroc 1980; mouse: Fields and Larkin 1980; cattle: Perezgrovas and Anderson 1982; Musah, Schwabe, Willham, and Anderson 1986). Morphological evidence confirmed an increase in the extracellular matrix, but biochemical evidence so far is restricted to a report of increased collagenase and PZ-peptidase activity in the relaxed pubic symphysis of mice after relaxin treatment (Weiss, Nagelschmidt, and Struck 1979). It is of interest, however, that human fibroblasts possess receptors for relaxin and the hormone may exert effects on collagen and proteoglycans through modulation of fibroblast activity. Relaxin also exerted a mitogenic influence on fibroblasts in culture and proliferation of these cells may participate in the cervical response to relaxin (McMurtry, Floerscheim, and Bryant-Greenwood 1980).

VIII   Conclusion

Technical difficulties beset fundamental work on the cervix in many species, but as long as management and clinical problems exist, these difficulties must be overcome. Empirical use of hormones and other drugs may partially alleviate abnormal cervical function, but a deeper appreciation of the mechanisms involved may lead to better (preventive) treatments.

In summary, changes in proteoglycan and collagen metabolism occur throughout pregnancy, reaching a crescendo in the periparturient period (Less dramatic changes occur during the oestrous cycle, permitting easier catheterization of the cervix at oestrus than during the luteal phase). These changes appear to involve a reduction in intact collagen concentration, accompanied by increased solubility, due to synthesis of new, less cross-linked collagen, as well as increased collagenolytic activity. Confirmation of this hypothesis should concentrate on molecular size of the hydroxyproline-containing molecules, as well as isolation, characterization and localization of the collagenase involved. Parallel with the changes in collagen, there is an increase in the amount of GAGs and a change in composition, with possible synthesis of a new population of proteoglycans. Further work on the proteo-glycans should be productive, because of their influence on the binding of collagen fibres, as well as on water retaining capacities, both of which alter the mechanical properties of connective tissue. Clearly, all these changes must take place via transcription of DNA within the cells of the stroma: un-fortunately, very little work has been centred to date on this altering popula-tion of cells, to say nothing of their regulation by hormones.

Acknowledgements

I am very grateful to Professor R. J. Fitzpatrick for arousing my interest in the cervix and for his encouragement throughout our joint studies. I am also grateful to Jill Lanham, Eleanor Blears, Jo Owen, Brenda Jamieson, David Spiller and David Horne for their technical help over the years. Similarly, to Patricia Hardman, James Owiny, John Anderson and John Raynes who have provided stimulating discussions. Thanks too to Sheila Nugent for typing the manuscript. None of the work would have been possible without the gener-ous financial support of the Lalor Foundation, the Wellcome Trust and the University of Liverpool, to whom I express gratitude.

References

Anderson, A. B. M. and Turnbull, A. C. (1969). Relationship between length of gesta-tion, cervical dilatation, uterine contractility and other factors during pregnancy. *Am. J. Obstet. Gynec.* **105**, 1207–1214.

Bassett, E. G. (1962). Infiltration of eosinophils into the modified connective tissue of oestrous and pregnant animals. *Nature, Lond.* **194**, 1259–1261.

Bishop, E. H. (1964). Pelvic scoring for elective induction. *Obstet. Gynaec.* **24**, 266–268.

Bryant, W. M., Greenwell, J. E., and Weeks P. M. (1968). Alterations in collagen organisation during dilatation of the cervix uteri. *Surg. Gynaec. Obstet.* **126**, 27–39.

Buckingham, J. C., Buethe, R. A., and Danforth, D. N. (1965). Collagen muscle ratio in clinically normal and clinically incompetent cervices. *Am. J. Obstet. Gynec.* **91**, 232–237.

—— Selden, R., and Danforth, D. N. (1962). Connective tissue changes in the cervix during pregnancy and labor. *Ann. N.Y. Acad. Sci.* **97**, 733–742.

Cabrol, D., Huszar, G., Romero, R., and Naftolin, F. (1981). Gestational changes in the rat uterine cervix: protein, collagen and glycosaminoglycan content. In *The Cervix in Pregnancy and Labour: Clinical and Biochemical Investigations.* (eds.

D. A. Ellwood and A. B. M. Anderson), pp. 34–39. Churchill Livingstone, Edinburgh.

Calder, A. A. (1981). The human cervix in pregnancy, a clinical perspective. In *The Cervix in Pregnancy and Labour: Clinical and Biochemical Investigations.* (eds. D. A. Ellwood and A. B. M. Anderson), pp. 103–122. Churchill Livingstone, Edinburgh.

Chang, W. C., Abe, M., and Murota, S. (1977). Stimulation by prostaglandin $F_{2\alpha}$ of acidic glycosaminoglycan production in cultured fibroblasts. *Prostaglandins* **13**, 55–63.

Conrad, J. T. and Hoover, P. (1982). Variations in the mechanical behavior of the rabbit cervix with endocrine state and anatomic site. *Am. J. Obstet. Gynec.* **143**, 661–666.

——and Ueland, K. (1976). Reduction in the stretch modulus of human cervical tissue by prostaglandin E[2]. *Am. J. Obstet. Gynec.* **126**, 218–223.

——— (1983). Physical characteristics of the cervix. *Clin. Obstet. Gynec.* **26**, 27–36.

Cullen, B. M. and Harkness, R. D. (1960). The effect of hormones on the physical properties and collagen content of the rat's uterine cervix. *J. Physiol., Lond.* **152**, 419–436.

Danforth, D. N. (1947). The fibrous nature of the human cervix and its relation to the isthmic segment in the gravid and non gravid uteri. *Am. J. Obstet. Gynec.* **54**, 541–560

—— (1980). Early studies of the anatomy and physiology of the human cervix and implications for the future. In *Dilatation of the Uterine Cervix.* (eds. F. Naftolin and P. G. Stubblefield) pp. 3–15, Raven Press, New York.

—— (1983). Morphology of the human cervix. *Clin. Obstet. Gynec.* **26**, 7–13.

——Veis, A., Breen, M., Weinstein, H. G., Buckingham, J. C., and Manalo, P. (1974). The effect of pregnancy and labor on the human cervix: Changes in collagen, glycoproteins and glycosaminoglycans. *Am. J. Obstet. Gynec.* **120**, 641–649.

Edqvist, S., Einarsson, S. Gustafsson, B., Linde, C., and Lindell, I. O. (1975). The *in vitro* and *in vivo* effects of prostaglandins E_1 and $F_{2\alpha}$ and of oxytocin on the tubular genital tract of ewes. *Int. J. Fert.* **20**, 234–238.

El-Bana, A. A. and Hafez, E. S. E. (1972). The uterine cervix in mammals. *Am. J. Obstet. Gynec.* **112**, 145–164.

Ellwood, D. A., Anderson, A. B. M., Mitchell, M. D., Murphy, G., and Turnbull, A. C. (1981). Prostanoids, collagenase and cervical softening in sheep. In *The Cervix in Pregnancy and Labour: Clinical and Biochemical Investigations.* (eds. D. A. Elwood and A. B. M. Anderson) pp. 57–73 Churchill Livingstone, Edinburgh.

Endell, W., Holtz, W., and Schmidt, D. (1976*a*). Histochemical characterisation of the uterine cervix in the pig at the time of parturition. *Berl. Münch. Tierartz. Wochenschr.* **89**, 313–316.

Endell, W., Holtz, W., and Schmidt, D. (1976*b*). Macro- and micro-morphological changes of the uterine cervix in the periparturient sow. *Berl. Münch. Tierartz. Wochenschr.* **89**, 349–354.

Evans, M. I., Dougman, M. B., Moawad, A. H., Evans, W. H., Bryant-Greenwood, G. D., and Greenwood, F. C. (1983). Ripening of the human cervix with porcine relaxin. *Am. J. Obstet. Gynec.* **147**, 410–414.

Fields, P. A. and Larkin, L. K. (1980). Enhancement of uterine cervix extensibility in oestrogen-primed mice following administration of relaxin. *J. Endocr.* **87**, 147–152.

Fitzpatrick, R. J. (1957). The activity of the uterine cervix in ruminants. *Vet. Rec.* **69**, 713–717.

—— (1977). Changes in the cervical function at parturition. *Annls. Rech. Vet.* **8**, 432–449.

—— and Dobson, H. (1979). The cervix of sheep and goat during parturition. *Anim. Reprod. Sci.* **2**, 209–224.

—— —— (1981). Softening of the ovine cervix at parturition. In *The Cervix in Pregnancy and Labour: Clinical and Biochemical Investigations.* (eds. D. A. Ellwood and A. B. M. Anderson) pp. 40–56, Churchill Livingstone, Edinburgh.

Flint, M. (1972). Interrelationships of mucopolysaccharide and collagen in connective tissue remodelling. *J. Embryol. exp. Morph.* **27**, 481–495.

Fosang, A. J., Handley, C. J., Santer, V., Lowther, D. A., and Thorburn, G. D. (1984). Pregnancy related changes in the connective tissue of the ovine cervix. *Biol. Reprod.* **30**, 1223–1235.

Gathercole, L. J. and Keller, A. (1975). Light microscopic wavefronts in collagenous tissue and their structural implications. In *Structure of Fibrous Biopolymers.* (eds. E. D. T. Atkins and A. Keller) pp. 153–175, Butterworths, London.

Gay, S. and Miller, E. J. (1978). *Collagen in the Physiology and Pathology of Connective Tissue.* Gustav Fischer Verlag, Stuttgart.

Golichowski, A. M., King, S. R., and Mascaro, K. (1980). Pregnancy-related changes in rat cervical glycosaminoglycans. *Biochem. J.* **192**, 1–8.

Harkness, M. L. R. and Harkness, R. D. (1959). Changes in the physical properties of the uterine cervix of the rat during pregnancy. *J. Physiol., Lond.* **148**, 524–547.

Hendricks, C. H., Brenner, W. E., and Kraus, G. (1970). Normal cervical dilatation pattern in late pregnancy. *Am. J. Obstet. Gynec.* **106**, 1065–1082.

Hillier, K. and Coad, N. (1982). Synthesis of prostaglandins by the human uterine cervix *in vitro* during passive mechanical stretch. *J. Pharm. Pharmacol.* **34**, 262–263.

—— and Wallis, R. M. (1982). Collagen solubilities and tensile properties of the rat uterine cervix in late pregnancy, and the effects of arachidonic acid and prostaglandin $F_{2\alpha}$. *J. Endocr.* **95**, 341–347.

Hollingsworth, M. (1981). Softening of the rat cervix during pregnancy. In *The Cervix in Pregnancy and Labour: Clinical and Biochemical Investigations.* (eds. D. A. Ellwood and A. B. M. Anderson) pp. 13–33, Churchill Livingstone, Edinburgh.

—— and Gallimore, S. (1981). Evidence that cervical softening in the pregnant rat is independent of increasing uterine contractility. *J. Reprod. Fert.* **63**, 449–454.

Hughesdon, P. E. (1952). The fibromuscular structure of the cervix and its changes during pregnancy and labour. *J. Obstet. Gynaec. Br. Commonw.* **59**, 763–776.

Huszar, G. B. (1981). Biology and biochemistry of myometrial contractility and cervical maturation. *Semin. Perinat.* **5**, 216–235.

Ito, A., Kitamura, J. H., Mori, Y., and Hirakawa, S. (1979). The change in solubility of type 1 collagen in human uterine cervix in pregnancy at term. *Biochem. Med.* **21**, 262–270.

—— Naganeo, K., Mori, Y., Hirakawa, S., and Hayashi, M. (1977). PZ-peptidase activity in human uterine cervix in pregnancy at term. *Clin. Chim. Acta* **78**, 267–273.

Jeffrey, J. J. and Koob, T. J. (1980). Endocrine control of collagen degradation in the uterus. In *Dilatation of the Uterine Cervix.* (eds. F. Naftolin and P. G. Stubblefield) pp. 135–145, Raven Press, New York.

Johnstone, F. D., Boyde, I. E., McCarthy, T. G., and McClure-Brown, J. C. (1974). The diameter of the uterine isthmus during the menstrual cycle, pregnancy and puerperium. *J. Obstet. Gynaec. Br. Commonw.* **81**, 588–591.

512   Hilary Dobson

Junqueira, L. C. U., Zugaib, M., Montes, G. S., Toledo, D. M. S., Krisztian, R. M., and Shigihara, K. M. (1980). Morphologic and histochemical evidence for the occurrence of collagenolysis and for the role of neutrophilic polymorphonuclear leukocytes during cervical dilatation. *Am. J. Obstet. Gynec.* **138**, 273–281.

Kitamura, K., Ito, A., Mori, Y., and Hirakawa, S. (1979). Changes in the human uterine cervical collagenase with special reference to cervical ripening. *Biochem. Med.* **22**, 332–338.

————————— (1980). Glycosaminoglycans of human uterine cervix: Heparan sulphate increase with reference to cervical ripening. *Biochem. Med.* **23**, 159–166.

Kleissl, H. P., van der Rest, M., Naftolin, F., Glorieux, F. H., and de Leon, A. (1978). Collagen changes in the human uterine cervix at parturition. *Am. J. Obstet. Gynec.* **130**, 748–753.

Koob, T. J., Stubblefield, P. G., Eyre, D. R., and Ryan, K. J. (1980). Connective tissue alterations associated with parturition in the rabbit. In *Dilation of Uterine Cervix* (eds. F. Naftolin and P. G. Stubblefield) pp. 203–218, Raven Press, New York.

Ledger, W. L., Ellwood, D. A., and Taylor, M. J. (1983). Cervical softening in late pregnant sheep by infusion of PGE_2 into a cervical artery. *J. Reprod. Fert.* **69**, 511–515.

—— Webster, M., Harrison, L. P., Anderson, A. B. M., and Turnbull, A. C. (1985). Increase in cervical extensibility during labour induced after isolation of the cervix from the uterus in the pregnant sheep. *Am. J. Obstet. Gynec.* **151**, 397–402.

Leppert, P. C., Keller, S., Cerreta, J., and Mandl, I. (1982). Conclusive evidence for the presence of elastin in the human and monkey cervix. *Am. J. Obstet. Gynec.* **142**, 179–182.

Leppi, T. J. (1964). A study of the uterine cervix of the mouse. *Anat. Rec.* **150**, 51–66.

Liggins, G. C. (1981). Cervical ripening as an inflammatory reaction. In *The Cervix in Pregnancy and Labour: Clinical and Biochemical Investigations.* (eds. D. A. Ellwood and A. B. M. Anderson) pp. 1–9, Churchill Livingstone, Edinburgh.

MacKenzie, I. Z. (1981). Clinical studies on cervical ripening. In *The Cervix in Pregnancy and Labour: Clinical and Biochemical Investigations.* (eds. D. A. Ellwood and A. B. M. Anderson), pp. 163–186, Churchill Livingstone, Edinburgh.

MacLennan, A. H. (1981). Cervical ripening and the induction of labour by vaginal prostaglandin $F_{2\alpha}$ and relaxin. In *The Cervix in Pregnancy and Labour: Clinical and Biochemical Investigations.* (eds. D. A. Ellwood and A. B. M. Anderson), Churchill Livingstone, Edinburgh.

—— Katz, M., and Creasy, R. (1985). The morphologic characteristics of cervical ripening induced by the hormones relaxin and prostaglandin $F_{2\alpha}$ in the rabbit model. *Am. J. Obstet. Gynec.* **152**, 691–696.

McMurtry, J. P., Floersheim, G. L., and Bryant-Greenwood, G. D. (1980). Characterization of the binding of ^{125}I-labelled succinylated porcine relaxin in human and mouse fibroblasts. *J. Reprod. Fert.* **58**, 43–49.

von Maillot, K., Stuhlsatz, H. W., Mohanaradhkrishan, V., and Greiling, H. (1979). Changes in the glycosaminoglycans distribution pattern in the human uterine cervix during pregnancy and labor. *Am. J. Obstet. Gynec.* **135**, 503–506.

—— and Zimmerman, B. K. (1976). The solubility of collagen of the uterine cervix during pregnancy and labour. *Arch. Gynaek.* **220**, 275–280.

Mitchell, M. D. and Flint, A. P. F. (1978). Use of meclofenamic acid to investigate the role of prostaglandin biosynthesis during induced parturition in sheep. *J. Endocr.* **76**, 101–109.

Mochizuki, M. and Tojo, S. (1980). Effect of dehydroepiandrosterone sulphate on softening and dilatation of the uterine cervix in pregnant women. In *Dilatation of the Cervix.* (eds. F. Naftolin and P. G. Stubblefield) pp. 267–286, Raven Press, New York.

Mori, Y., Ito, A., Hirakawa, S., and Kitamura, K. (1981). Proteinases in the human and rabbit. In *The Cervix in Pregnancy and Labour: Clinical and Biochemical Investigations.* (eds. D. A. Ellwood and A. B. M. Anderson) pp. 136–143, Churchill Livingstone, Edinburgh.

Mullins, K. J. and Saacke, R. G. (1982). A study of the functional anatomy of the bovine cervix. *J. Dairy Sci.* **65**, *Suppl.* **1**, 184.

Murota, S., Abe, M., and Otsuka, K. (1977). Stimulatory effect of prostaglandins on the production of hexosamine containing substances by cultured fibroblasts. (3) Induction of hyaluronic acid synthetase by $PGF_{2\alpha}$. *Prostaglandins* **14**, 983–991.

Musah, A. I., Schwabe, C., Willham, R. L., and Anderson, L. L. (1986). Influence of relaxin on induction of parturition in beef heifers. *Endocrinology* **118**, 1476–1482.

Norstrom, A. (1984). Acute effects of prostaglandins on the biosynthesis of connective tissue constituents in the non-pregnant human uteri. *Acta obstet. gynec. scand.* **63**, 169–173.

Owiny, J. R. (1986). *Factors affecting softening of the ovine cervix at parturition.* Ph.D. Thesis, University of Liverpool.

——Fitzpatrick, R. J., Spiller, D. G., and Appleton, J. (1987). Scanning electron microscopy of the wall of the ovine cervix uteri in relation to tensile strength at parturition. *Res. Vet. Sci.* **43**, 36–43.

Parry, D. S. and Ellwood, D. A. (1981). Ultrastructural aspects of cervical softening in sheep. In *The Cervix in Pregnancy and Labour: Clinical and Biochemical Investigations.* (eds. D. A. Ellwood and A. B. M. Anderson) pp. 74–84, Churchill Livingstone, Edinburgh.

Perezgrovas, R. and Anderson, L. I. (1982). Effects of porcine relaxin on cervical dilatation, pelvic area and parturition in beef heifers. *Biol. Reprod.* **26**, 765–776.

Postlethwaite, A. E. and Kang, A. K. (1976). Collagen and collagen peptide induced chemotaxis of human blood monocytes. *J. exp. Med.* **143**, 1299–1307.

Putnam, M. R., Rice, L. E., Wetteman, R. P., Lusky, K. S., and Pratt, B. (1985). Clenbuterol (Planipart ™) for the postponement of parturition in cattle. *Theriogenology* **24**, 385–393.

Raynes, J. G., Anderson, J., Fitzpatrick, R. J., and Dobson, H. (1988). Collagenase activity is not increased during cervical softening in the ewe. *Coll. rel. res.* (in press).

——Clarke, F., Anderson, J., Fitzpatrick, R. J., and Dobson, H. (1988). Collagenase inhibitor concentration in cultured ovine cervical tissue is increased in late pregnancy. *Biochem. Biophys. Acta* (submitted).

van der Rest, M. (1980). Collagen and its metabolism. In *Dilatation of the Uterine Cervix.* (eds. F. Naftolin and P. G. Stubblefield) pp. 61–78, Raven Press, New York.

Rimmer, D. M. (1973). The effect of pregnancy on the collagen of the uterine cervix of the mouse. *J. Endocr.* **57**, 413–418.

Rorie, D. K. and Newton, M. (1967). Histologic and chemical studies of the smooth muscle in the human cervix. *Am. J. Obstet. Gynec.* **99**, 466–469.

Scott, S. E. and Dorling, J. (1965). Differential staining of acid glycosaminoglycans (mucopolysaccharides) by alcian blue in salt solutions. *Histochemie* **5**, 221–233.

Soden, P. D. and Kershaw, I. (1974). Tensile testing of connective tissues. *Med. Biol. Eng.* **12**, 510–518.

Steinetz, B. G., O'Bryne, E. M., and Kroc, R. L. (1980). The role of relaxin in cervical softening during pregnancy in mammals. In *Dilatation of the Uterine Cervix.* (eds. F. Naftolin and P. G. Stubblefield) pp. 157–177, Raven Press, New York.

Stys, S. J., Clark, K. E., Clewell, W. H., and Meschia, G. (1980). Hormonal effects on cervical compliance in sheep. In *Dilatation of the Uterine Cervix*. (eds. F. Naftolin and P. G. Stubblefield) pp. 147–156, Raven Press, New York.

Toole, B. P. and Lowther, D. A. (1968). Dermatan sulfate protein: isolation from and interaction with collagen. *Biochim. Biophys. Acta* **128**, 567–578.

Uldbjerg, N., Malmstrom, A., Ekman, G., Sheehan, J., Ulmsten, U., and Wingerup, L. (1983a). Isolation and characterisation of dermatan sulphate proteoglycan from human uterine cervix. *Biochem. J.* **209**, 497–503.

—— Ulmsten, U., and Ekman, G. (1983b). The ripening of the human uterine cervix in terms of connective tissue biochemistry. *Clin. Obstet. Gynec.* **26**, 14–26.

Varner, D. D., Hinnichs, H., Garcia, M. C., Osborne, H. G., Blanchard, T. L., and Kenney, R. M. (1985). A comparison between cervical dimensions of pregnant and non-pregnant *Santa Getrudis* and *Bos taurus* cows. *Theriogenology* **24**, 109–118.

Wallis, R. M. and Hillier, K. (1981). Regulation of collagen dissolution in the human uterine cervix by oestradiol-17β and progesterone. *J. Reprod. Fert.* **62**, 55–61.

Ward, W. R. (1968). *Structure and function of the ovine cervi uteri* (*with special reference to the condition of ringwomb in the sheep*). Ph.D. Thesis, University of Liverpool.

Wedmore, C. V. and Williams, T. J. (1981). Control of vascular permeability by leucocytes in inflammation. *Nature, Lond.* **289**, 646–650.

Weiss, M., Nagelschmidt, M., and Struck, H. (1979). Relaxin and hormone metabolism. *Horm. metab. Res.* **11**, 408–410.

Index